THE OVERH
PICTURE

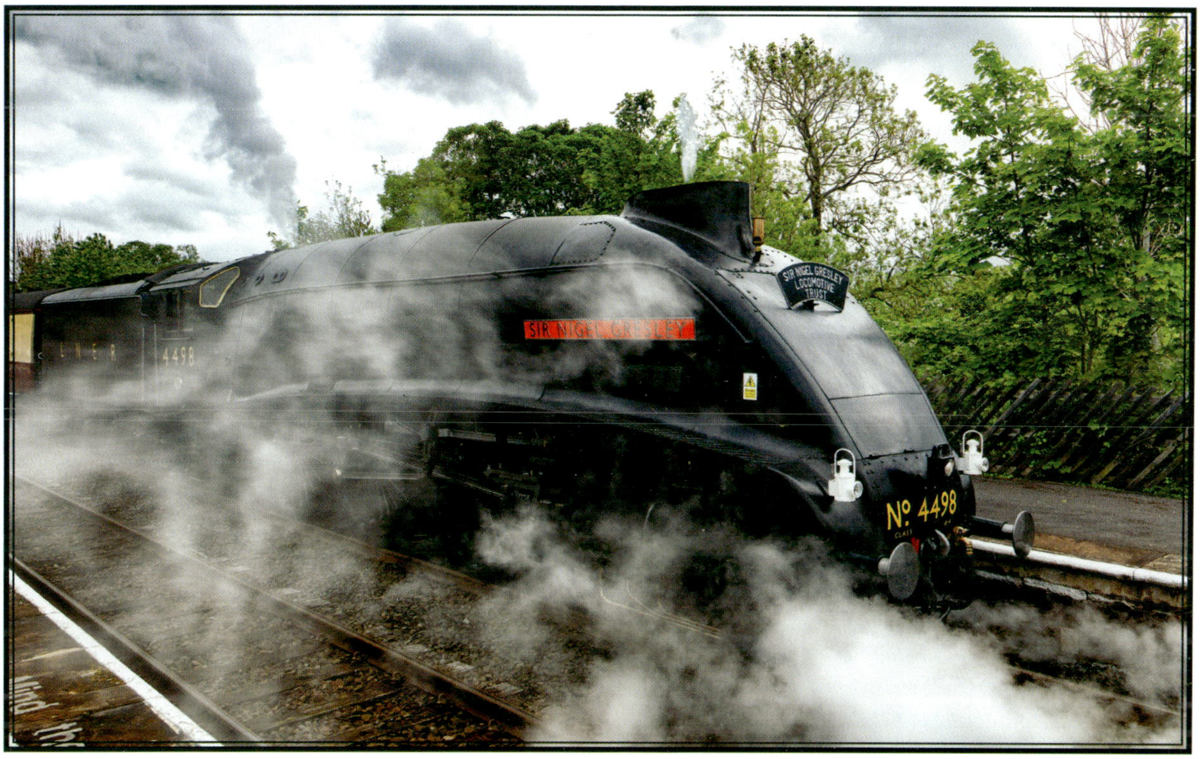

TREVOR CAMP

The Overhaul Picture by Trevor Camp

Published by the Sir Nigel Gresley Locomotive Trust

Registered Charity No.1079591
Registered offices: Lloyds Bank Chambers, Hustlergate, Bradford BD1 1UQ
Registered in England : Company N°901015
A Company Limited by Guarantee
VAT N° GB 740 843437

www.sirnigelgresley.org.uk

Artwork, design and layout: Trevor Camp

ISBN: 978-1-7395885-0-2
Printed by: WTTB - London, Unit 15, Thames Gateway Park, Dagenham, RM9 6FB

First Edition

Cover image: Sir Nigel Gresley during the first trial run at Bridgnorth on the Severn Valley Railway March 23rd 2022

Page i image: Sir Nigel Gresley at Appleby on the Inaugural Run May 21st 2022

Back cover: Sir Nigel Gresley stands beside his locomotive (credit SNGLT//Mike Read Collection)

FOREWORD

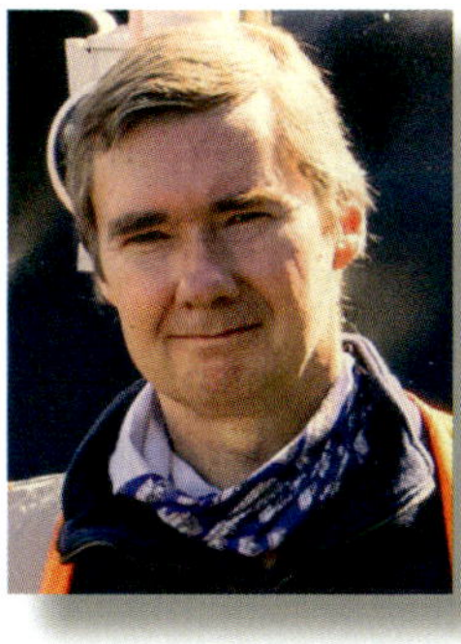

Nigel Wilson

At the beginning of 2015 iconic steam locomotive 60007 *Sir Nigel Gresley* was approaching a much-needed major overhaul with an uncertain future: its longstanding locomotive manager was on the verge of retirement and in poor health, and there were question marks about whether enough people could be found to progress the necessary work in a timely manner. The nadir was reached in May when the locomotive was unavailable to haul a dining special. A number of things were in the locomotive's favour, however: the Sir Nigel Gresley Locomotive Trust, which has owned *Sir Nigel* since 1966, had always found a way to keep it running, and a newly appointed locomotive engineer was taking up the reins. The events in May triggered a decision to undertake the overhaul in York which, as it transpired, allowed the building of a fantastic team of engineers who carried out the most comprehensive overhaul ever undertaken on 60007. Another key advantage was that responsible for the engineering activities was a volunteer Chief Mechanical Engineer who had been wielding his spanner around *Sir Nigel* since 1971. And the final advantage was that the Trust had cash in the bank together with the support of a loyal membership who continued to contribute to the costs during the next seven years - without whom the work could not have been completed to the standard that it has. This book is a tribute to all the team who took the locomotive apart, measured, repaired, polished and painted every nut, bolt, rivet and spring – not to mention the miles and miles of pipes – before lovingly putting it all back together.

The spectacular images that Trevor has captured in this book paint a picture which is, as far as I know, unique in documenting an overhaul from inside the team carrying it out. I think you'll find it interesting as well as entertaining.

In 1967, after its first overhaul in preservation. *Sir Nigel Gresley* hauled a railtour from Crewe to Carlisle - its first tour in present ownership. In 2022, as you will read, history was recreated when, again fresh from overhaul, the locomotive triumphantly hauled a train with 253 SNGLT members onboard over the same route - a remarkable tribute to the team you will read about in these pages.

Nigel Wilson

Chairman, The Sir Nigel Gresley Locomotive Trust

June 2022

ACKNOWLEDGEMENTS

I would like to say a special thank you to all of you who responded to my appeal to pre-order a copy of

THE OVERHAUL PICTURE

by doing so you helped fund the publishing and printing of this first edition.

Below are the first 150 names, thank you so much.

Malc Bateman, Peter Brackstone, Doreen Pawson, John Garton, Philip Hourd, Dave Lee,
Ollie Dutton, Peter James, Barry Horton, Chris Whiteley, Dave Whiteley, Mike Page,
Andy Barwick, Phil Slater, Peter Ellis, John Wilkinson, Richard Hawkhead, Ian Hawkhead,
Mathew Hawkhead, Rachel Hawkhead, Craig Hawkhead, Andy Evans, Nigel Franklin,
Roger Turnbull, Stephen Turnbull, William Stafford, Nigel Wilson,
Richard Swales, David Camp, Graham Walker, Adam Camp, Stuart Hill,
Christopher Humpleby, Dean Cummins, Nicolas Clennett, Tony Bickerstaffe, Bob Shearman,
Tom Bell, Darrin Crone, Doug Blyth, Ray Thompson, Rod Thomas,
Andrew Jacques, Ian Jacques, Mike Bradley, Paul Aston, John Charles Doubleday,
Peter Bowes, Paul Carrigan, Nigel Hoskison, Dave Goodey, Anthony Holt, Peter Webb,
Nigel Rayner, Vida E Lyne, Eric Taylor, David Thornhill, Ian Mercer, Jim Hunt,Penny Aspey,
Richard Spoors, Ian Jefferson, David Daniels, Brent Orridge, Alex Bown, M F Sarney,
Bob Matkin, S Reddin, Robert Richardson, Douglas Burcham,
James D Burcham, Philip S L Burcham, Peter Bainbridge, Anne-Marie Lawn, Mrs A Miller,
Alison Flood, Sidney Horton, Alan Pitt, Ray Thompson, David Francis,
Luke Perry, Scott Middlemiss, Ron Lawson, James Porter,
John Atkinson, Donald J Frank, Robert Riddick, Martin Windeler, William Elliot, Pam Gillard,
Barrie Luck, S C Hall, Gavin Shepherd, J I Croft, H N Collingwood, P J Haynes, S D McNaughton, A P Gardner,
J S Silcock, G M S Carter, John Kirsopp, Teresa Jackson, Sam Perry,
Mike Smith, D W Clark, Jeff Greenwood, Robert Goulding, Richard Anstis, Jackie Butterworth,
Linda Vincent, Mick Hammond, Andrew Rampley, Jane Acaster, Philip John Salter,
Brenda Ward, Charles Newton, John Harris, Andy Furnell, Paul Rowell, Andrew Cole,
E L Peacock, Ian Holdsworth, Matthew Crampton, Graham Smith, Barbara Lanelzzy Lane,
Christopher Allen, Phil Cameron, David Brooks, Dawn Rogers, Albert E Blower, Rob Slaughter,
Robert Ince, Ken Bristow, Martin Joyce, Eric Lummas, Daniel Hindley, J V Vickers, James Porter,
Yvonne Tramm, Stuart Mann, Kayleigh Stanton, Peter John Bowen, Matthew Kay,
Anthony Endsor, Ross Chalmers, Rick Roper, Dennis Gregory,
Graham Southway, Peter Nicholas Sabini, C W Brooks,

CONTENTS

Thank you to everyone who has aided the production of this book. A special thank you must go to my wife, Janice, who has supported me, helped me and endured me discussing 'the book' through this long journey!

INTRODUCTION

I have two intended goals.

Firstly, and primarily, this is a photographic record of what turned out to be a six year overhaul of class A4 'Pacific' steam locomotive 60007 Sir Nigel Gresley. The overhaul was originally planned for 4-5 years but this was significantly delayed by the Coronavirus pandemic which started in 2020.

Every ten years, the boiler insurance certification must be renewed which requires a complete strip down and overhaul to ensure safe operation for the next ten years. Whilst that was being undertaken it was an opportunity to thoroughly overhaul the rest of the locomotive (affectionately known as 'No.7'). The owner of the locomotive, the Sir Nigel Gresley Locomotive Trust (SNGLT) negotiated an arrangement with the National Railway Museum (NRM) in York to rent space in their excellent engineering workshops so the Trust's engineering team could carry out the work. No.7 was duly moved there from Grosmont on the North Yorkshire Moors Railway in November 2015, after coming out of traffic in September that year but due to the museum having redevelopment plans No.7, along with the Trust's volunteer engineering team, was moved in September 2021 to the workshops of Locomotive Services Limited (LSL) in Crewe for completion. A logical move as the Trust had negotiated a contract with LSL to haul some of the SAPHOS mainline excursions which operate from the Crewe depot, so No.7 would spend a good deal of time there.

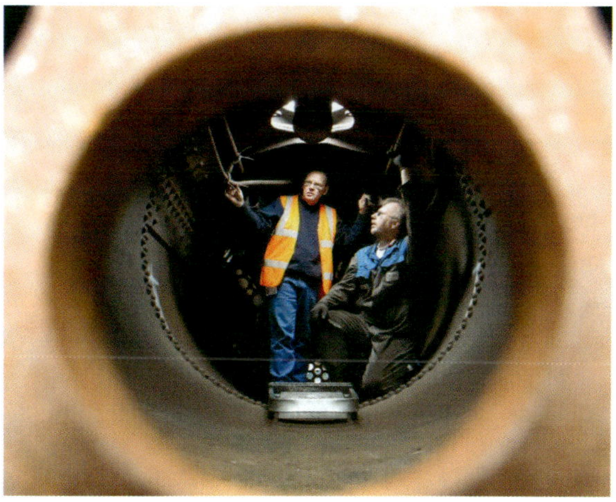

With the boiler tubes removed Darrin Crone & Roger Turnbull inspect the inside of the boiler

Secondly, I am a landscape photographer and as such my photography reflects the attributes of that style which I have applied to many of the images in the following pages. I was asked to document the progress of the overhaul, but it became clear to me, from day one, that this was an excellent opportunity to put an artistic take on the subject as well as documentary and technical shots.

This is not a technical diary, but my personal story of the overhaul from my many days spent on the project and is quite possibly a unique first time any steam locomotive overhaul has been covered in this way. Initially I anticipated this to be purely a portfolio of images with extended captions but after being encouraged to write the story behind the pictures it has evolved into what I hope you find an enjoyable read.

A note for photographers: the engineering workshop lights are not well suited for photography and hand-held shots require a high iso, wide apertures and longish exposures. Therefore, most workshop shots have been done on a tripod to overcome these problems and people movement is often blurred. A photographic technique often utilised to portray action (and I admit, not planned here) but welcomingly adopted. I like my images to be bright with high contrast and many have also been shot in High Dynamic Range (HDR): three exposures of different amounts merged together to give a full range of tones.

So, this portfolio is for steam train enthusiasts, photographers, and hopefully art lovers as well as many others!

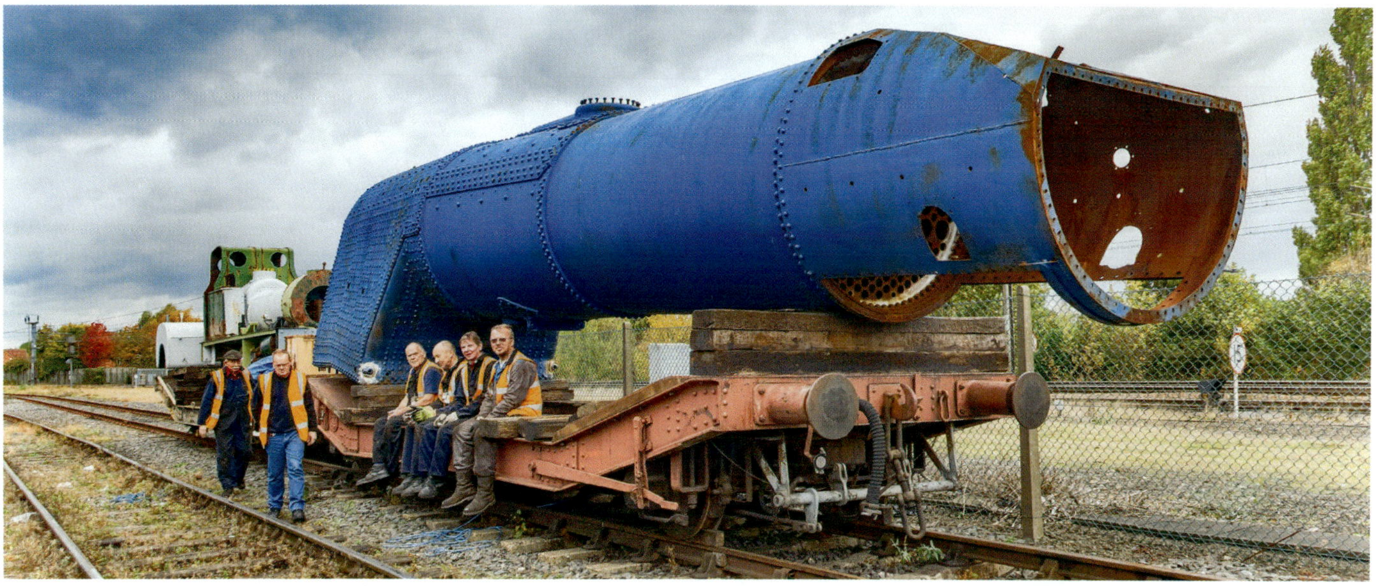

The boiler on a rail wagon, a temporary home before being moved by road to Llangollen for overhaul

TOP TEAM

Working your way through the following pages you will be introduced to the key players on the overhaul team at a point that I first met them. However it is only fitting that the two most important members of the team are introduced here at the beginning.

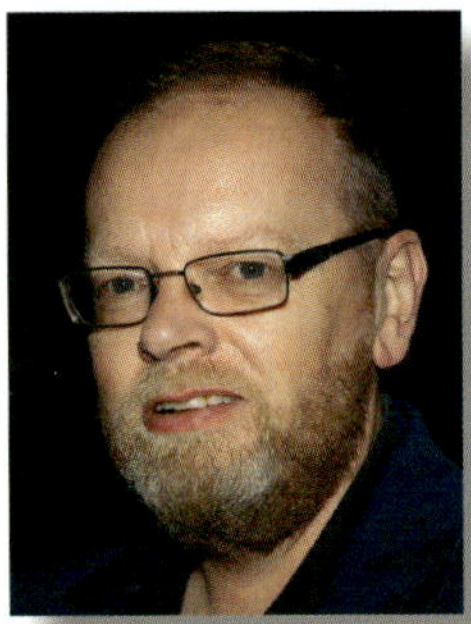

Darrin Crone

Locomotive Engineer Darrin Crone was appointed by the Trust to oversee what was to become the most thorough overhaul of Sir Nigel Gresley in preservation and possibly even before then in the LNER and British Railway days. His unique style of man management was to make him very popular with the rest of the volunteer engineering team; keeping together a group of volunteers for six years is no mean feat particularly when Covid was to interrupt the overhaul.

From leaving school Darrin's future was definitely set in engineering and after working for British Aerospace on Typhoon fighter jet aircraft in Preston he moved back to Teesside to take up a job with HNL engineering where he stayed for 24 years rising to be the Head of Engineering. He left in 2015 to take up his position with SNGLT which perfectly dovetailed his love of the railways with his engineering background.

Not that Darrin knew it but he was following in his father's footsteps when he volunteered with NELPG (North East Locomotive Preservation Group) and the NYMR (North Yorkshire Moors Railway), progressing from cleaner to fireman and eventually to driver. His father was a guard on the railways and knowing of the love he still held for trains, Darrin invited him along for a footplate ride one day and was amazed when he declined the invitation. "Why not?" was Darrin's bemused response. The reply was a family story the young Darrin had not heard before. Prior to Darrin entering this world, his father had been a fireman on the railways and after a break for army service had to rejoin the railways as a guard as dieselisation had meant there was no longer any positions left for firemen. Darrin's father, it seems, had had more rides on footplates than Darrin had had hot dinners.

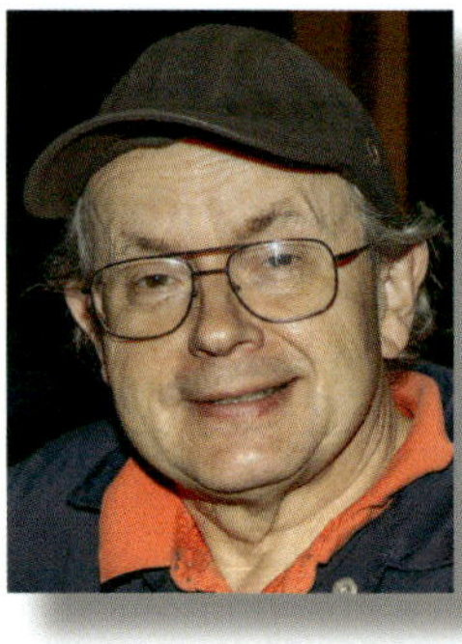

Richard Swales

Richard is the Trust's Chief Mechanical Engineer (CME) and has been involved in all the overhauls since becoming a Trust member in 1971. His association started as a schoolboy when No.7 was housed at the County Durham Philadelphia NCB shed where Mick Ennis, then head of maintenance, appointed Richard as honorary apprentice. This stimulated an interest in engineering which led to a degree in Mechanical Engineering from Newcastle University and a 40 year career at NSK Bearings Europe Ltd in Peterlee, during which time he achieved Chartered Engineer status with the Institute of Mechanical Engineers in 1988.

Richard carried on working on No.7 even after the move to Carnforth in 1977, together with spending time on the NYMR in Grosmont and the Monkwearmouth Station Museum in Sunderland, working on the overhauls which were completed in 1983, 1992 and 2006. During the last of these he was Assistant CME to John Graham, from whom he later took over as CME.

I think it is fair to say nobody knows No.7 as well as Richard, whose expertise and knowledge has been an invaluable contribution over the last 50 years, but who, nonetheless, is very generous in sharing the credit with Darrin and the rest of the team.

"The current overhaul has been the most thorough of them all, and what gives me most satisfaction is that the backbone of the project has been provided by a volunteer engineering team working to a very high professional standard, for which our Locomotive Engineer, Darrin Crone, deserves all the credit for managing the people and the processes."

IT STARTED WITH A PING

1

December 2015 – January 2016

My story began with a ping, just a short, sharp iPhone ping but enough to jolt me from my half dozing, half watching tv state and reach for my phone. Who is wanting to alter my timeline through this universe now and on Christmas Eve, I'm thinking. Well, I don't know this guy but what he has to say has suddenly woken me up big style. Would I be interested in photographing the overhaul of steam locomotive 'Sir Nigel Gresley'? Not that Malc, my new texting friend, knew at the time, but No.7 happens to be my favourite locomotive, and the one which I had only said farewell to a few months ago on the North Yorkshire Moors Railway, when it was on its last weekend of running before going out of traffic for several years whilst it was overhauled. It didn't take me long to say, "I would be delighted to be involved."

Christmas and New Year went, and after crossing a few Ts and dotting a few Is, my first visit to the workshops of the National Railway Museum was arranged for January 20th 2016. I met up with Malc for the first time and he introduced me to Darrin Crone, the Locomotive Engineer, and the only paid member of the team, who was in charge of the overhaul. Darrin gave me a tour of the locomotive showing the progress to date. All of a sudden, from previously being happy just to be standing on a platform next to this iconic and famous locomotive, I found I was wearing a hard hat down in the pit underneath it, standing on the footplate and finally climbing a ladder to stand inside the smokebox with my hat in the space vacated by the now removed double chimney. "Well, what do you think?" said Darrin, adding "it's going to be three years or more, so are you up for it and will you stay for the duration?" Silly questions but he had to ask, and I had to reply "Absolutely, yes."

Time to get my cameras out. The general lighting in the workshop is adequate for its purpose but is not studio level lighting for photography and to compensate I could have used additional lighting, but this was not a practical solution, so a lot of shots were taken with the camera on a tripod allowing a low noise iso of 100 to be used. This slowly changed as we got further into the overhaul and I simply had not the luxury of setting every shot up on a tripod, I had to handhold and subsequently the iso went north and I had to deal with the noise when editing.

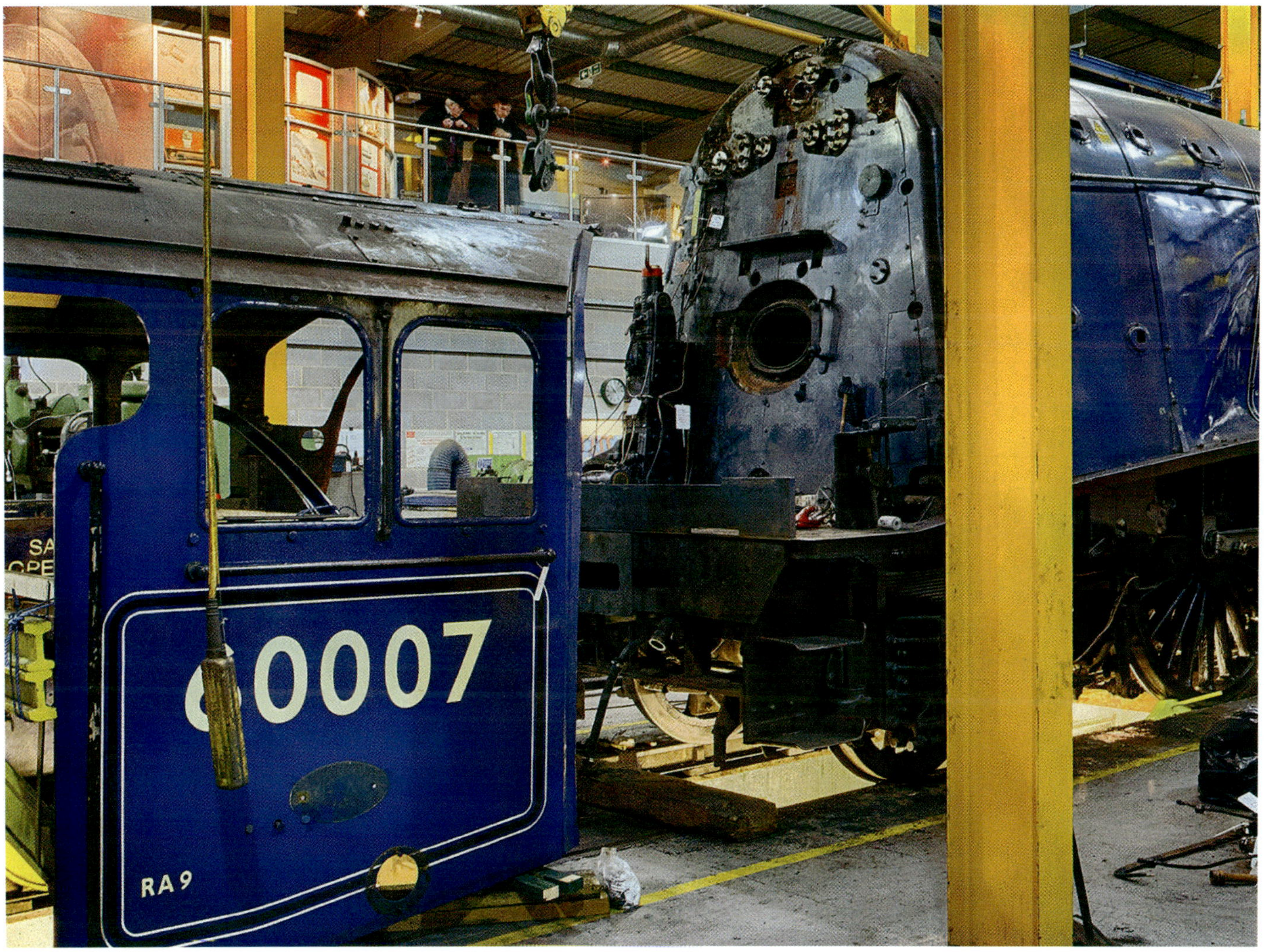

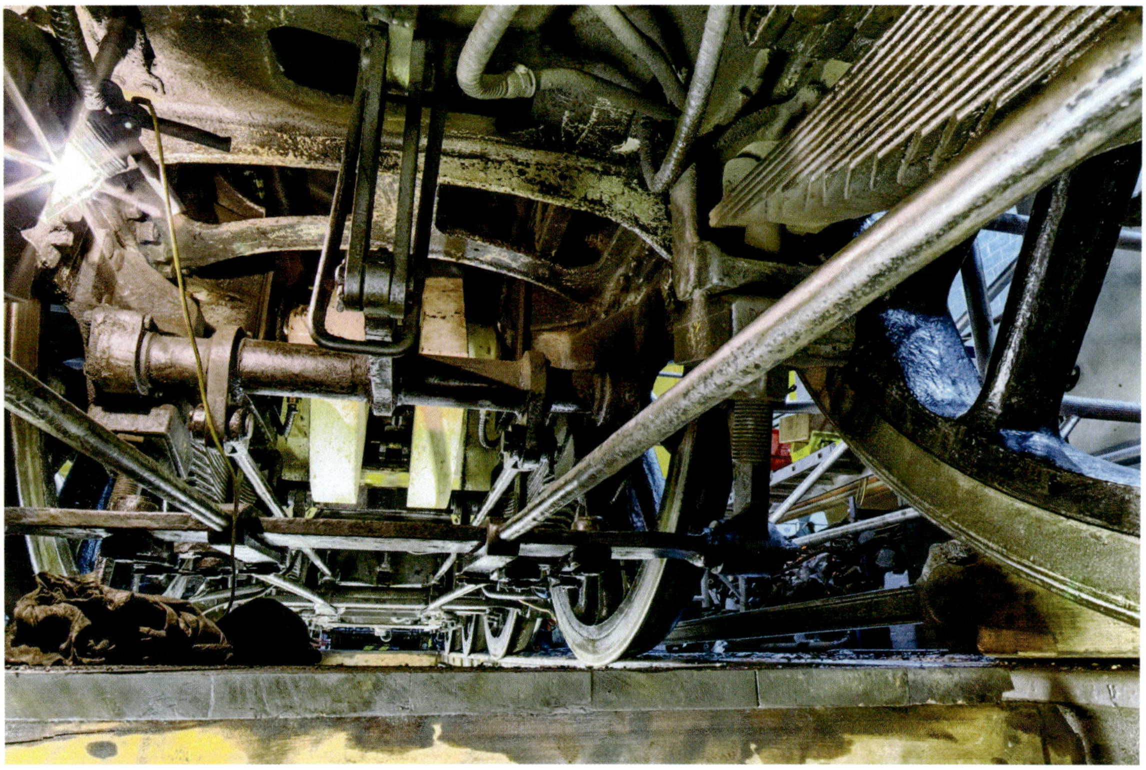

My first triple exposure High Dynamic Range (HDR) image taken in the gloomy surroundings of the pit below No.7

My first shot had to be back in the pit, dark and gloomy as it was, but I knew that shooting in raw format on a tripod I could extract some wonderful hidden colours, and I wasn't disappointed. I must admit that being down there alone with a hundred or so tons of a world famous A4 locomotive above my head, I was having to pinch myself. Smitten you're thinking - yes, I was! It would take a few more visits before familiarity would settle that feeling down. I was back to being a young excited teenager again!

Rising safely out of the pit, I went back to the location of my acceptance of the challenge and found Darrin perched on top of the smokebox that I had previously been standing inside. With the iconic front streamlining now removed, No7 had lost its familiar sleek look but it was clearly still looking like an imposing locomotive. Piece by piece she was being meticulously disassembled by the Trust's team of volunteer engineers with all the parts tagged and catalogued. The rest of the streamlining body panels were also being removed by Malc and Peter Brackstone. Peter was to become a key player in the volunteer team; a retired heating engineer whose knowledge, ability and crucially important strength,

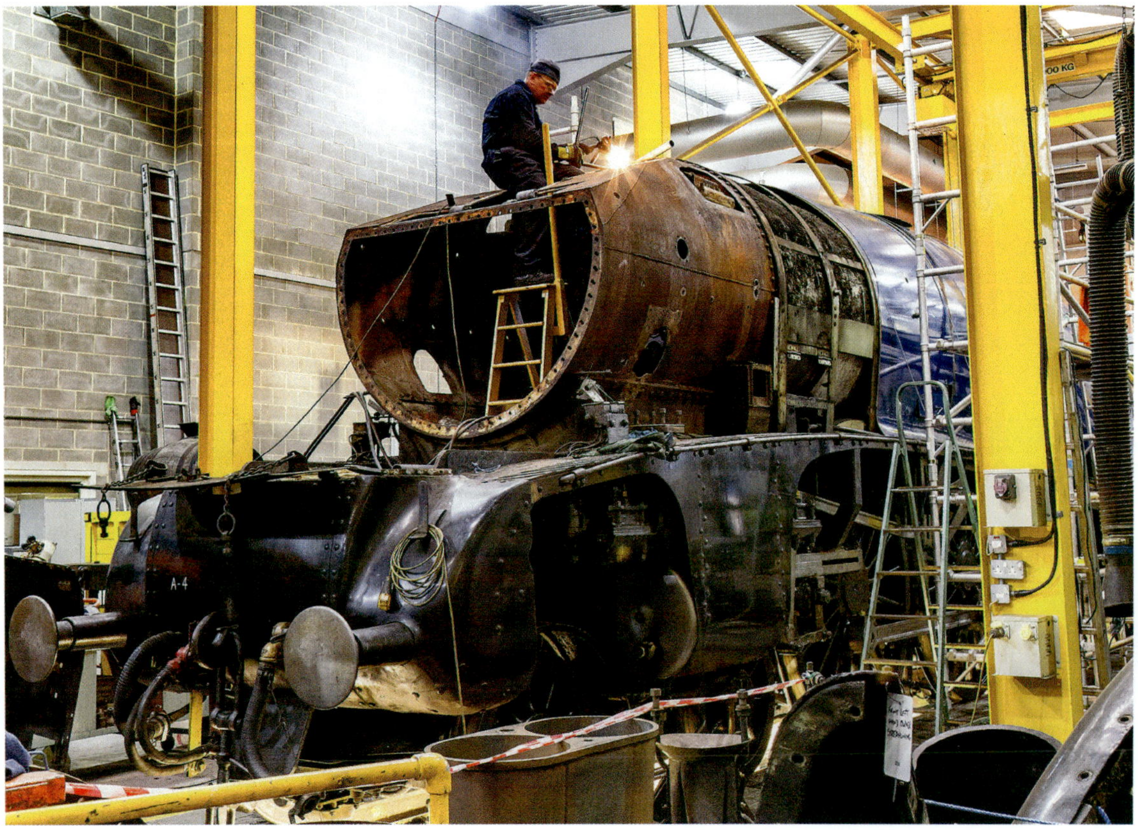

Darrin Crone on top of the smoke box with an oxy-acetylene burner

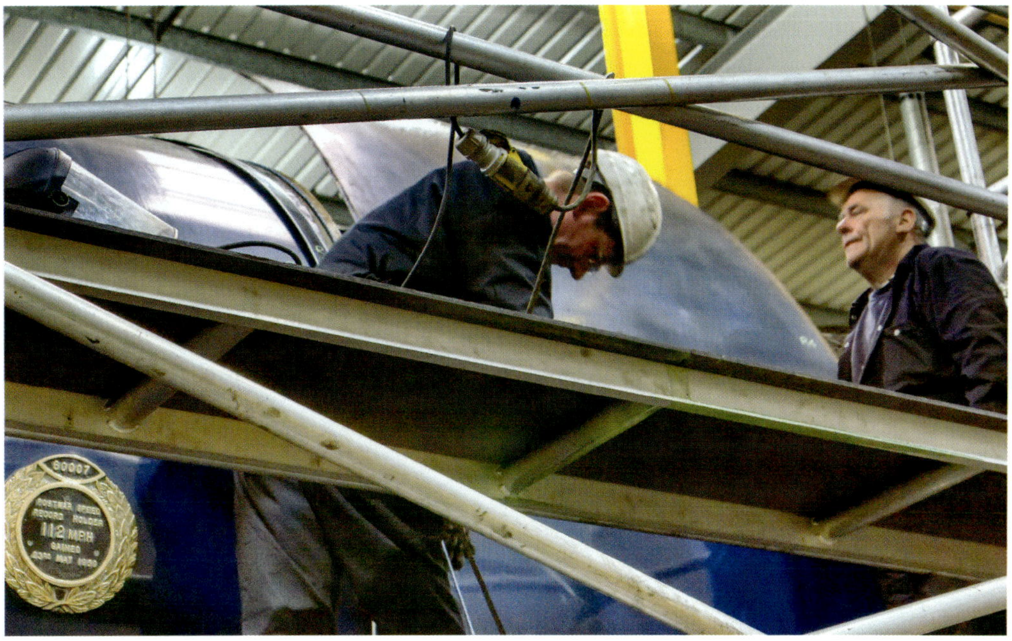

Malc Bateman and Peter Brackstone removing streamlining panels from the boiler sides.

Peter Brackstone

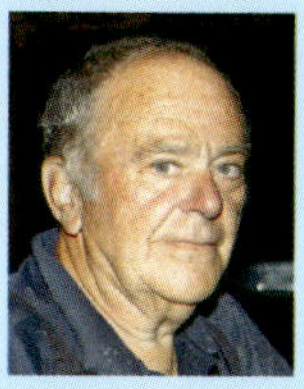

Pete (Bracko), a retired production engineer from the boiler industry, joined as a founding member of the volunteer team and after soon realising he had made a good choice in devoting his time to the project went on to become the leading volunteer in attendance days. "After all," he said " it was only a three year project, wasn't it!" Of the multitude of jobs he willingly undertook it was the first choice blue cabin tea maker that his comrades appreciated the most. He was also the leading member of the painting team and Darrin's 'appointed volunteer' when strength was needed.

"I feel very privileged indeed to say I was one of the volunteers who was involved in the most in depth overhaul that Sir Nigel Gresley has received in private ownership."

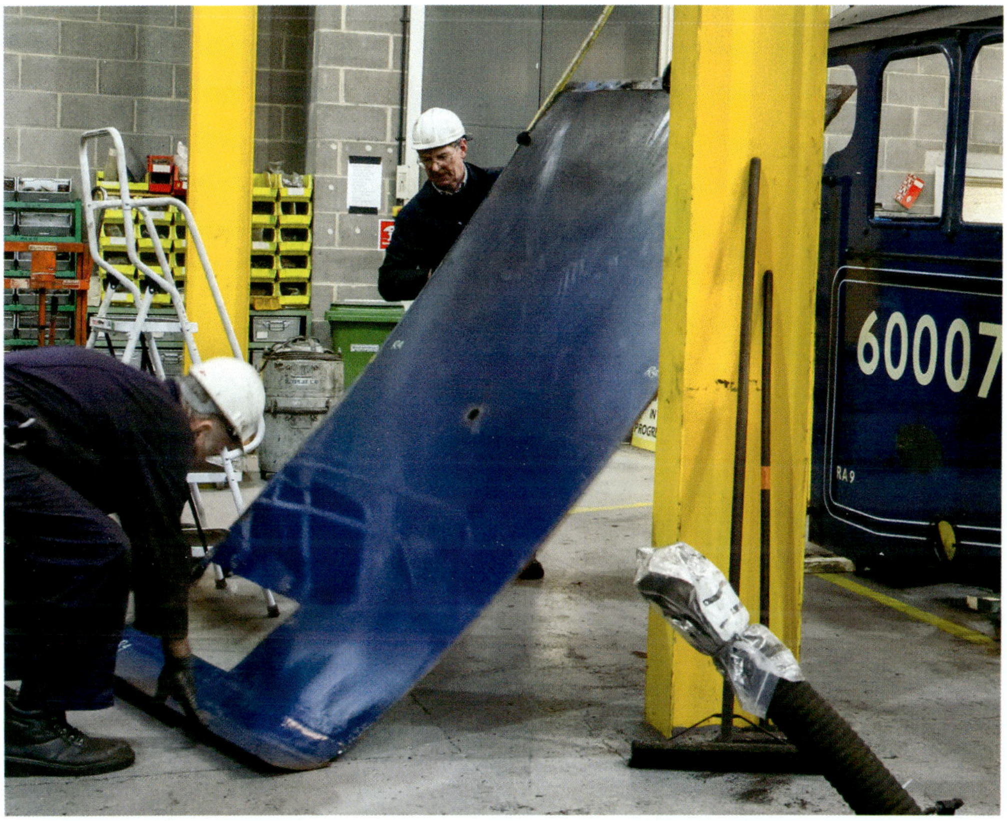

With the overhead crane taking the weight Peter Brackstone and Malc Bateman move a streamlined section of cladding to a storage location

Malc Bateman

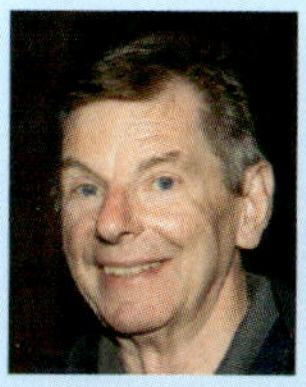

Malc has a long association with No.7 having been a volunteer working member of the Trust for many years. Now retired from a long career as an IT manager in the automotive industry he shares his time as a Trust volunteer with his much loved canal boat, the latter resulting in his nickname as Captain Pugwash. He is a Trust board member and also editor of the Trust magazine Chime; his hobbies include live steam model railways and motor racing.

"As the son of an LNER/BR 'Passed Fireman' at Darlington (51A) Shed it was inevitable that I would be fascinated by steam. That fascination started in 1955 when my father introduced me to an LNER A8 Class footplate as he arrived with a local passenger service at Darlington Station. Fast forward 54 years and I found myself, having retired, part of the weekly 'Thursday Gang' cleaning No.60007 at Grosmont on the NYMR. It would lead to becoming part of the SNG Engineering Team. The rest 'is history' as they say. I'm proud to be part of the Team that delivered this amazing overhaul under incredible professional guidance almost 67 years after my first footplate adventure!"

particularly in making the tea, were to make him into a VIP in the engineering team, but more of that later.

Day 1 was coming to an end, I now had a good idea of the work ahead and I had met a team of focused, enthusiastic, talented individuals who over time I would become good friends with in a strong 'band of brothers' led by Darrin who I was to discover to be a great man manager.

Day 2 a couple of weeks later was to see sparks flying as I was looking down the barrel.

LOOKING DOWN THE BARREL

February 2016

My second visit to the NRM. I was still feeling my way around this project, so Darrin and the team were getting on with their tasks and leaving me very much to get on with mine. They were always available if I had a question but were not directing me to take this picture or that, something I was very grateful for as it allowed me a greater time to think and scope to be creative. I think it was then, when I was wandering around the workshop looking at angles, that I realised the unique and iconic lines of A4 'streaks' could still be captured and be identifiable even during a major strip down of the locomotive. I was to return to this theme in future visits but today something spectacular was happening.

Before the boiler was lifted from the chassis and transported by road to the boilersmiths in Llangollen, the boiler tubes

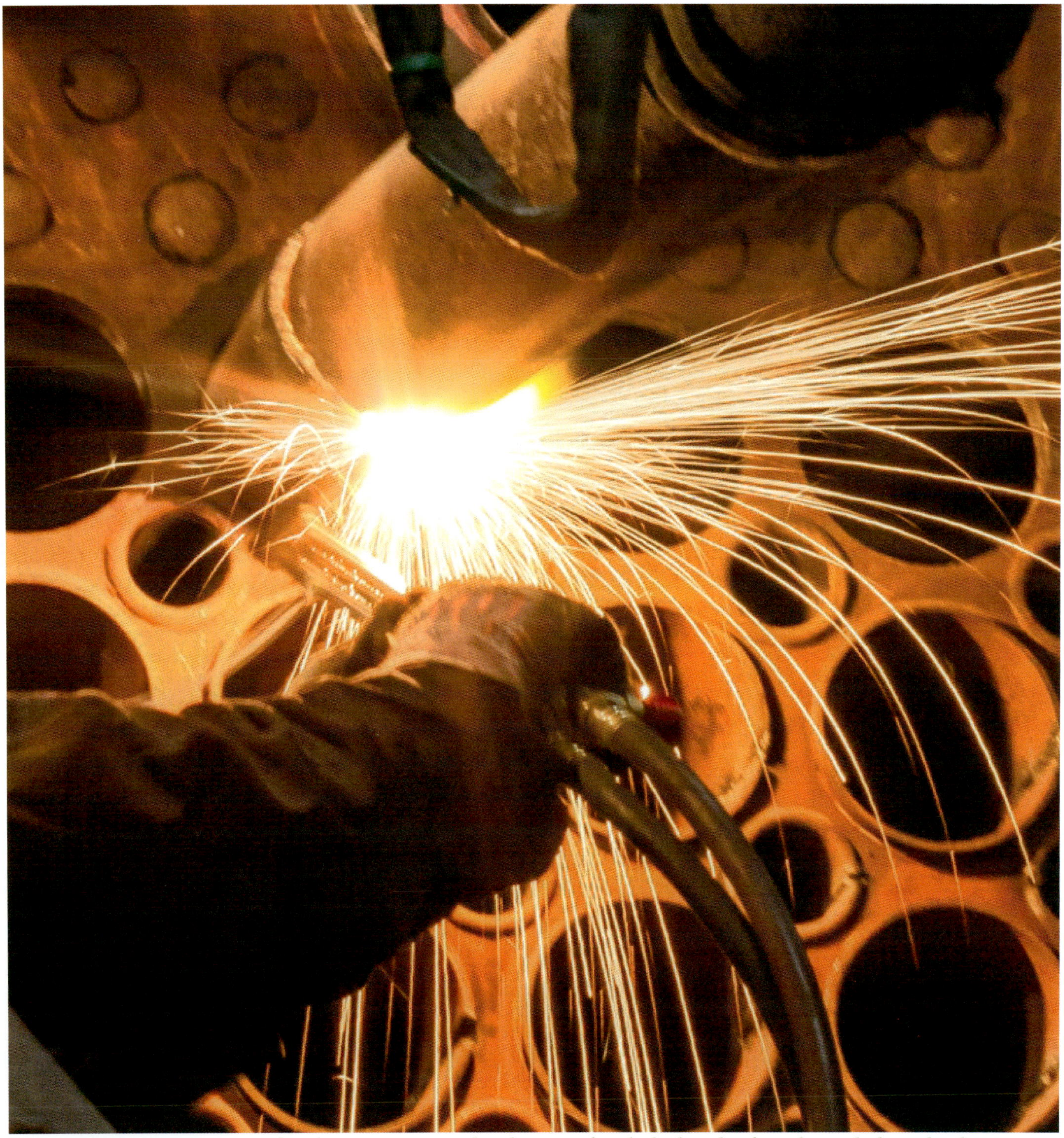

Darrin causing sparks to fly as he uses an oxy-acetylene burner to free the boiler tubes from the smoke box tube plate

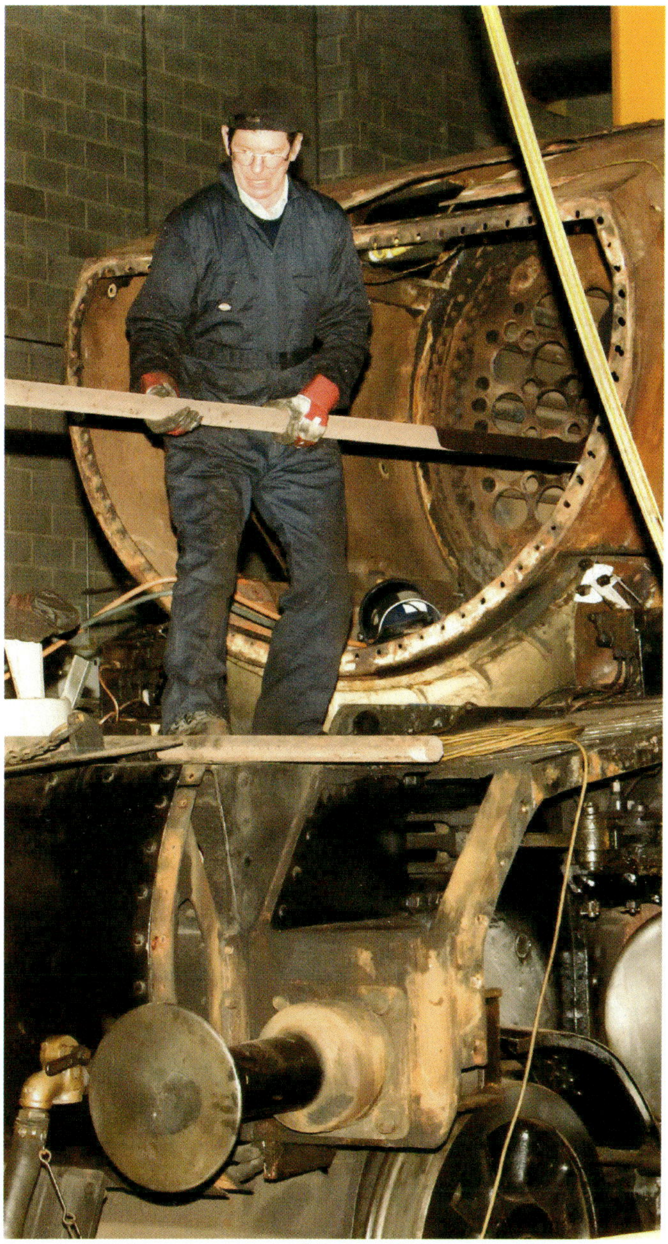

Malc removes one of the small diameter tubes from the boiler

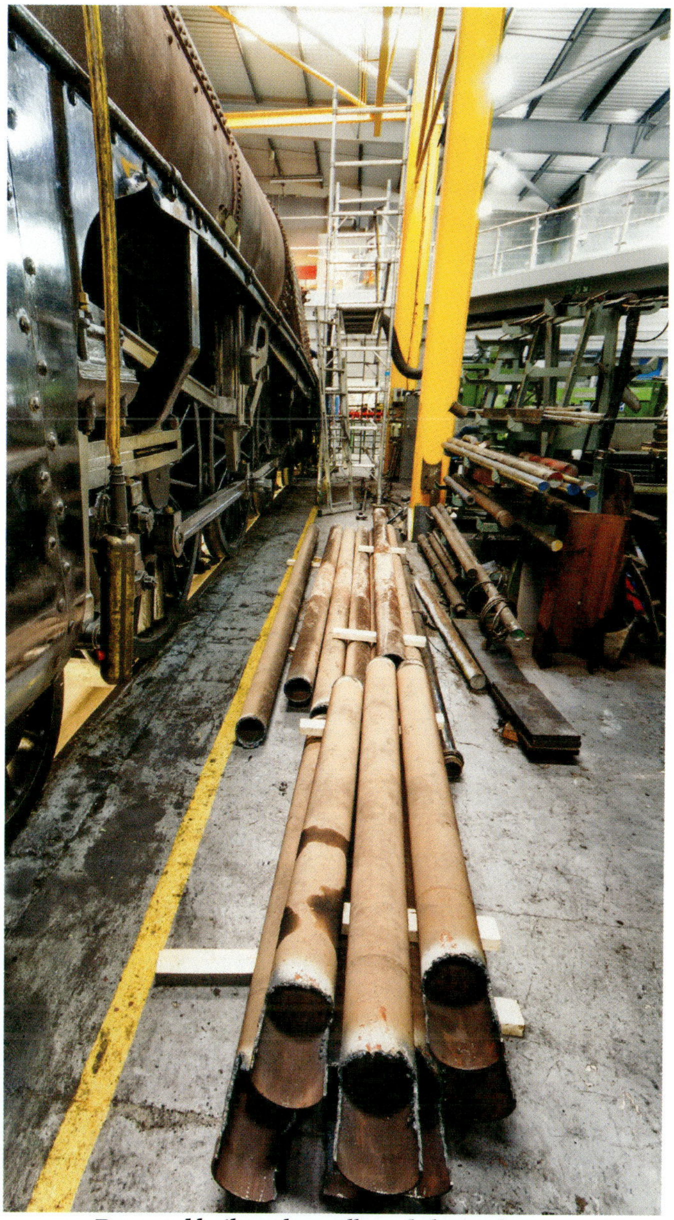

Removed boiler tubes collected alongside No.7

needed removing. To free the tubes from the boiler would require oxy-acetylene burning and I wasn't going to miss this opportunity to see sparks flying. It was a little bit of trial and error with the camera settings here - slow enough shutter speed to capture light trails but not too slow to avoid blurring the people in the shots who would inevitably be moving around. In the end the light trails won and the technique of deliberately blurring people to show action was, I admit, accidently employed now and in future similar shots.

Working from the outside of the boiler to remove the top tubes was technically difficult but once they had been removed there was another way to remove the lower tubes as will be revealed.

PASS ME A HAMMER

February 2016

"Are you coming in?"

"Trevor pass me a hammer," Darrin called, as I was contemplating my next shot. "Yeah sure," I replied but where was I passing it to and where was the hammer? The hammer I found quickly along with several others, and having a close friend who is an engineer, the decision on which one was easy - just pick up the biggest. Now to find Darrin. Homing in on his voice I found him inside the firebox. "Will this do?" "Perfect" was the reply, followed by "Are you coming in?" "Er no", I replied. Those who know me will know I would probably not get through the firebox doorway and more of a worry, if I had, would I ever get out. But I didn't need to get in anyway, as the best shot was through the firebox door, of course it was, just have a look. It turned out to be one of my favourite shots with Darrin wielding this Thor like hammer at the end of the boiler tubes to encourage them free from the tube plates.

Now, if being inside the firebox was claustrophobic, Dave Lee was to go one better. The top boiler tubes were released from the tube plates by burning free with an oxy-acetylene burner from outside of the boiler, but after they had been removed it freed up space inside the boiler, not a lot, just a little bit, for Dave to climb inside the boiler through the steam dome on top. Taking the burner equipment with him the task of burning free the tubes from inside was considerably easier, I was told. Another of my favourite shots resulted from this - Dave working with the burning equipment behind the smokebox tube plate inside of the boiler. You won't get me inside there I thought. Hmm -that's what I thought way back then.

A couple of other favourite shots on this visit came from Bob, who was busy freeing the diaphragm bolts at the firebox end of the boiler. One shot was from back down in the pit where Bob had very little light to work with and had to use the aid of a high powered torch, and the other came when it was discovered some bolts could not be freed and a grinder was used to cut through them. Long exposure was used on this latter shot to highlight the sparks but resulting in a blurred Bob, or maybe that was just the speed he was working at.

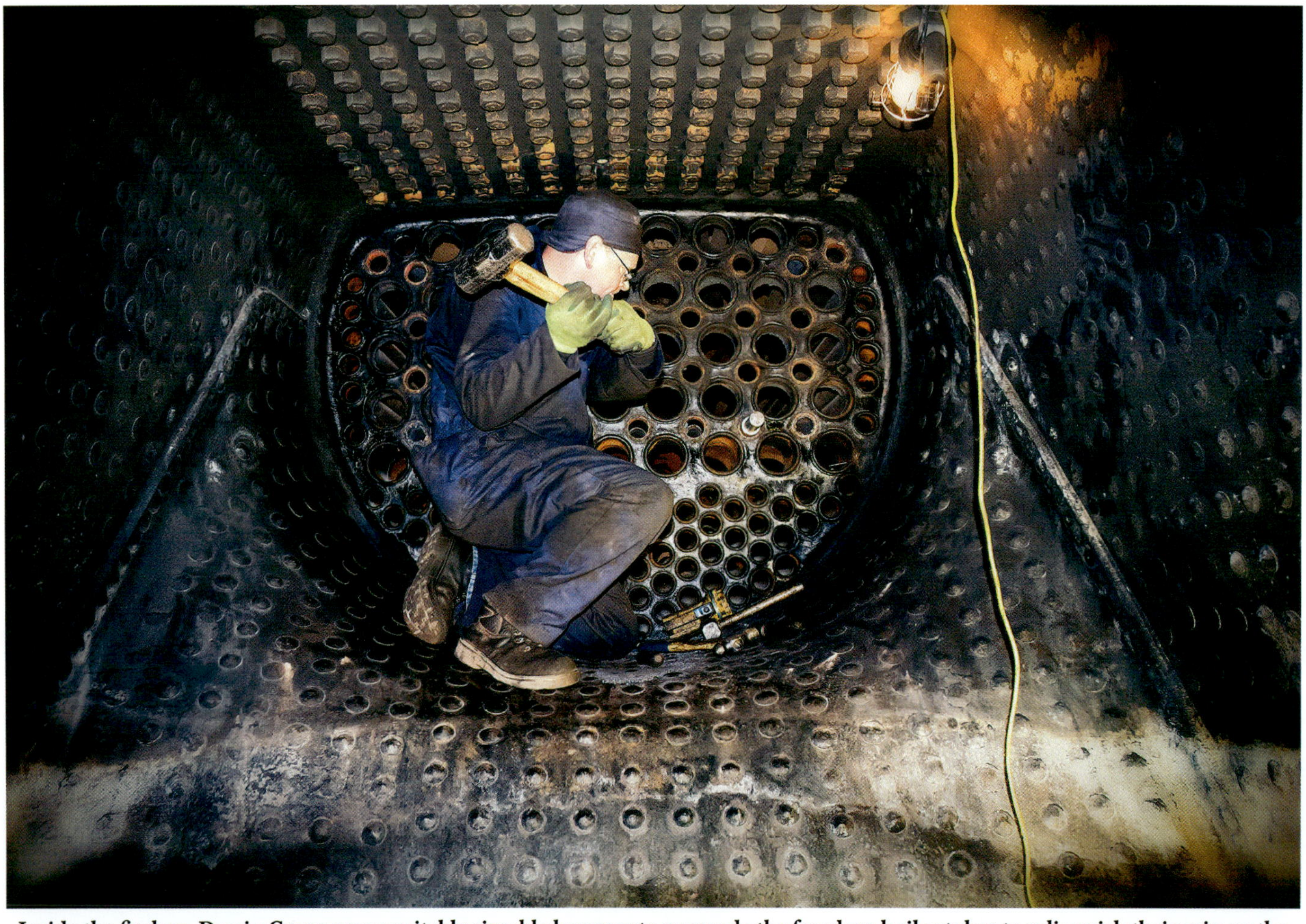

Inside the firebox, Darrin Crone uses a suitably sizeable hammer to persuade the freed-up boiler tubes to relinquish their grip on the firebox tube plate

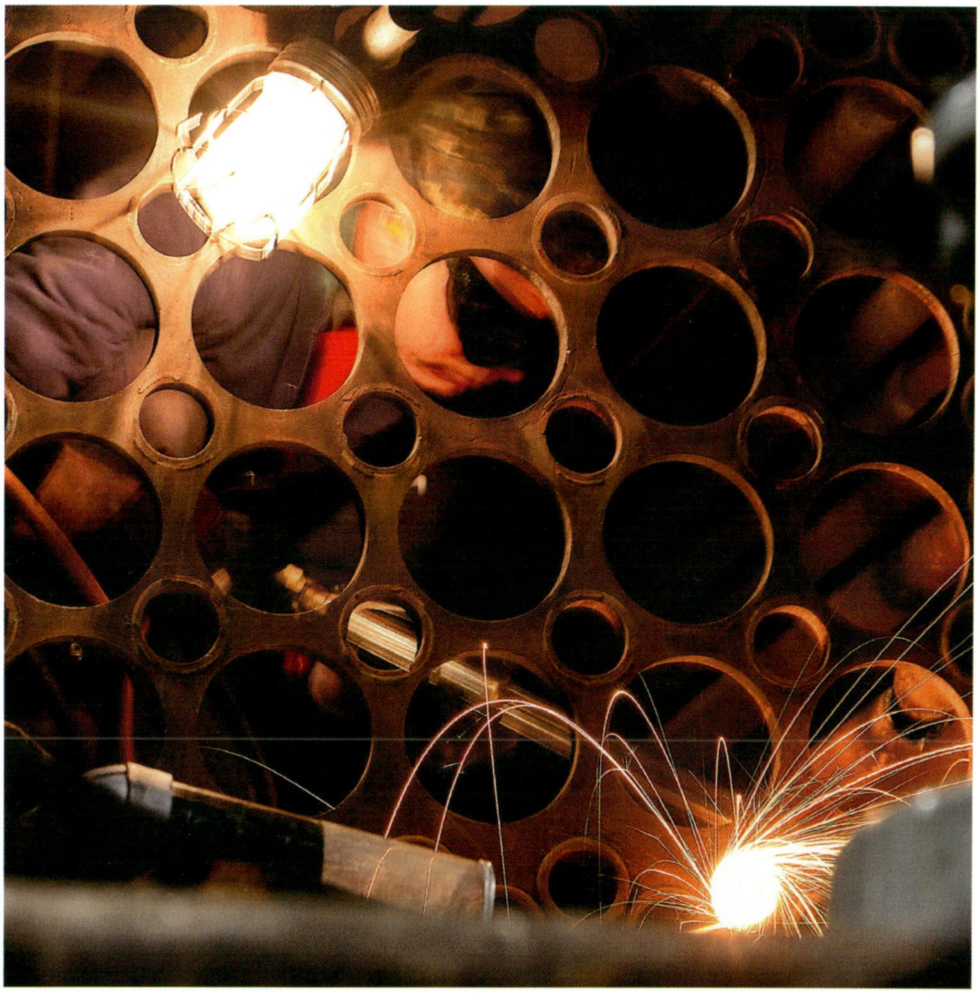

Dave Lee

I served my apprenticeship as a Road Transport Engineer with GPO/Post Office/British Telecom. I started volunteering on preserved steam railways in 1980, becoming loco driver and Head of Loco Dept. on Pontypool & Blaenavon Railway. I joined NYMR as a full time fitter on paid staff in 1990, being involved in repair and maintenance of steam and diesel locos, plus running Armstrong Oilers for 13 years, supplying Axle box Oiler Pads to British steam preserved railways and even some overseas. I retired in 2013, after 23 years full time on NYMR, and started volunteering on the SNG overhaul in September 2015. It gave me much enjoyment to be around steam locos again, and the pride of contributing towards bringing it back to working order, being part of a team of once new acquaintances who have now become friends, has been very satisfying.

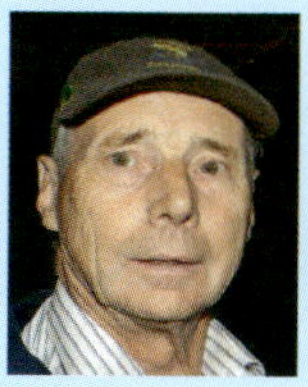

Dave Lee uses an oxy-acetylene burner to release the boiler tubes from the smokebox tube plate – note that Dave is inside the boiler, gaining access through the steam dome

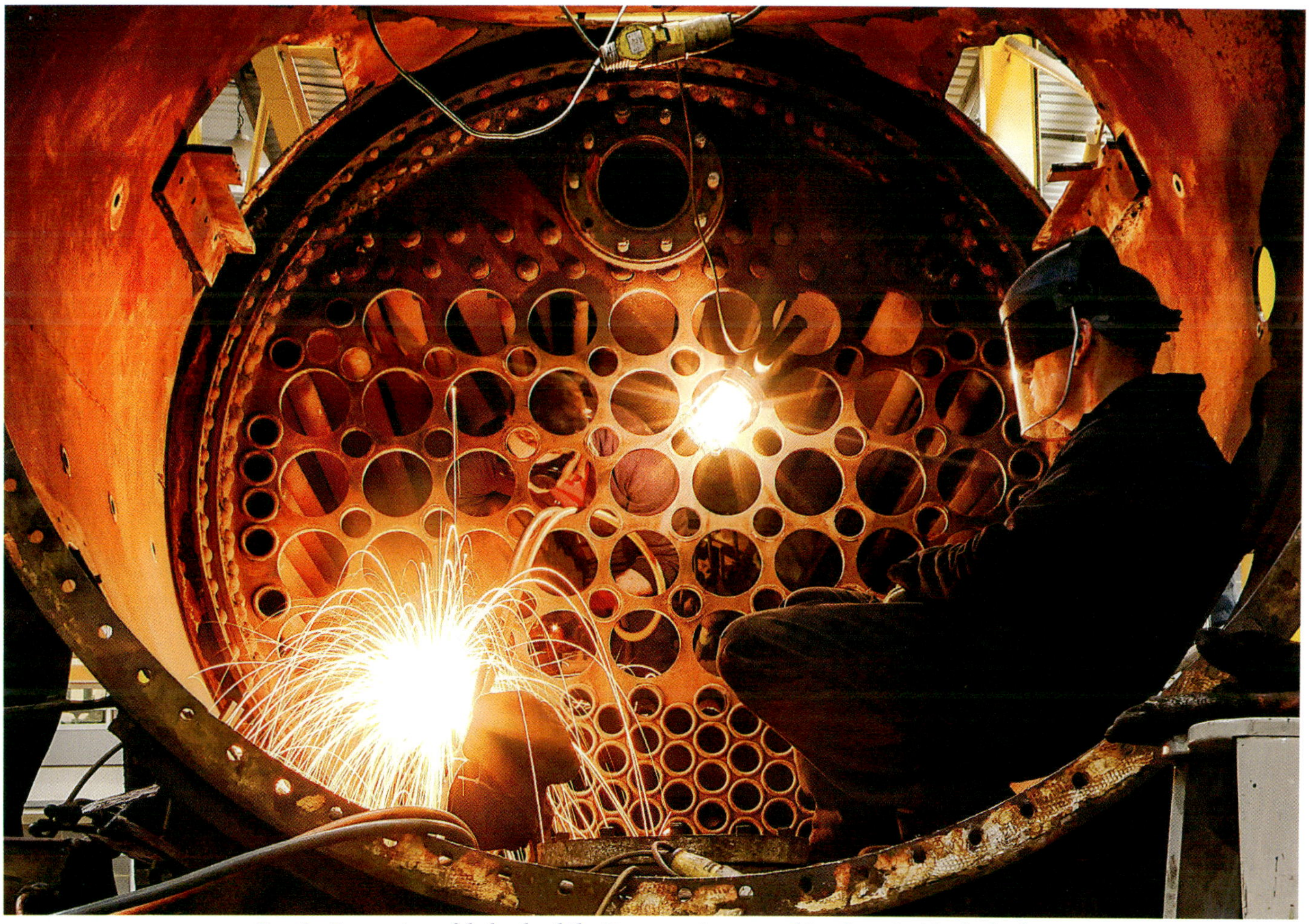

Darrin Crone sits safely back whilst Dave continues working from inside the boiler

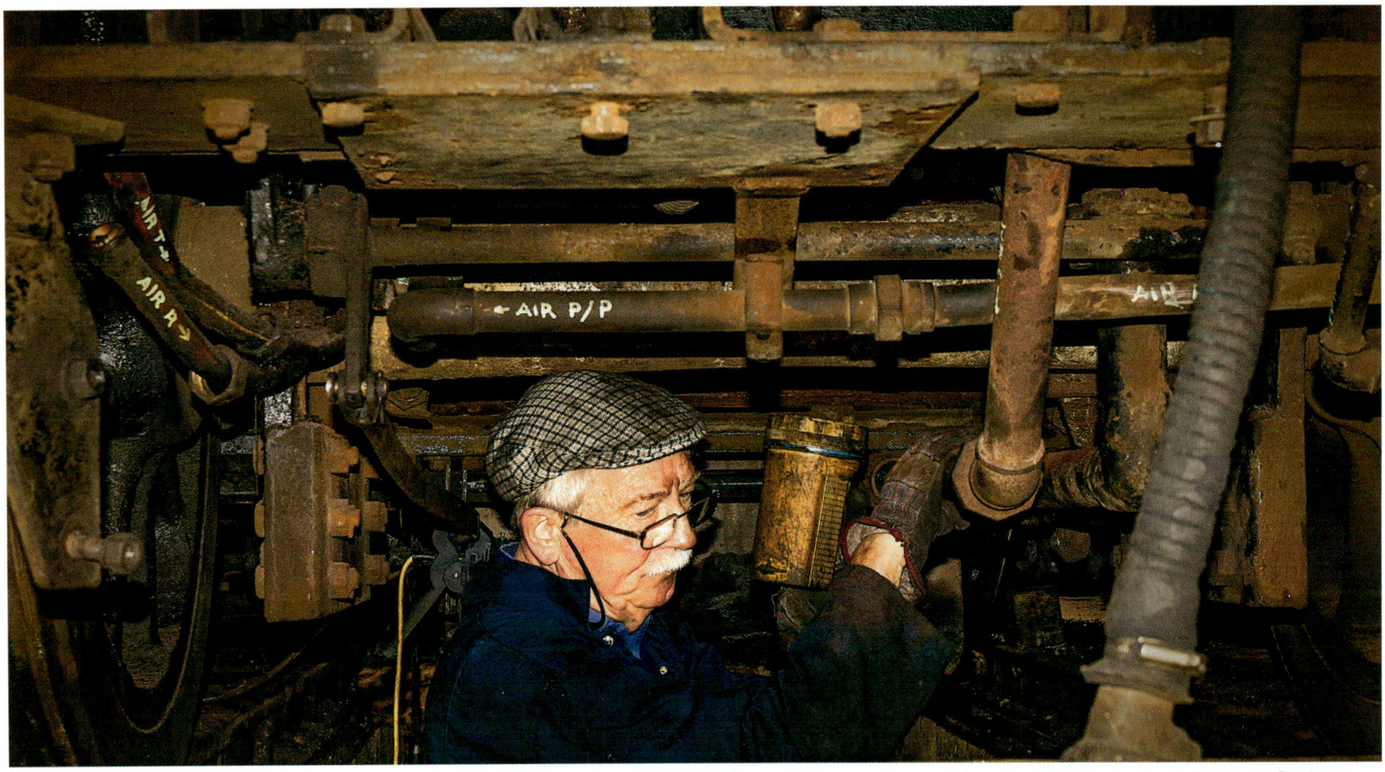

Bob Shearman, in the pit below No.7, uses a high-powered torch to locate the position of the diaphragm bolts below the firebox

Bob Shearman

I spent my entire career in the Meteorological Office but never worked as a forecaster. My hobby was mechanical work on cars including building a kit car. In retirement, I worked as a volunteer in the workshop at the Mid Hants Railway. The opportunity to work on Bittern's restoration began my love of A4s. When we decided to move north, a transfer was arranged for me to the Gresley Thursday gang at Grosmont. I looked forward to the overhaul for five years and the reality has been ever better than anticipated - an opportunity to brush up on old skills and learn lots of new ones as part of a happy, friendly engineering team. The best part is the challenge of restoring a very special A4 to a high standard. Being involved in solving the problems which frequently occur in this work creates the buzz for me.

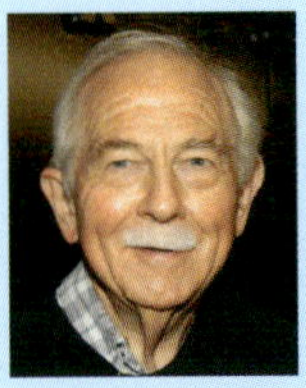

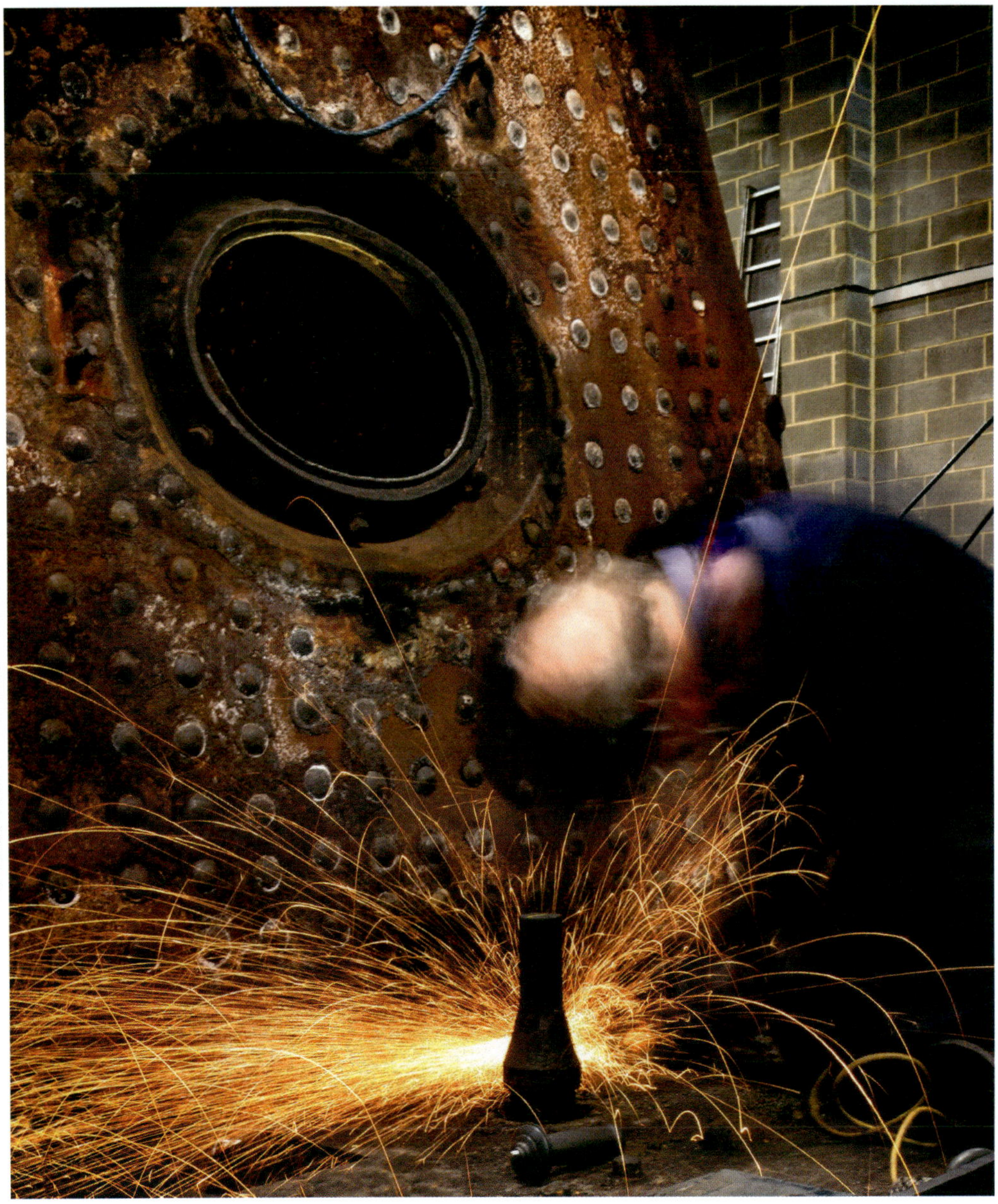

Bob uses grinding equipment to cut through stubborn diaphragm bolts below the firebox. The long exposure capturing the light trails of the sparks but also Bob's movements

THE BOILER LIFT

2

March 2016

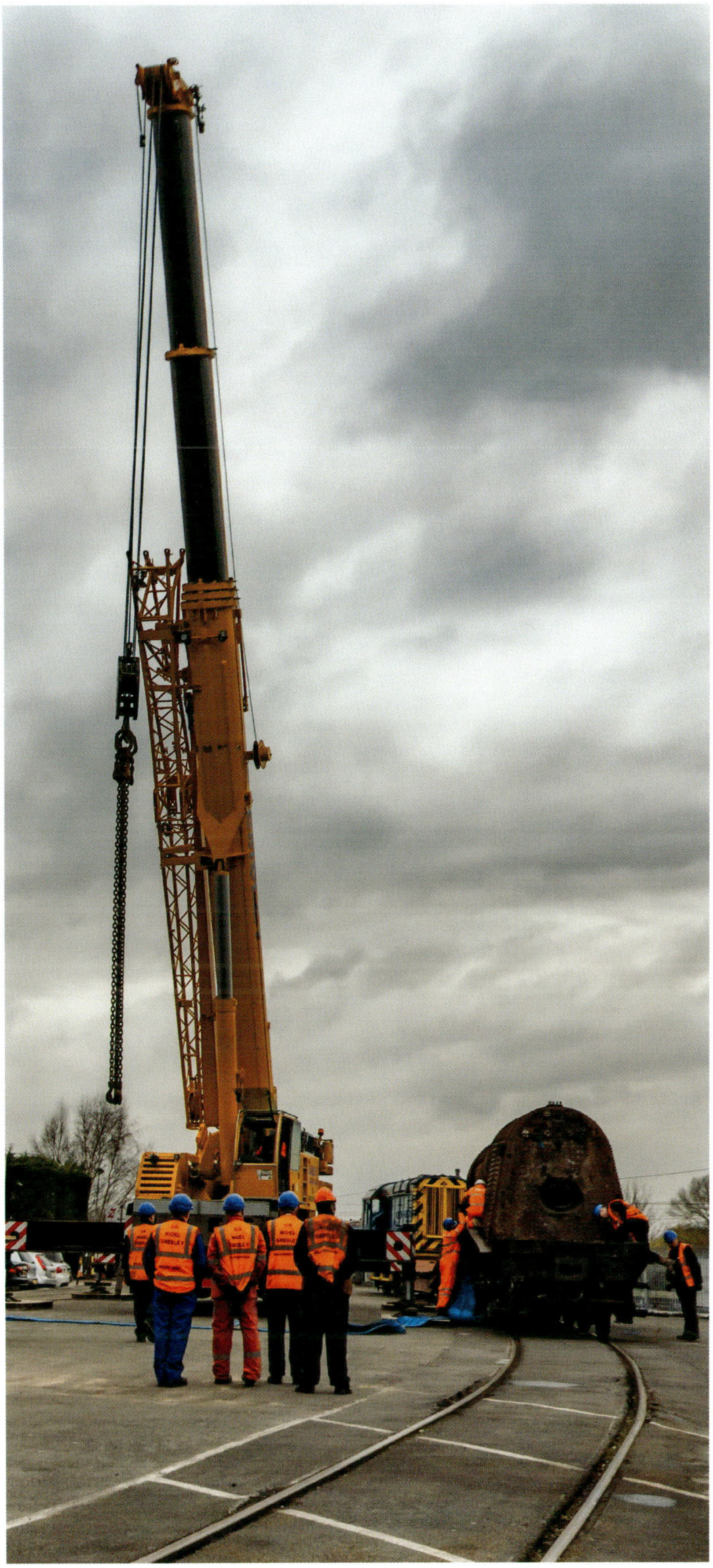

Up until now the dismantling of No.7 had been just the pre-amble - removal of the boiler tubes, cab, streamlined casing and all the associated components. Now was the first big event, what I refer to as 'Milepost' events. With the help of the NRM's diesel shunter, No.7 was moved into the North Yard so the boiler could be lifted from the frames and moved onto an accommodation rail wagon for further work prior to it being lifted again onto a heavy load lorry for transportation to the Boilersmiths at Llangollen.

The boiler lift took place on March 24th, 2016. It was a chilly, overcast day when a mobile crane met No7 and the assembled crew in the NRM's North Yard. The crew slowly got to work; nobody wanted to be responsible for making a mistake here, and none were. After some adjustments to the straps in order to find the correct balance position, the day went without a hitch and the boiler was safely moved onto its temporary home on the rail wagon. No.7, now with a considerable weight removed from the frames, was a lot lighter, and shunting the remaining wheels and frames had to be done with care as the derailment risk was greater, but this again was successfully completed.

One last job for the crane crew was to lift the ashpan from between the frames. The enormous reach of the crane boom made light work of this task, lifting and moving it back over the fence and down onto the rail side of the North Yard.

Most of my photography so far had been carefully set up in the workshop, but this day needed a different approach. I found I had to shoot quickly when the action required and to try to be in several places at once - all whilst trying to not get in the way of the work. This was not something I was used to doing on this assignment and it became a learning curve for similar days to come.

The team watch on as the crane is extended ready for the big lift

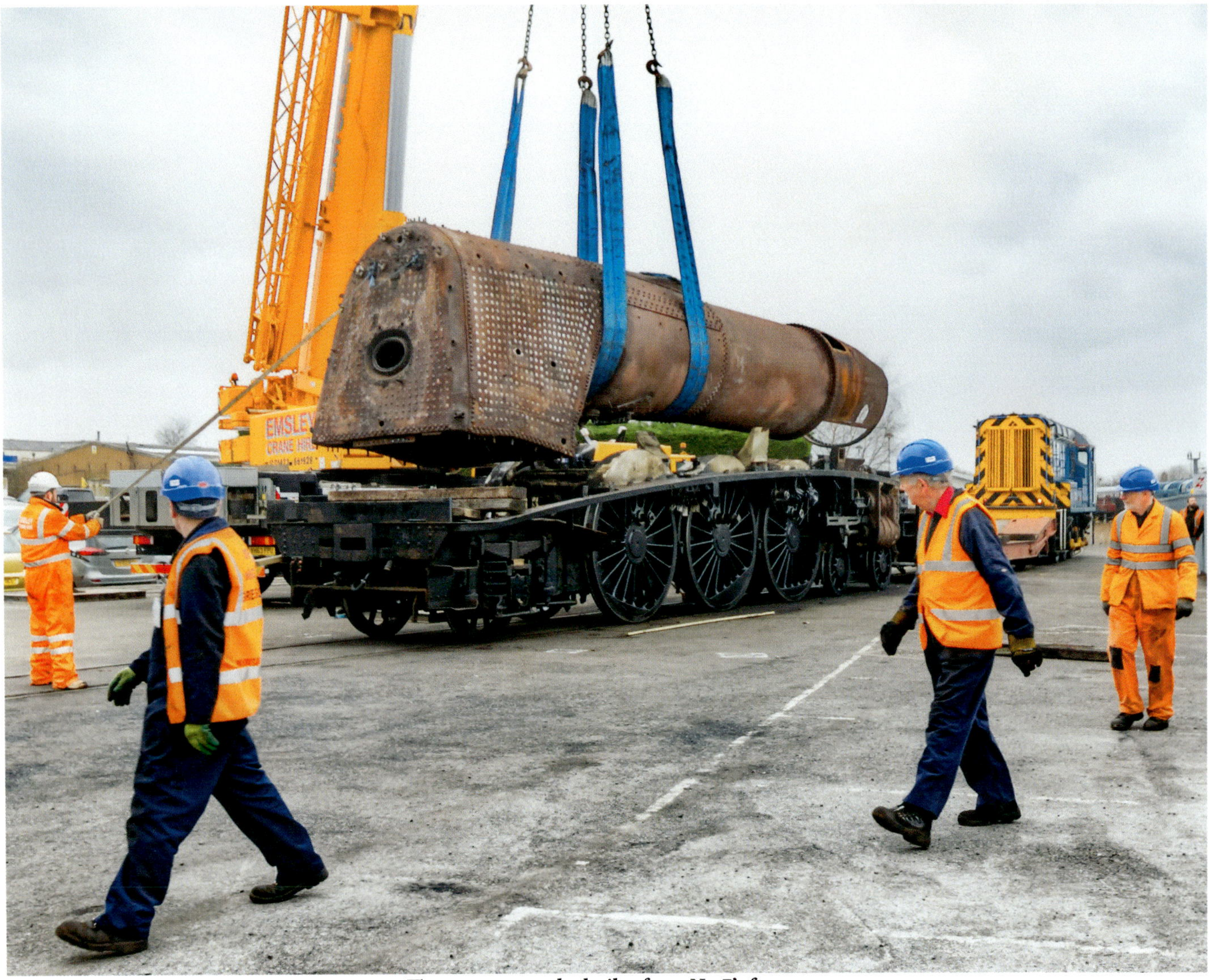

The crane eases the boiler from No.7's frames

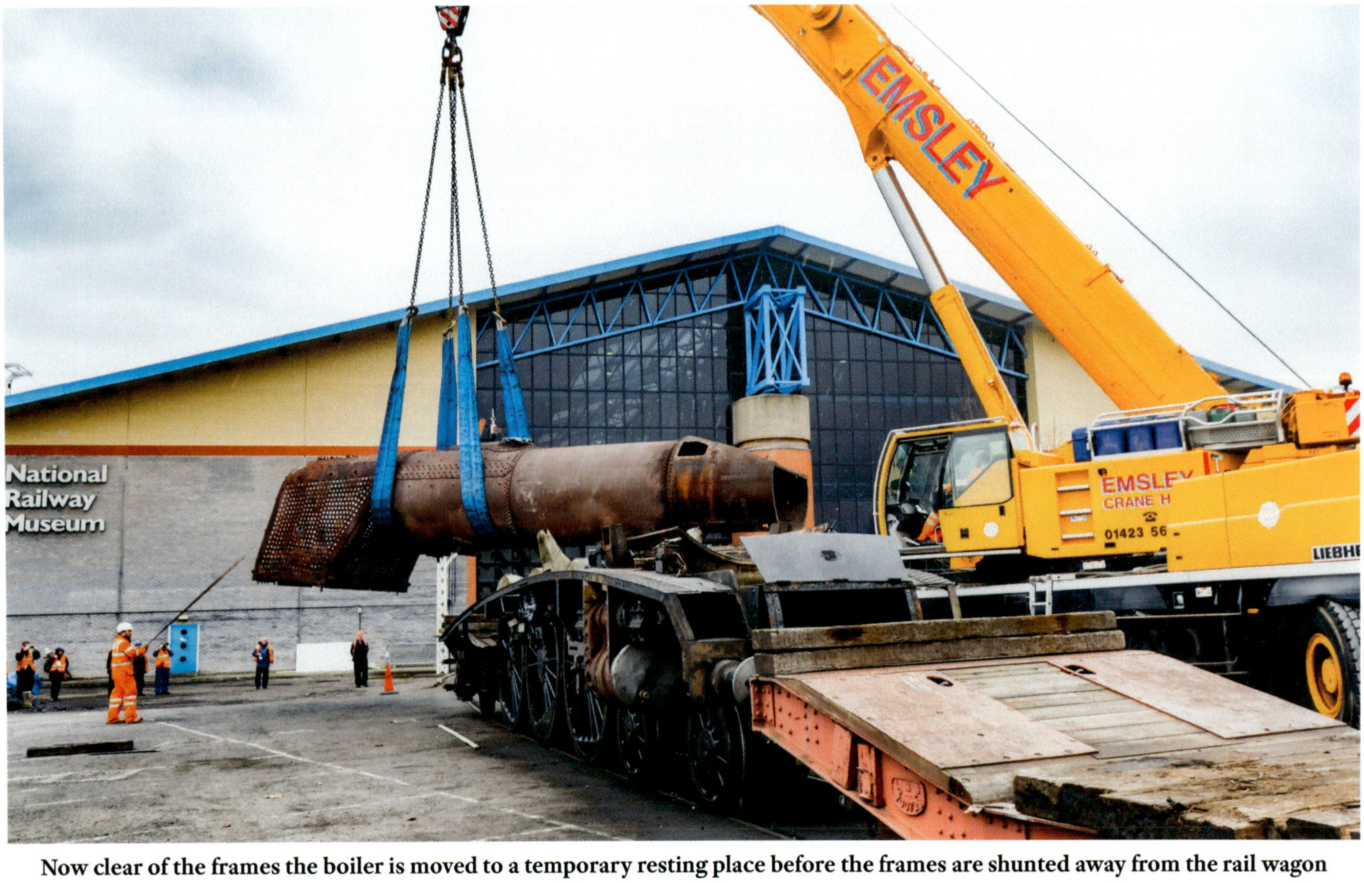

Now clear of the frames the boiler is moved to a temporary resting place before the frames are shunted away from the rail wagon

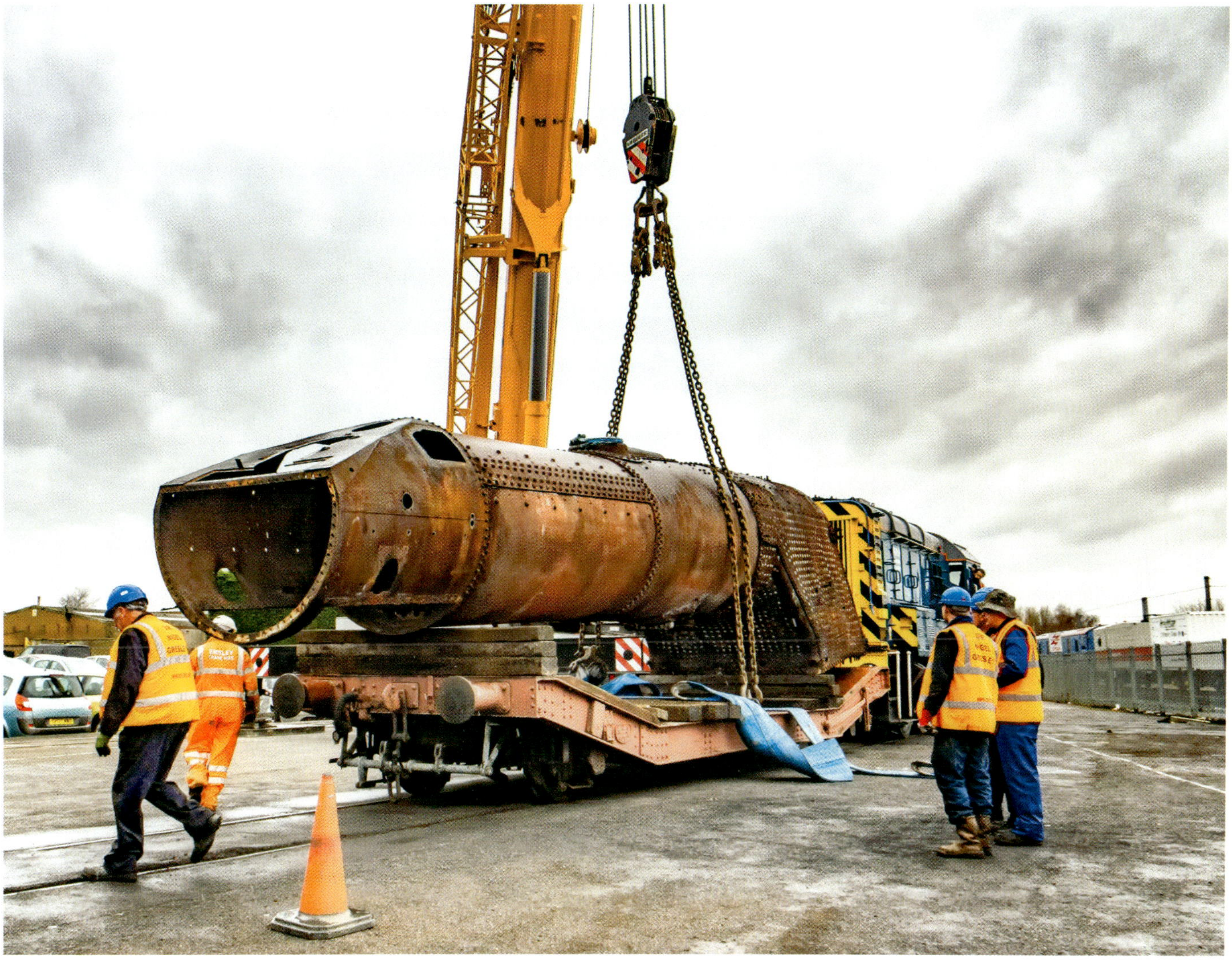

The crane relinquishes its grip on the boiler as it is safely in position on the low loader rail wagon

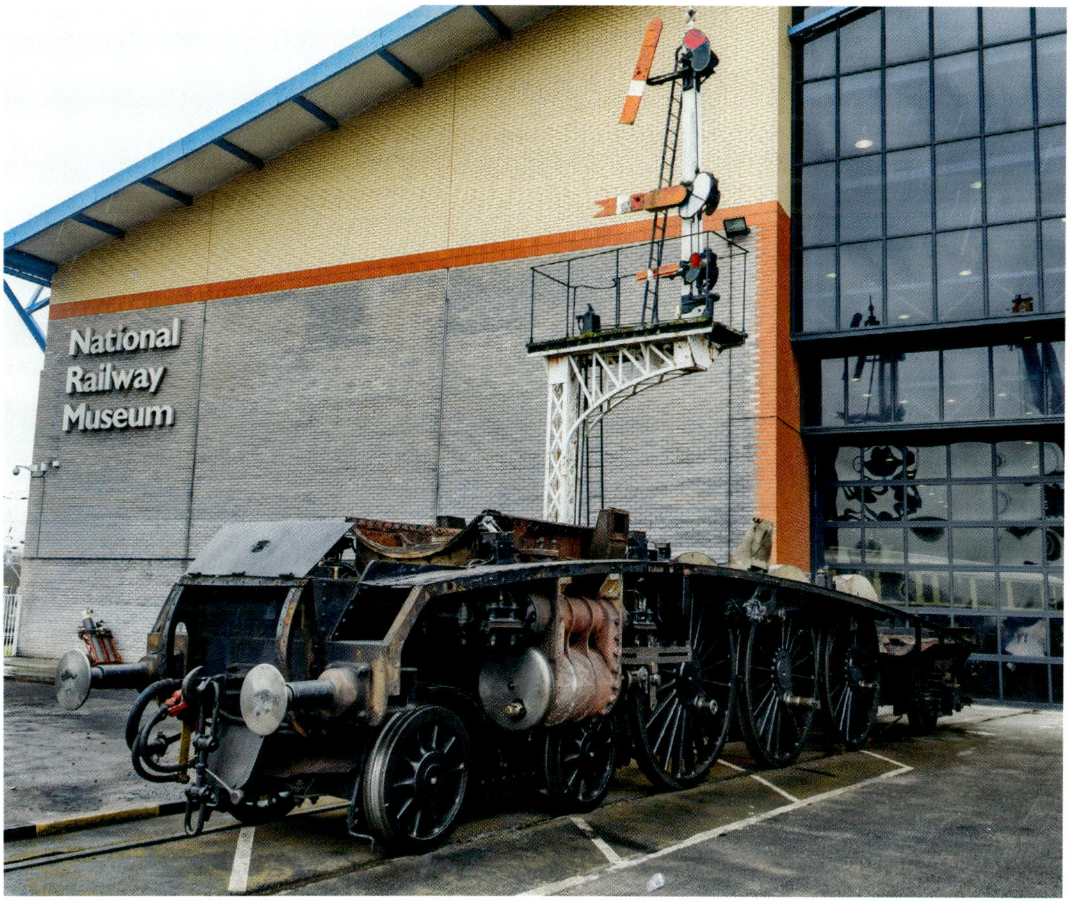

The frames, devoid of boiler, are shunted up to the NRM Great Hall windows

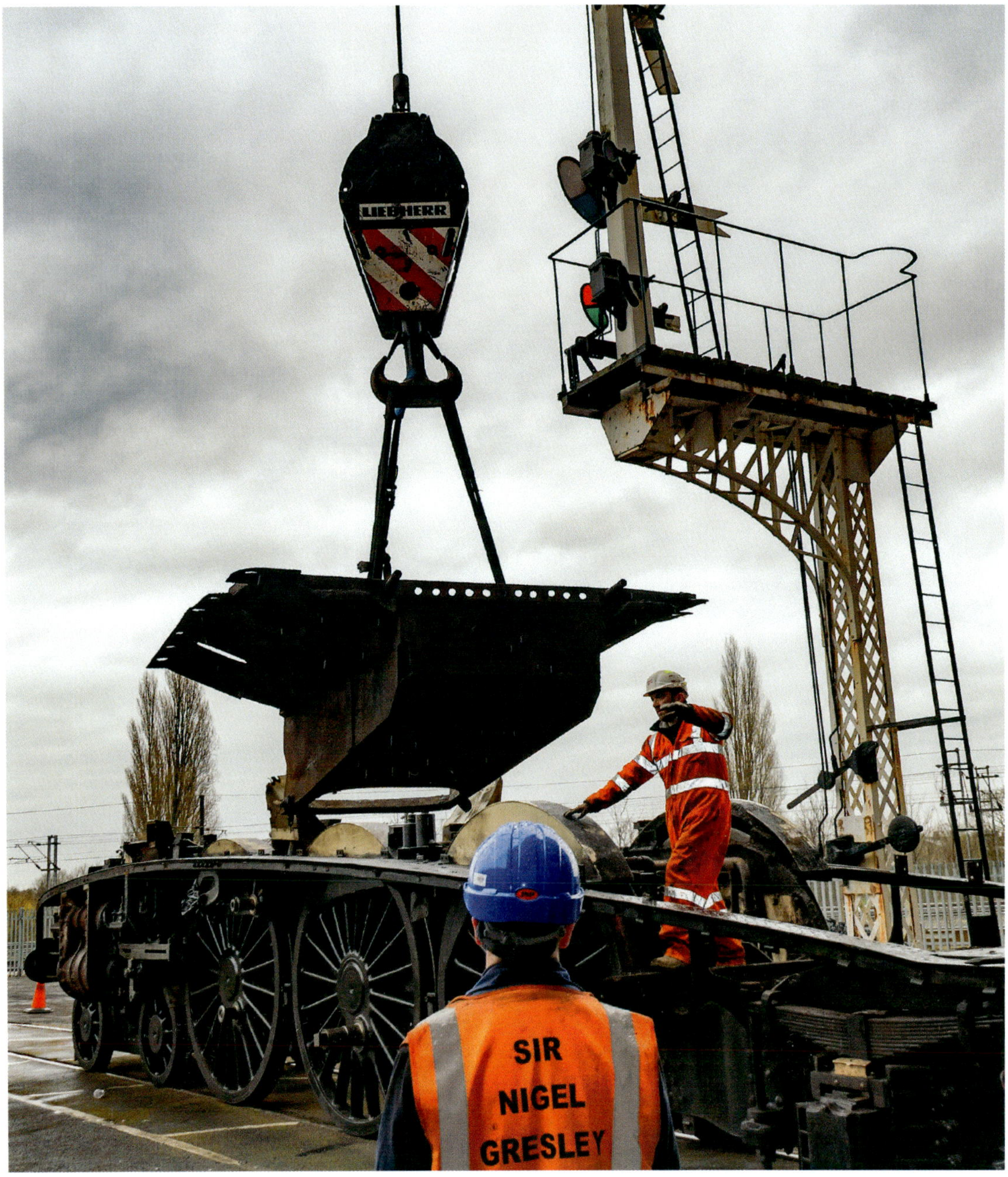

With the boiler removed from the frames the crane used its enormous reach to remove the ashpan

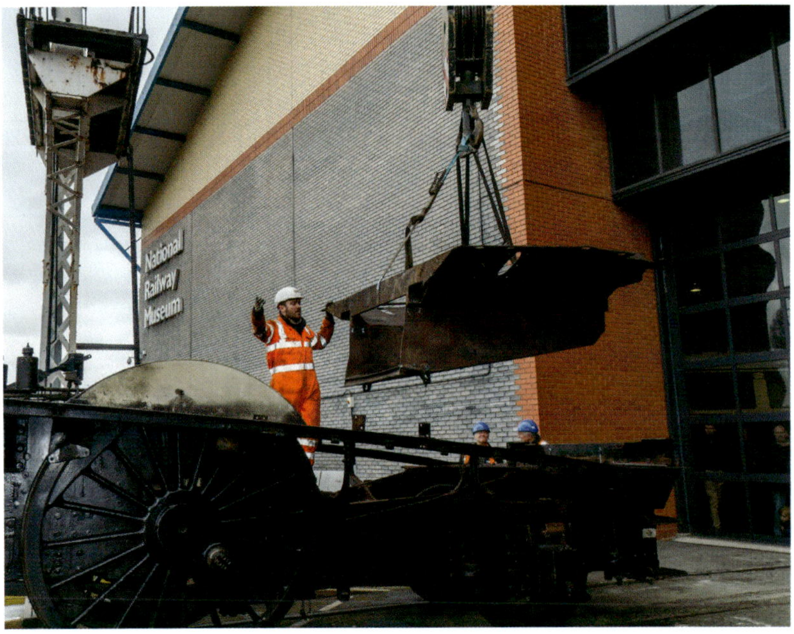

The ashpan is lifted clear of the frames

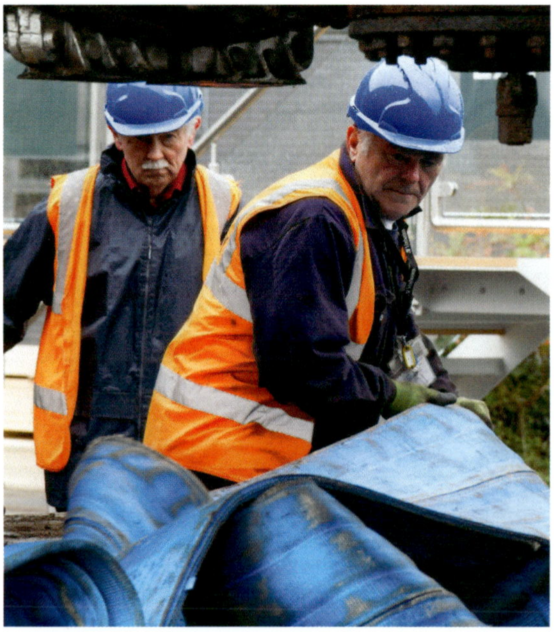

Bob Shearman (left) and Peter Brackstone remove the crane harness from under the boiler

A SHOT IN THE DARK

April 2016

Photographic equipment and a photographer who suffers from a dust allergy don't mix well with a shot blasting crew, but hey-ho I have plenty of lens cleaner fluid and antihistamine tablets. The boiler, now separated from the frames and sitting on a rail truck in the North Yard, was ready to be painted, but first required the rub down to bare metal which wasn't going to be done with emery paper but by professional shot blasters. Wearing what looked like a space suit, one of the team slowly worked his way around the boiler. But it wasn't just the outside - the scaling on the inside from ten years of operation also needed removing and to do that, our spaceman went inside to what must have been an incredibly noisy, dark and claustrophobic world. I wondered if this was a career choice when he was growing up. He certainly earned his money and my gratitude, as he rewarded me with a couple of great shots. Anybody passing by on the mainline would probably have been wondering why a sudden fog had descended over their train.

After the shot blasting was completed and following an overnight wait for the dust to finally settle, the shot blasting team applied a sprayed on protective coating of paint, which naturally was blue.

The team's attention now returned to the wheels and frames in the workshop. It would be June before focus returned to the boiler when the NDT (Non-Destructive Testing) team paid a visit to help assess the repairs needed to the boiler.

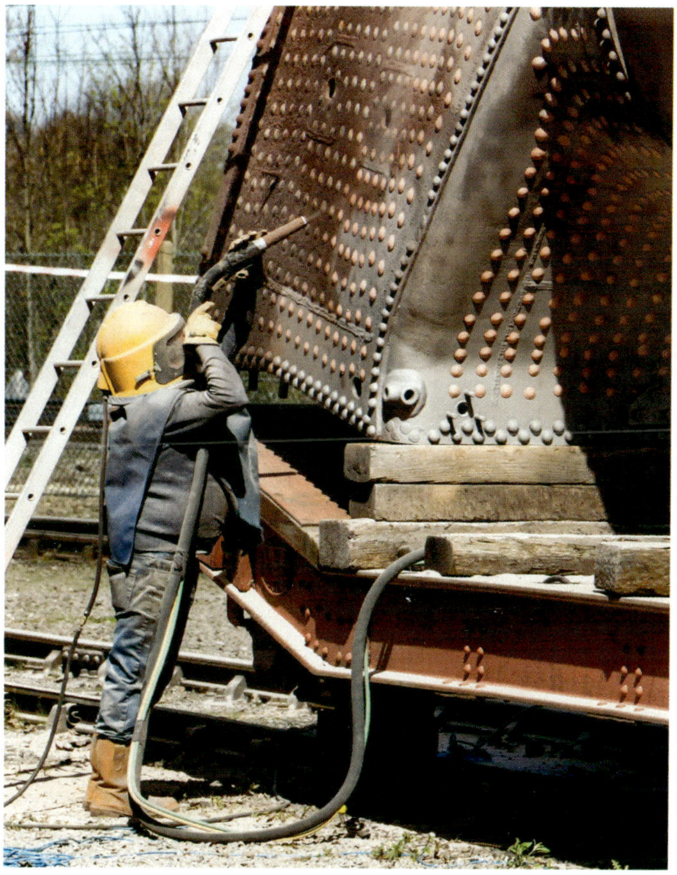

Shot blasting the outside of the boiler

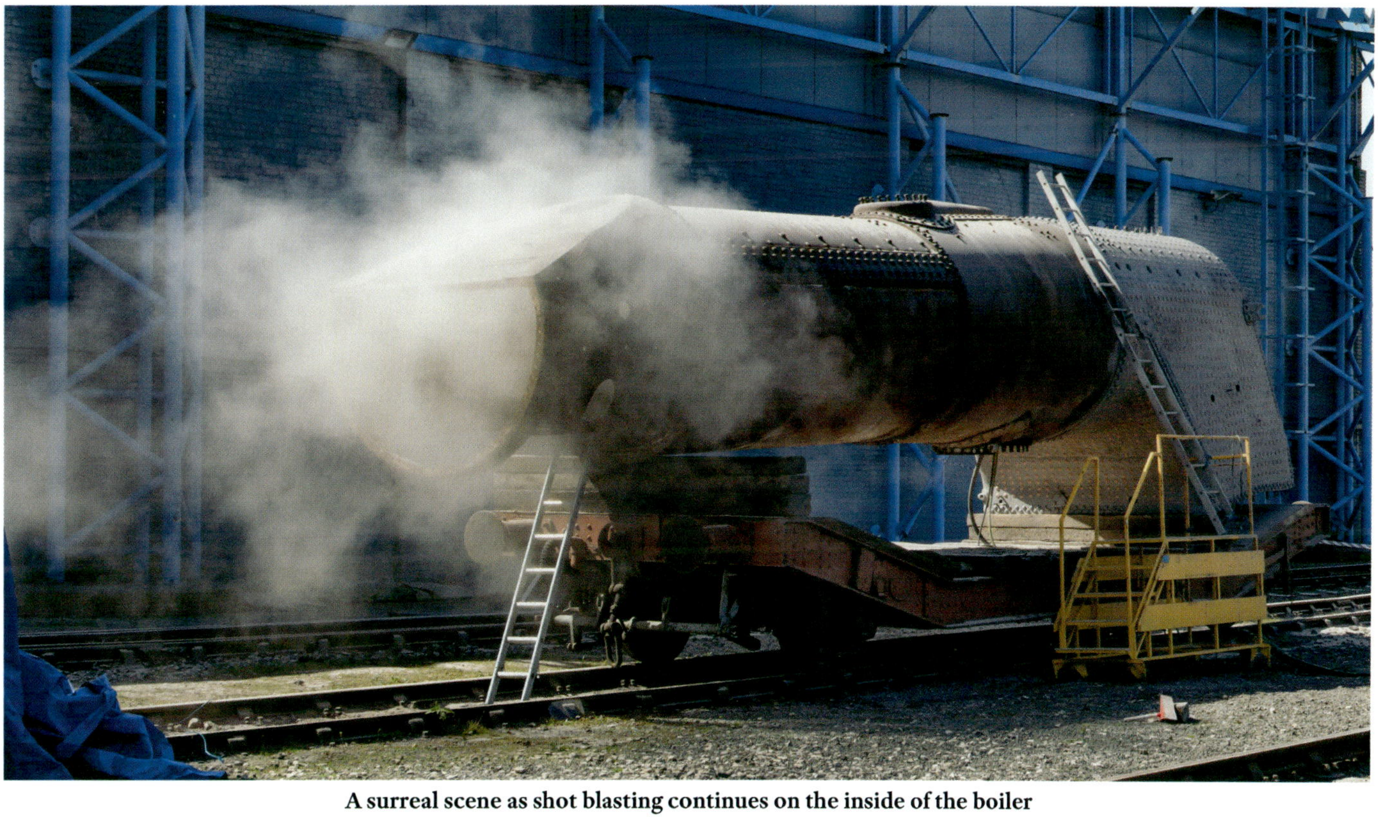

A surreal scene as shot blasting continues on the inside of the boiler

3

WHEELS DOWN

May 2016

So far, I had seen the boiler separated from the frames and wheels by using a large mobile crane, so separating the wheels from the frames would involve lifting the frames off the wheels - wrong! Let me introduce you to the 'Wheeldrop'. It operates like a normal lift except it's called a drop - it is the railway industry after all! With a locomotive positioned above the wheeldrop (which is a section of railway track that can be lowered into the ground, like a lift going to the basement), the wheels are released from the locomotive frame and lowered by the lift, sorry I mean drop

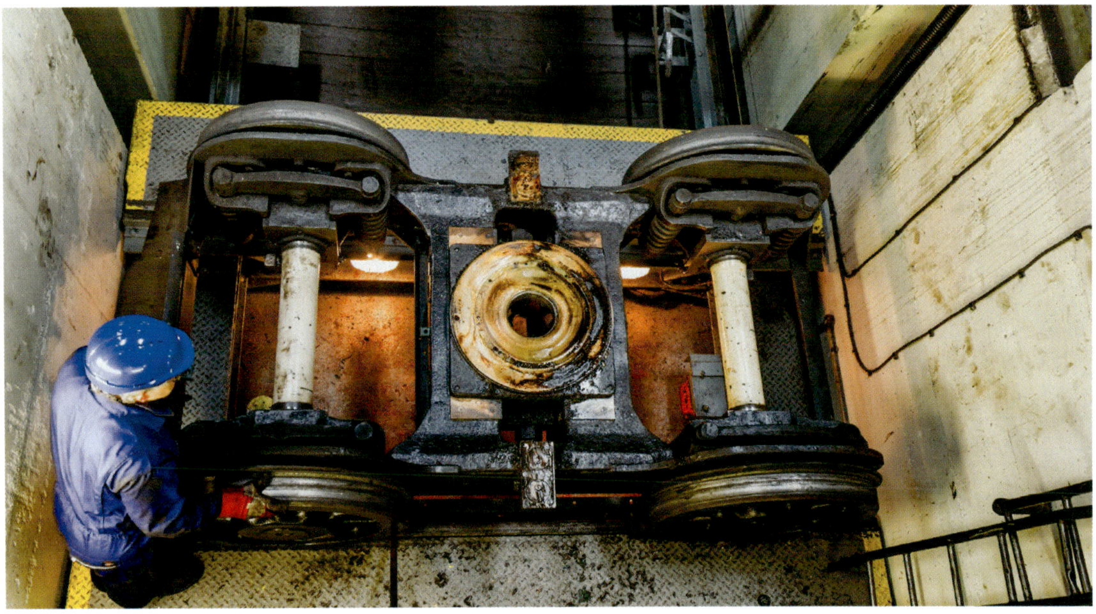

The bogie sits on the platform at the bottom of the wheeldrop

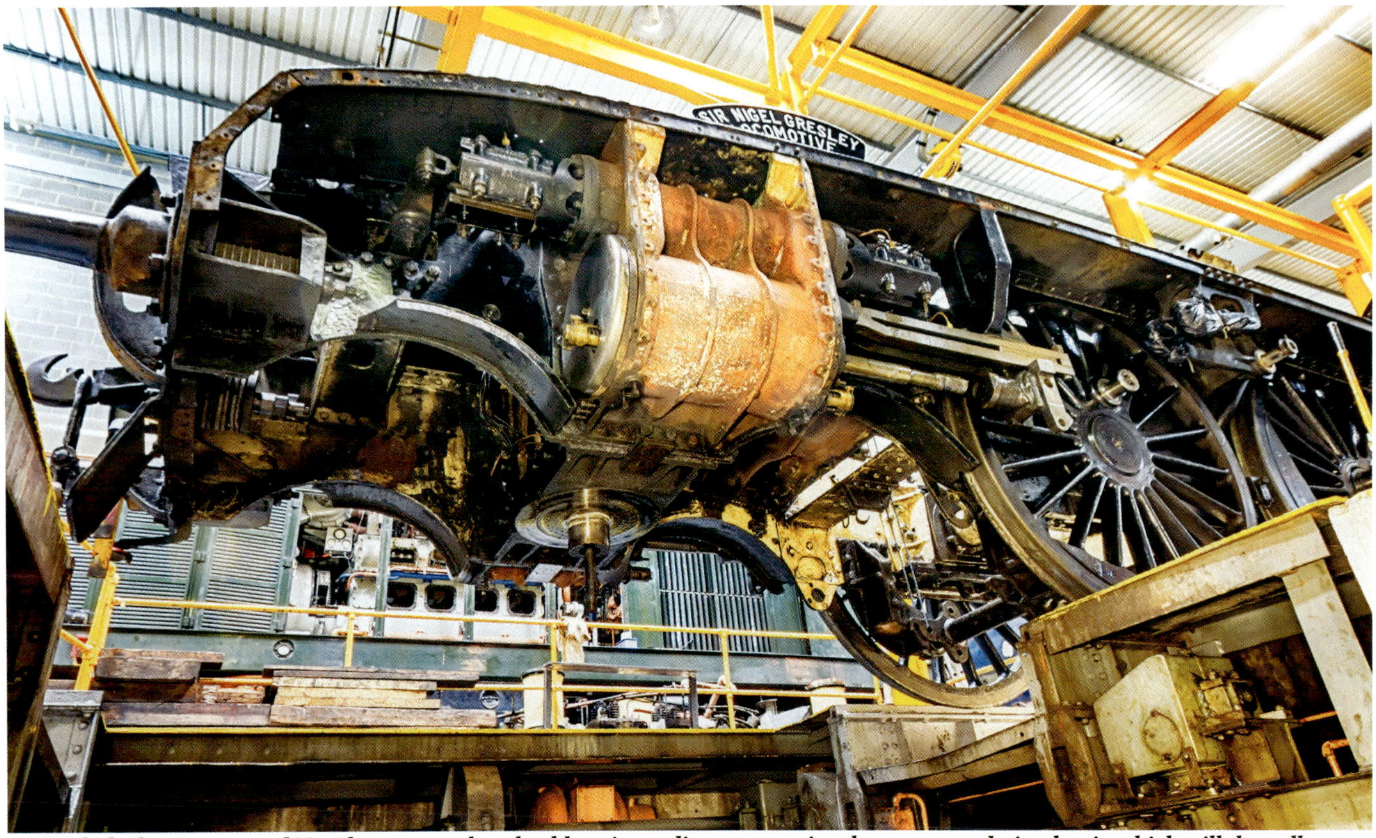

With the bogie removed No.7 hangs over the wheeldrop in readiness to receive the accommodation bogie which will then allow removal of the drive wheelsets

Andy Barwick

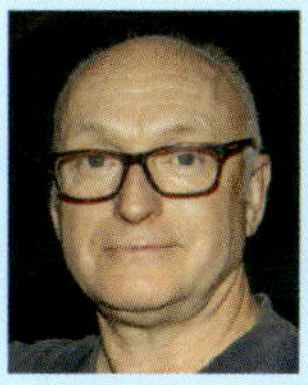

As a teenager living close to Pickering Andy followed in his father's footsteps as a volunteer on the NYMR in the 1970s. He is an avid collector of railway photographs and has an extensive collection. After a telephone conversation with the late Roger Barker from the Trust and a meeting in Grosmont he was invited to join the engineering team in 2011. In 1978 Andy worked on a North Sea oil rig as a 16 year old Instrument Apprentice before moving on to a long career with BP in 1980 where he was involved with testing and commissioning control and safety systems and eventually becoming a Maintanance Planner. With these abilities Andy is a key member of Darrin's engineering team particularly after retirement and being able to be a more regular volunteer.

Andy says "Will be able to make a greater commitment to the engineering team now that I am retired and look forward to learning more new skills during the loco rebuild under the guidance of Mr Crone," adding he "enjoys drinking Peter Brackstone's tea and listening to tall stories in the cabin."

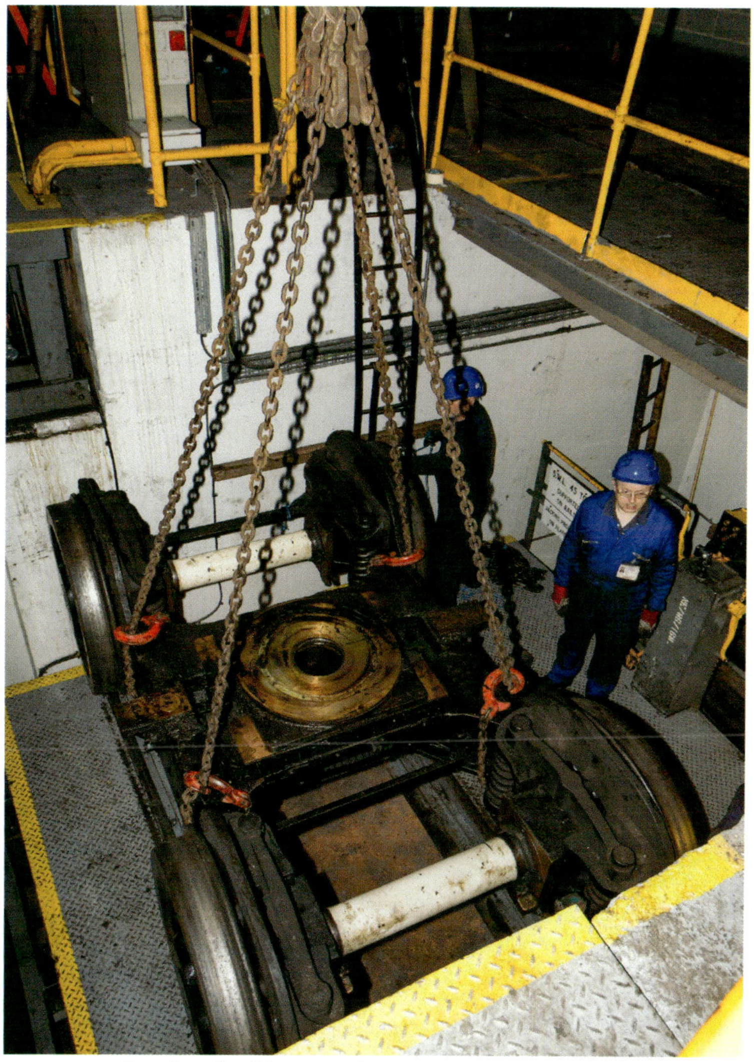

After the wheeldrop platform traverses to a position parallel to No.7 the overhead crane picks up the bogie and commences lifting it clear

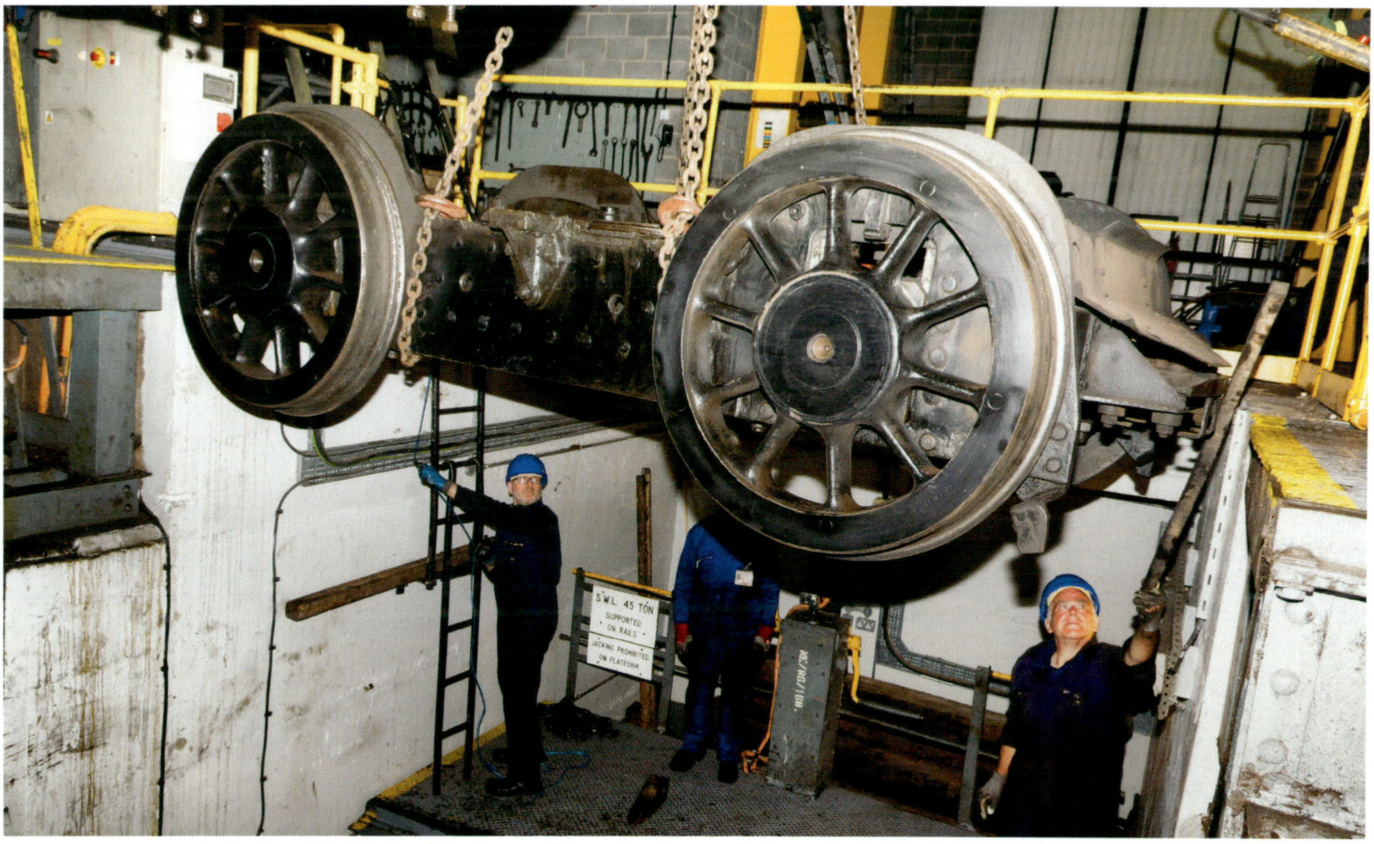

Andy Barwick, Richard Swales and Peter Brackstone help guide the bogie clear of the wheeldrop

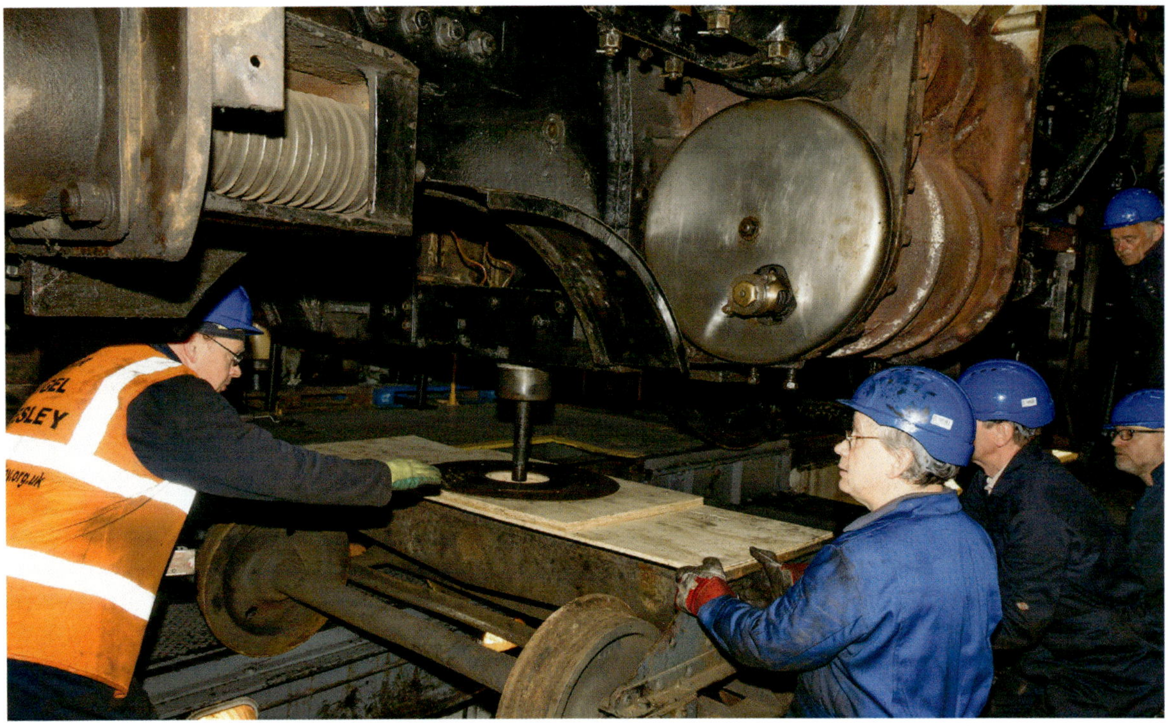

The accommodation bogie is positioned under the frames

(not literally), into the ground. They are then mechanically moved laterally to a position parallel to the track, where an overhead crane can lift them clear. Hopefully, my images will make this clearer.

"Are we pushing this all the way to Kings Cross?"

The bogie was first, and with it safely lowered to the bottom of the wheeldrop it was time for me to climb down and take an image of the frame overhanging the now open section of track above. Being under a train is not to be recommended, but the angle offered gave me an exceptional and dramatic shot of No.7. It was then time to climb out and take the shots of the team using the overhead crane to lift the bogie clear and park it on a parallel track.

"A bit more"

With the bogie now removed, the front of No.7 would need supporting before any of the 6'8" driving wheels could be removed, so an accommodation bogie was rolled in to take the original bogie's place. This then paved the way for

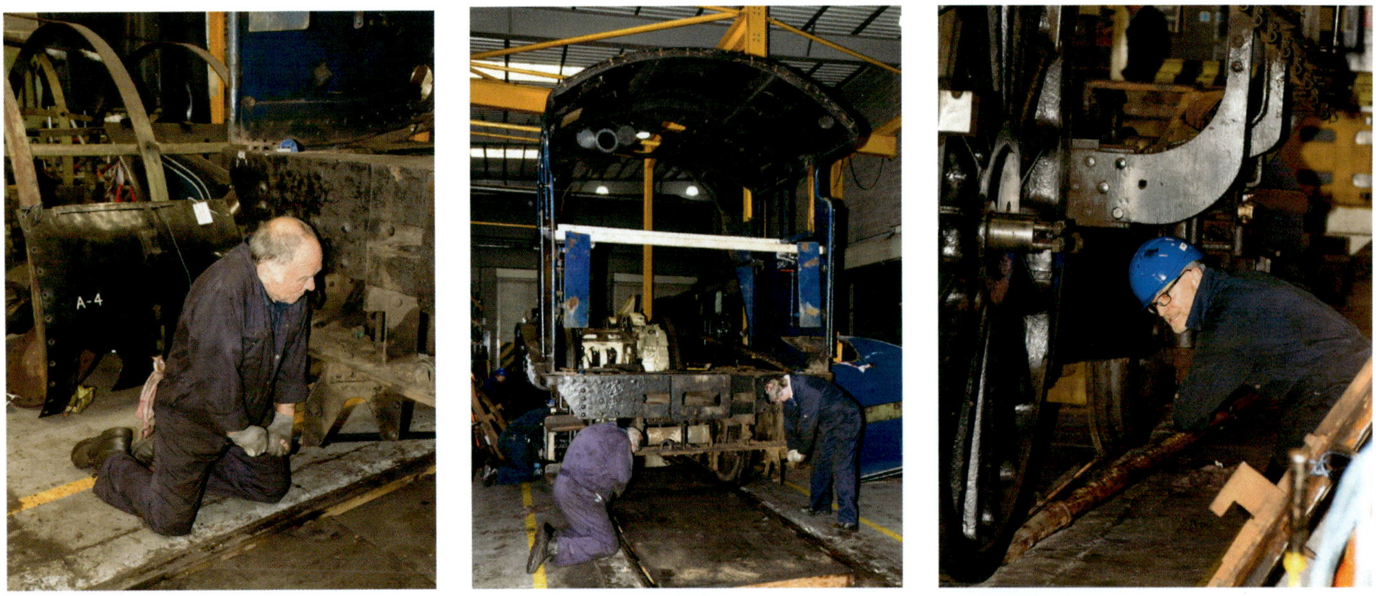

Peter Brackstone, Andy Barwick and Malc Bateman inch No.7 forward using pinch bars

removal of the driving wheels, but first No.7 needed moving forward to position the front drive wheels over the wheeldrop. How are we going to do that, I wondered? Muscle power, that was how. Well I have pushed a car before but pushing No.7 down railway tracks? I was pleased I was just the photographer, as up stepped the engineering team of the day. They gathered along the sides and rear, some armed with pinch bars to persuade the wheels to inch forward, and with all responding to the team leader's calls like a

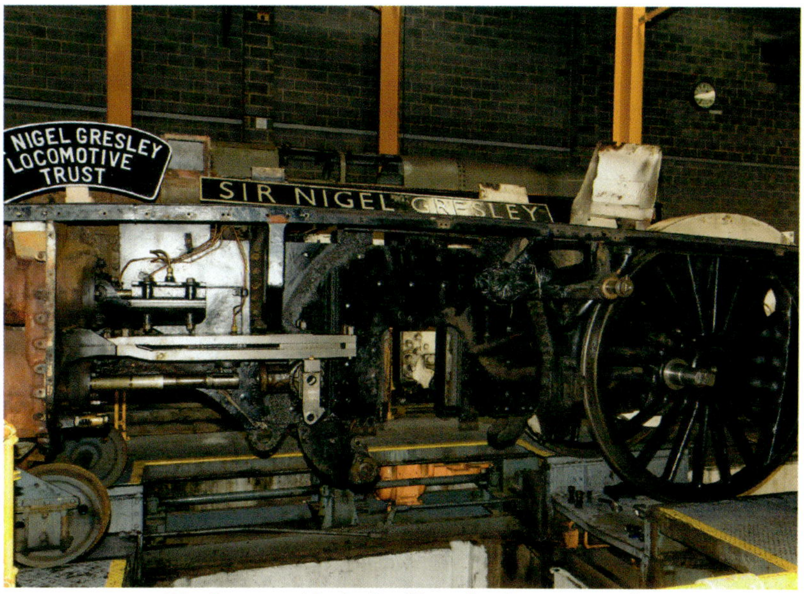

The frames with the leading wheelset removed

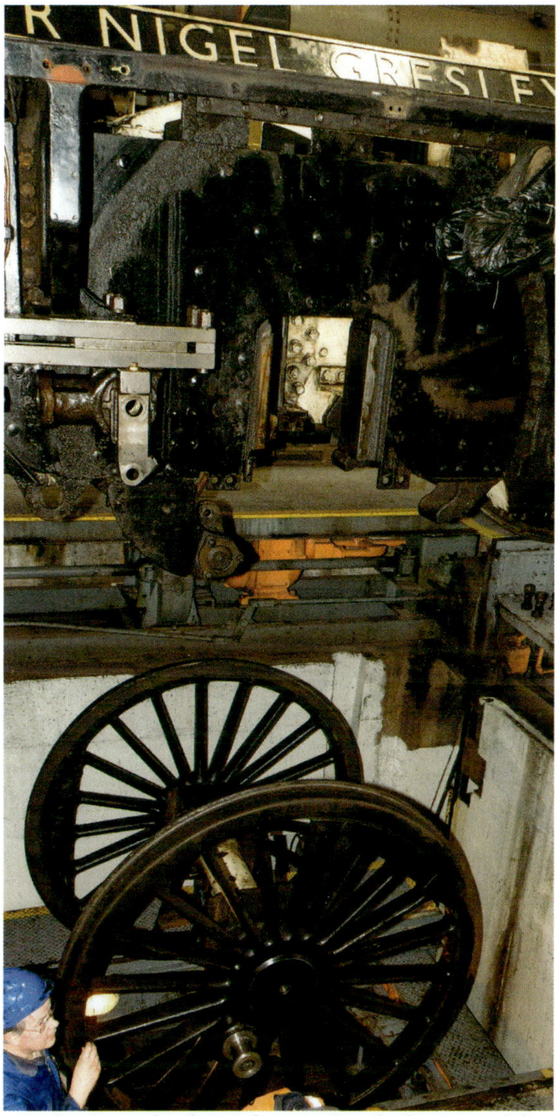

The leading wheelset is lowered to the bottom of the wheeldrop

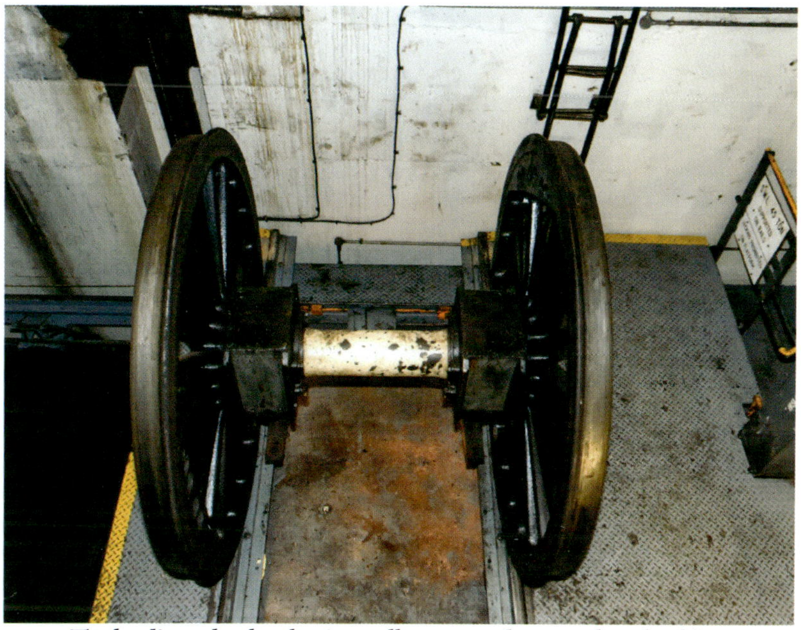

The leading wheelset horizontally traversed across the wheeldrop

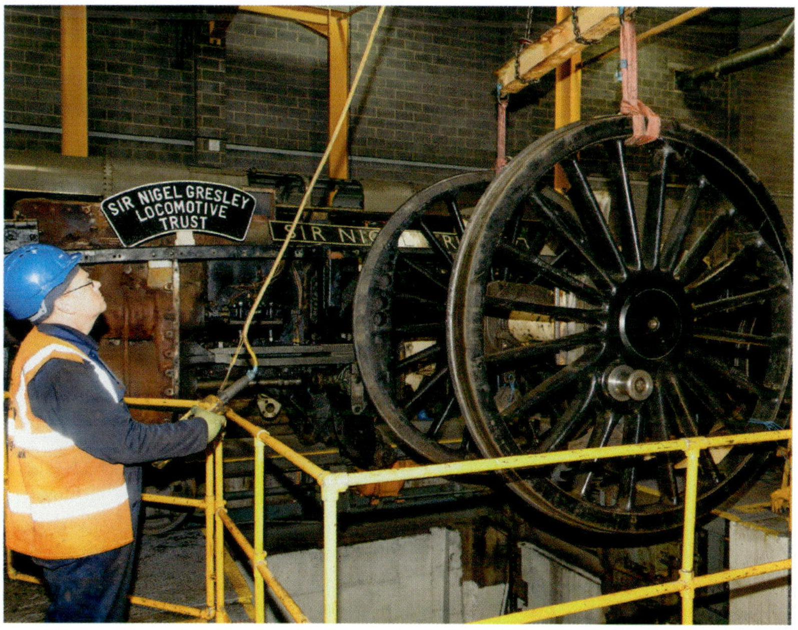

The leading wheelset is lifted from the wheeldrop to be positioned on a parallel track

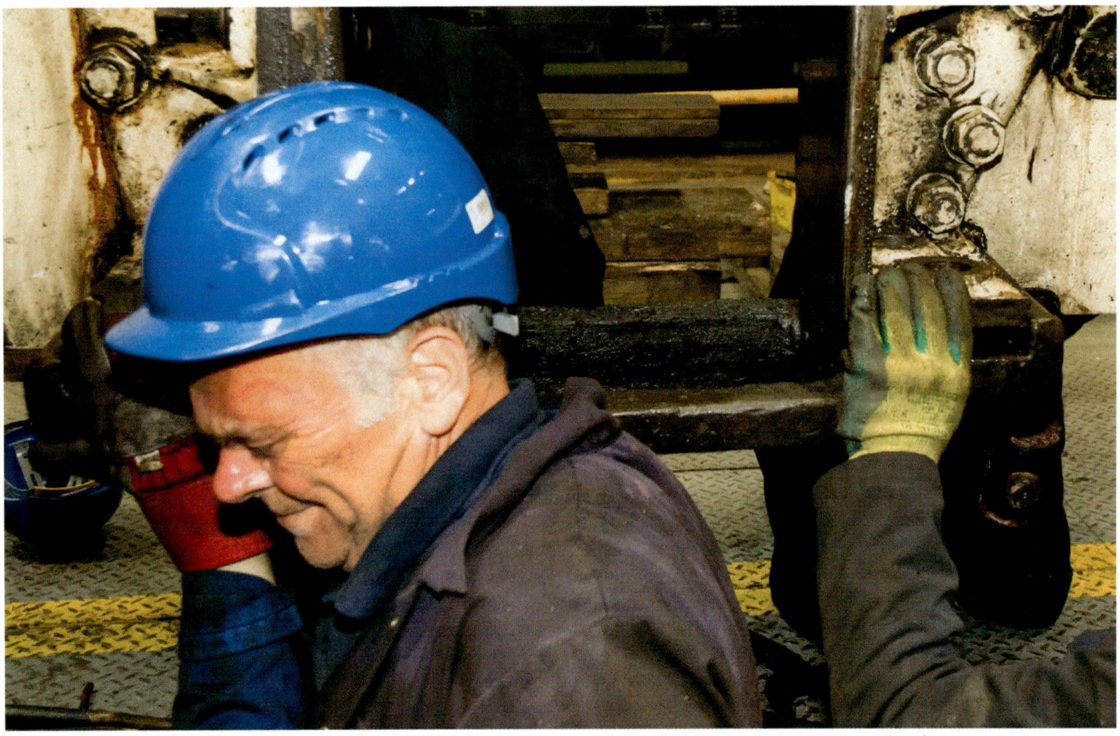

It takes the combined effort of five people to reattach a heavy hornstay to the frames

Red Arrows display, No.7 slowly moved forward over the wheeldrop. It wasn't long before one 'a bit more' cry was met with the reply of 'are we pushing this all the way to Kings Cross?'. Eventually No.7 reached the correct location and the team collapsed exhausted but in great spirits.

Now it was the turn of the leading wheelset to be lowered into the wheeldrop. The heavy hornstays, which secure the axle boxes in the frames, were removed and the lowering of the wheeldrop platform began. Following the same procedure as the bogie, the wheelset was traversed horizontally to be clear from under No.7 and then lifted clear by the overhead crane.

To avoid unecessary stress on the frames the hornstays were now refitted - a difficult and strenuous job with the cast steel being very heavy, awkward to handle and a tight fit. Muscle power was again called upon with Peter Brackstone once more excelling in that department, using his shoulder to hold the hornstays in position whilst the other team members set about fastening them onto the frames. Words of, er, encouragement were jokingly provided by Darrin

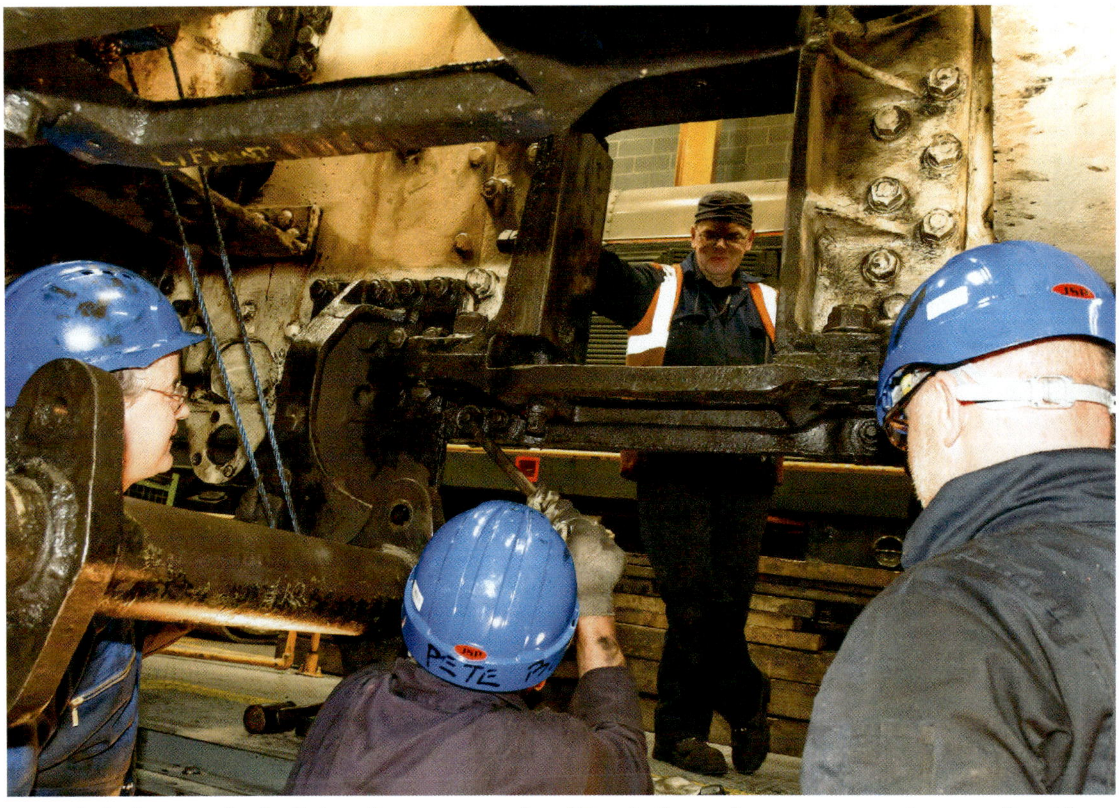

With the heavy task of refitting a hornstay completed Darrin Crone shares some amusing remarks with (left to right) team members Richard Swales, Peter Brackstone and Andy Barwick

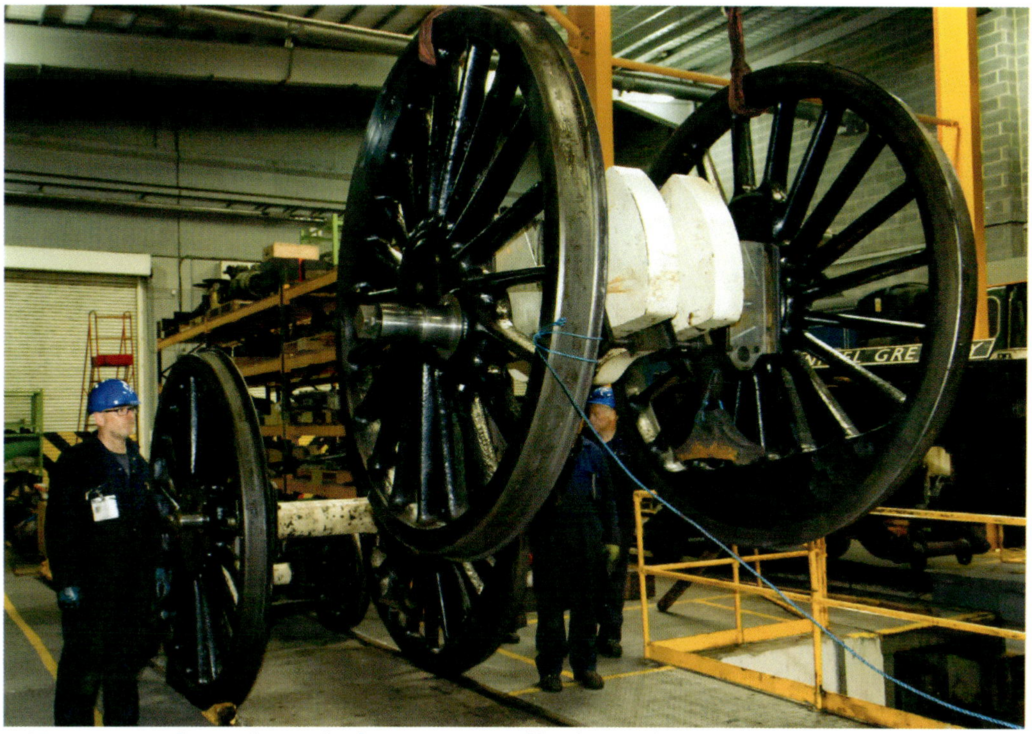

The centre and main drive wheelset is positioned by the overhead crane onto the parallel track.

as the exhausting task was carried out. Having discovered the difficulty of this task, a better procedure was devised by the engineering team to be used in the future - a policy of evolving procedures at which the team were to become very adept.

Taking a well earned break, the team set about repeating the whole procedure again on the centre wheelset. This went equally well and the day's efforts came to a close when the bogie and the two driving wheelsets were safely positioned on the parallel track. Only the trailing wheelset remained in the frames and that could wait until tomorrow as time, and not to mention the energy of the now exhausted team, was running out.

It had been a long tiring day for everyone, including myself, and I was only taking the pictures! As I set off home the thought crossed my mind that I could not help but be impressed by the efforts of the not so young volunteer engineers in the team.

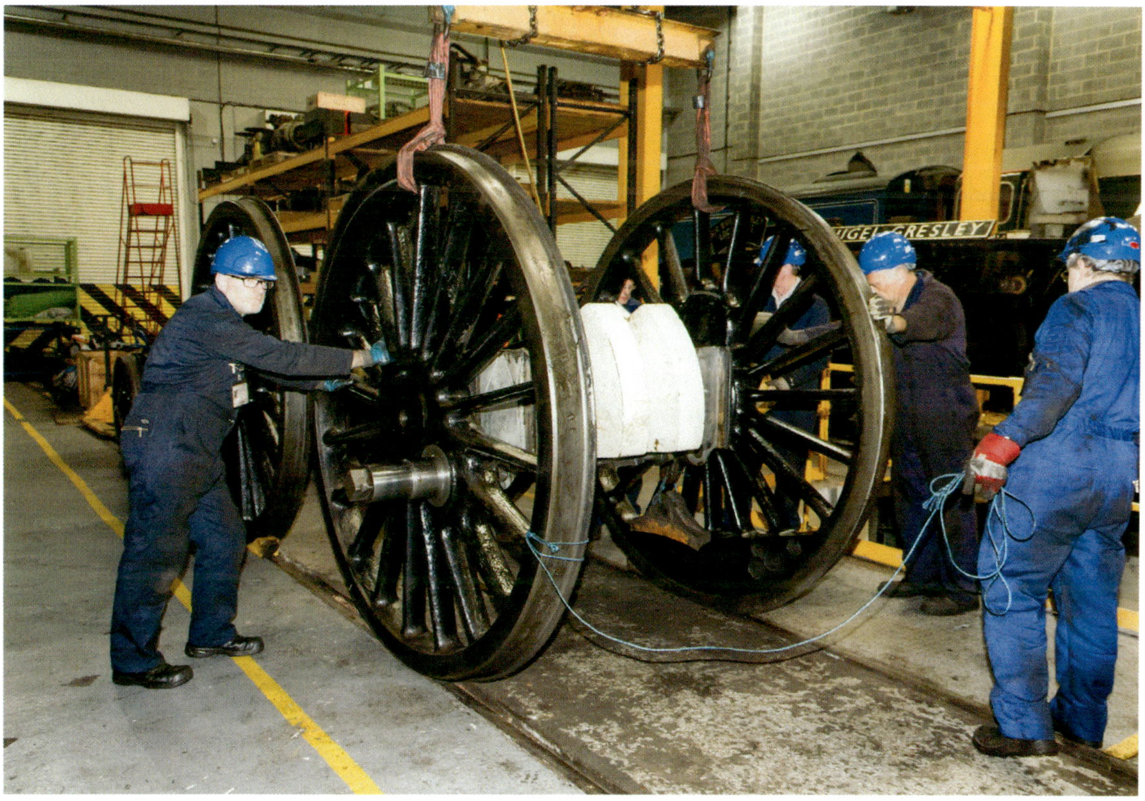

The days work comes to conclusion when the centre drive wheelset is placed on the parallel track along with the bogie and leading wheelset

GOOD VIBRATIONS

June 2016

It was now more than six months since the overhaul had started. The wheels and boiler had been separated from the frames and the emphasis of the work was moving from dismantling to refurbishment. I was introduced to a new tool in the armoury of paint stripping equipment - the needle gun. The needles are a collection of fine chisels which are grouped together and pneumatically vibrated at high speed in a handheld device against a metal surface to remove paint or scale. This was an important piece of kit given the amount of paint No.7 needed removing, but prolonged operation by any individual was not a good idea, so the work was shared around everybody and this time everybody included myself. I don't know why, but the Beach Boys song 'Good Vibrations' was constantly in my head during these sessions!

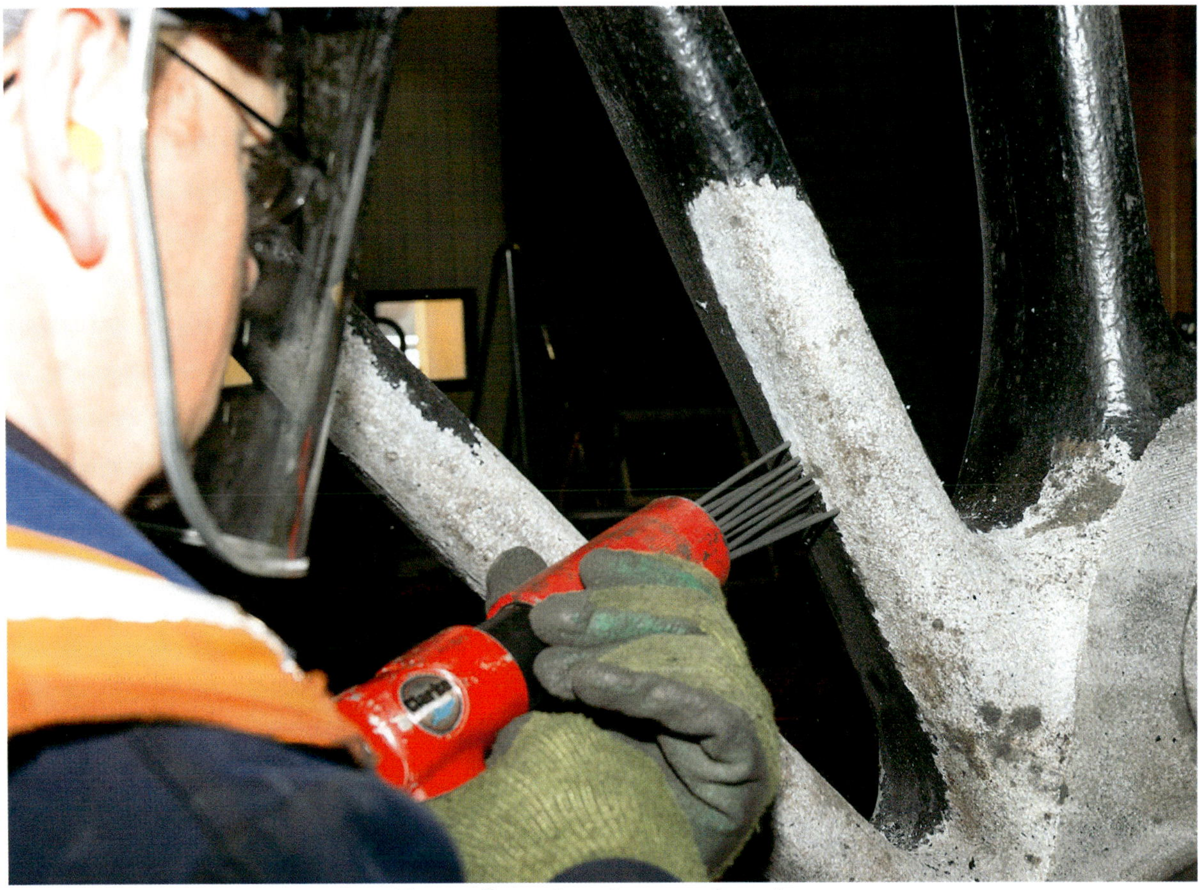

The needle gun in action removing paint

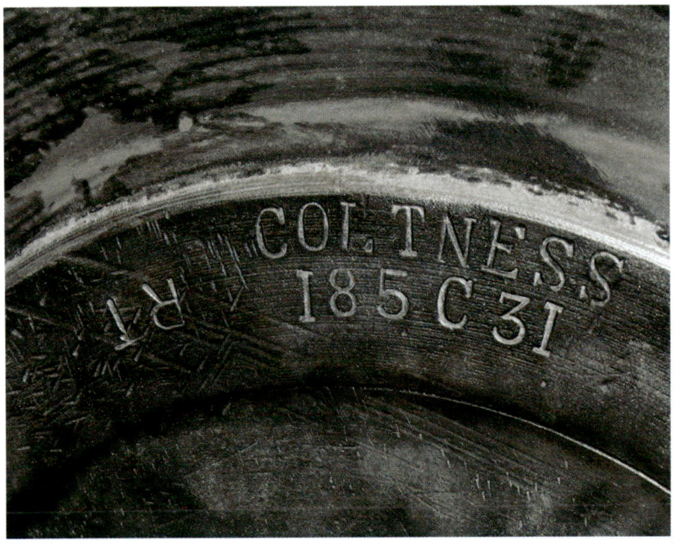

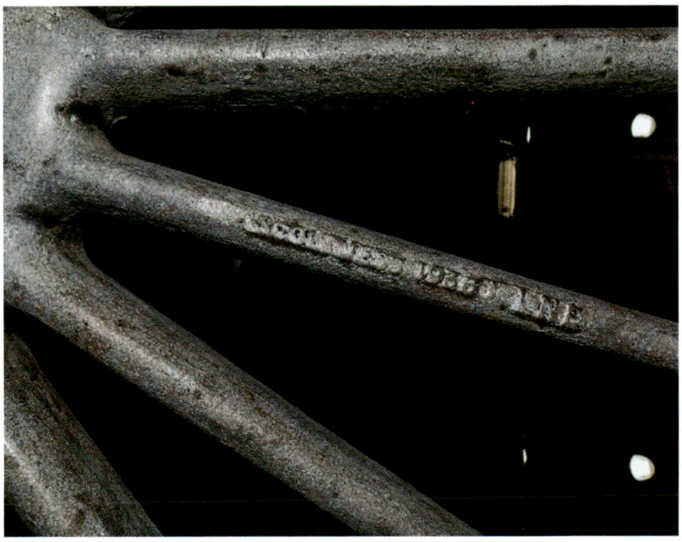

Stampings are revealed after paint removal

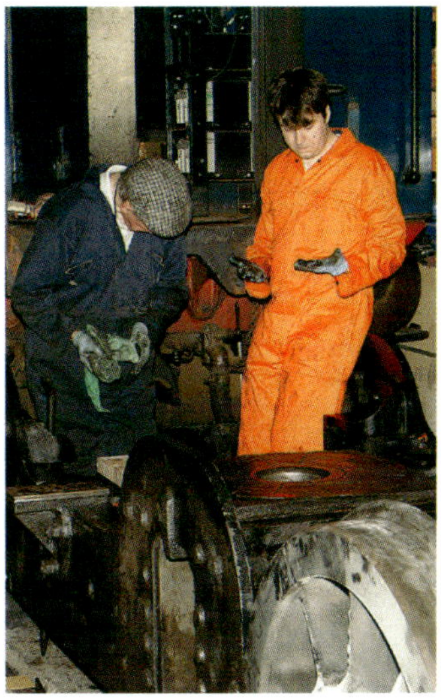
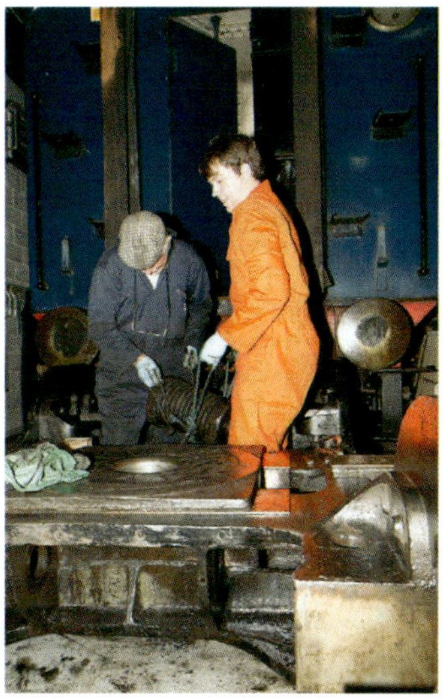
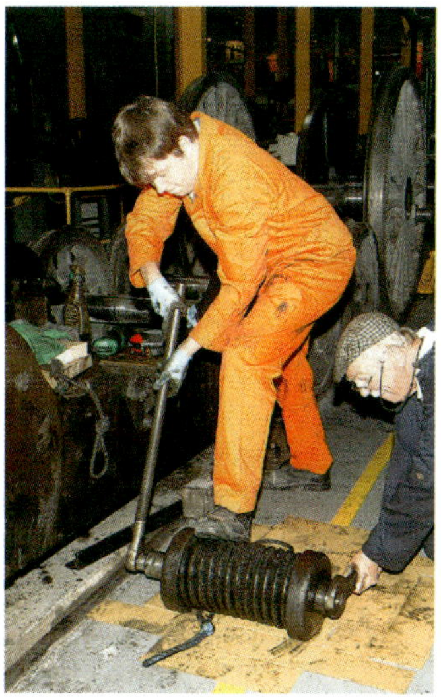
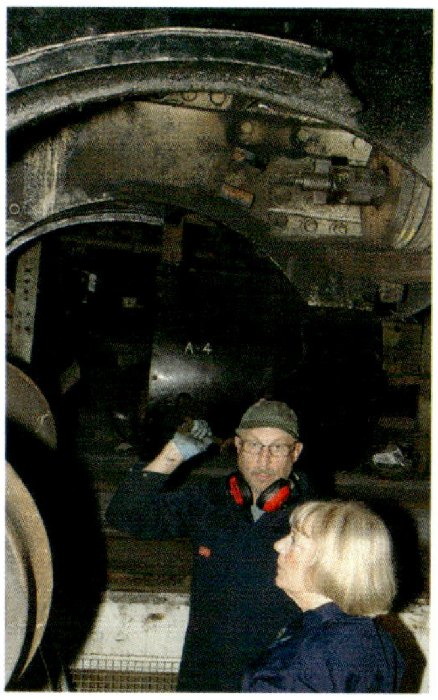

Bob Shearman and Tom Bright get to grips with a bogie spring whilst Mike Page and Jane Skinner work on the lubricating system

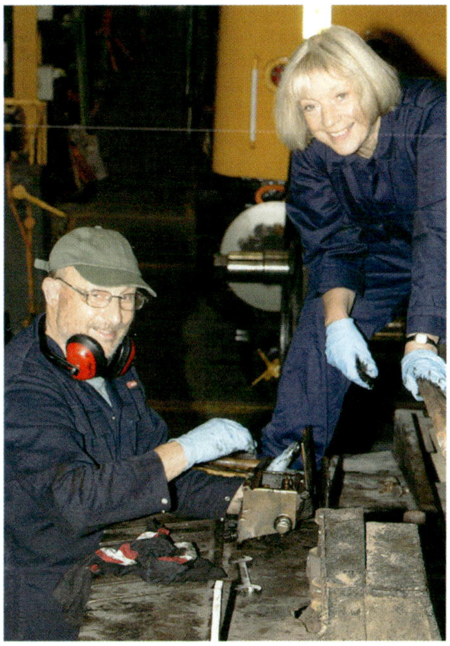
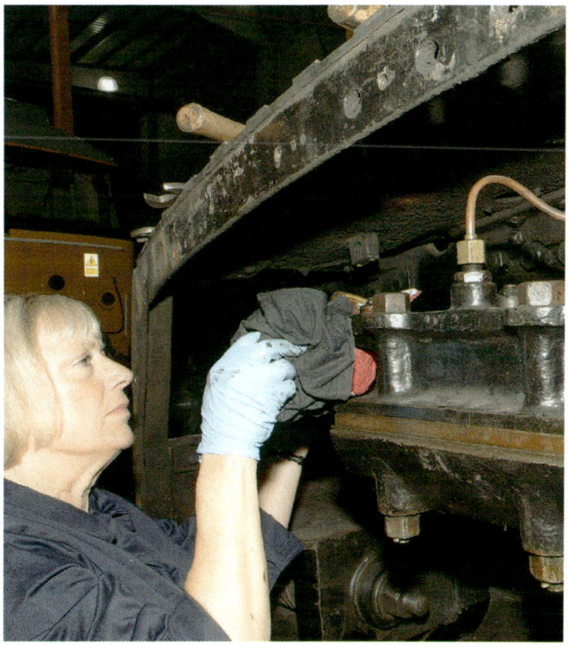

Mike Page

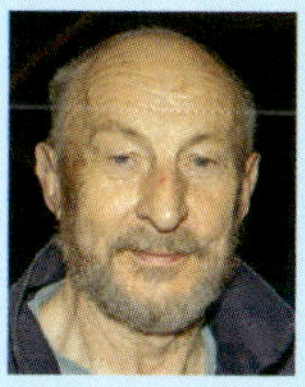

Mike is a long term member of the volunteer engineering team who has worked on No.7 at Grosmont on the NYMR before the start of the overhaul and is one of the six members who has worked throughout the overhaul from coming out of traffic in September 2015.

"Following retirement from an engineering background in both the RAF and Nuclear Research, I joined the team at Grosmont, under the watchful eye of Roger Barker, helping to keep No.7 running and subsequently moving into this overhaul, guided by Darrin."

Removal of the paint started to show some history of No.7 with old markings coming to light. I have included several shots here displaying these, many of which are the markings of the Coltness foundry manufacturer dating back to 1936.

This period of the overhaul saw many of the 1000 plus components which had been removed receiving a lot of tender loving care. Steam Railway magazine's Tom Bright and Jane Skinner joined up with engineering team members Bob Shearman and Mike Page as they worked on a bogie spring and the lubricating system respectively. Tom clearly has a similar view to me when getting his hands dirty, but a great day was had by all and their support was very much appreciated.

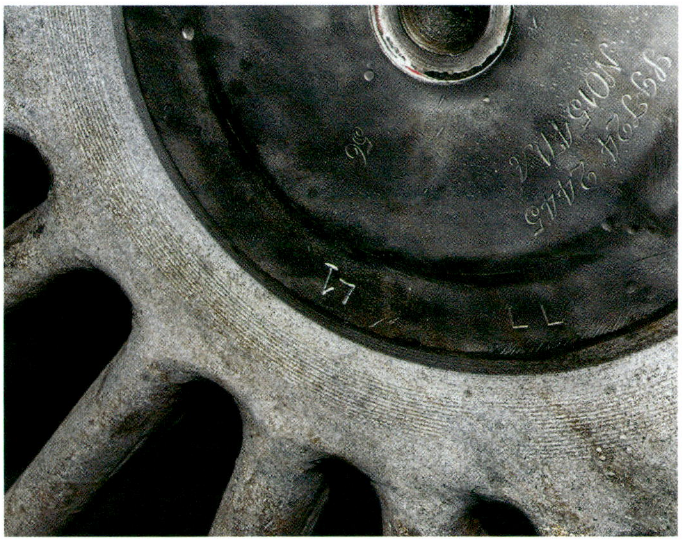
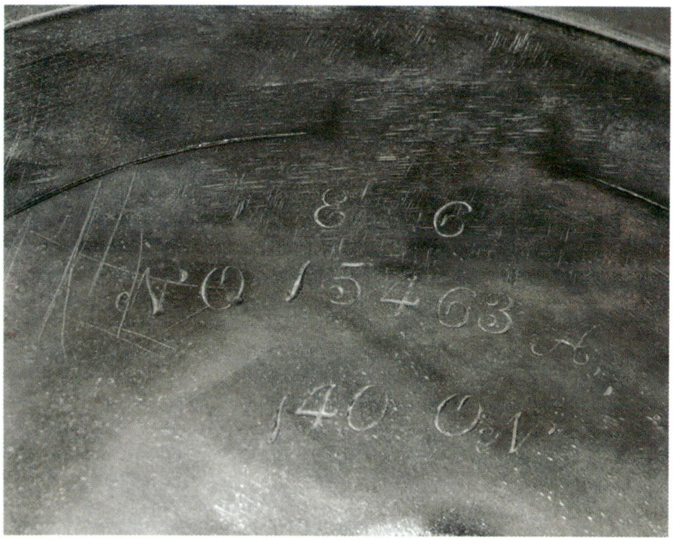

A TESTING TIME

June 2016

While work continued at a pace in the workshop on dismantling the lubricating system and removing paint from the wheels, my next visit was back outside when the NDT (Non-Destructive Testing) team arrived to check the integrity of the boiler using MPI (Magnetic Particle Inspection) and Ultrasound equipment. I'll pass on explaining the technical aspects of this as the internet has some far better explanations than I could provide. I was happy with these images, being outside with more light than the gloom of the workshop was a bonus. Another victim was swallowed up inside the firebox and boiler which offered up some humorous shots. One set of legs are easy to spot passing through the firebox door, but there are also legs entering the boiler through the steam dome in another image. You won't find me getting in there, I kept telling myself.

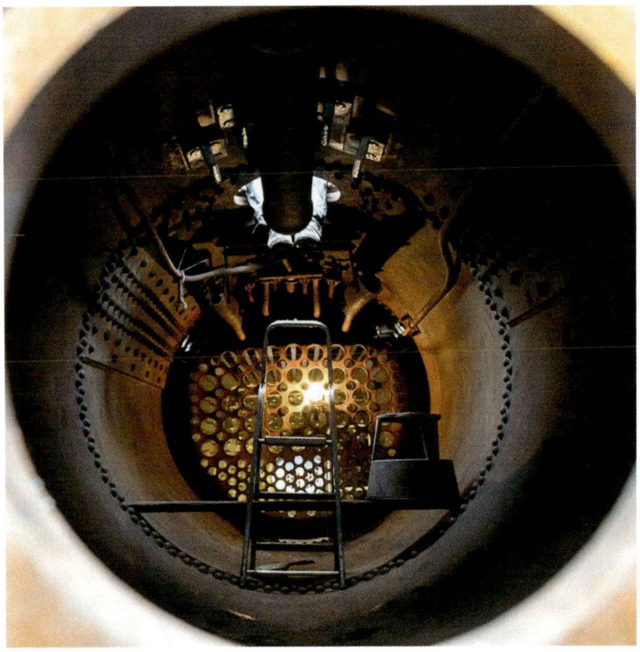

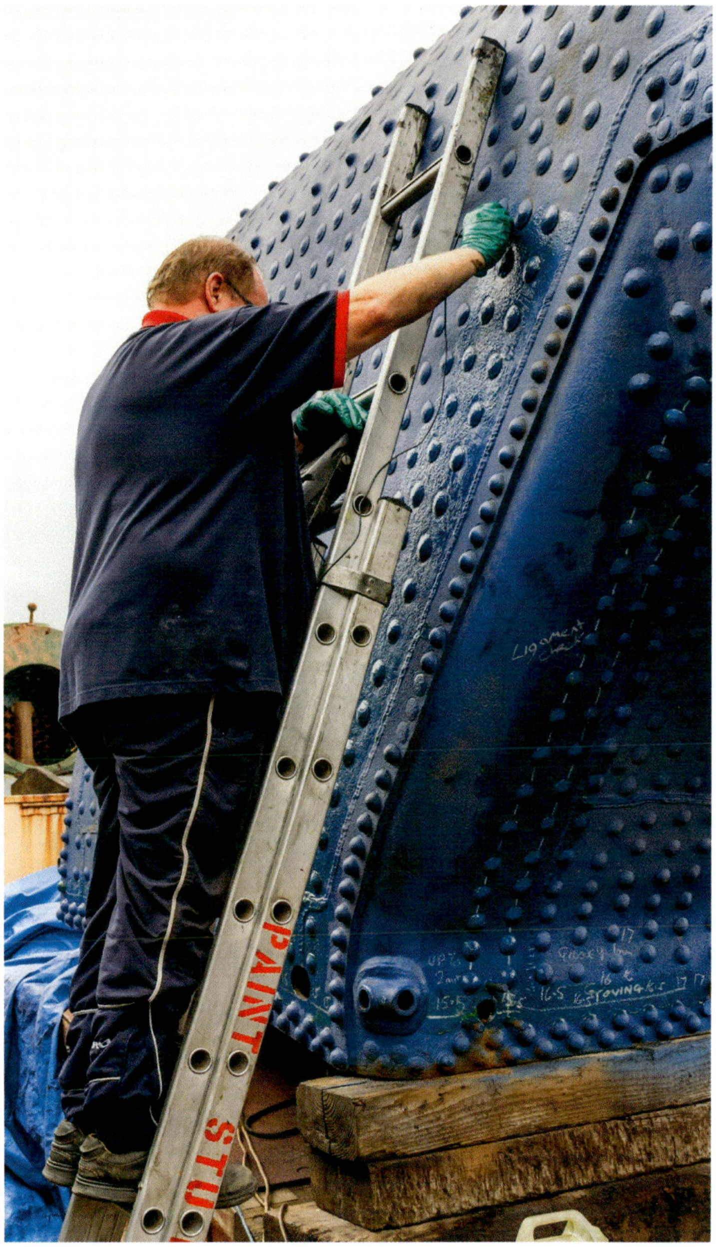

ABOVE - A pair of legs descend into the boiler as entry is gained through the steam dome.

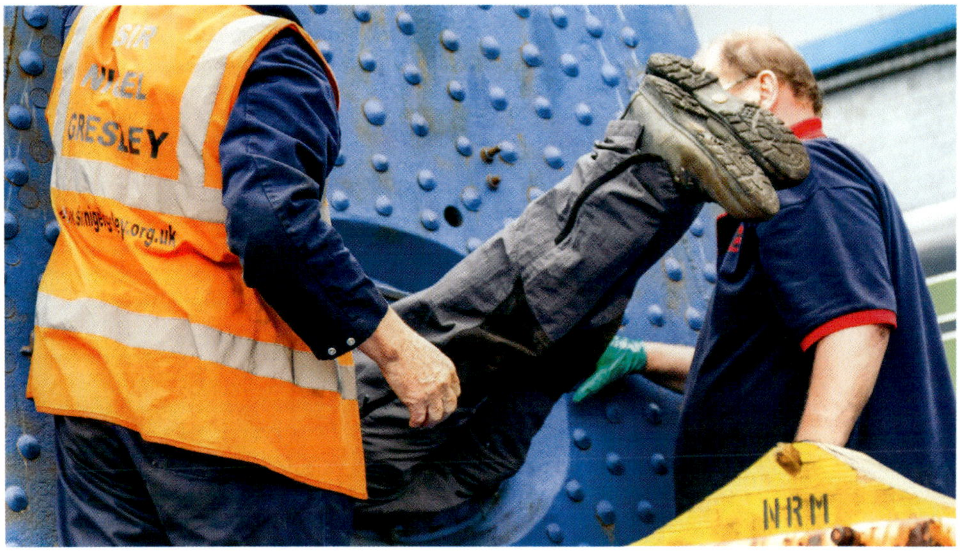

TOP RIGHT & RIGHT - The thickness and integrity of the boiler and firebox required Non Destructive Testing (NDT) to be carried out externally and internally. Entry to the firebox is head first via the firebox door

WHEELS ON THE MOVE

4

July – August 2016

Paint removal on the wheels continued during the summer in preparation for their move to the workshop of the South Devon Railway (SDR) in Buckfastleigh. There the wheels would be retyred - which doesn't mean they are going to get a well-earned rest after a lifetime of work. It means the well-worn metal tyres around the circumference of the wheels would be replaced with new ones, just like a Grand Prix car at a pit stop. Except this pit stop would take considerably longer than a few seconds; we would not see No.7's wheels again until September 2017.

Archers Transport were contracted to move the wheels to Devon and driver Paul duly arrived with his rigid lorry and trailer on the 10th August 2016. Whilst the access to the workshop is good, there was not enough room for both lorry and trailer, so the lorry was used to ferry some wheels to the trailer before it returned to be loaded with the remainder.

An interesting feature of the NRM workshop is the drawbridge on the gallery walkway. When raised this allows the overhead crane access to the workshop side exit doors and thus our wheels could be moved directly from the workshop onto the back of Paul's lorry. With trailer and lorry fully loaded and secured the long drive to Devon began.

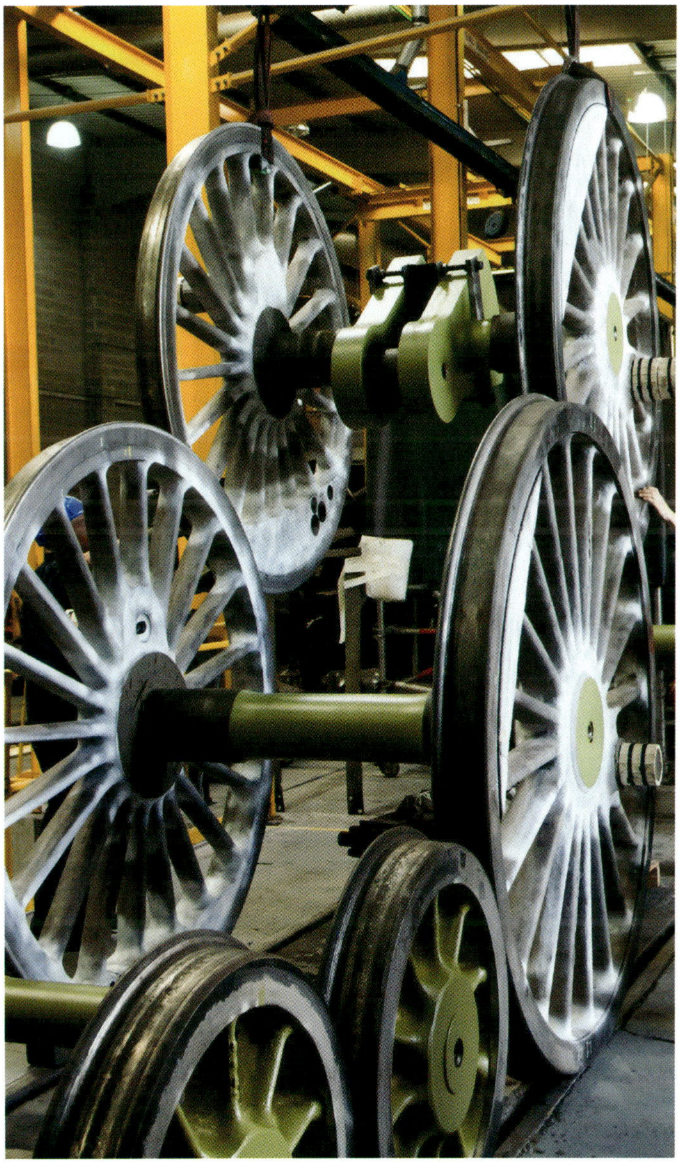

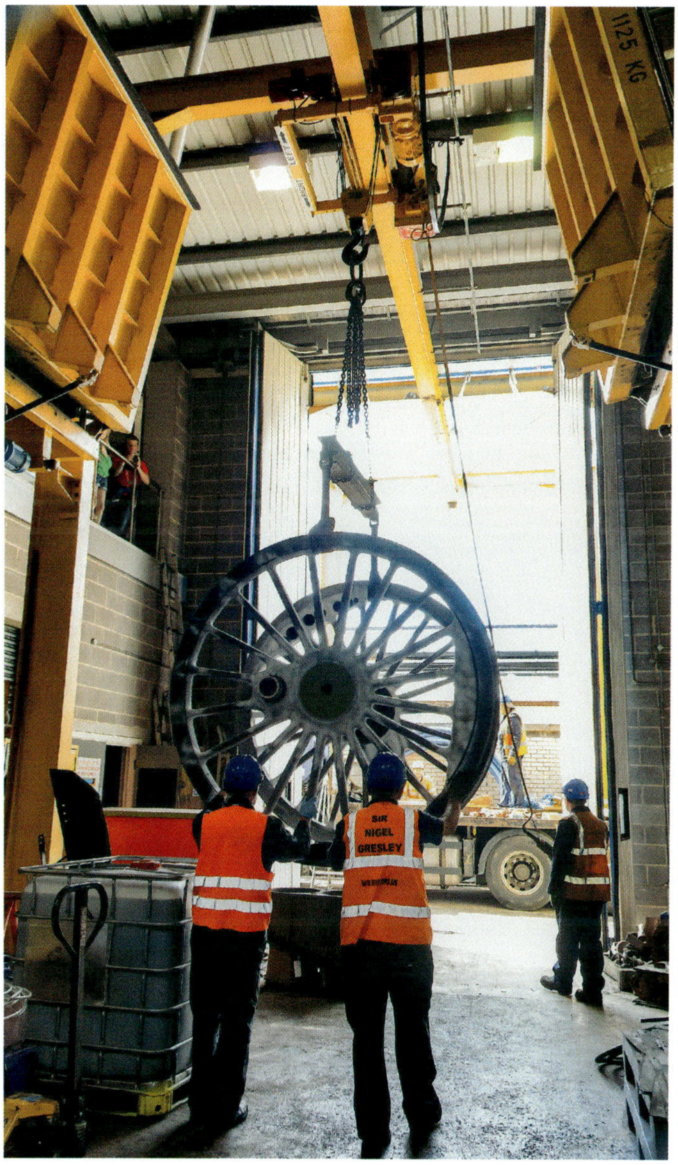

The first wheelset is lifted and moved through the raised viewing gallery drawbridge by the overhead crane to the waiting lorry

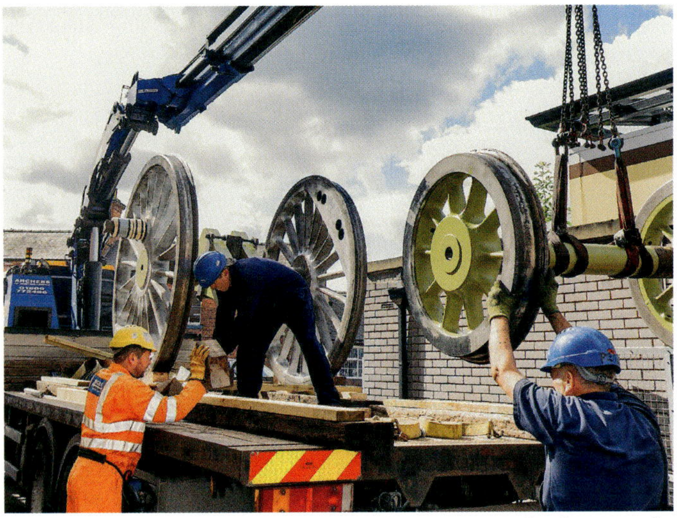

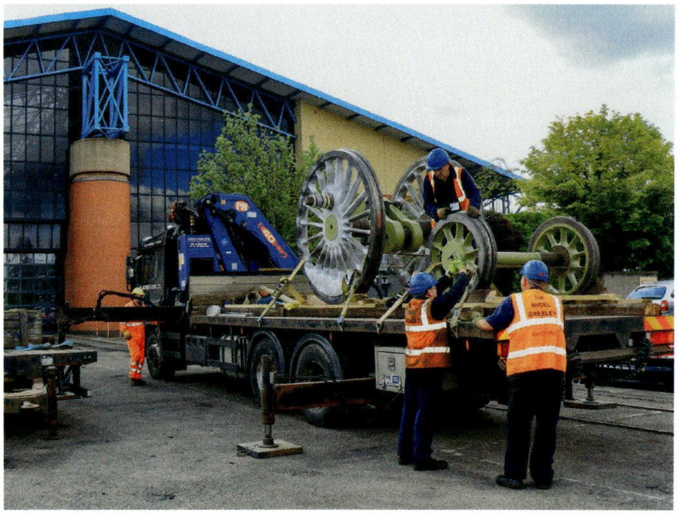

Wheelsets are loaded on to the lorry

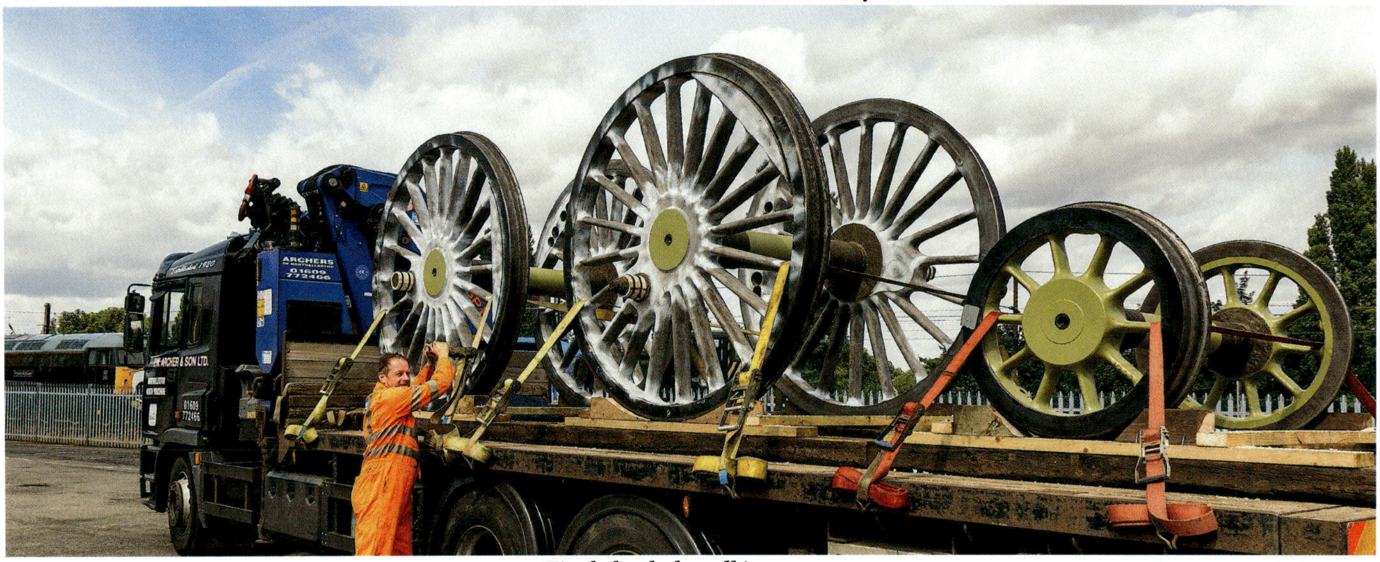

Final check that all is secure

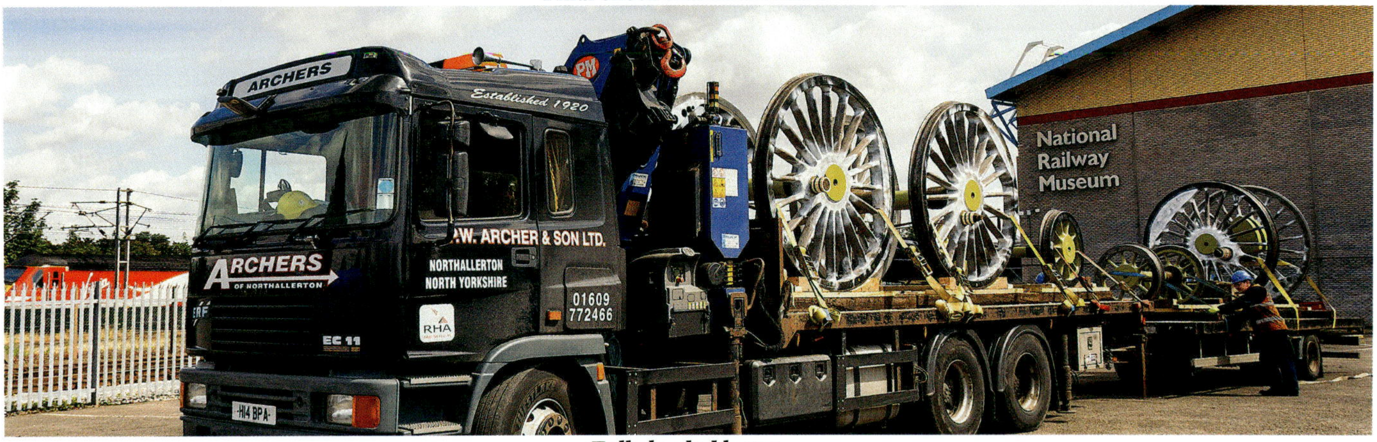

Fully loaded lorry

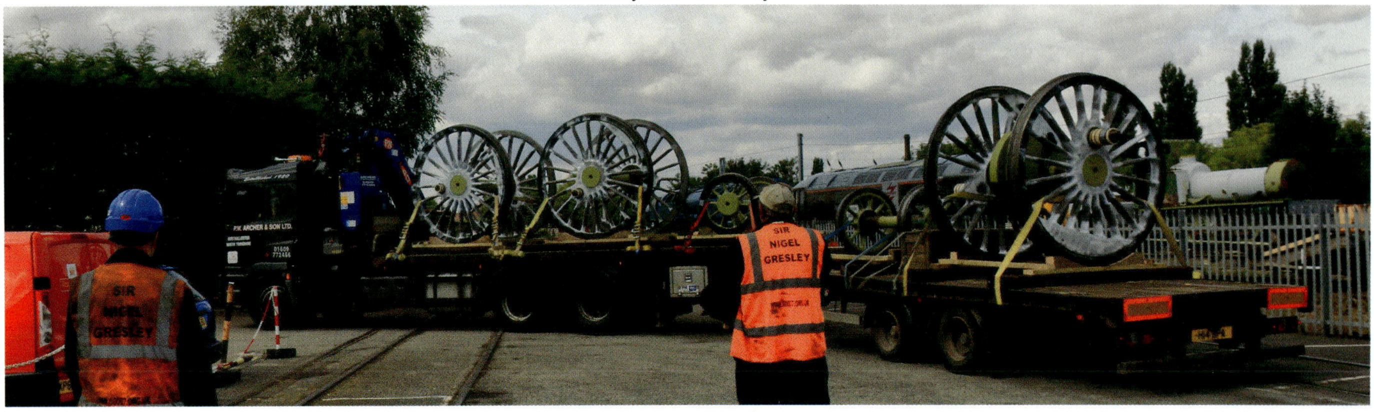

The precious cargo departs

September 2016

No.7 was now well and truly stripped down. The boiler was in the NRM's North Yard, the wheels were at the SDR in Devon, the cab was on the floor in the workshop and the frame was devoid of its working parts. This was a time for the paint strippers, and any decorator will tell you that preparation is the key (I use the word 'key' in more ways than one) to a good finish and it's hats off to all the team who prepared the surfaces for re-painting on No.7. Hour after hour and day after day was spent on this task with some extremely difficult nooks and crannies not to be missed.

Having already prepared the wheels, before they left for retyring at the SDR in Buckfastleigh, the team was becoming very proficient at the task.

The images I took here were to make some interesting comparisons with the ones taken when fresh paint was applied. More markings became visible – the most notable one on the right-hand inside frame above an axle box showing BR for British Railways, as opposed to LNE for London North Eastern.

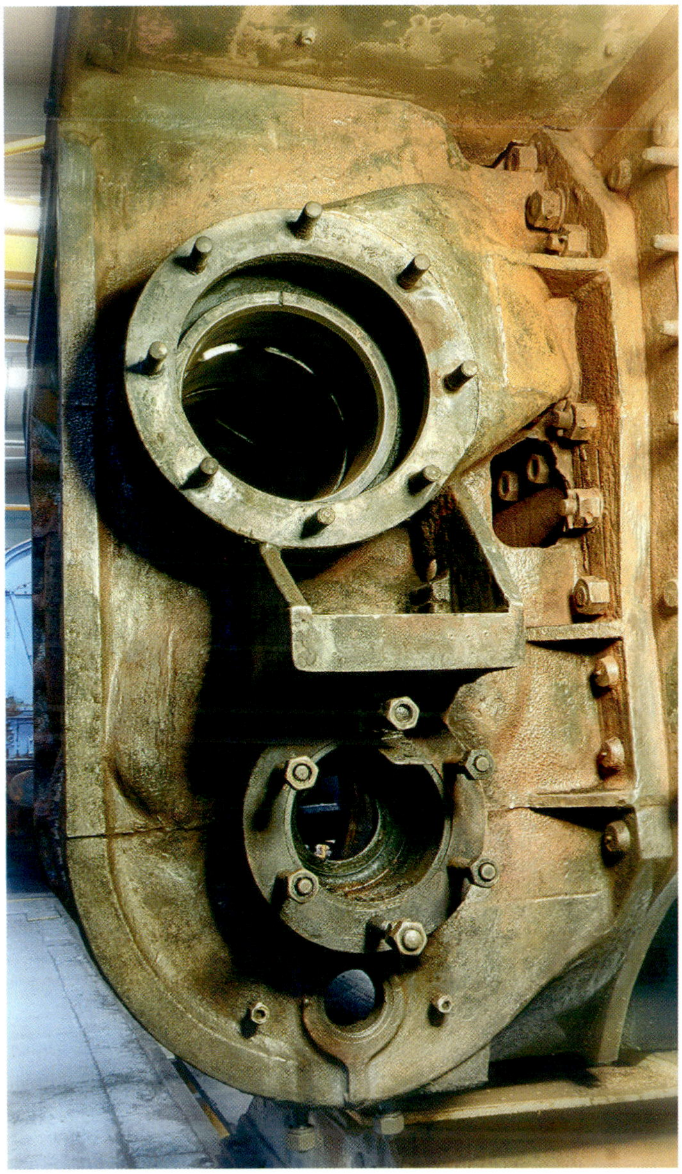

Old paint has all been removed in preparation for repainting

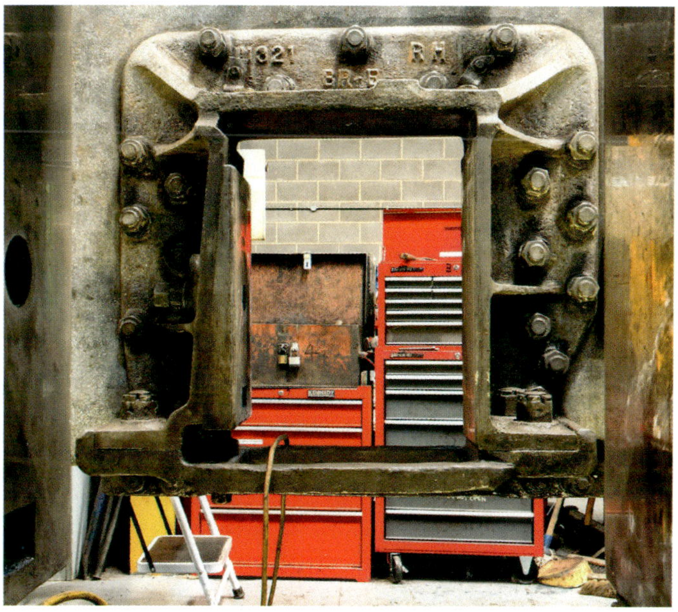

Removal of paint reveals BR markings (above) and LNE markings (below)

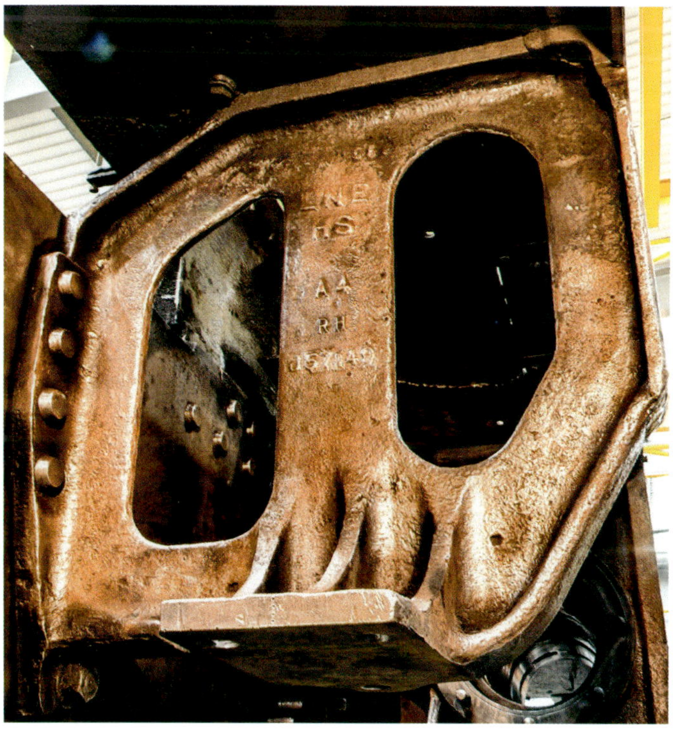

A BIG COVER UP

October 2016

Back outside with the boiler it was time to cover up and hide from the gaze of passing travellers on the mainline; no, sorry, covering to protect from the elements after the NDT team had finished their work.

It was overcast, chilly and a bit breezy when Darrin assembled the team, equipped with a huge tarpaulin, colour suitably blue. After Darrin gave Roger a quick tour of the inside of the boiler the cover up commenced. Now, if you have ever tried putting up a tent in the wind you will have some idea of the task ahead - but only some. This tarp was considerably larger than most tents. Planning was important, so with Darrin positioned on top of the boiler issuing instructions and looking very happy with his ringside seat as well, the unfurling began. The cry to take care came. After all we didn't fancy wrapping the tarp around the front of an ECML class 91 express heading to Edinburgh and blinding the driver, or indeed incinerating the tarp on the high voltage overhead lines. I jest of course, the likelihood of that happening was very remote, as the ECML high protective fencing and the professional manner of the assembled team would ensure no such mishap occurred but the humorous replies to the cry were in line with the great spirit the engineering team held in such high regard.

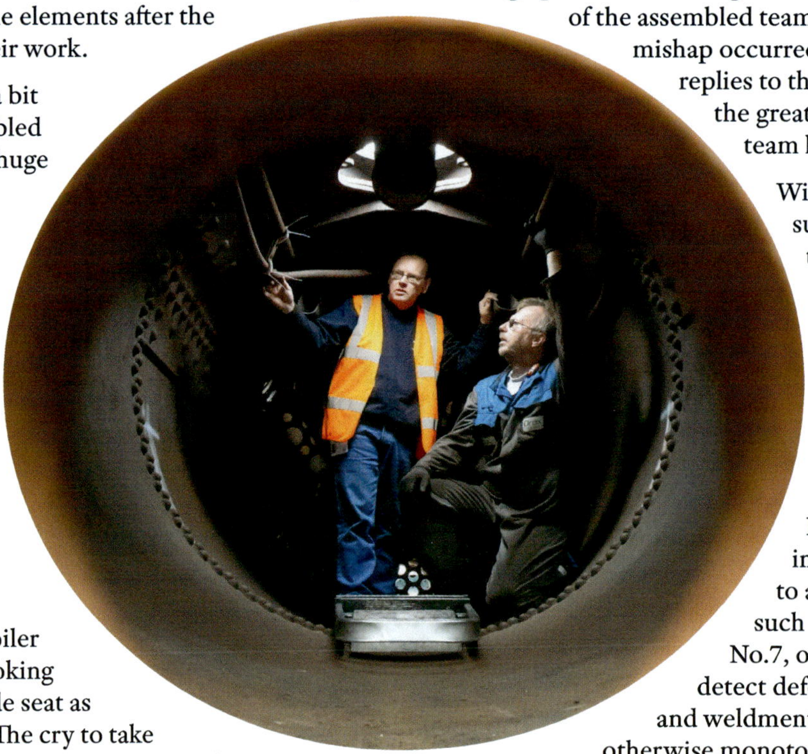

With the job duly successfully completed, the team returned for one of Peter's special brews of tea before it was back to the workshop to continue work on the frames, where some colourful shots were to be taken.

Dye penetrant inspection can be applied to all non-porous materials such as the metal frames of No.7, on which it is used to detect defects in castings, forgings and weldments by highlighting the otherwise monotone surfaces. Mike and Roger cracked on with this task (although 'cracked' is probably not the best word to use here) and, indeed no serious cracks were found, just some minor

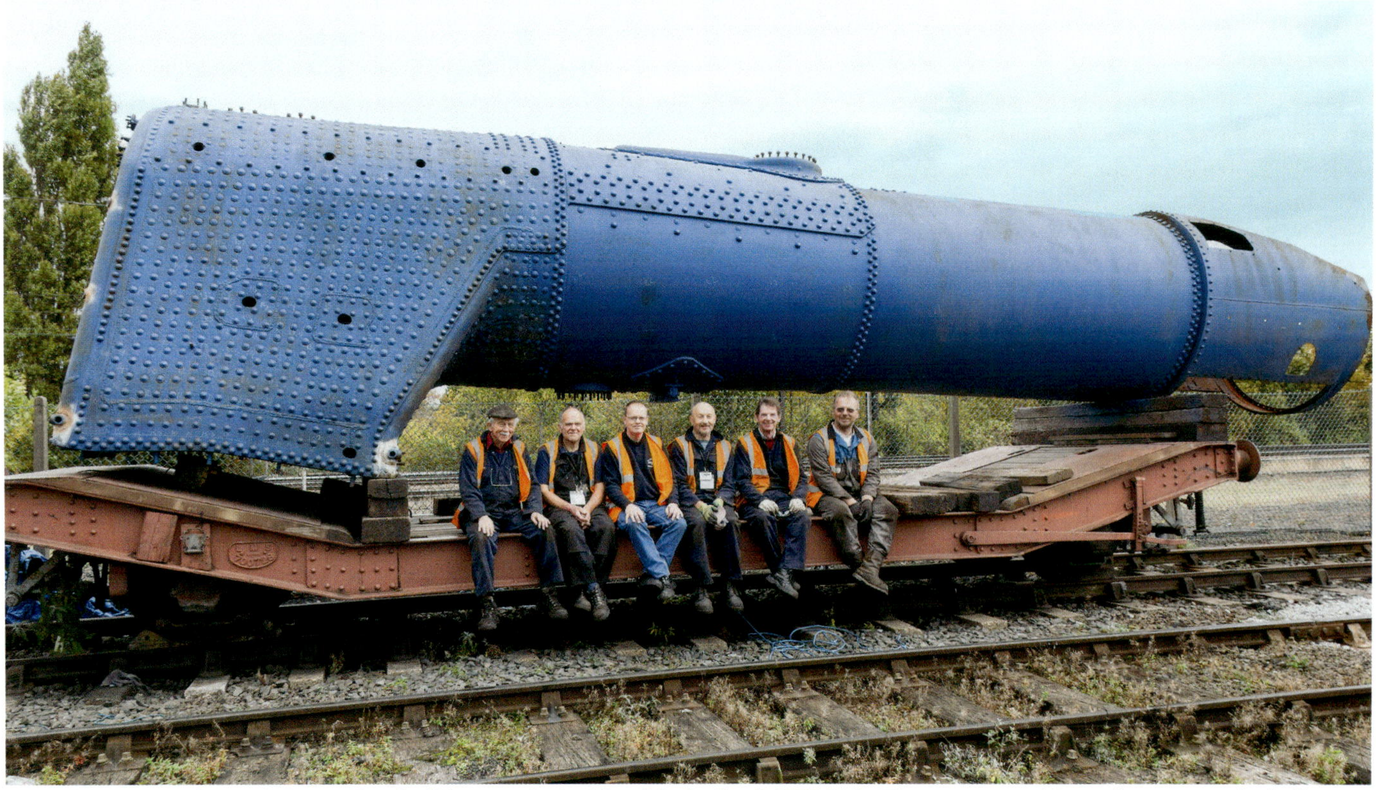

TOP - Darrin Crone & Roger Turnbull inspect the inside of the boiler BOTTOM - The team assemble ready to cover the boiler

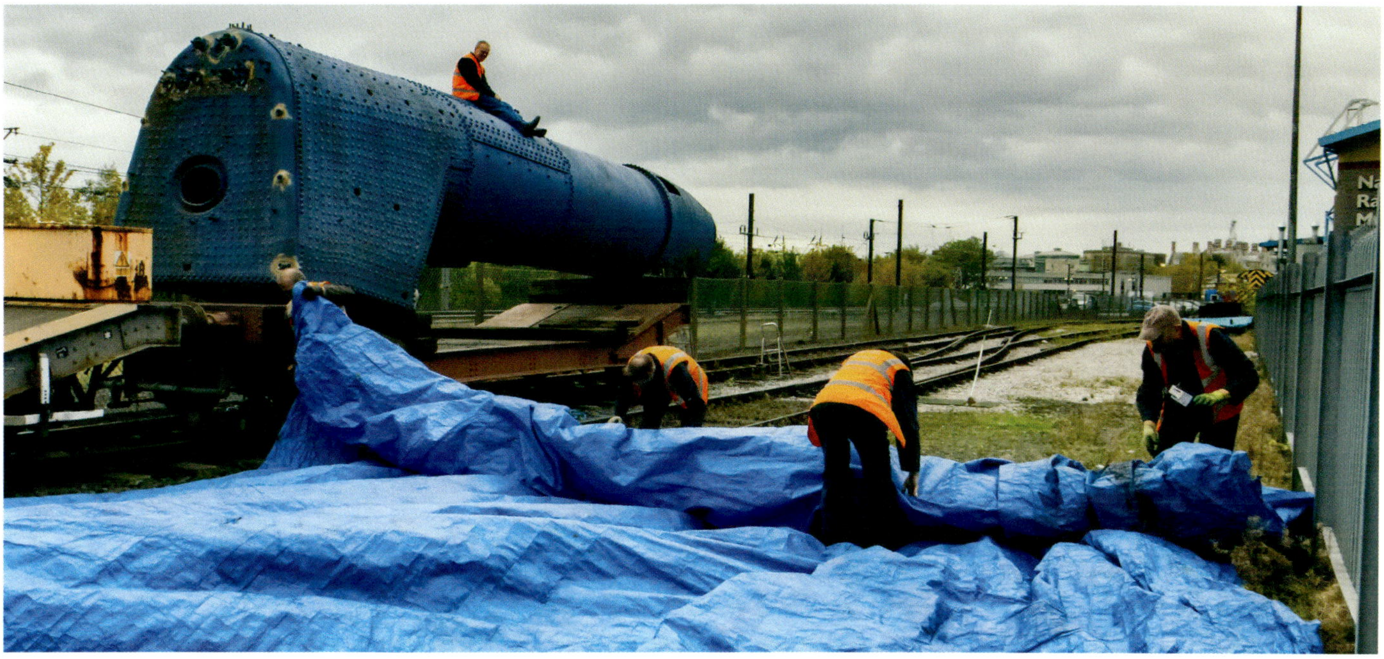

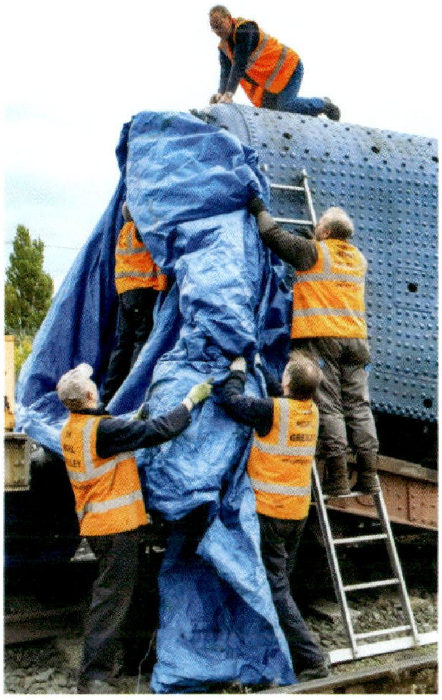

P26 TOP Darrin Crone introduces Roger Turnbull to the inside of the boiler

P26 BOTTOM Left to right - Bob Shearman, Peter Brackstone, Darrin Crone, Mike Page, Malc Bateman, Roger Turnbull take a minute for the camera before starting the big cover up

TOP With Darrin Crone orchestrating procedure from his lofty perch on top of the boiler the team unwrap the huge cover sheet

LEFT Working from left to right the sheet is placed over the boiler

BOTTOM Job done, the sheet is secured around the boiler

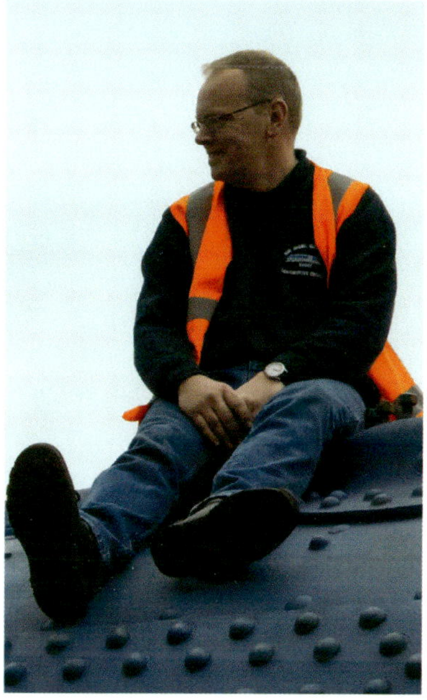

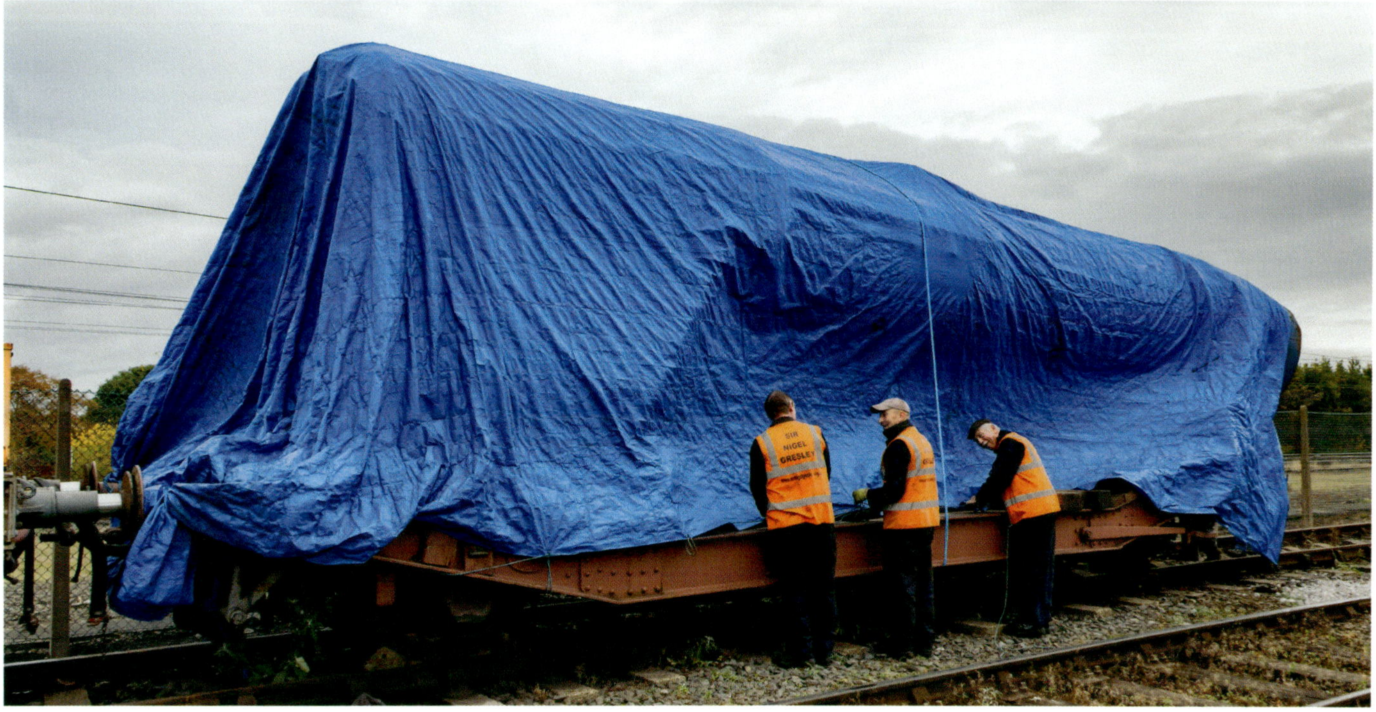

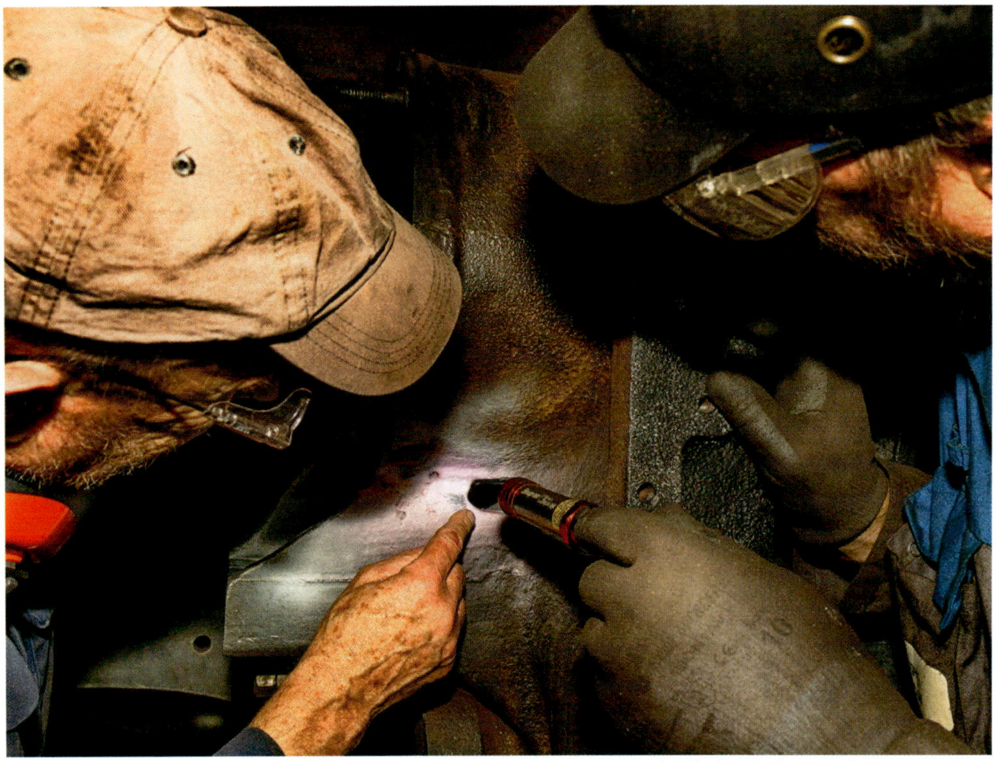

Using dye penetrant as a highlighter Mike Page & Roger Turnbull check for cracks in metal surfaces

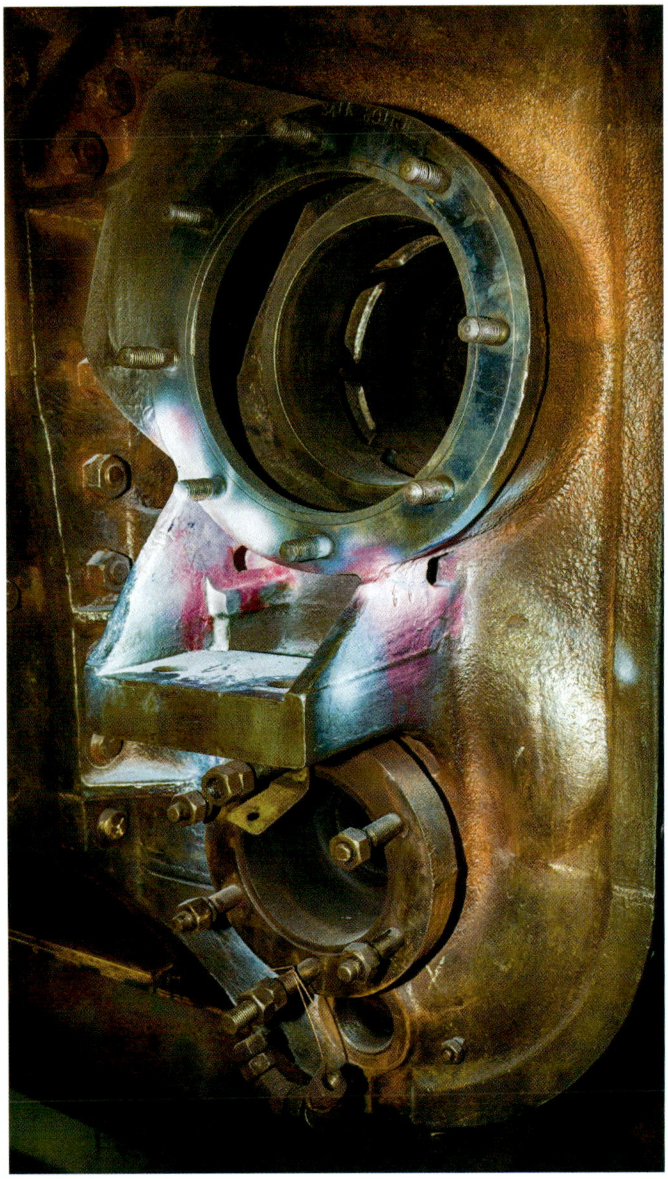

Roger Turnbull

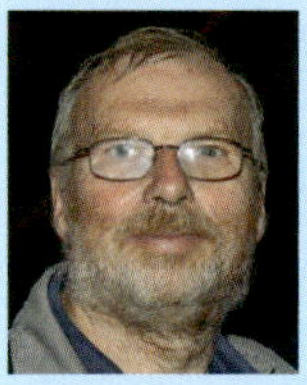

I have stood on the NRM workshop balcony over the last twenty odd years, looking down at the projects that have been done in there and thinking that it would be interesting to get involved, if only I had time. Following early retirement from Drax Power Station near Selby, where I had worked for 36 years, mainly on supervising and managing the operation of six, one million horsepower steam engines and boilers one hundred and fifty feet high, I saw on the SNGLT website that I could volunteer to help with the rebuild. I worked part time in the last six months of my employment to enable me to attend the NRM one day a fortnight before doing two days a week after I retired. The overhaul of Sir Nigel Gresley is a big project with a big budget and a bunch of volunteers who all get on well, providing there are enough seats in the blue cabin at tea break time! This is an enjoyable project to do for someone who is interested in steam engines, railways and hands on engineering, which suits me down to the ground.

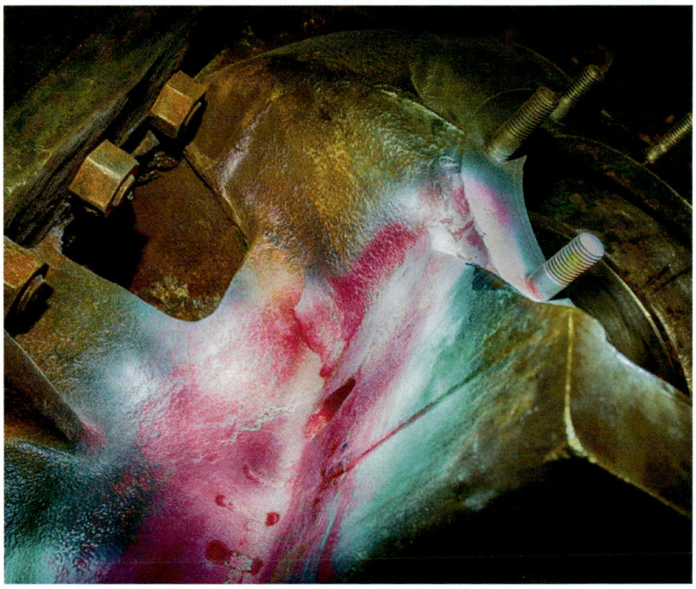

LEFT & ABOVE Dye penetrant is applied to surfaces as a highlighter for cracks

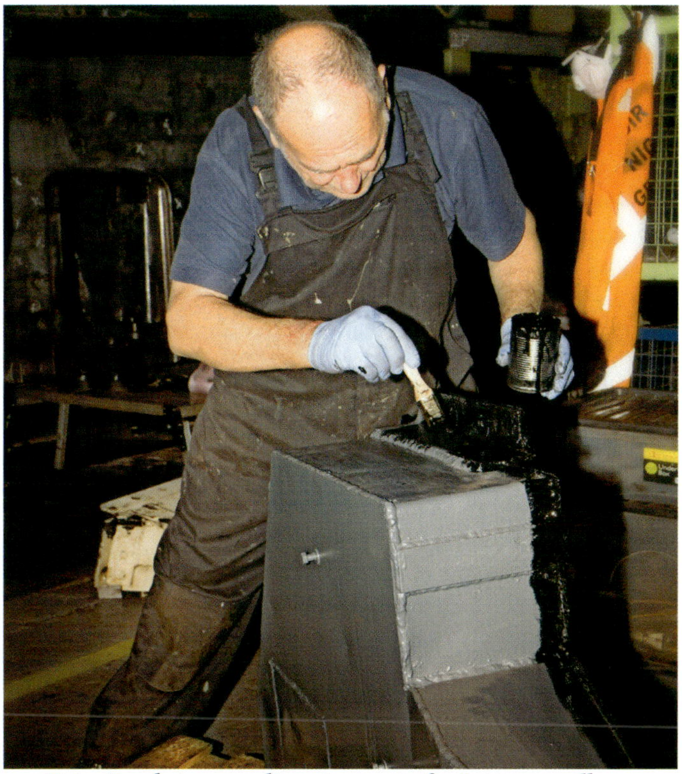

Peter Brackstone applies a top coat of paint to a sandbox

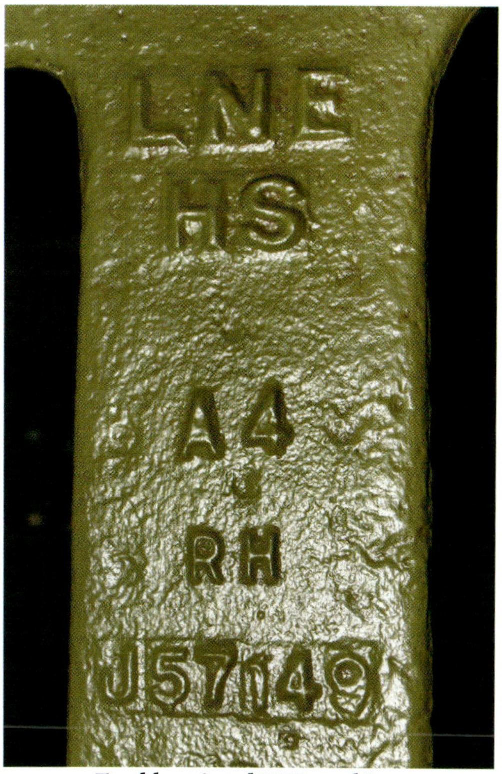

Freshly painted LNE markings

repairs- a nice bonus for the team.

Painting was now in full swing. Peter took charge of these duties and he took great pride in his work which was nearly as good as his tea making skills - the man's talents are endless. It was great to see primer and undercoat being applied after all the needle gunning and preparation work done by Peter and the team, (with even a little bit by me. Have I already mentioned that?) The newly painted sections of frames were starting to highlight markings identifiable with No.7's original build.

Along with stubborn to remove nuts and bolts, there are hundreds, possibly thousands of rivets and boiler stays on a steam engine and over time some will come loose. So, after

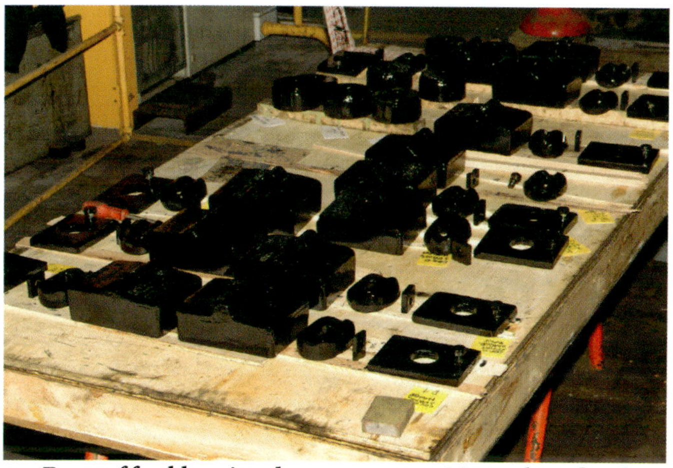

Rows of freshly painted components waiting to be refitted

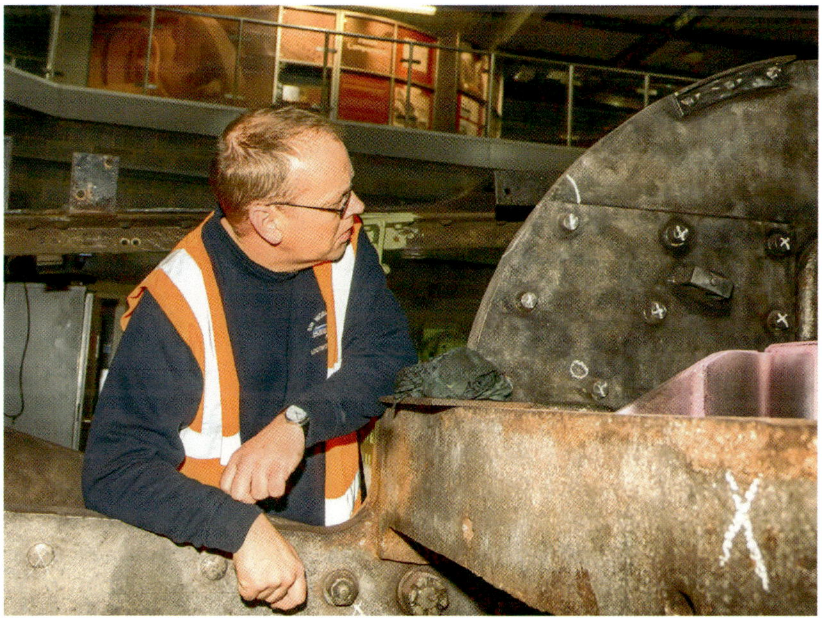

Darrin Crone marks up the rivets with chalk

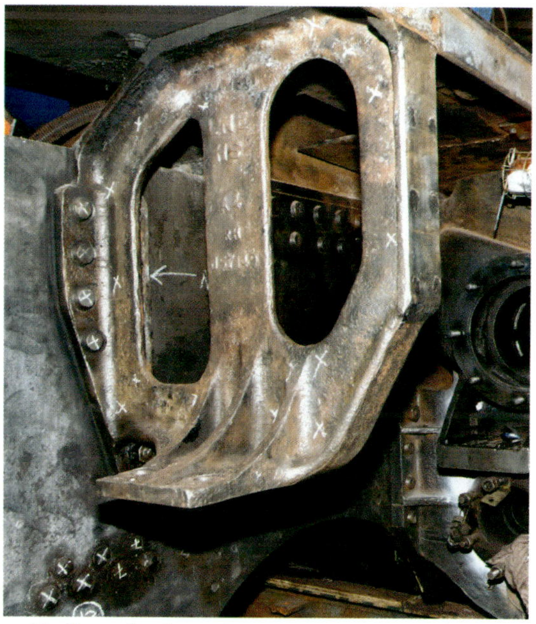

Chalk marks show rivets as secure or needing to be replaced

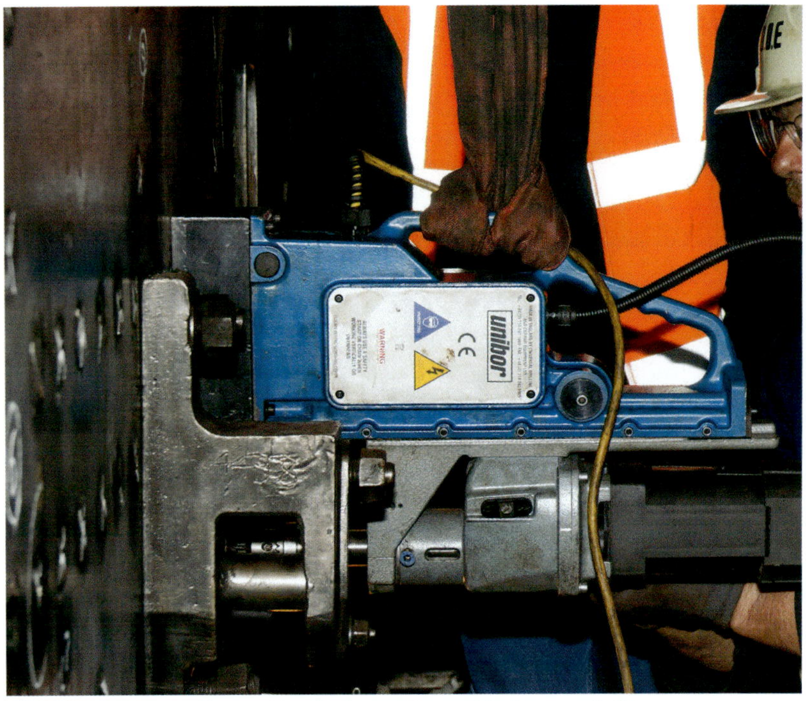

Roger Turnbull working with the mag drill

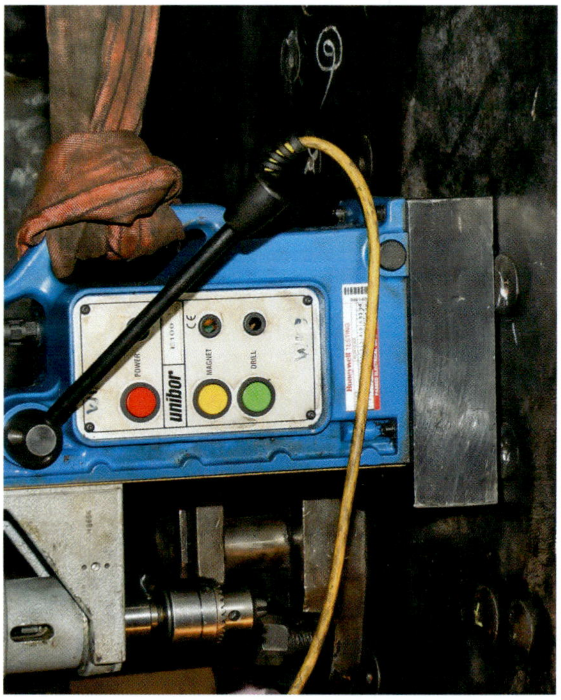

The mag drill magnetically held to the side of the frames and a strop from the overhead crane providing extra support for when the power is off

being identified by Darrin using chalk crosses and circles, those that needed replacing were removed by drilling through their centres. Whilst a standard heavy-duty drill will suffice for the smaller bolts the rivets are a different proposition. Let me introduce you to the mag drill. As it is too large to hold, it is held in position, as its name suggests, by an electromagnet. The images show Roger at the controls of the mag drill, and Malc Bateman holding a torch whilst Bob Shearman uses the drill on a bolt.

Another day at the workshop came to an end and I headed back the 60 miles home where my day was only half-way through. Transferring the images onto my computer and applying individual edits to each selected file usually took up the rest of the evening. Just as well the 14 or so hour shift was very much a labour of love. A couple of weeks later I would have an even longer shift as the boiler was to be transported to the boilersmiths in Llangollen, with Malc and I following behind, returning the next day, when winter arrived early.

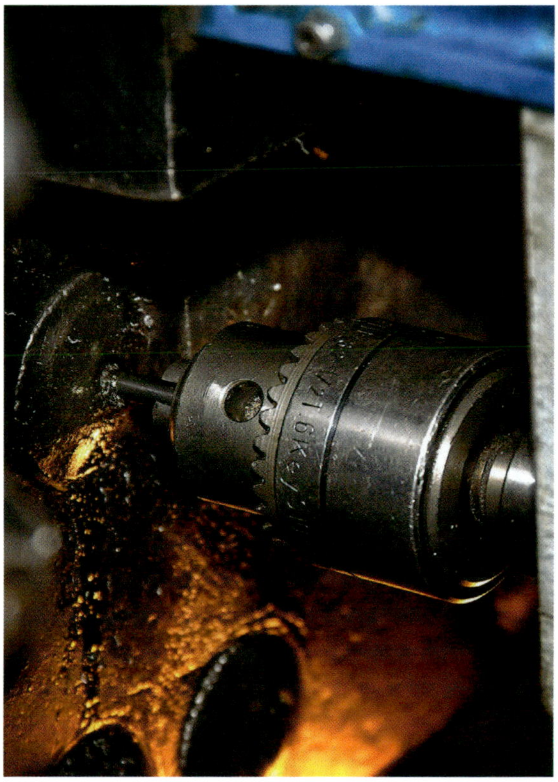

The mag drill in action

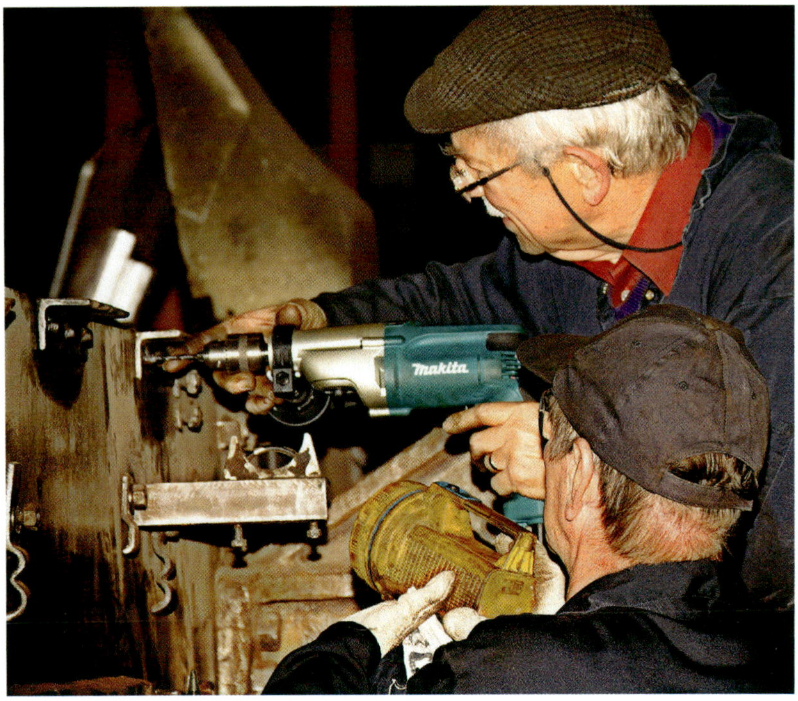

Bob Shearman and Malc Bateman drilling out stubborn bolts

007 GANG

October 2016

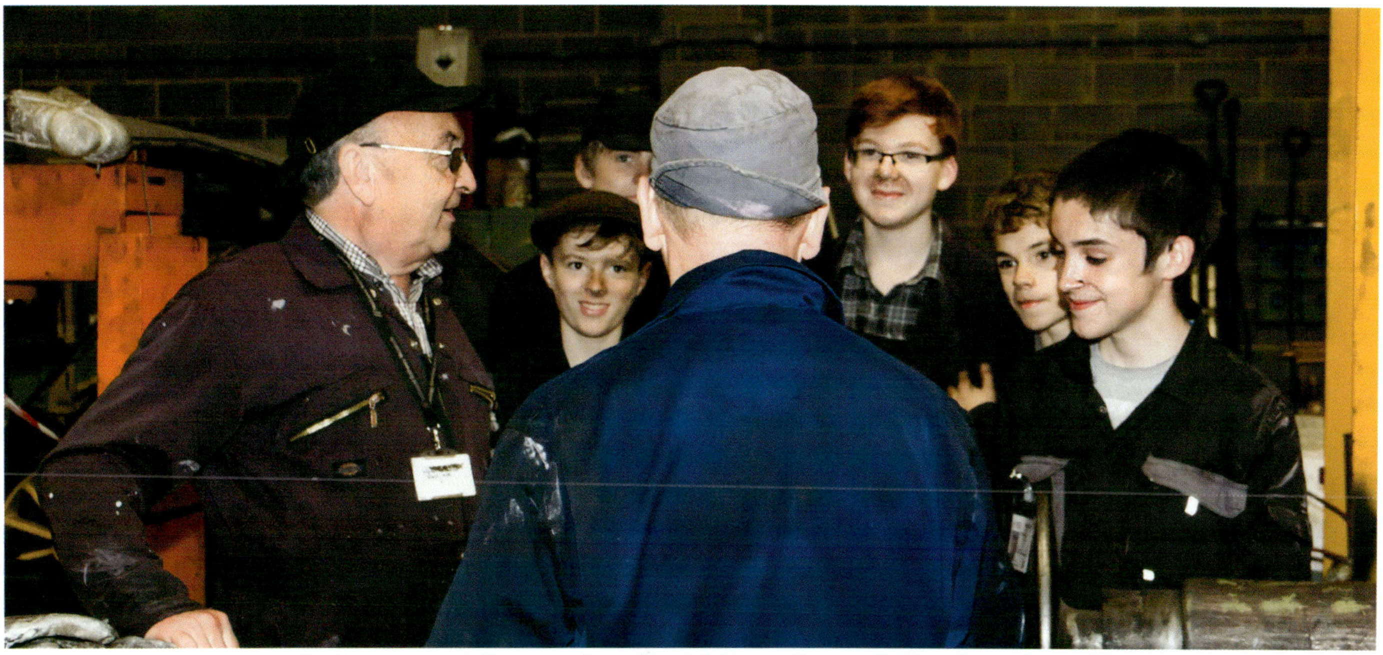

Darrin Crone (back to camera) discusses the tasks with Bryan Orange (left) and members of the 007 Gang

One of my visits to the workshop this month was specifically to record the work of some of the junior volunteer members who usually are only able to attend on Saturdays or other non-school days.. The group, currently numbering around 20 members, was formed in 2016 by Bryan Orange who arranges their monthly trips to work on No.7. These youngsters are, and will continue to be, of huge benefit to SNGLT and No.7 now and in the future and the '007 Gang' is equally a huge benefit to them in so much as they are learning great skills from the senior members as well as gaining good life skills. There was, indeed, a great bond between them all.

These days were great opportunities for the juniors. The nearest I got to a steam engine when I was their age was just standing at the end of a station platform or tip-toeing around the steam sheds of Thornaby, Gateshead, Hartlepool or York until we got caught and chased out by the foreman. I had waited much longer for my official opportunity to come around and I reminded myself that I was as equally pleased to be doing what I was now doing.

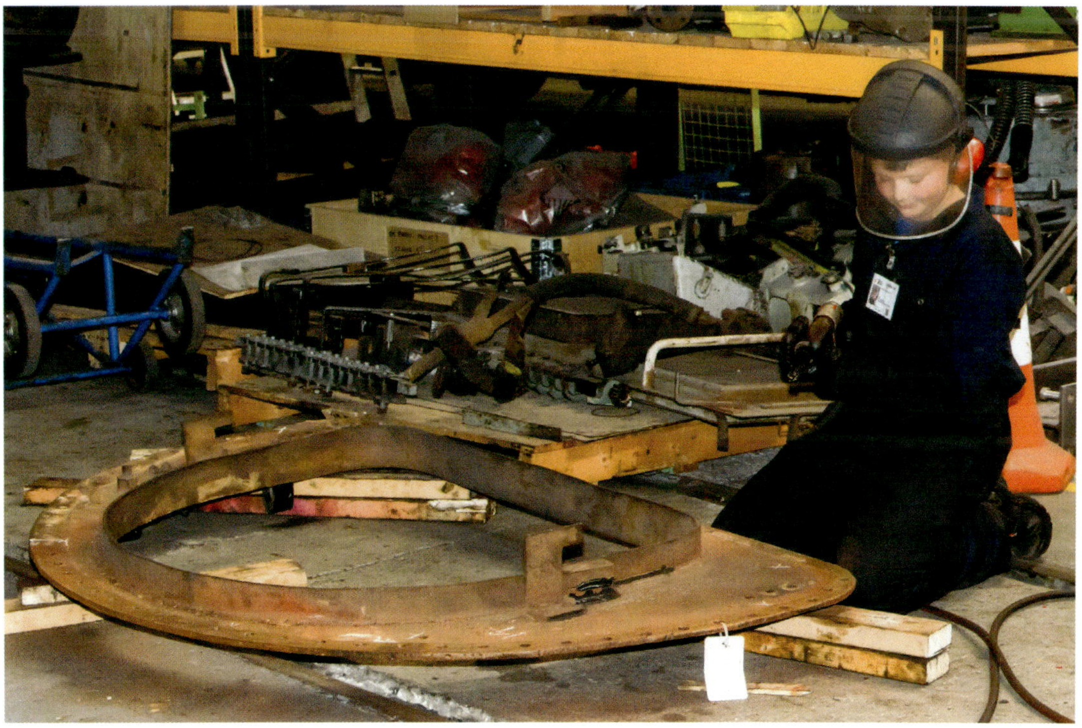

Junior volunteer Henry Pritchard working with the needle gun

RIGHT - Junior volunteers Tom Houseman and Finn Allen working on top of the frames next to the boiler saddle

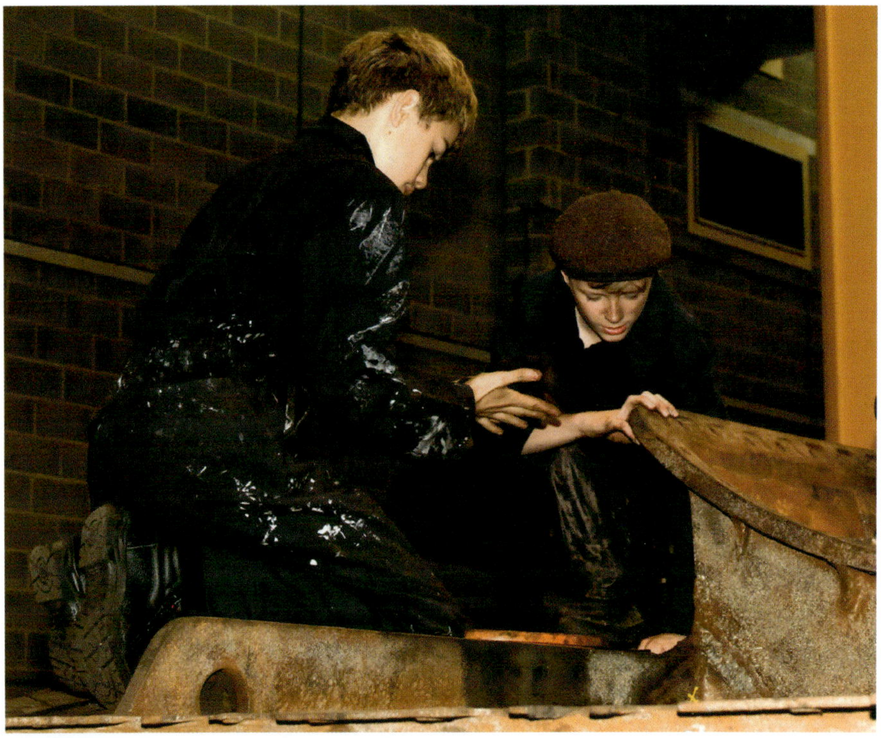

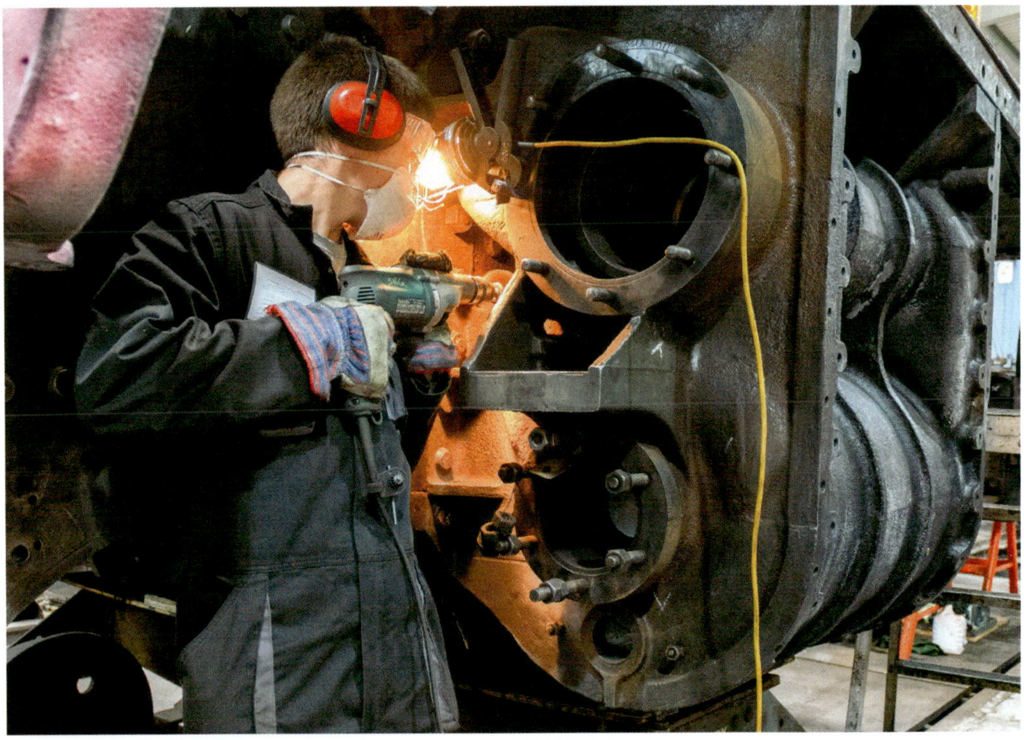

LEFT - Junior volunteer Ethan Humble uses a power drill with a wire wool attachment on a cylinder housing

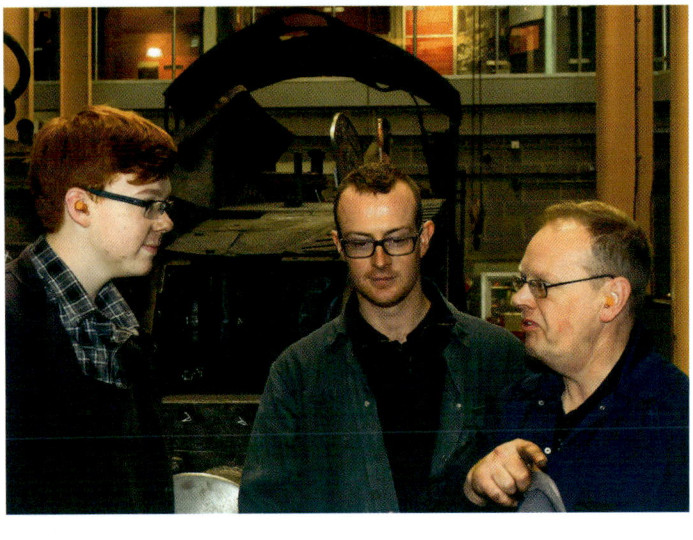

LEFT - Junior volunteer Tom Noble (left) with Darrin (right) and Jono Taylor

BOILER DEPARTS

5

November 2016

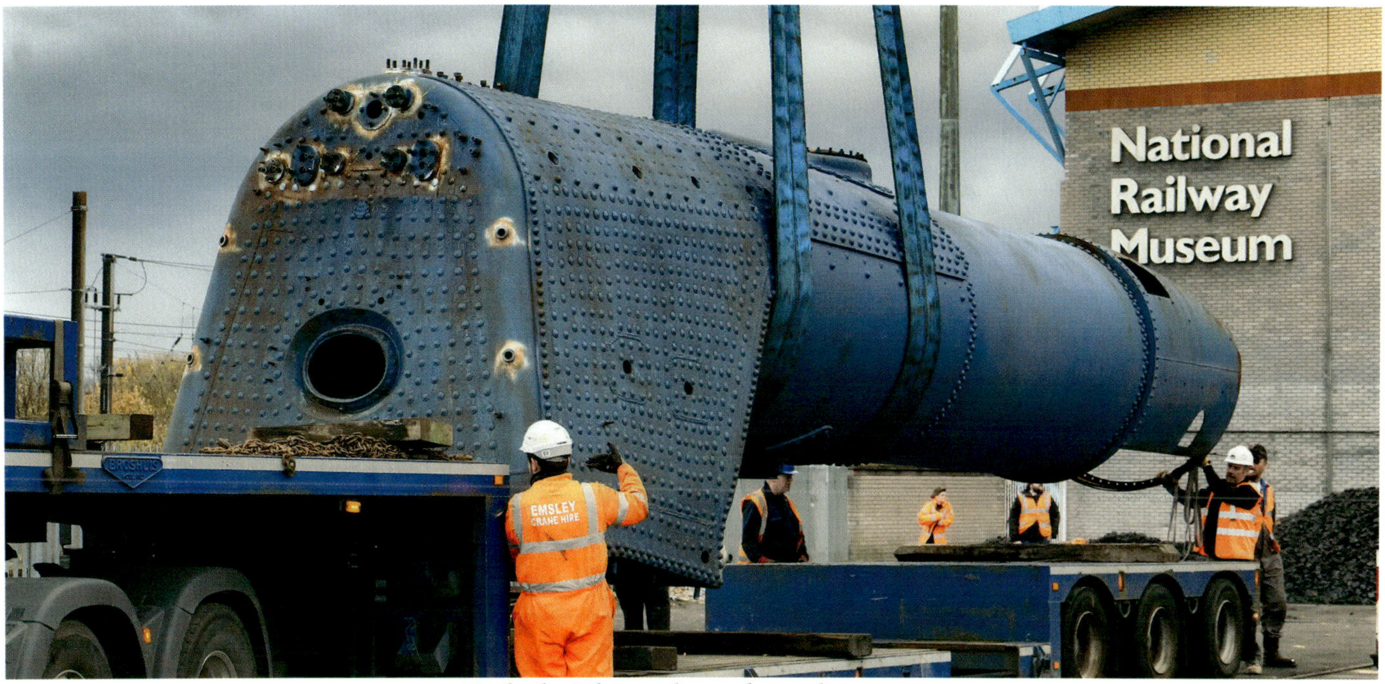

The boiler is lowered onto the road transport

It was nearly a year from the start of the overhaul and the team were now well and truly into the swing of things but there were plenty of nerves around when the boiler started on its journey to the Llangollen boilersmiths. With a second visit of one of the huge Emsley mobile cranes and specialist lo-loader transport from Reid Freight, the nerves would eventually be calmed as the boiler departed the NRM on route to north Wales.

I decided I wanted to cover this journey from York to Llangollen so with Malc as my driver, we did just that or tried to anyway, as we were in Wrexham before we caught up with the boiler, which did have a head start. It was important to me to get pictures of the lorry and its enormous load navigating their way through the city of York, and it was worth it as can be seen with the complicated manoeuvre passing through the city walls twice to avoid tight turns,

TOP & RIGHT - The boiler is lifted into position on the road transport in the NRM's North Yard

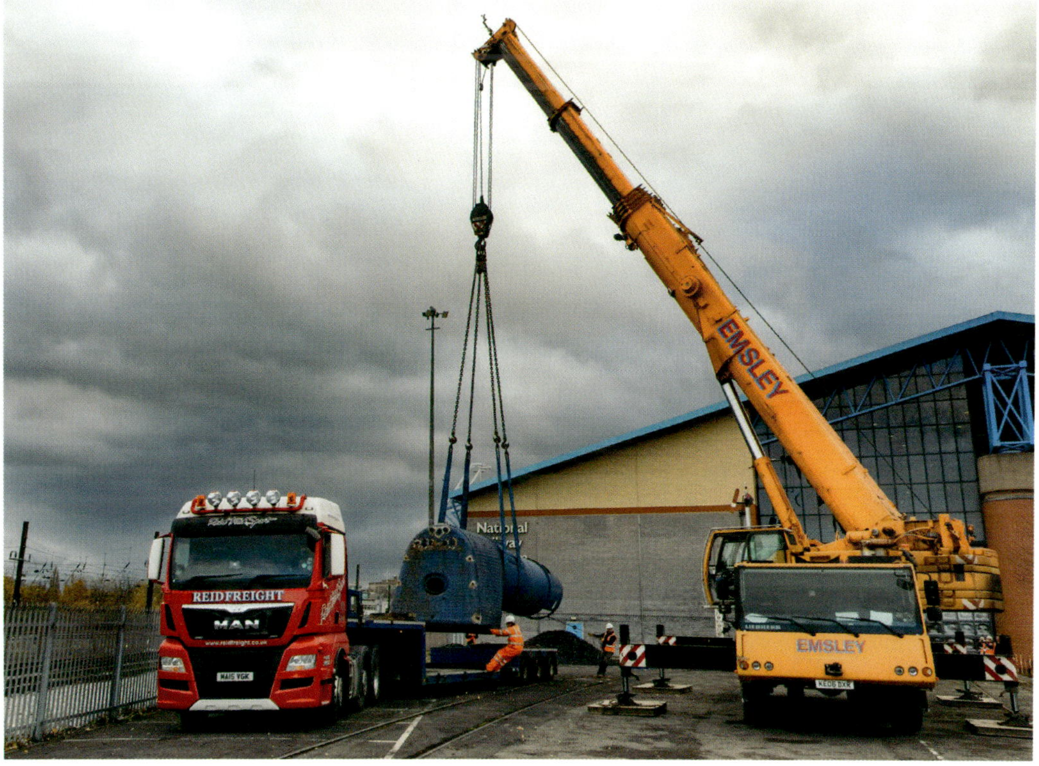

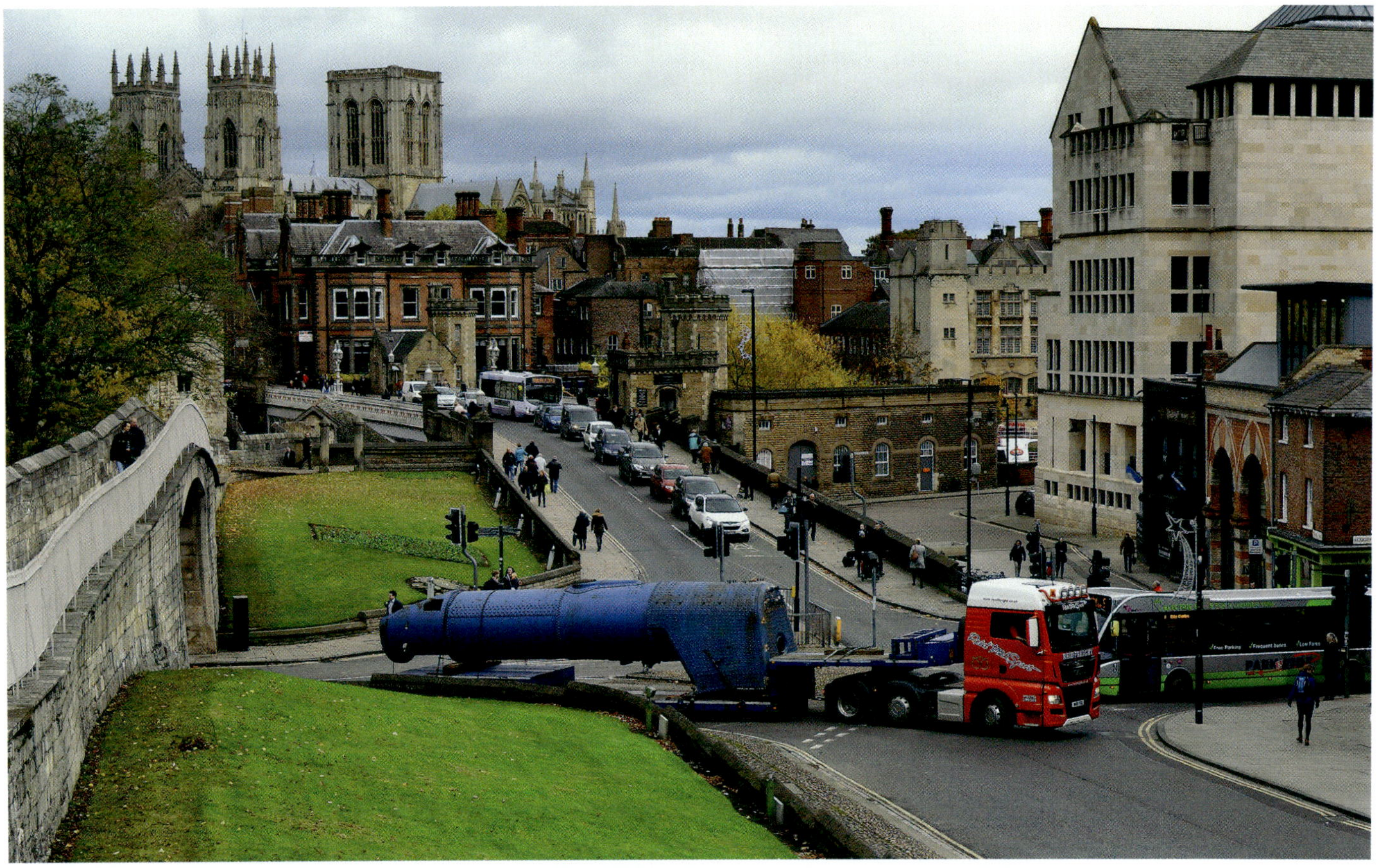

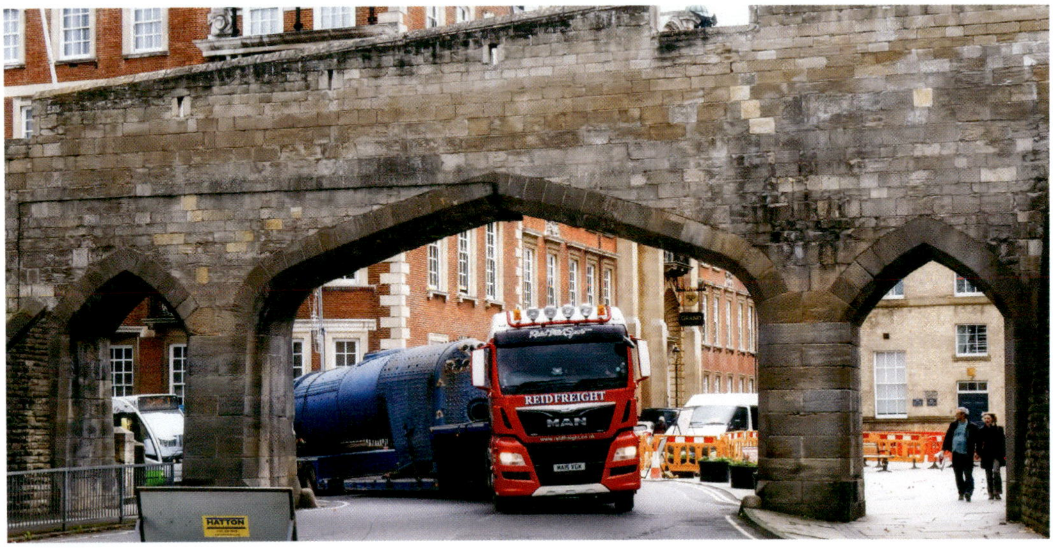

ABOVE & LEFT The boiler transport winds its way through York's city walls

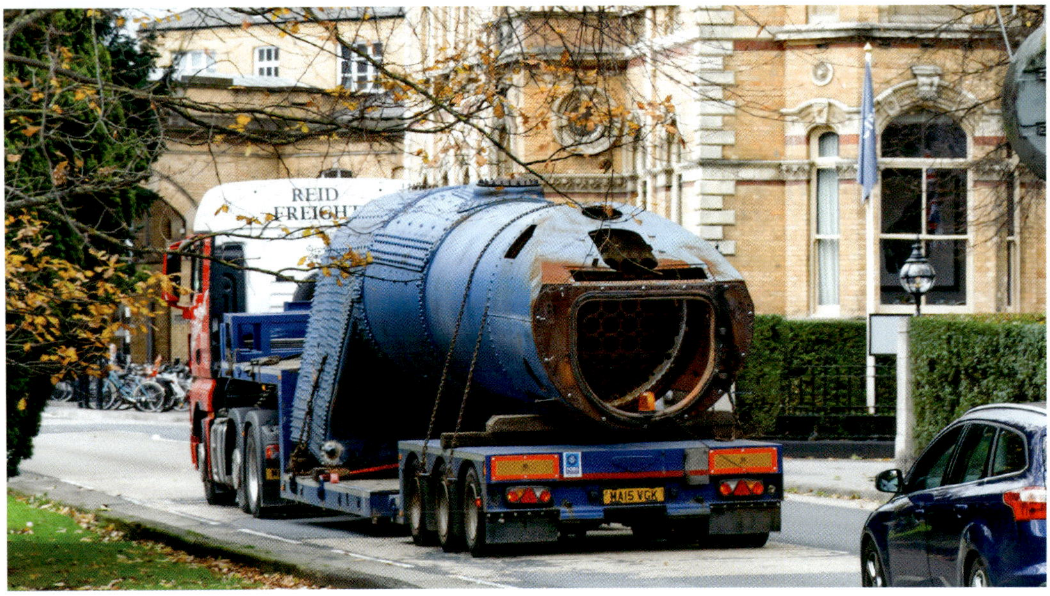

LEFT - Passing York station en route to Llangollen

"Oh, that's easy. You wait around 30 seconds and then you put the reversing lights on, and the traffic soon scatters."

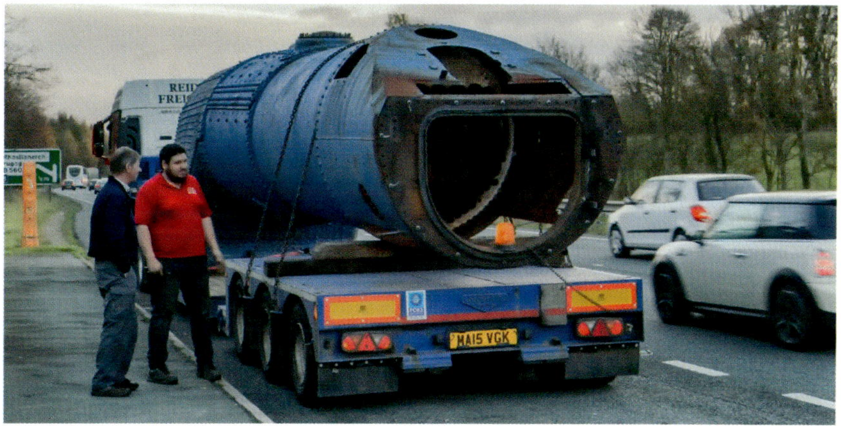

Malc Bateman and myself catch up with the transport near Wrexham and discuss the next part of the journey

which would have probably demolished some of York's fine flower beds. I needed to be in more than one place at once here, so my trusted friend Malc quickly turned into my apprentice and was duly rewarded when he took the classic shot from the city walls looking back over Lendal Bridge to York Minster in the background.

After seeing the lorry disappear into the sunset beyond the railway station we quickly got back to the car and set off in pursuit. It shouldn't be a problem catching up with the lorry and its exceptional load; wrong. As mentioned above it was Wrexham before we did and that was only because the driver had pulled into a layby and was waiting for us. Hats off to these heavy haulage guys - they certainly know how to get on with the job which was set to get even more challenging.

Our stop in the Wrexham layby was for the driver to ask us to go ahead of him into Llangollen as he said you will get some good shots when I arrive on the bridge. Intrigued, I asked why he needed to turn left onto the River Wye bridge when the Llangollen Railway workshop was straight on. "I can't get in that way. I have to reverse down the last couple of hundred yards!" This I had to see. Next question, "How do you reverse a huge load like this around the t- junction off the bridge when you will have queuing traffic behind you?" "Oh, that's easy. You wait around 30 seconds and then you put the reversing lights on, and the traffic soon scatters." "Oh right," I said, this I really did have to see. The pictures tell the story.

It had been a long day and we were ready for a bit of chill

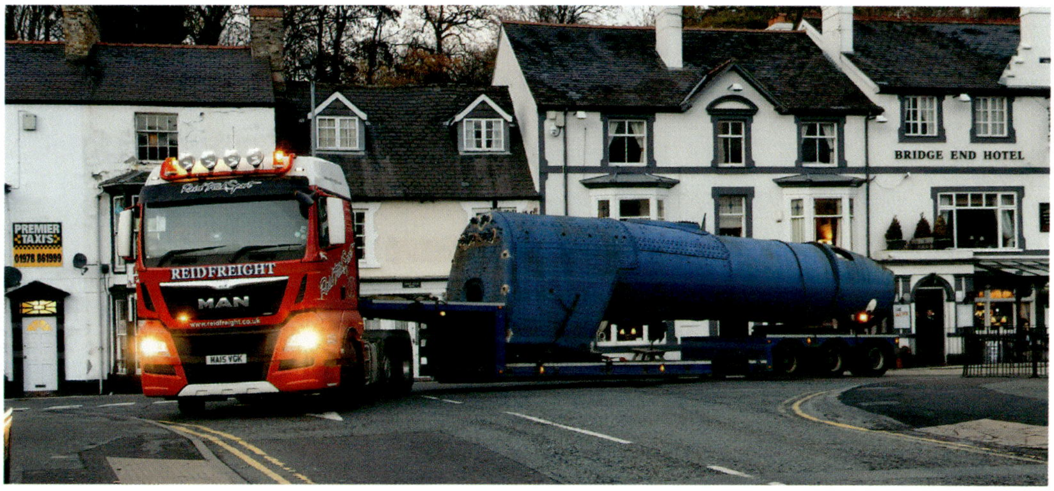

The boiler transport swings onto the bridge in Llangollen ready to reverse to the workshops

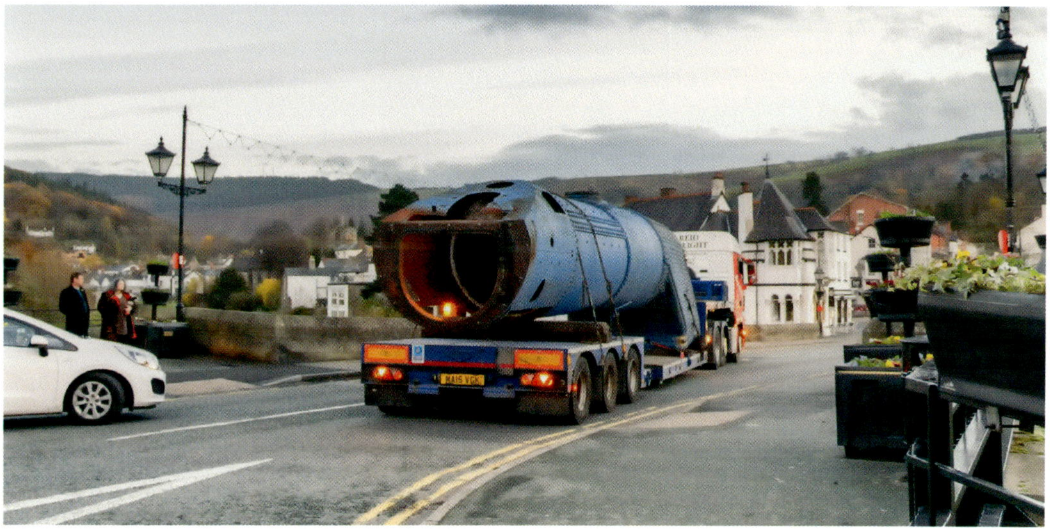

The boiler transport starts reversing around the T junction corner towards the workshops in Llangollen

time. Trust chairman, Nigel Wilson, had joined up with us and after a couple of hours' relaxation at our hotel, we had an evening meal at one of the excellent Llangollen establishments before retiring for the night. Tomorrow would see the formal handover of the boiler and the arrival of winter with a vengeance.

Suitably rested overnight the three of us enjoyed a hearty breakfast before heading off to the workshops of the Llangollen Railway where the boiler was to be unloaded from the lo-loader. Paul Aston, SNGLT engineering team representative in Llangollen met us at the workshops along with representatives from the Llangollen Railway. It was November and the weather had deteriorated into heavy rain; it was icy cold rain the kind that soaks you in seconds. Keeping the rain drops off the camera lens was an impossible challenge and as uncomfortable and challenging as it was, I doubt it came near to the task the Llangollen team were undertaking in moving the boiler from the lo-loader and onto an adjacent rail wagon. Aided by their wonderful steam crane the task was completed with great skill and to everybody's relief the wet, very cold and bedraggled team were rewarded with hot tea in the station café.

It was time to head back home which would mean crossing the Pennines to North Yorkshire and County Durham. Both Malc and I are none too keen on the M62, preferring to head north on the M6 and then to cross the Pennines on the A66, so that's what we agreed to do. The rain hadn't eased when we stopped for lunch at Ravenstonedale and by the time we left it had turned into snow, transforming the already pretty village into a Christmas card scene. It wasn't that bad we thought, as we stopped for fuel in Kirkby Stephen, only to be advised that the snow gates had been closed on the A66. Plan B, (logic wasn't working very well here) – if the best route over the Pennines was closed due to snow let's try the worst route - south to Nateby and then over the top to Keld into Swaledale - very pretty in summer and very pretty now. We climbed up following the tyre tracks of another intrepid explorer before finally giving in when the other vehicle turned around in the distance. One last attempt, further south again down the Eden valley and over the top to Hawes before descending into Wensleydale, and finally we were home and dry - well maybe not dry.

Paul Aston

Paul has been an important member of the team, living closer than most to Llangollen and Crewe he was ideally situated to oversee the work on the boiler at Llangollen and to volunteer in the final stages of the overhaul at Crewe. Paul is also a former chairman of the Trust.

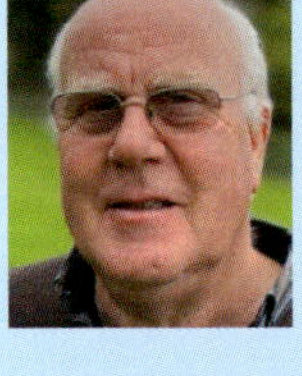

"At the last overhaul at Grosmont, where the chassis and boiler were overhauled together, I took 5 years off work to be able to be involved in every aspect of the work. Following my 7 year project in the NHS, I was delighted to be invited back to the SNG Engineering Team to manage the boiler overhaul. The key moments were: - the extent of the severe wastage of the 16mm steel plates newly fitted above the foundation ring at the last overhaul and the replacement of this platework; the removal of the copper tubeplate (which was welded in) and the intricate work involved in its replacement; and the challenges encountered in successfully fitting the refurbished superheater header. Over the three years there were many technical issues encountered, but the team at Llangollen, led by Chief Boilersmith, Graham Beddow, were great to work with, and Darrin and Richard were great to liaise with to resolve problems as they arose. The establishment of these new working relationships together with the renewal of old ones was a particular pleasure and most rewarding. It has also been great to keep in touch with the rest of the overhaul from Darrin's comprehensive weekly reports."

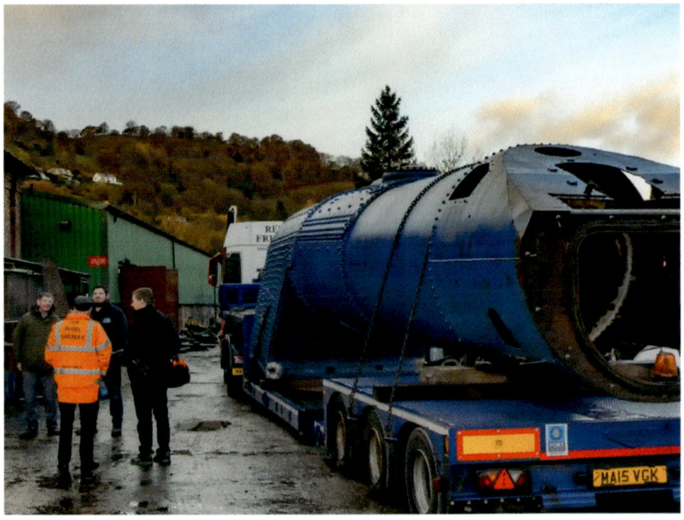

Boiler arrives at Llangollen workshops

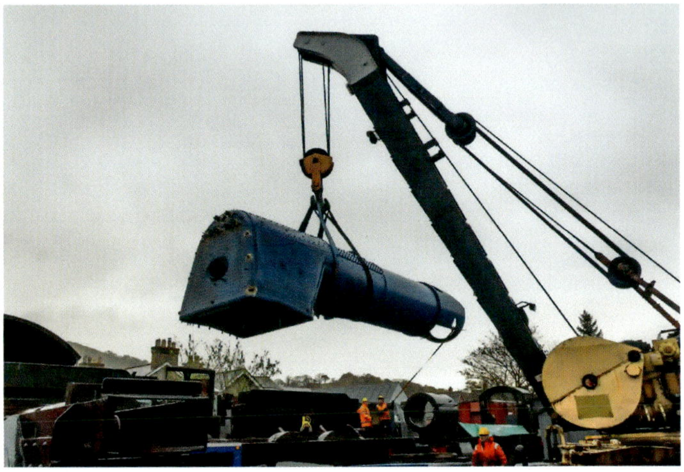

The steam crane lifts the boiler from the road transport

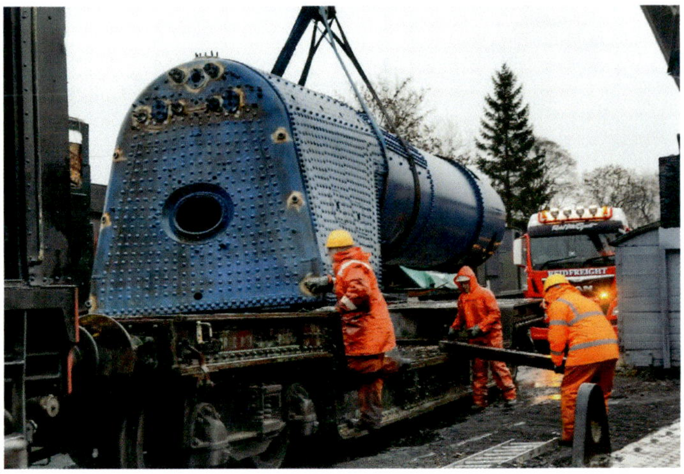

The boiler is lowered onto the rail wagon

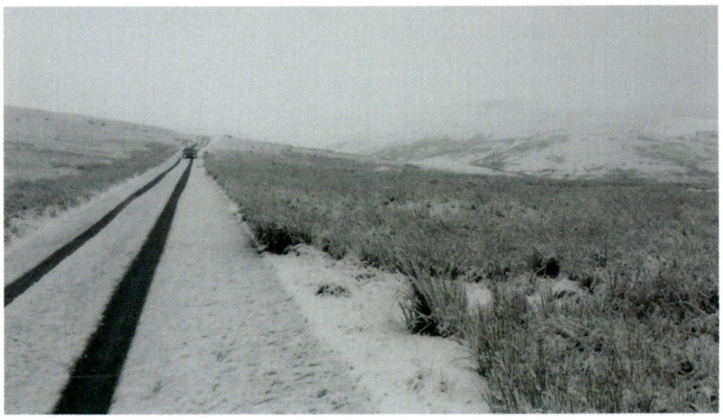

Time to abort Plan B's attempt to cross the Pennines

REBUILD BEGINS

6

December 2016

Year one had become year two and No.7 was now scattered far and wide - the wheels in Devon and the boiler in north Wales, whilst the frames remained in York. My focus would remain with the frames in York where a corner had been turned. There was a lot going on but the most visible signs of progress, to me, were like the green shoots of spring, as large areas of bare metal were now being primed with new paint and in some cases undercoat and topcoat. The team were no longer just stripping down No.7. The rebuild, if only in a very quiet way, had started. The first component to be refitted was a sandbox, although Darrin did add, with his customary smile, that it wasn't the first as some bolts had got that honour, but I'm sticking with the sandbox claim. The sand from these boxes can be released onto the track to help the wheels get traction when conditions are slippery, owing to leaves on the track etc. This image also shows coloured dyes used for checking for metal cracks, (no it isn't fancy undercoat), and at the top right you can see the orifices of the middle cylinder which is normally hidden from view under the boiler between the frames.

The poor light in the workshop and the numerous dark areas meant I was continuing to use the tripod and triple exposures of the High Dynamic Range (HDR) technique in a lot of shots, which was lifting out different colours and giving great contrast as a result. The two shots I took of the frames, taken from in front of No.7, do, I think, emphasise this point and coupled with using a wide-angle lens, I was happy with the resulting images. Whilst juggling the swapping of lenses, as I only had one camera body with me, I was looking around for a temporary shelf (as I haven't got three hands) and it was only after I had retrieved the unused lens that I realised I had only used the buffer beam from Flying Scotsman as a shelf. Yes, we now had two of the most famous locomotives together in the workshop as Flying Scotsman enjoyed a winter break from a demanding schedule of rail tours and can be seen at the top of the picture taken from the gallery walkway.

The Christmas break was upon us and when we returned in the new year it would be a riveting day as things were hotting up.

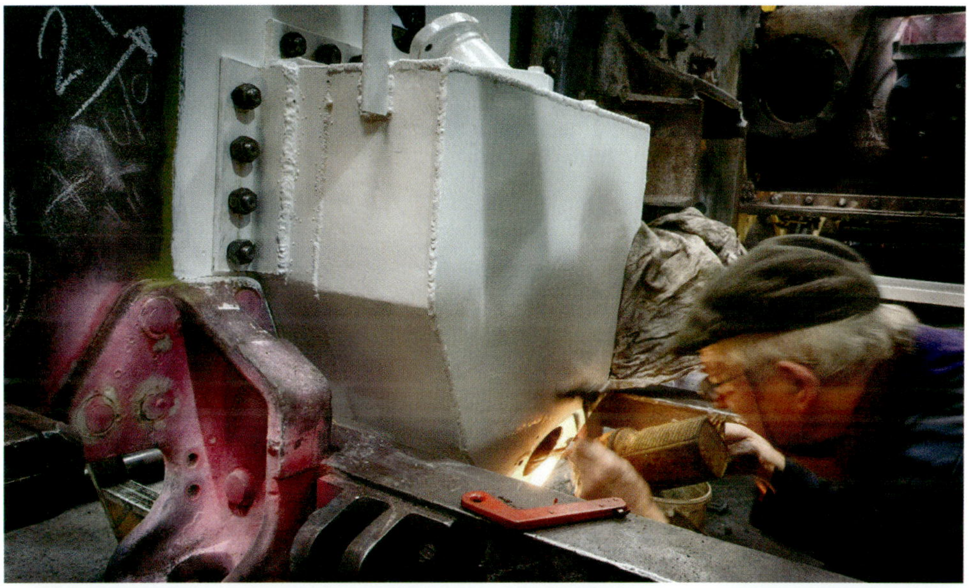

Bob Shearman checks there are no blockages into the newly fitted sand box, the first component refitted

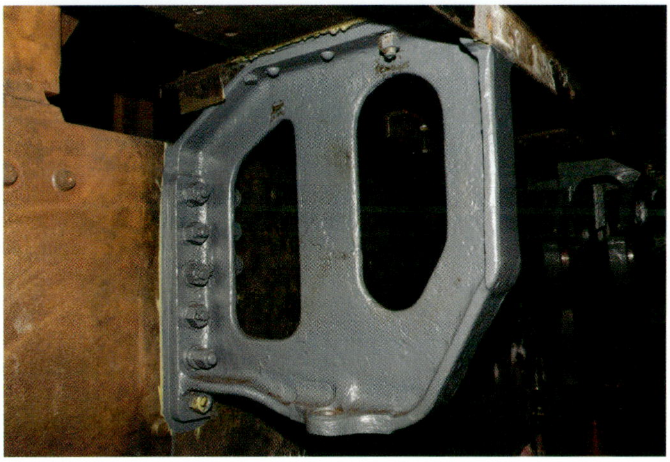

Fresh primer applied

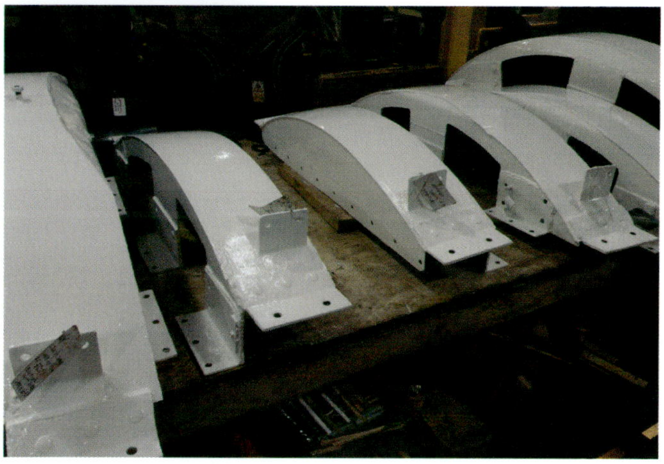

Freshly painted wheel splashers.

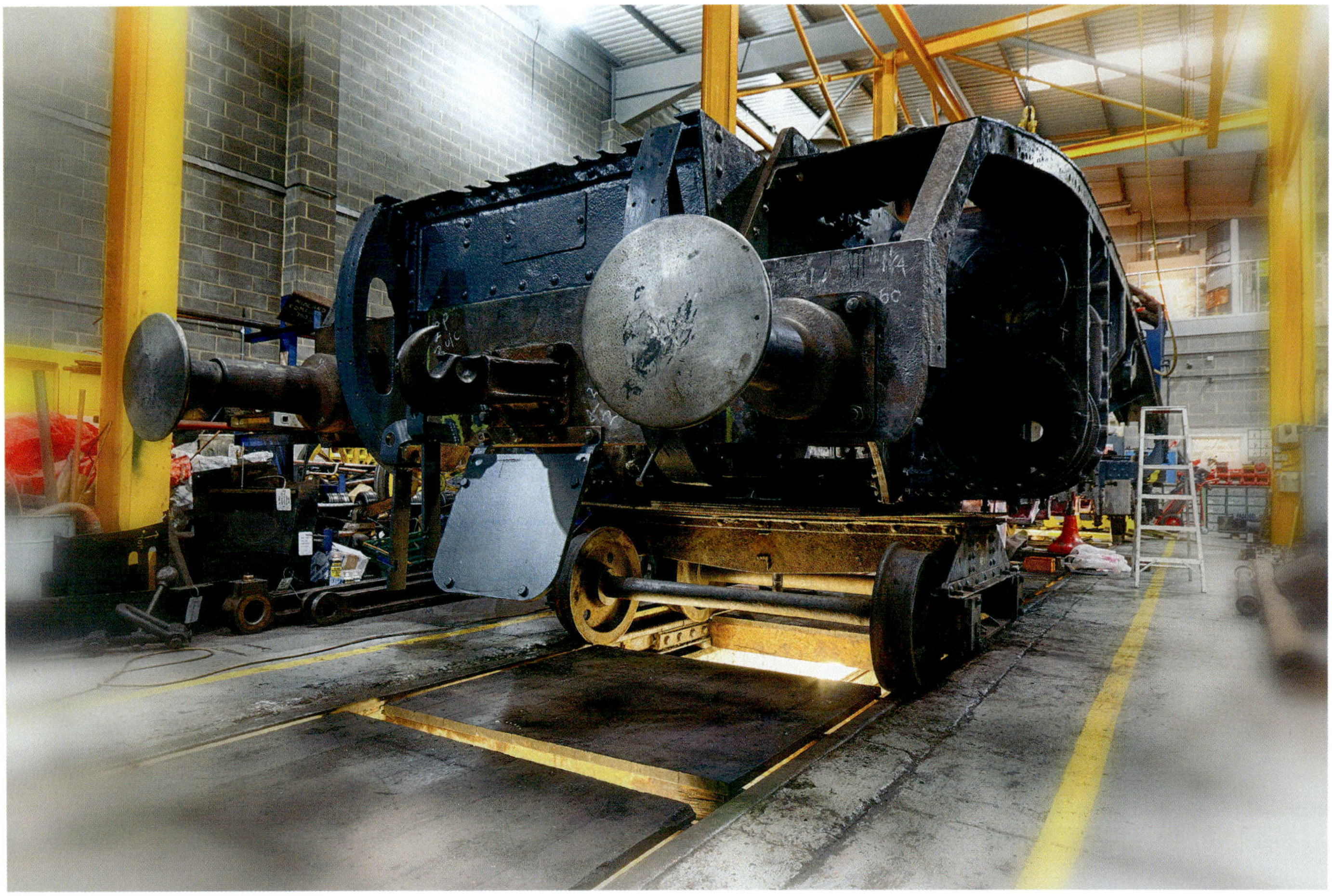

ABOVE & BELOW High Dynamic Range (HDR) images, each picture is created by merging three differently exposed images into one image thus capturing a wider spectrum of colours and tones

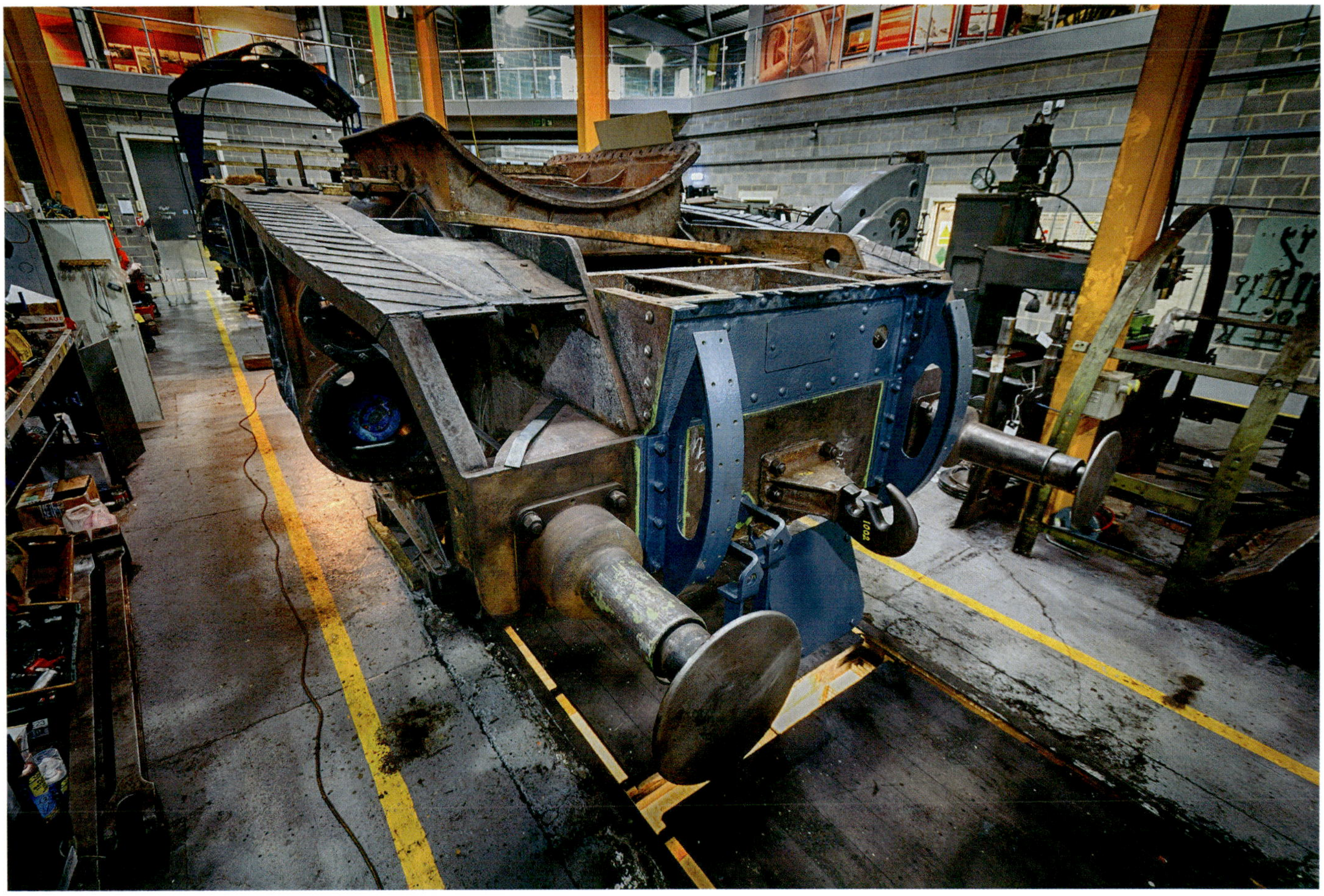

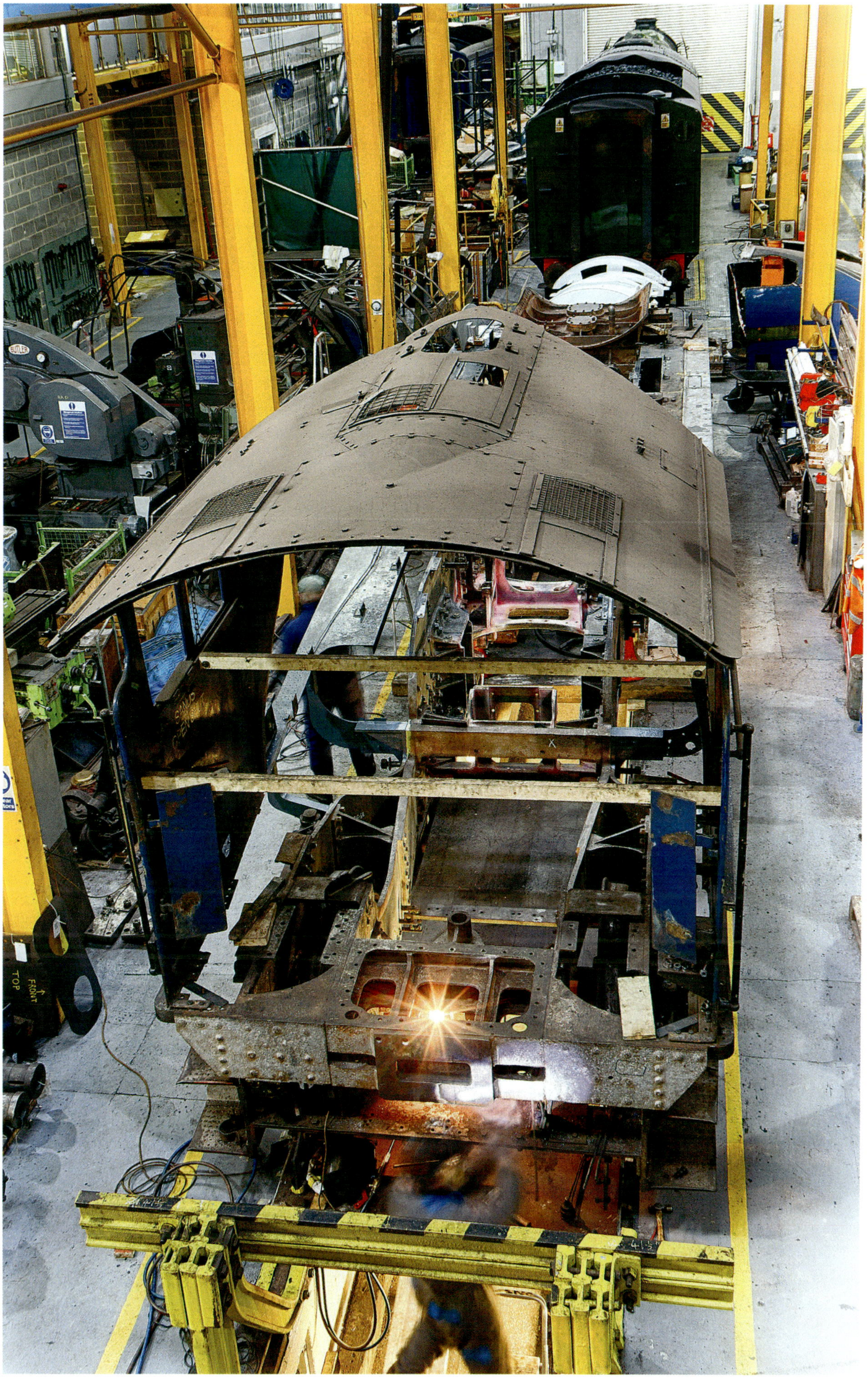

Resembling a scene from the early 1950s with A3 Flying Scotsman keeping No.7 company in the workshop. Note the long exposure resulting in a blurred Roger Turnbull working from the pit on the dragbox and Darrin Crone down the left hand side of the frames

A RIVETING DAY

January 2017

The Christmas break was over, and it was back to work for the team with a new form of expertise required. No.7 had lots of rivets but nobody on the team with riveting skills. Back in the day riveting was a common trade, particularly an hour or so north of York in the ship building north-east of England, but those days are long gone. However, the ever -resourceful Darrin managed to seek out a contractor with the necessary capability and so the riveting began.

Riveting support came in the form of new volunteers, Trevor Wilford and Peter Ellis, who also had previous experience and the team quickly grew when it became apparent other team members wanted to learn and practice this old trade.

The picture sequence here portrays the first of many rivets being heated up, transported and then fitted in No.7 frames. This was hot work and great care had to be taken, so, to stay at a safe distance, I needed to use a longer lens, allowing the lens to zoom in on the close-ups.

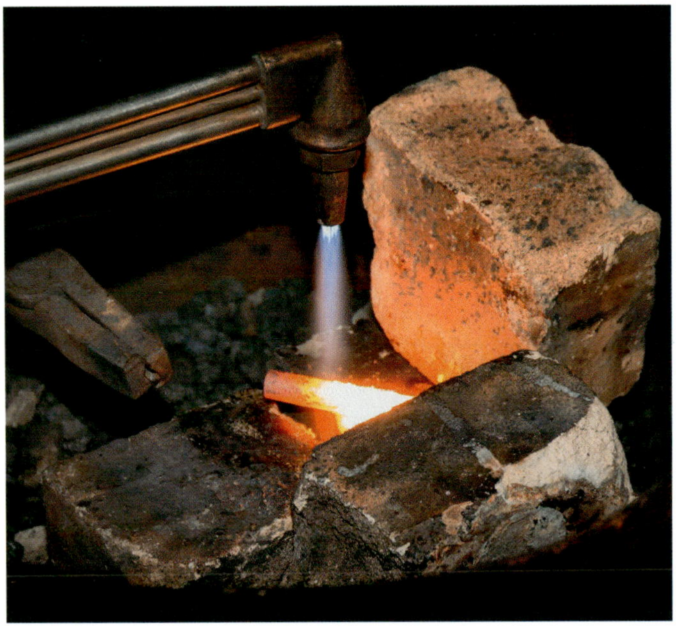

Heating up a rivet

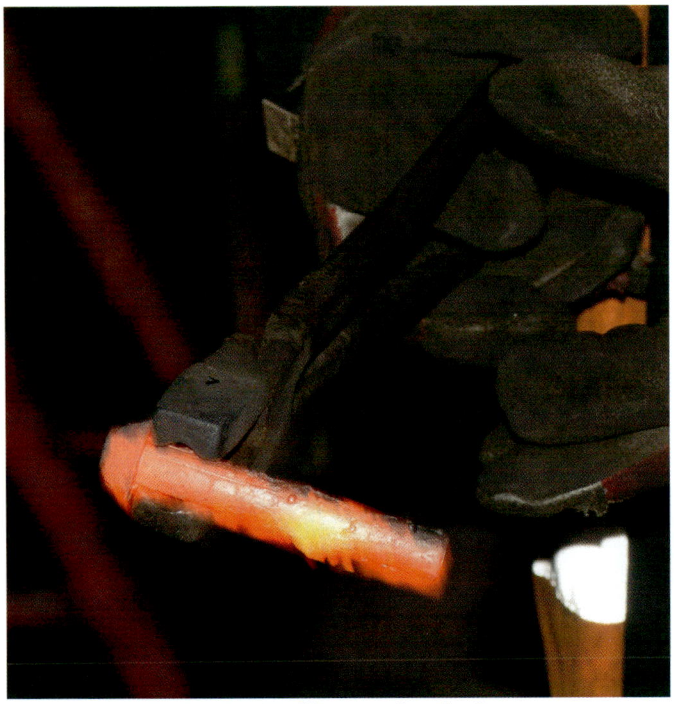

The rivet is quickly moved to the frames

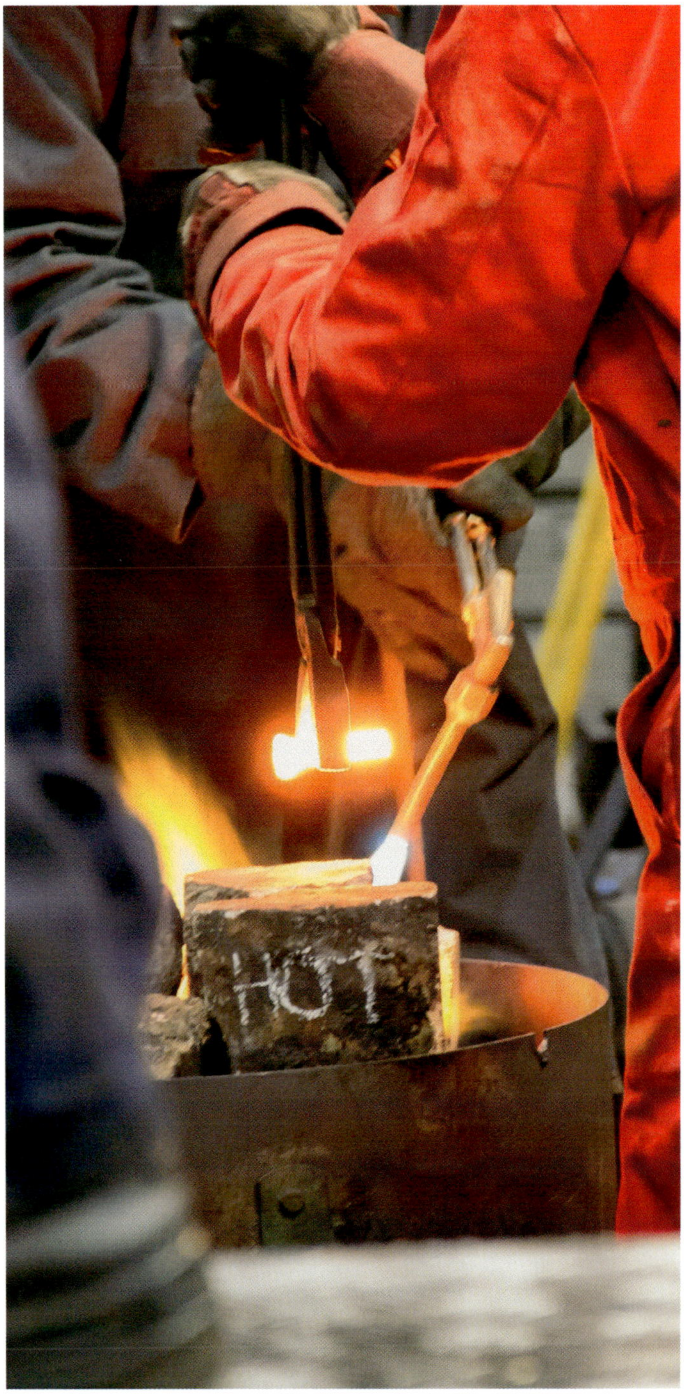

Another hot rivet is lifted carefully and moved to the frames

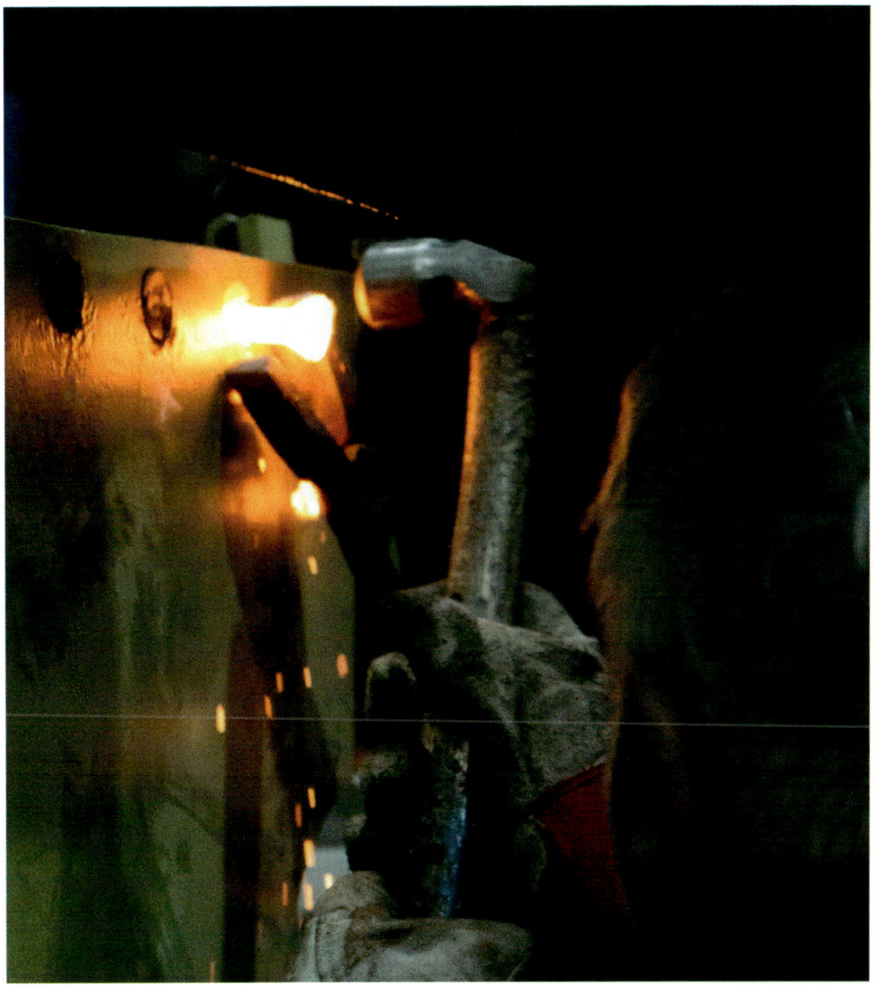

Peter Ellis

A.K.A 'the skilled artisan, the pipe strangler, the silver fox'

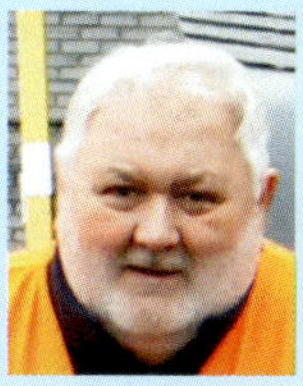

Born, bred and living in Stockton all my life, my earliest memories of the A4s were trainspotting from the platforms at Darlington Bank Top Station in the late 50s and early 60s. I served my time as an industrial plumber/ pipefitter at ICI Billingham and then spent a number of years on the tools before leaving in 1975 to join the Local Authority as a Building Inspector – a position I held for 37 years before taking early retirement in 2012. Having been headhunted by Darrin in 2016 to give some practical guidance to the lads about annealing, silver soldering, pipe bending etc., I was conned into staying with the overhaul until the end. My work has been to refit, repair or replace as necessary all the pipework on the engine and tender. Together with a number of lads who were happy to work with me as a team, we have spent many a happy hour cursing and swearing in the pit!

Peter Ellis hammers a rivet into the frames

BELOW & RIGHT - The riveting gun is applied to form the head whilst the rivet is still hot enough

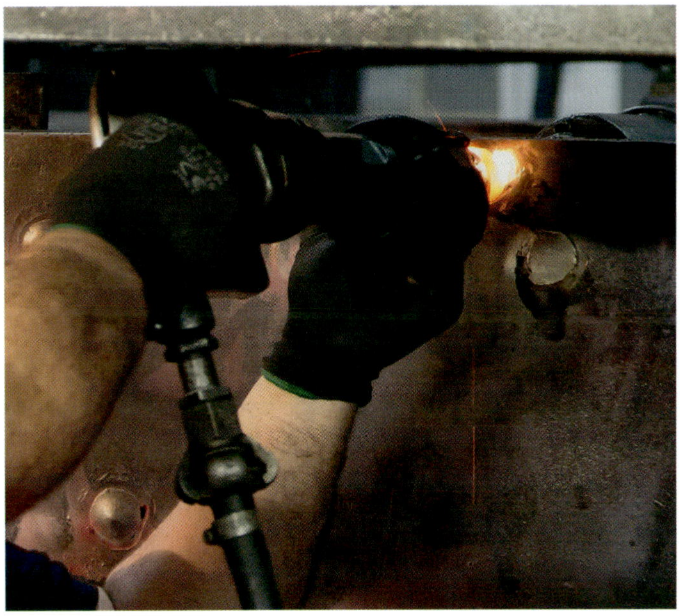

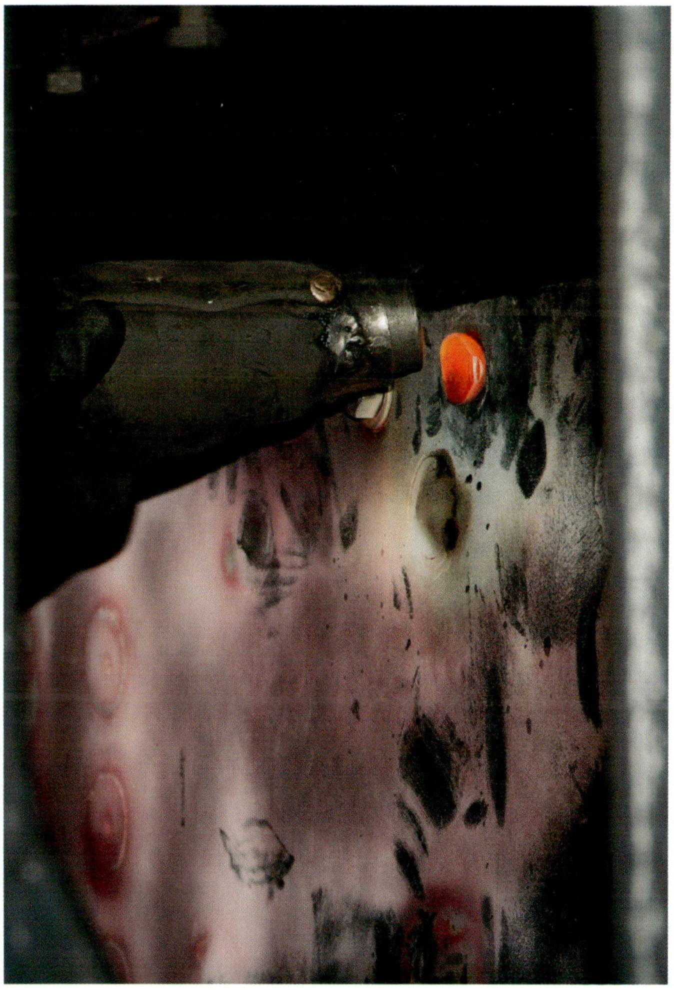

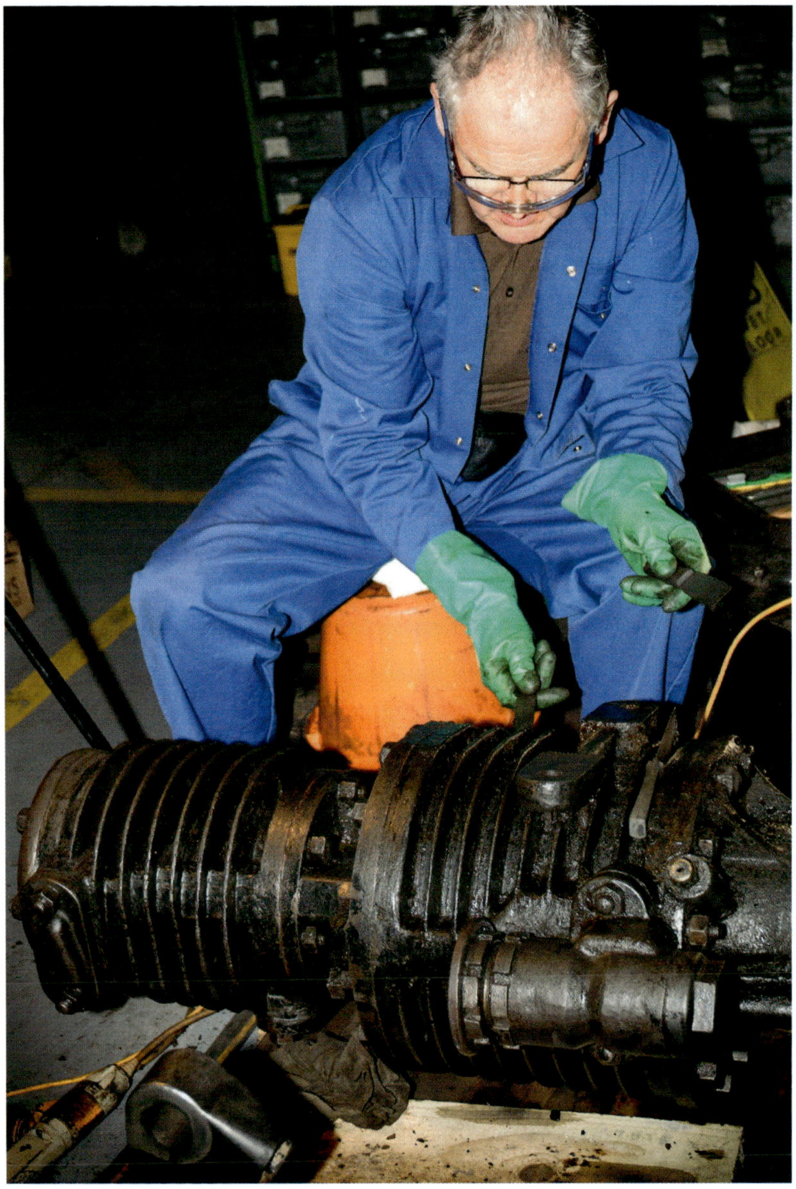

Rod Thomas gets to grips with cleaning an air brake cylinder.

Roderick Thomas (Rod)

Christened by the team as Dr Ebola as I use green chemical gloves when cleaning!

Christened as Rod the Hod because I worked in construction albeit not as bricklayer but as a Quantity Surveyor!

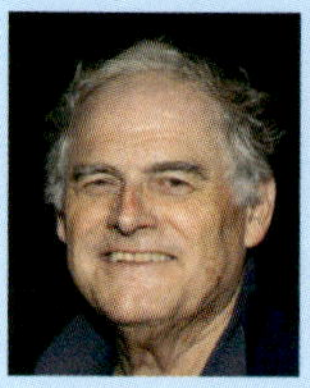

In the 50s my Dad gave me a Hornby clockwork O gauge tinplate set (still got it!) and I have been a steam fan ever since!

An armchair member of NYMR but regularly following 60007 so when Darrin asked for volunteers and allowed me to join the team in May 2016, aged 69, I was chuffed to bits!

Dr Ebola started cleaning and after two years joined the piping team as gofer/ apprentice plumber's mate to a silver medallist plumber which has been enlightening in more ways than one and I am still serving my time!

Proud to have been involved in the overhaul of 60007 with bits of me on the loco, pipe lagging my speciality!

Proud to have been part of a great team with its comaraderie and banter which is something I will miss greatly when we are finished!

My thanks go to my wife in supporting and encouraging me to have a go!

"It's a dirty job but someone has to do it"

Whilst the riveting was keeping several team members occupied others had less glamorous tasks and Rod Thomas's role for the day was certainly less appealing. The brake air cylinders had 10 years of grime and sludge encasing them, and armed with rubber gloves, scrapers and pure dedication, Rod was up to the challenge. As they say, 'It's a dirty job but someone has to do it'.

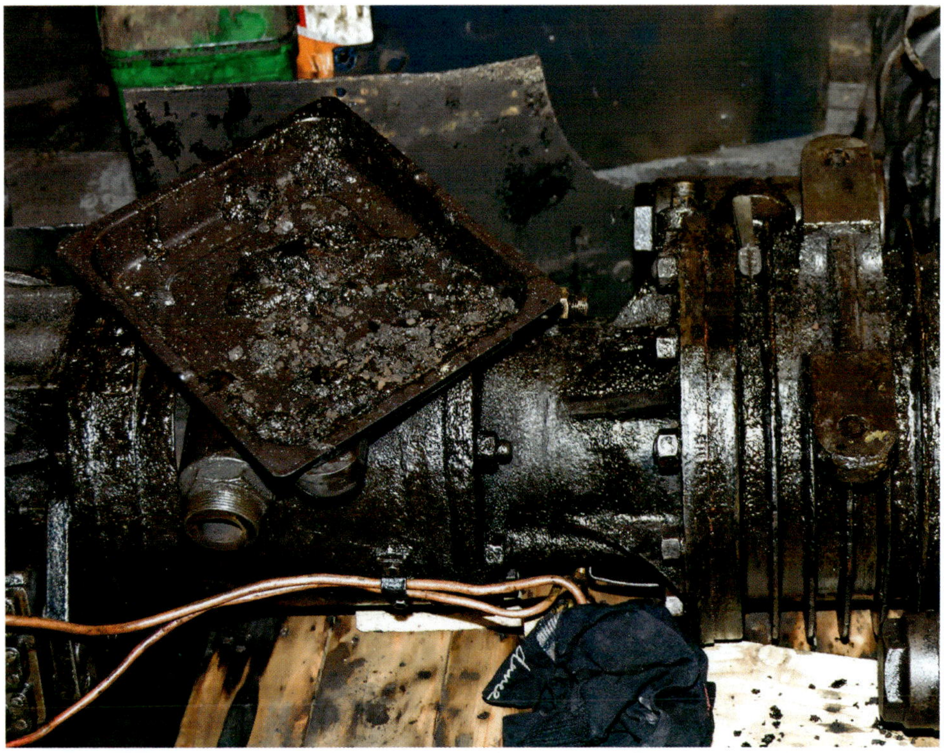

Years of sludge composed of various materials is removed from an air brake cylinder

FRESH PAINT

February 2017

We were barely in the second year of the overhaul but the sight of fresh paint being applied to a lot of the meticulously prepared metalwork gave, for me at least, a feeling of progress. The many experienced engineers in the team would be labouring hard with sparks flying and hammers hammering; careful measurements would be being taken and many other technical points would be addressed, but it was the paint which provided that visible progress. I include a few images here showing the signs of progress together with a picture of SNGLT Chief Mechanical Engineer (CME), Richard Swales taking measurements on the horns. The horns sit either side of the axle boxes in the frames, and if you look carefully, you will notice that they have a diagonal fit against each other, allowing them to slide against themselves. This either decreases or increases the overall thickness and thus they can be set to form a good fit against the axle box when it is refitted.

Richard had the overall responsibility for the overhaul on behalf of the SNGLT Board and members, so everything was being done to Richard's methodical and meticulous plan, and I was now beginning to see just how good his excellent attention to detail was. I was to make a point of focusing, pardon the pun, on Richard's work every time our visits to York coincided. Darrin was the day to day manager onsite and I was becoming very aware of not only his engineering prowess but also his team building skills. Darrin's team of volunteers was moulding into a great team, - reliable, trustworthy, experienced, knowledgeable and

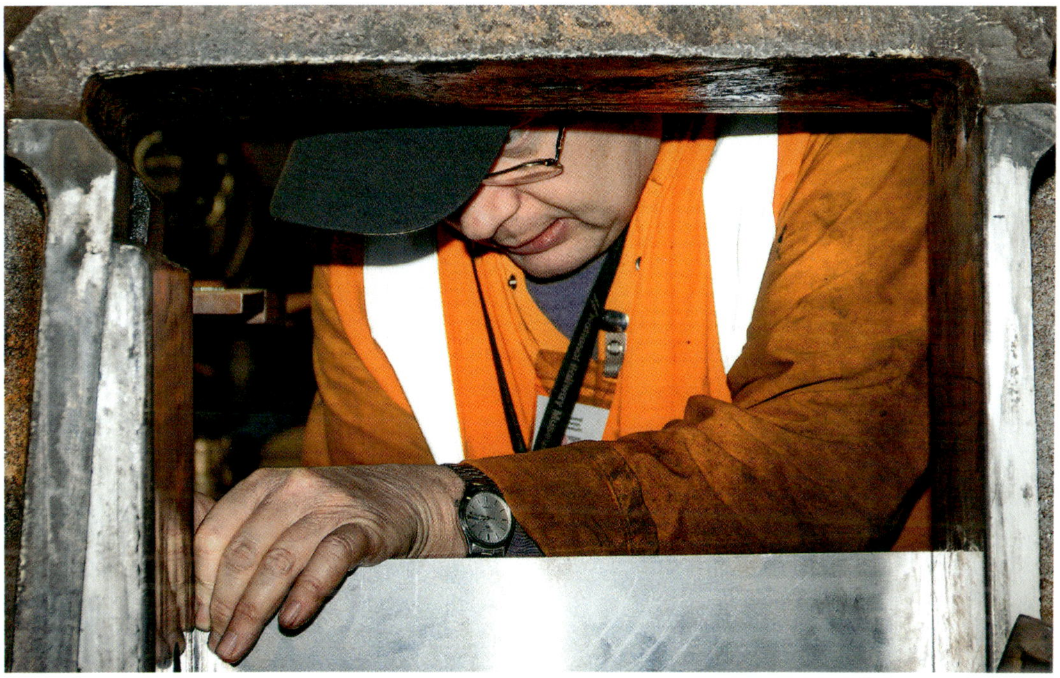

Richard Swales taking measurements from the adjustable horns

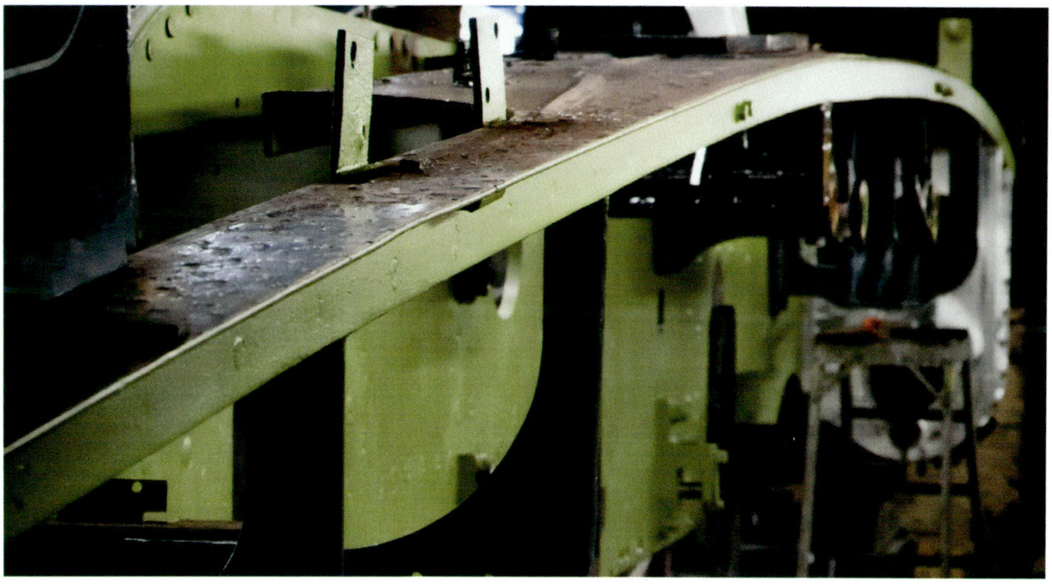

The unmistakable lines of an A4 even with the boiler and wheels removed and now with fresh primer and undercoat paint applied

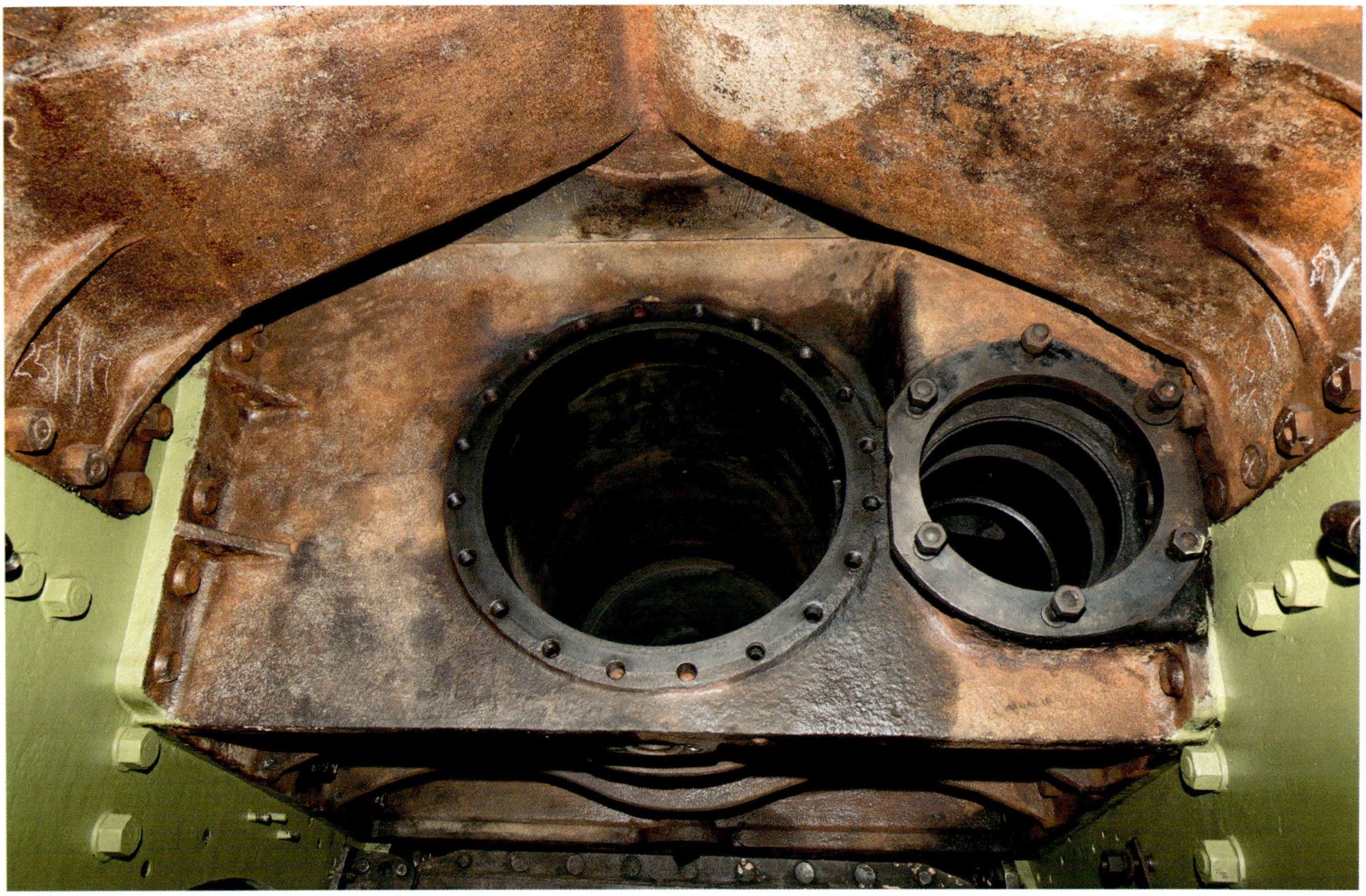

A rare opportunity to photograph the horizontal middle cylinder in its normally inaccessible position between the frames

very dedicated to the project. They came from all walks of life – mainly, but not all, retired. Weatherman Bob, IT manager Malc, central heating boiler engineer Peter, oil rig engineer Andy, National Park Ranger Mike, power station engineer Roger, and myself, a newspaper man are just a few of the diverse team members. Now, joining as riveters Peter Ellis, later to be awarded a silver medal by Darrin for pipe gymnastics and Trevor Wilford had joined up to this ever-increasing Band of Brothers.

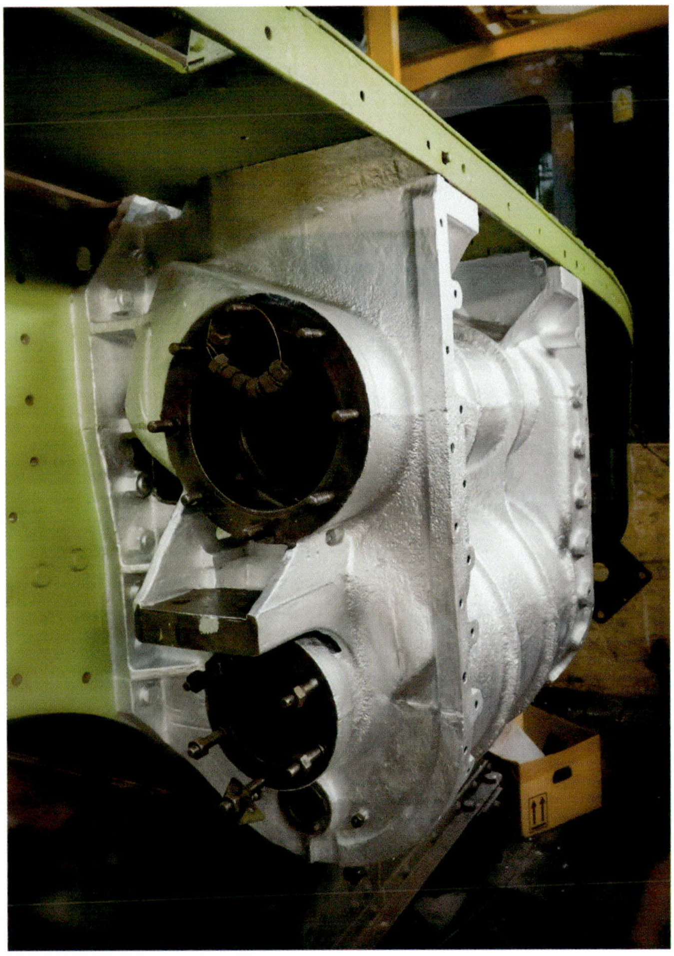

The right hand side cylinder freshly painted with primer and undercoat

My last visit this month was back in the pit below the frames, as the middle cylinder held a fascination for me. No.7 was a three-cylinder locomotive and, unlike the very much on view outer cylinders, the third was between the frames underneath the boiler. Before the boiler had been lifted, Darrin had tried to show me this from in the pit, but without taking lighting with me it was impossible to see clearly. It was literally pitch black and a bit too hostile an environment for my camera and tripod, so I opted to wait until the boiler had been lifted for this shot. You will notice this cylinder looks different; this is because it is in a horizontal plane as opposed to the vertical plane of the outer cylinders. I was keen to capture this shot as it was a rare opportunity to photograph something that is normally well out of the reach of a camera lens.

HOOKED

March 2017

I normally try to arrive at the workshops around 11am, just in time for one of Peter's cuppas. Well not just that - I leave home after the morning rush-hour on Teesside and my subsequent arrival in York then fortunately coincides with brew time in the affectionately named 'blue cabin'. This is a portakabin used by the team as a mess room and is appropriately coloured blue like No.7. However, today I was late, and having missed out on a cuppa, I went straight into the workshop. A new face greeted me - or was it new? Tony Bickerstaffe had visited as a guest of Darrin's a few weeks earlier as he was building his own scaled down working model of No.7. and wanted to look over some original drawings. But the magical attraction of the real thing, or Darrin's persuasive powers, had done the trick and another first-class volunteer engineer had joined the team. I got my cuppa at the lunch time break and I asked Tony what got him hooked and the reply was "I don't know really. I only came for some measurements." The team was growing. No.7, far from being out of traffic and out of the limelight, was very much still an extremely popular locomotive with which people wished to be associated.

With components all stripped from the frames, painting was again marching on and I now had the opportunity to get between the frames and photograph the newly painted middle cylinder, which had proved impossible to do when the boiler was in situ and would be impossible to do again when re-assembly took place.

Tony Bickerstaffe

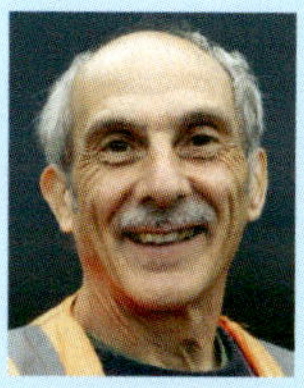

I have had a varied career in engineering. Having a keen interest in steam locomotives, one of my hobbies is making models of them at about one eighth of full size. Around the year 2000 I decided to build an LNER A4 and had acquired copies of the original LNER drawings which could be scaled down to produce an accurate model. Then, as though it had been arranged by the Gods of Steam for yours truly, Sir Nigel Gresley was moved to the National Railway Museum's workshop in York for its 10 year overhaul, and it was even over a pit which allowed access underneath to get photographs! I sought permission from both Darrin Crone, SNG Locomotive Engineer, and the NRM for access to the locomotive and after an in depth safety briefing, I was let loose to take my pictures. I was invited to join the volunteers for tea break time, felt really at home and had soon volunteered, paid my subscriptions and was attending on a weekly basis. I remember thinking, "I only came to get some photographs!" There's a lot of pride goes into what the volunteers do, which is contagious, and it rubs off on each other, with a shared feeling of satisfaction and achievement.

"I only came for some measurements"

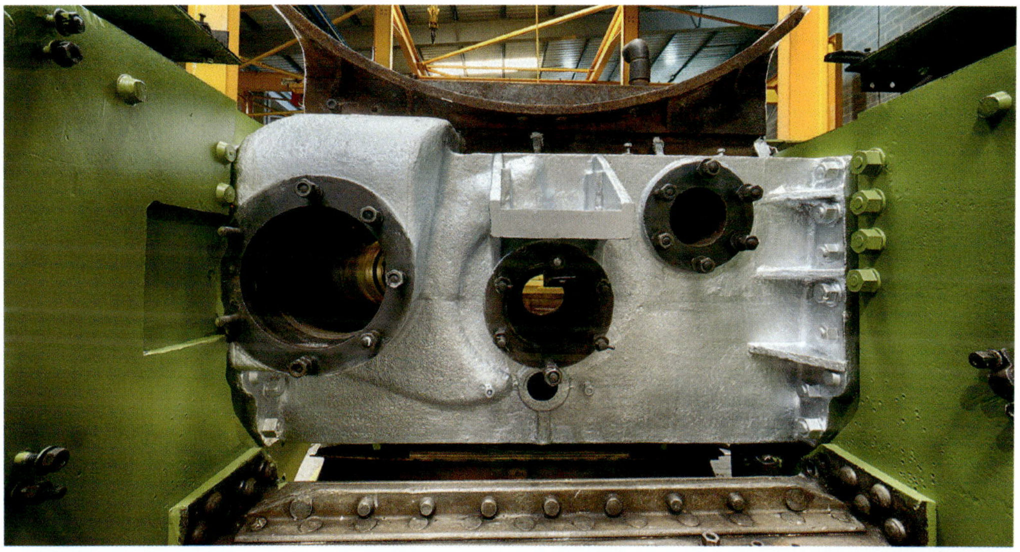

The freshly painted middle cylinder viewed looking forward from inbetween the frames

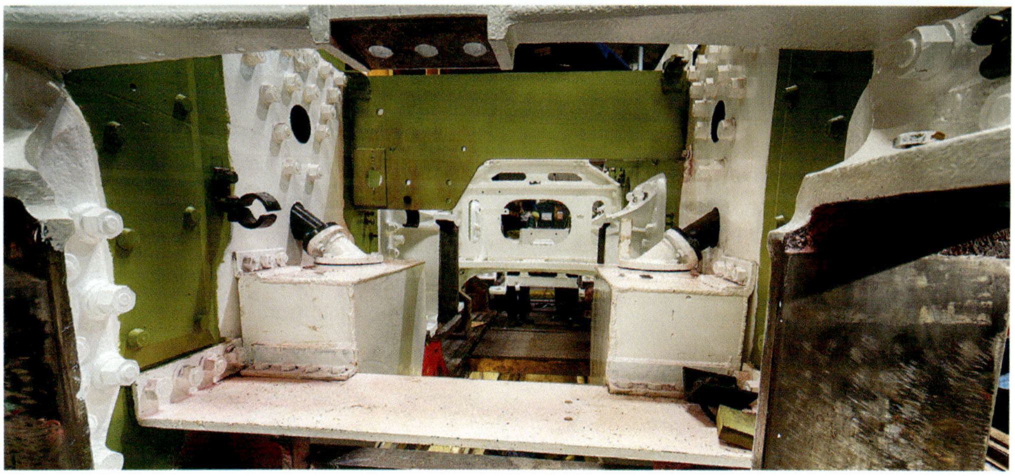

Freshly painted surfaces between the frames, not easily visible when the boiler is in place

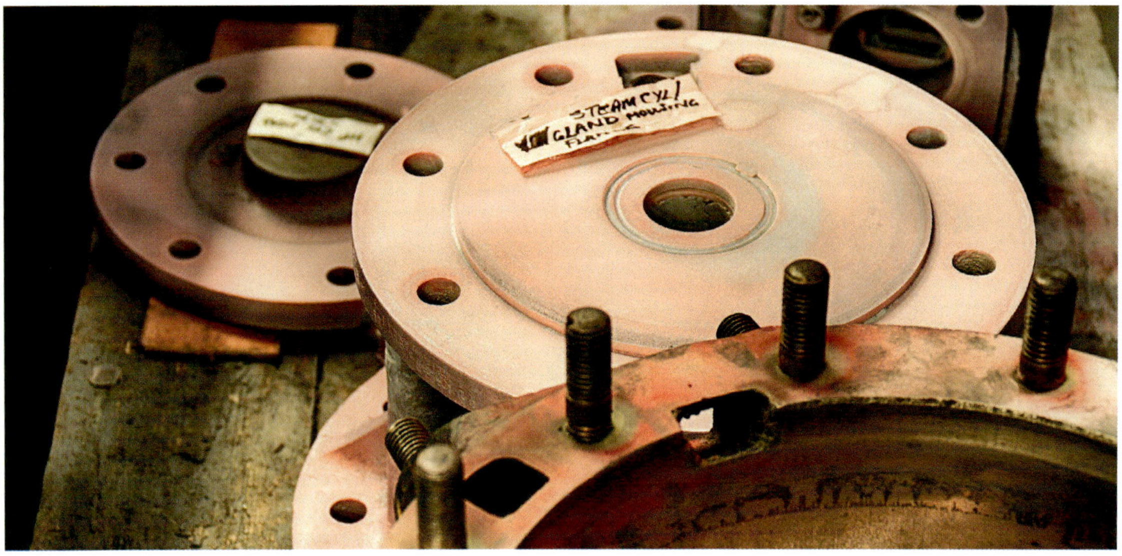

Air cylinder brake components, cleaned and restored

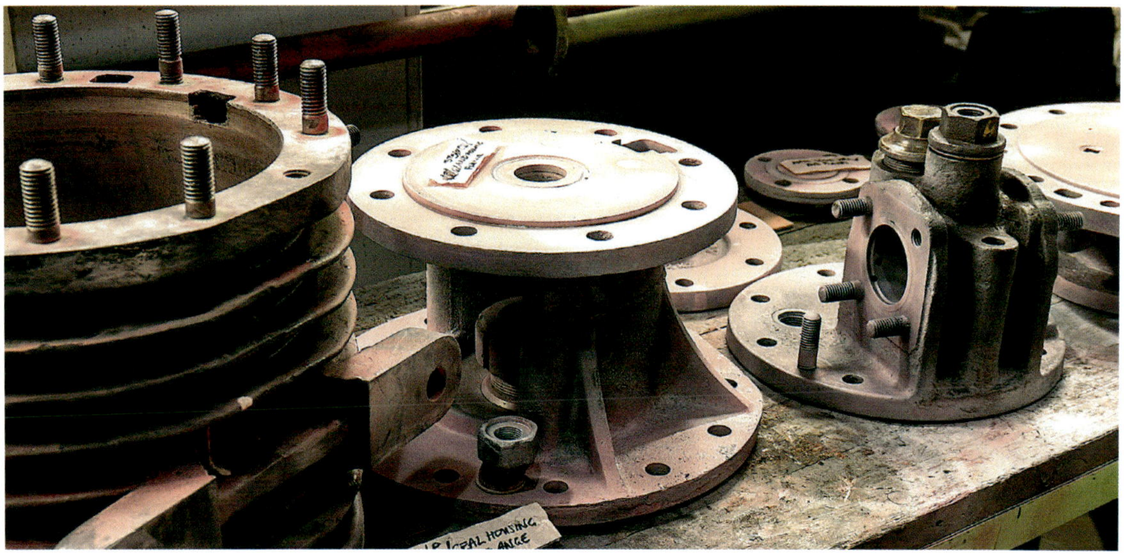

The brake cylinders had now been cleaned thoroughly which enabled them to be stripped down for mechanical overhaul and that gave me the chance to photograph the multitude of components which make up the cylinders.

With shots now in the bag I left in time to avoid the evening Teesside rush-hour traffic and wondered as I drove away how much impact Tony joining the team would have on his own version of No.7.

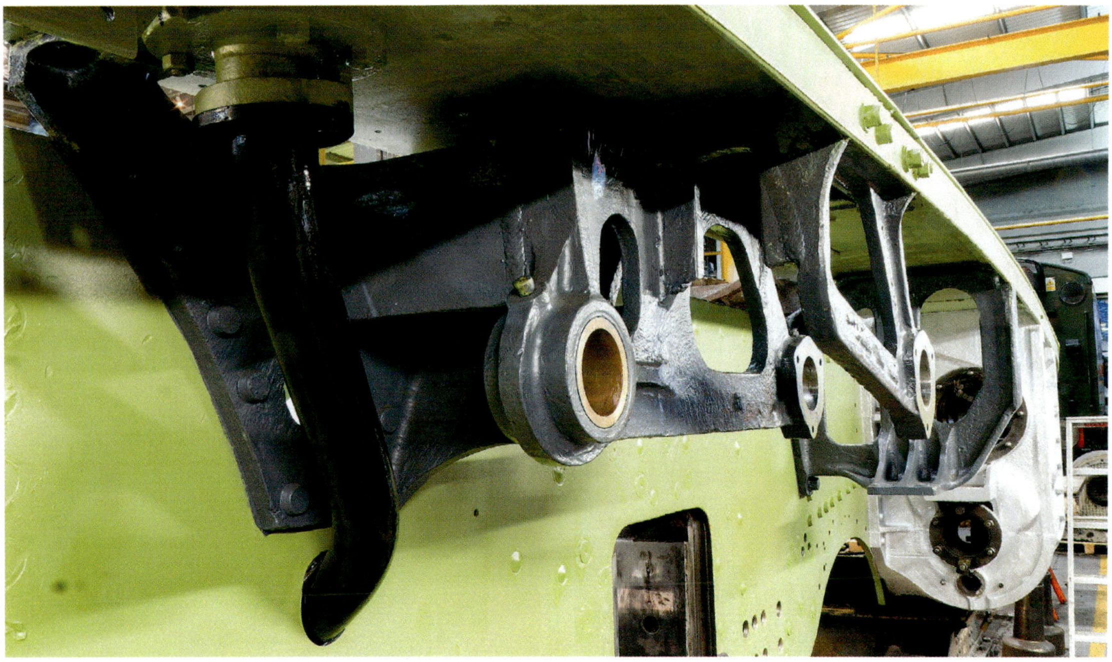

Starting to look like a brand new locomotive immaculately renovated and freshly painted components on the right hand side frame

MORE COATS

March 2017

Stand around the workshop long enough and you will end up with a paint brush in your hand or be next in line for using the needle gun. Painting was in full swing and it was good to see fresh paint being applied in so many areas. Peter Brackstone's other duty, besides making the tea, was to head up the painting team and he had now set up a command centre near the front of No.7 from which to organise operations. Most, if not all, removable components had been removed from the frames and many of these were now sitting in orderly lines proudly showing off their newly painted surfaces. I was keen to get more images between the frames for no other reason than that whilst the frames were devoid of boiler and wheels it was a great opportunity to do so.

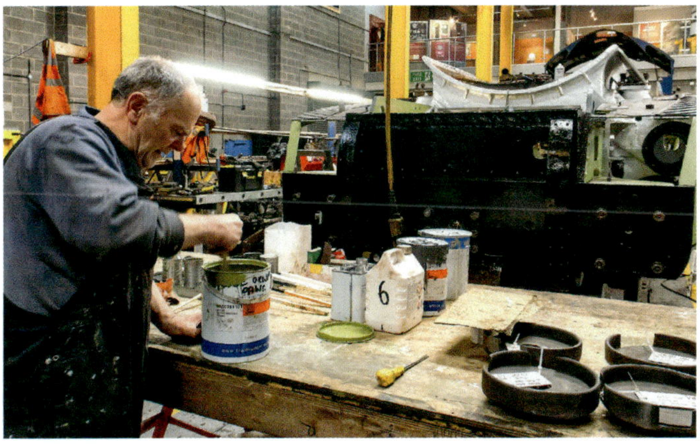

Paint team leader Peter Brackstone preparing paint at the paint station in front of No.7

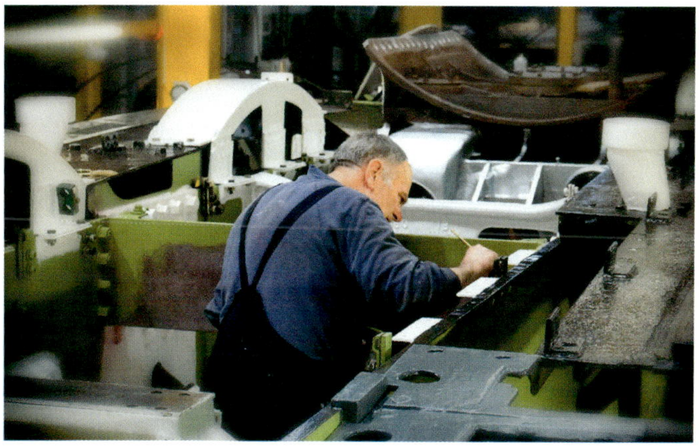

Painting continues between the frames

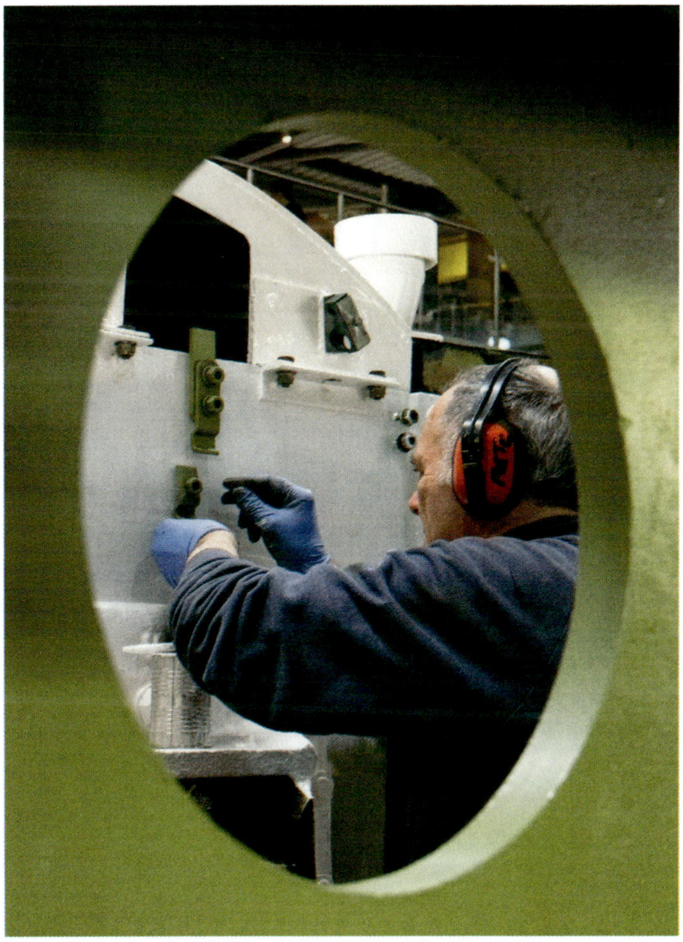

Attention to detail, brackets being replaced after temporarily being removed to allow painting behind them

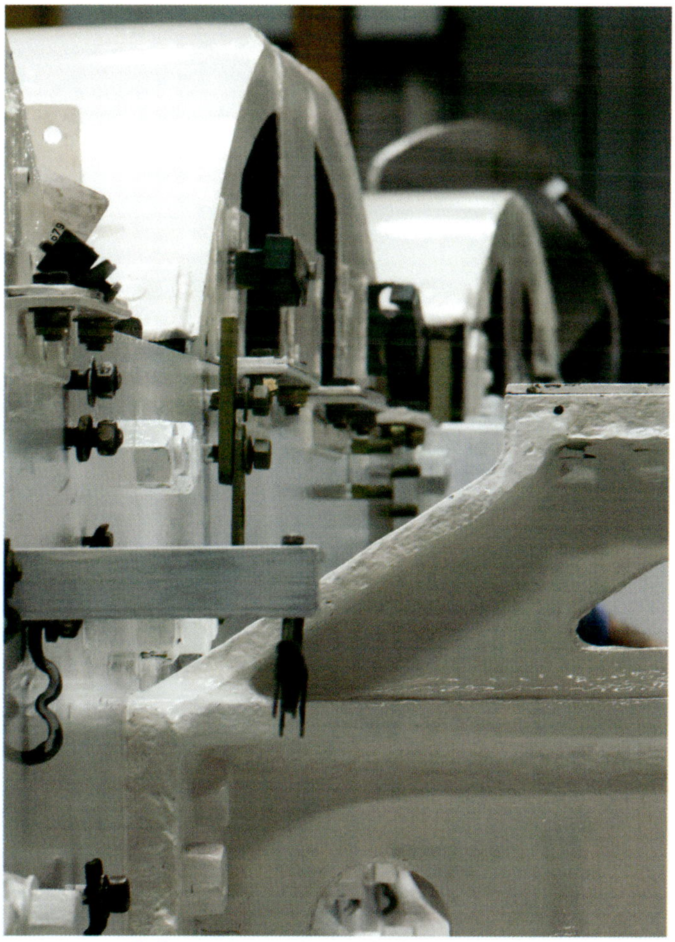

Top coats of white paint on the newly fitted wheel splashers and the inside of the frames

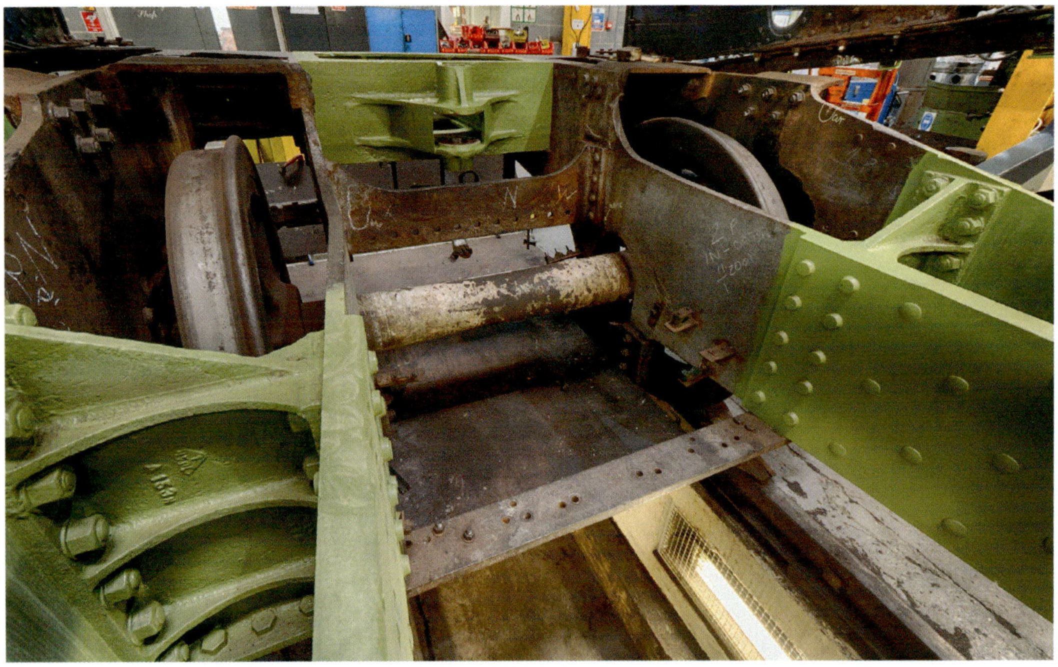

With the cab floor not in place a wide angle lens captures the freshly painted Cartazzi

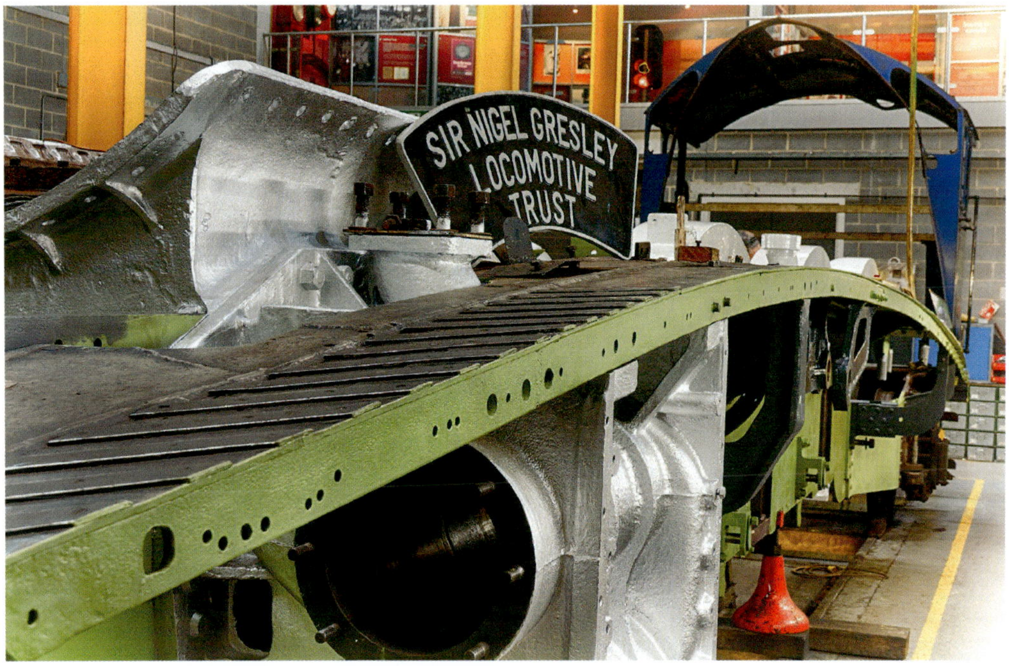

Freshly painted surfaces on the outside of the left side frames

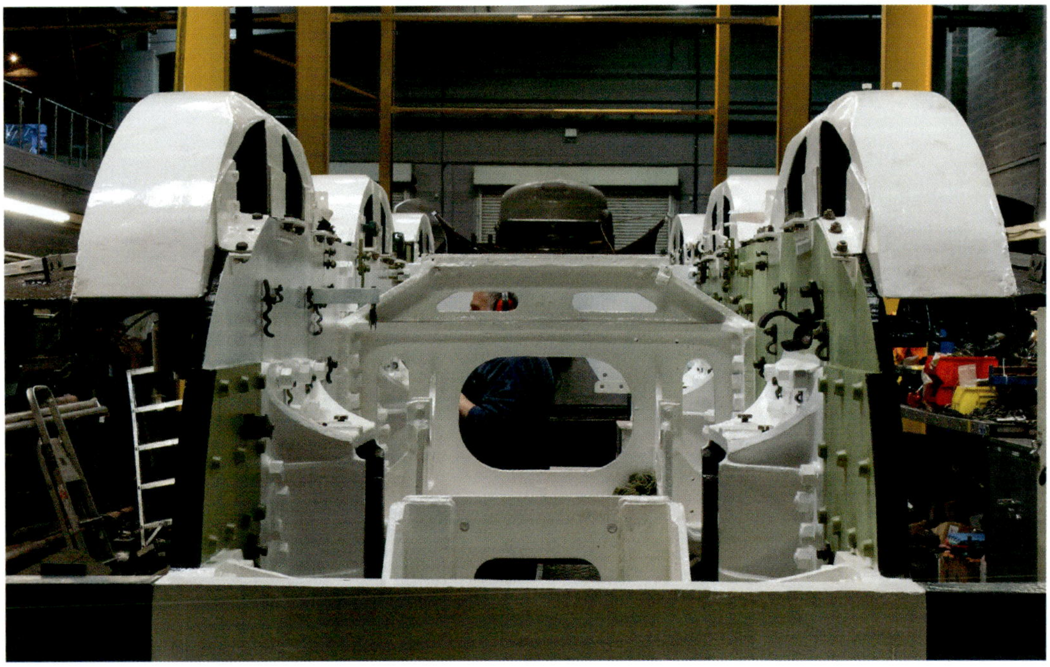

Peter Brackstone applies a top coat of white paint between the frames

RETURN TO LLANGOLLEN

April 2017

It was time for another trip to Llangollen to check on the progress with the boiler and it was an educational trip for me as I started to understand a bit more about the boiler workings and especially what a stay is. My stay was in a very nice hotel a couple of miles up the valley from the workshops, but the stays I refer to here are what hold the boiler inner cylinder to the outer casing. I hadn't even realised that there was an outer and inner until now. I understood that the hot air from the firebox passed through the boiler along the boiler tubes to the smokebox, heating the water in the boiler along the way, but what I hadn't realised is that the water also ran all the way around the outside of the boiler between an outer and inner casing which were held together with 'stays'. These stays from the outside look much the same as rivets which is what I originally thought they were. Now, I almost certainly should have known that, but I am just the photographer I reminded myself. I was happy to have a better understanding of what I was taking these images of, especially as the stays are items that a lot of generous SNGLT members and supporters of No.7 were sponsoring. The stays are of varying length with some reaching vertically and others horizontally, between the outer and inner casings and several of the horizontal ones pass down the entire length of the boiler. They are threaded at both ends to fit into threaded holes in the casing. The type of ends on the stays vary but those that I thought were rivets do resemble rivets as the ends are shortened and then riveted over in the same way. Most of the stays are steel but around the firebox they are copper due to the higher temperatures.

Removing and replacing the stays required drilling through their centres as can be seen in the images. There are around 2000 stays, not all of which required replacing but the best part of 200 did, highlighting one of the many time demanding procedures undertaken in the overhaul which make you appreciate why the boiler overhaul takes so long.

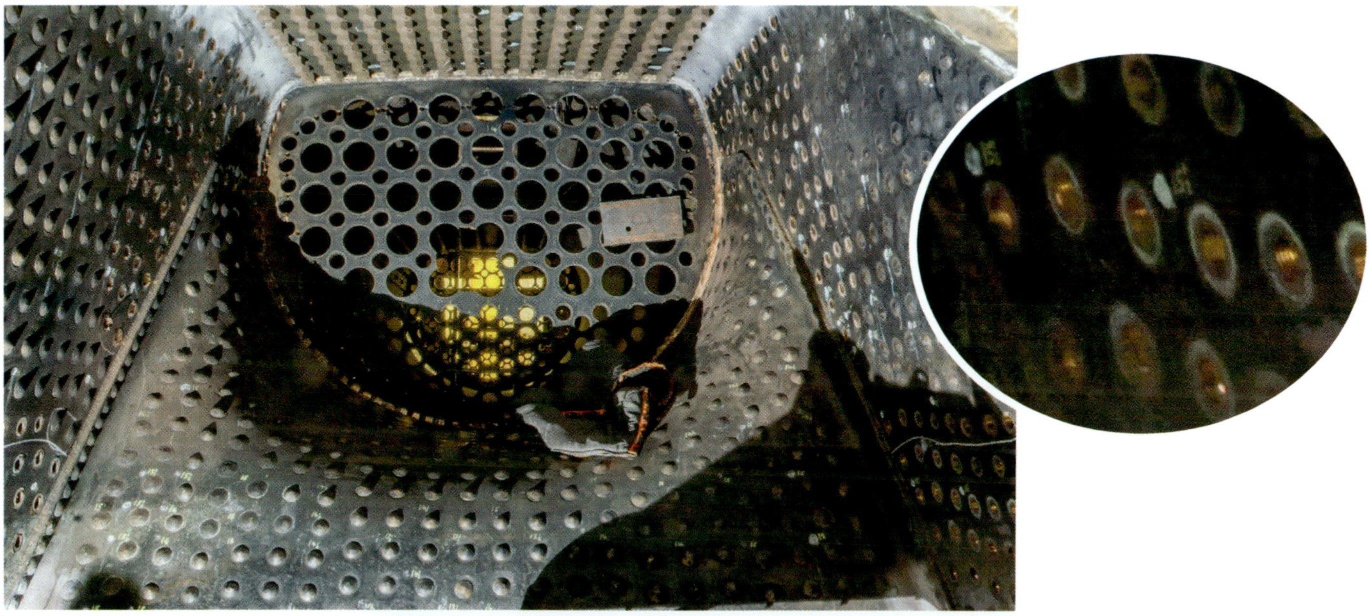

ABOVE & INSET - Inside the firebox work on the tube plate is evident with sections cut out. Drilled out copper stays can be seen on the left and right hand side The steel stays at the top of the firebox are held by nuts as opposed to the riveted over ends on the copper stays

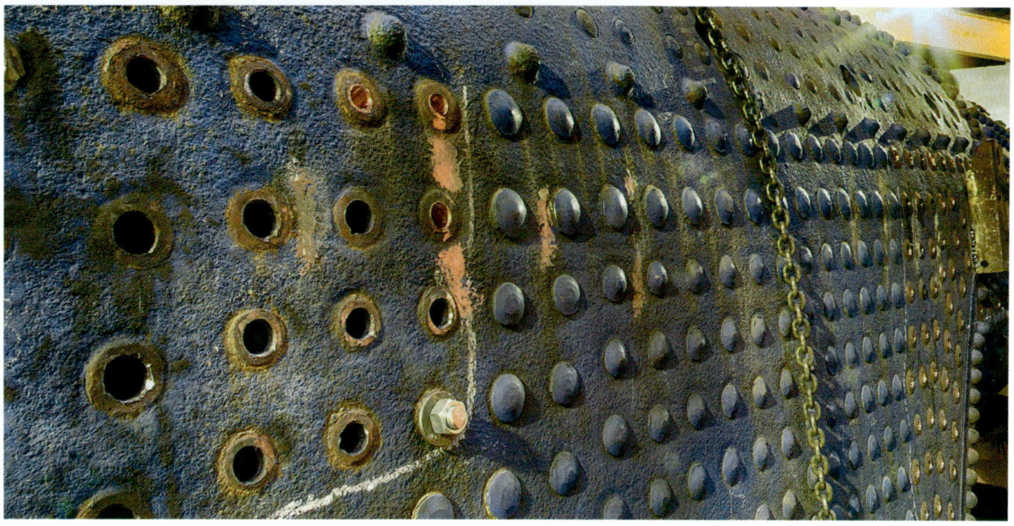

RIGHT - drilled out copper stays from the outside of the firebox end of the boiler

SPANNER IN THE WORKS

May – June 2017

Work continued on the wheels and boiler in Buckfastleigh and Llangollen respectively, but my focus was on the Trust's engineering team of volunteers in York where progress was being made on the frames and ancillary components.

With the cartazzi (that's the wheels under the cab), and the accommodation bogie at the front being the only wheels supporting the frames, a little extra support was needed. This was provided by two old, discarded frame stands which had been rescued from undergrowth on the trackside of the North Yorkshire Moors Railway, donated to, and now beautifully restored by No.7's engineering team. With shiny new red paint applied the two stands had a new lease of life as they sat proudly at the front of No7.

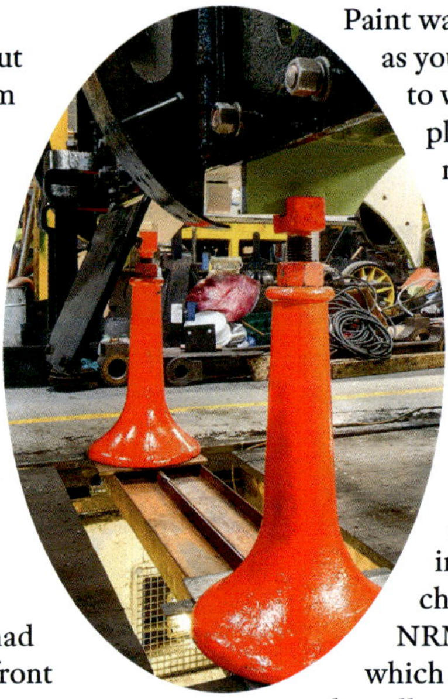

Paint was very much, for me, still the pictorial theme as you could clearly see the progress from week to week but lots of less obvious work was taking place, not least of which was the renovated and refitted pipework. Peter Ellis, who now had the nickname of Peter the Pipe Strangler, was excelling in his role of pipe strangling, and some of his new work was clear to see against the freshly painted white frames.

One concern with No.7 being 'out of traffic' for several years was that interest would wane but doing the overhaul in the shop window of the NRM and its viewing gallery was a massive help to keep No.7 in the limelight. SNGLT is an educational charity so, building on that the Trust and the NRM arranged some 'Object Handling Days', which involved setting up some display tables on the gallery and undertaking some talks on the work and the history of No.7 which were given by Malc Bateman. Supported by other Trust members including myself, Bob Matkin, Mike Page, Jackie Shearman and Julie Crone, the days generated quite a lot of interest, particularly in one of the very large workshop spanners.

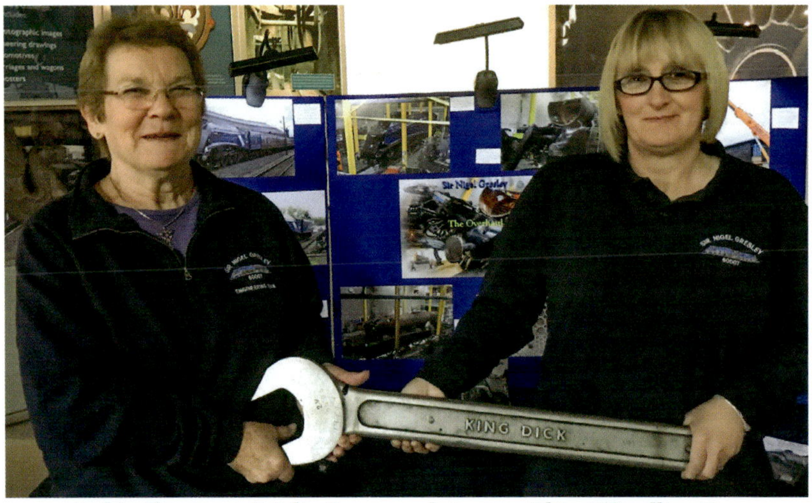

Jackie Shearman & Julie Crone on duty at an Object Handling Day

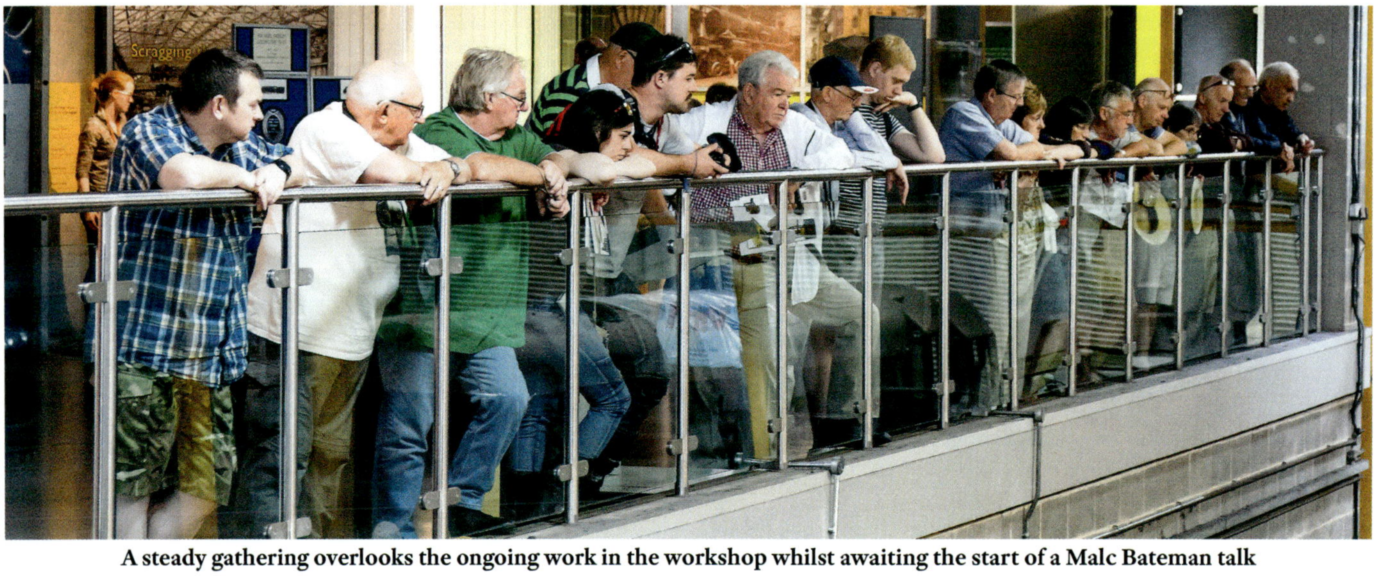

A steady gathering overlooks the ongoing work in the workshop whilst awaiting the start of a Malc Bateman talk

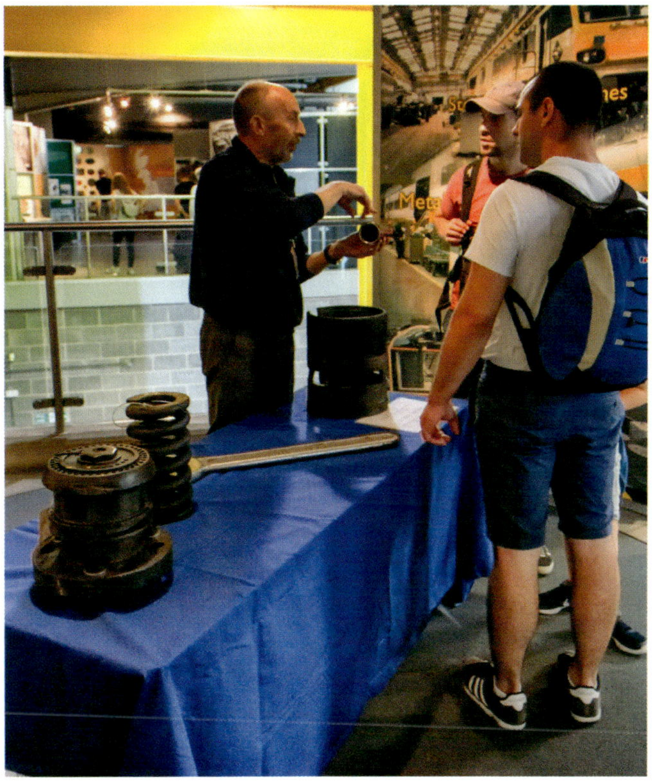

Mike Page discusses the overhaul work with visitors at an Object Handling Day event

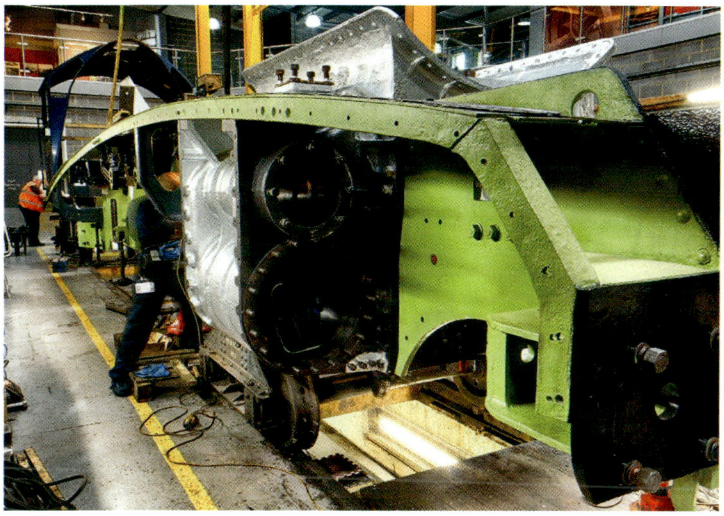

Fresh paint highlights the iconic lines of No.7's running board

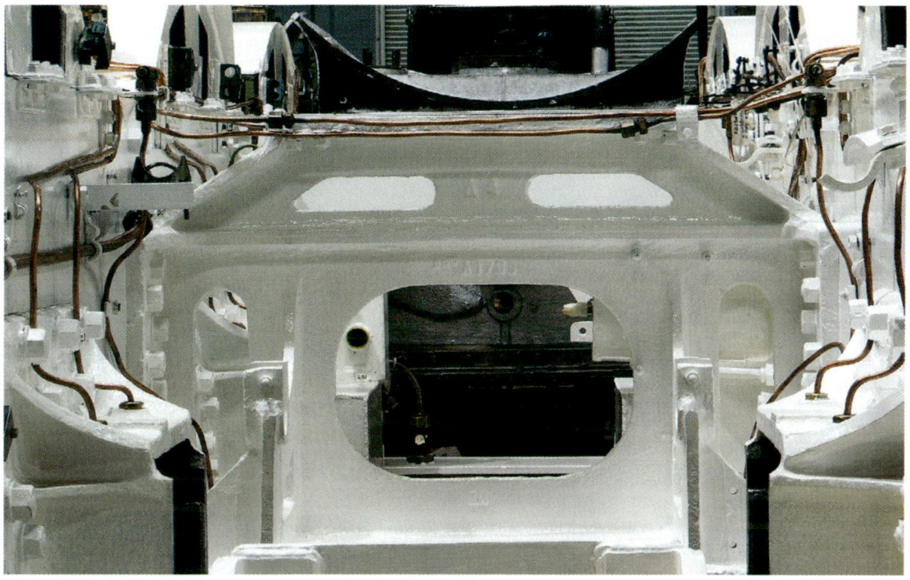

Peter Ellis's newly installed pipework stands out against the freshly painted frames

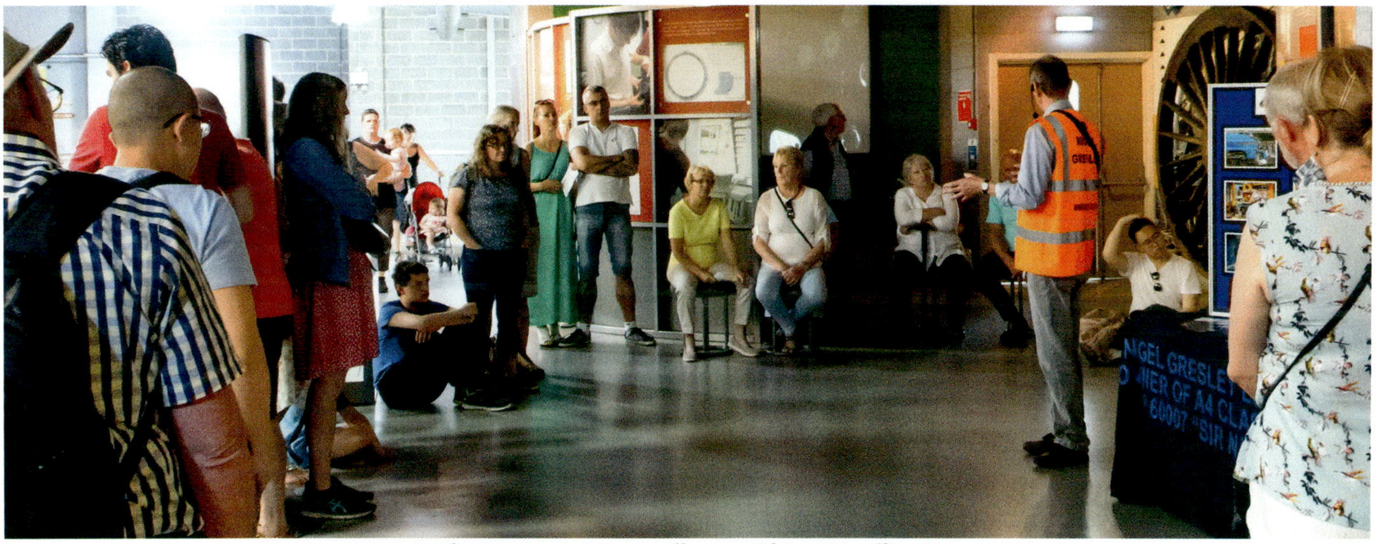

Malc Bateman giving a talk at an Object Handling Day

SPARKS FLYING

August 2017

I had two consecutive days of attendance this month; the first on Wednesday 16th when I was supporting Malc Bateman and Mike Page on the gallery in another of the 'Object Handling Day' events. However, I always still took my camera with me as I made sure I could spend a little time getting a few shots in the workshop be it only briefly.

I wasn't disappointed today as work was underway on the tender. Trevor Wilford provided me a spectacular shot whilst he was using burning equipment to help release the tank from the frames. As always with trying to capture the light trails of the sparks you need a longer than normal exposure and it is very much hit and miss, so lots of shots and then selecting the best later. Thank heavens we no longer have to buy rolls and rolls of film!

It was planned for the specialist liquid nitrogen contractors to attend on Thursday 17th as their services would be required for the fitting of the valve liners, so some final preparation work was taking place with the three liners. As I took some pictures of one of them, Darrin commented that there would be some good shots to be taken tomorrow when the contractors arrived, and he wasn't wrong.

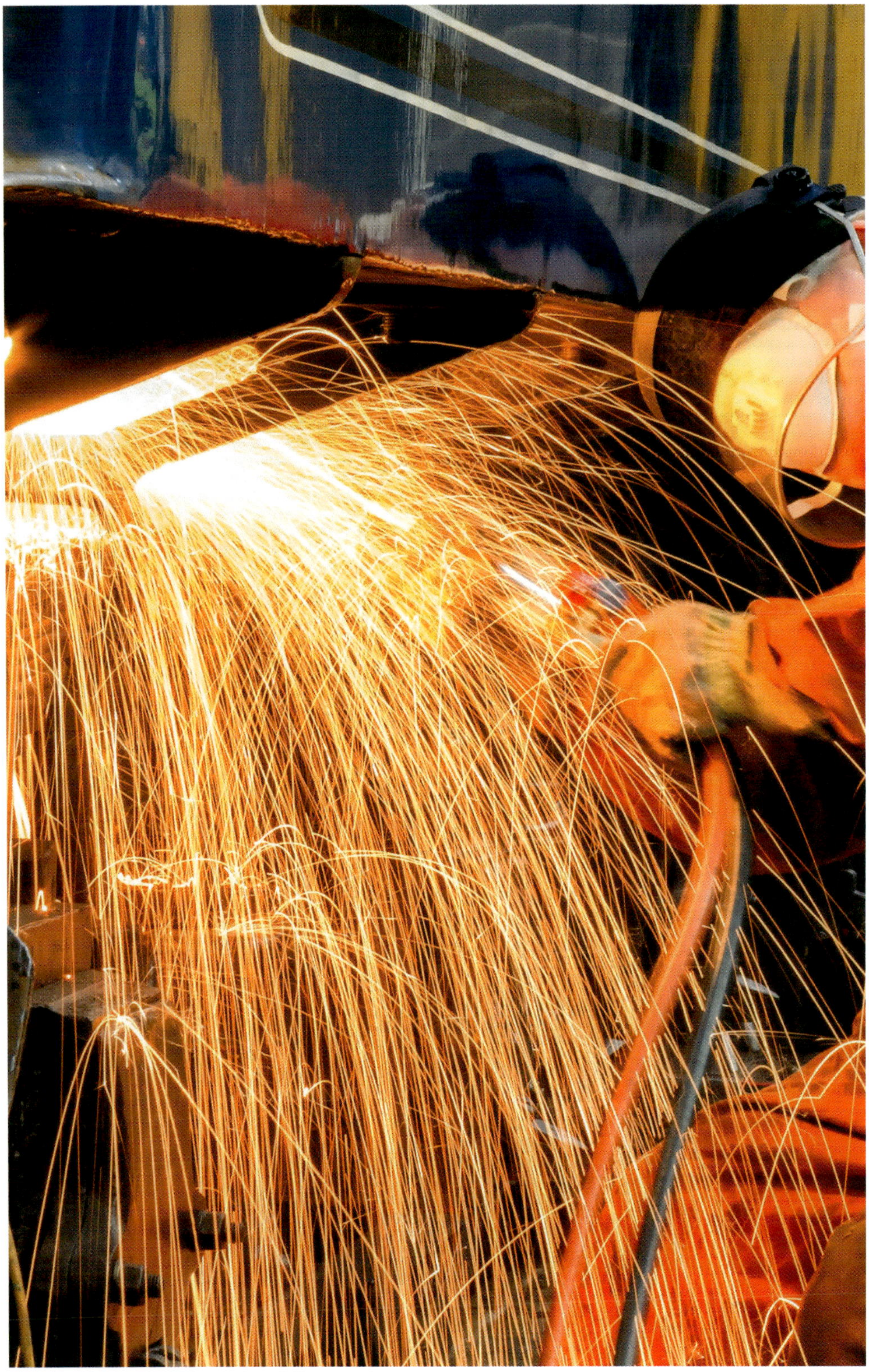

Trevor Wilford creates a dramatic shot as he works with the burning equipment to release the tender from its frames

SMOKE ON THE WATER

August 2017

I had taken the decision that using any kind of studio lighting in the workshop was totally impractical and I would have to deal with the lighting in other ways, but when the workshop floor started resembling a Deep Purple concert and everybody's lower legs and feet were disappearing into the mist that was clinging to the floor, I was wishing I had a few laser lights to add to the atmosphere created by the use of liquid nitrogen.

The three valves which supply the steam to the cylinders have liners fitted into their housings. These liners are a tight fit and immersing them into cooling liquid nitrogen contracted them to a smaller diameter, and a quick transfer to the casing when cooled enabled easier fitting into the valve housings, where, when they returned to the ambient temperature, they expanded, making a tight fit.

I think the pictures tell the story. It was amusing to see apparent steam coming from No.7 without a boiler to generate any, and all we needed now was the PA system to start playing 'Smoke on the Water'.

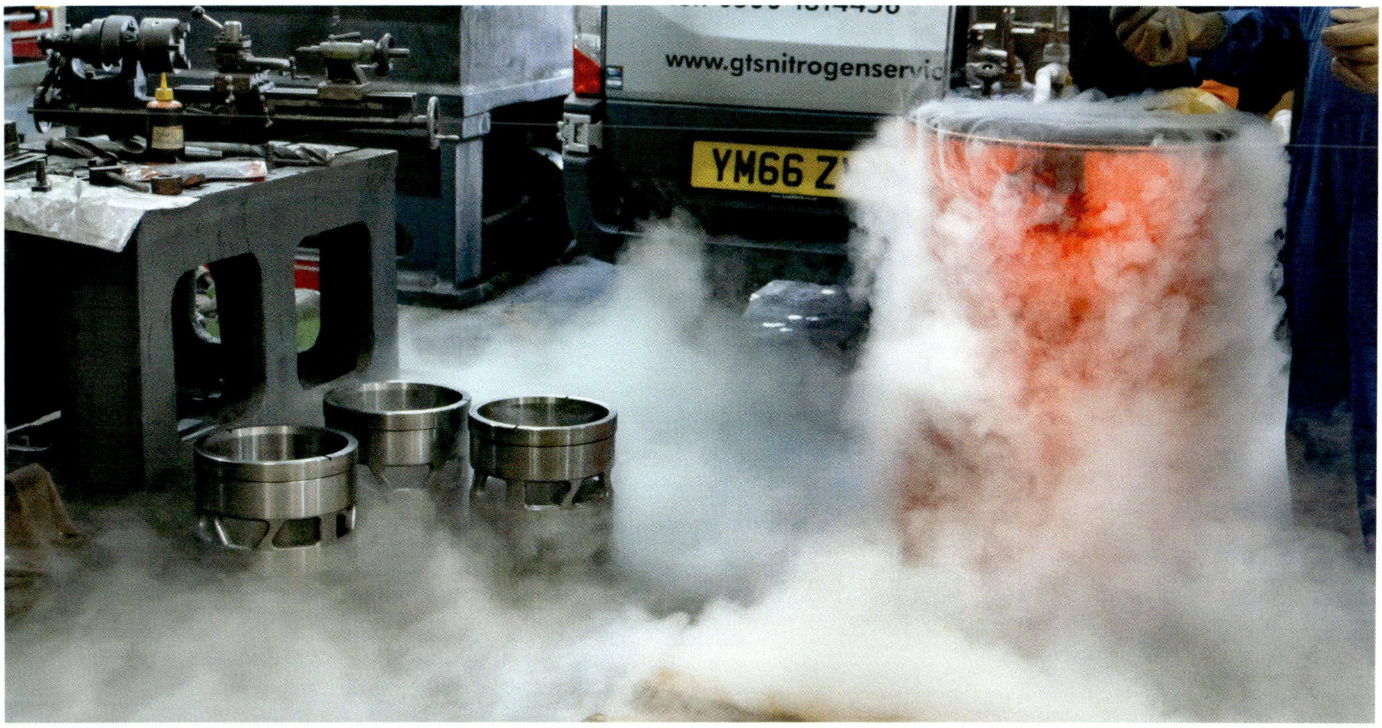

ABOVE - The liners ready to be immersed into the liquid nitrogen tank

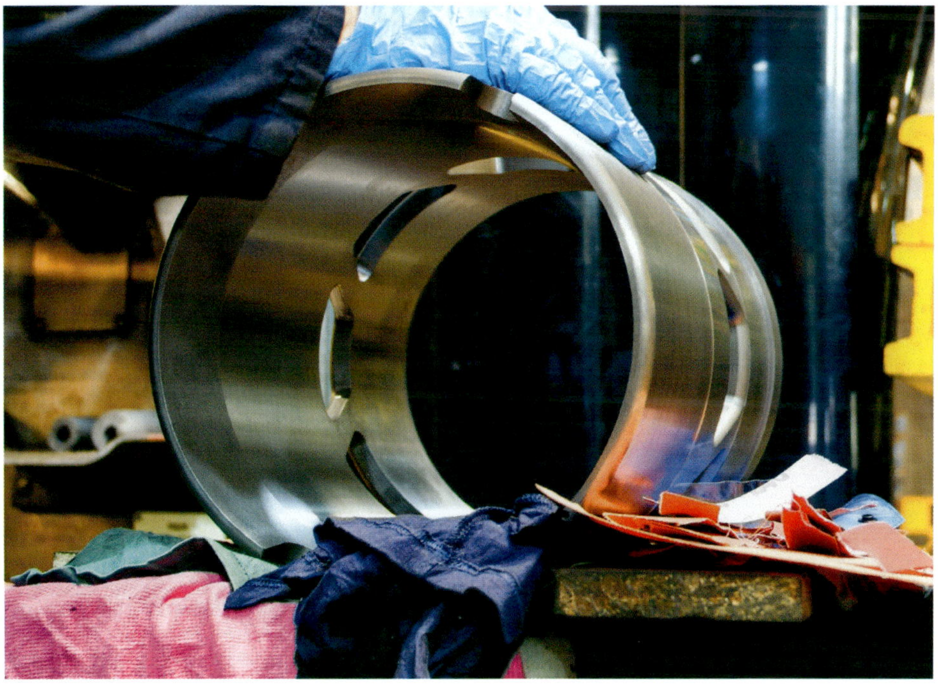

LEFT - One of the three valve liners being prepared for immersion in the liquid nitrogen

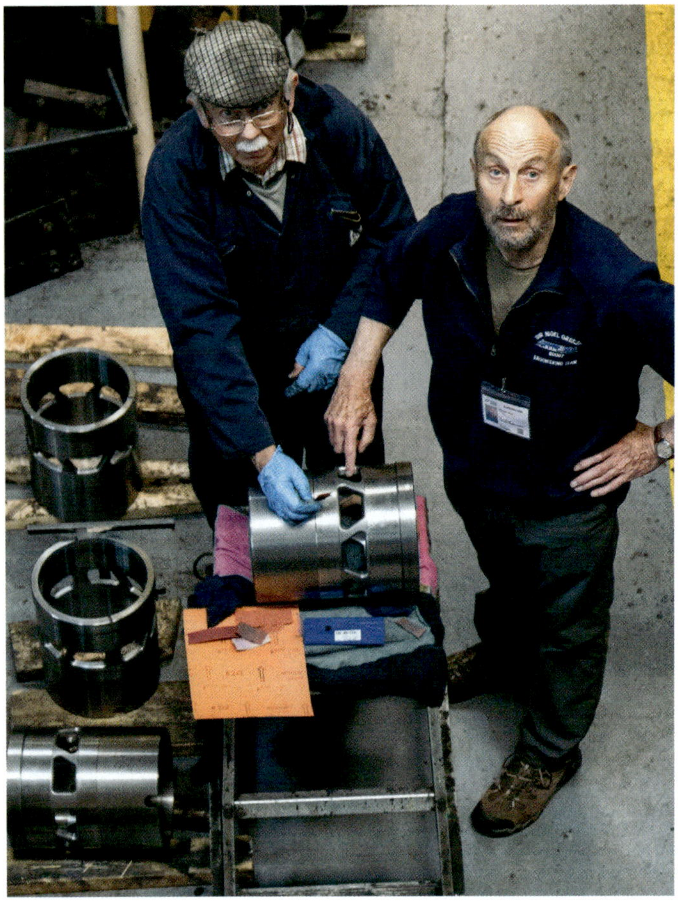

Bob Shearman & Mike Page carry out a final inspection of a valve liner before it is shrink cooled in the liquid nitrogen

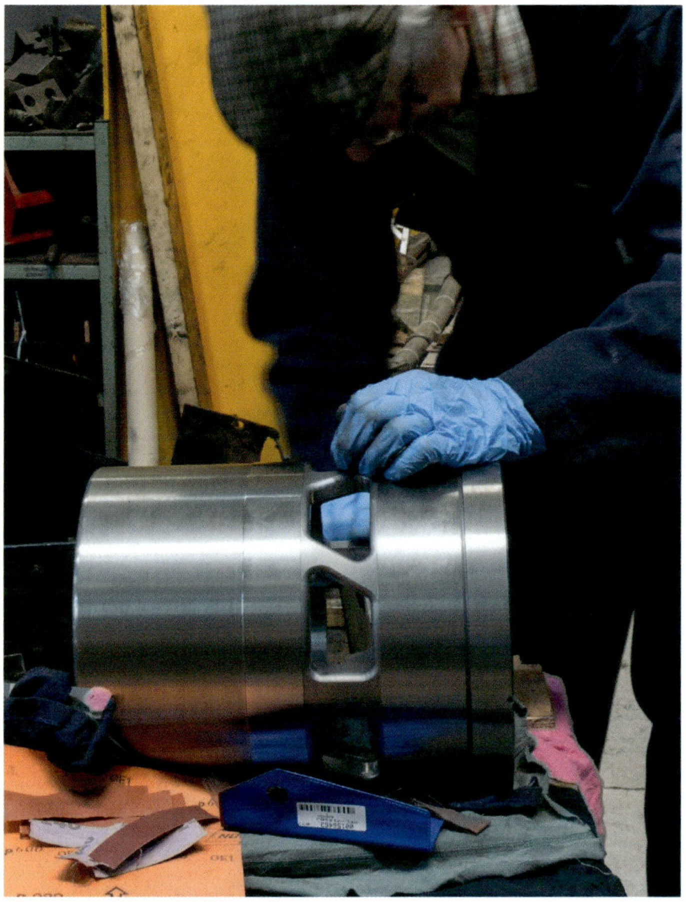

The valve liner is ready to be shrink cooled in liquid nitrogen

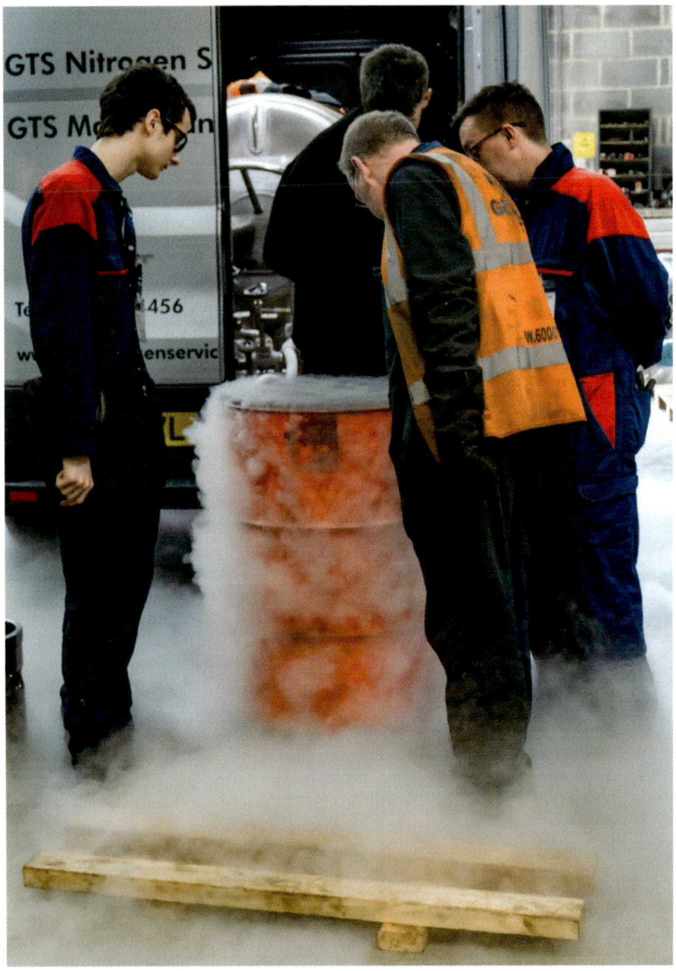

The team look on as one of the valve liners is cooled in the liquid nirogen tank

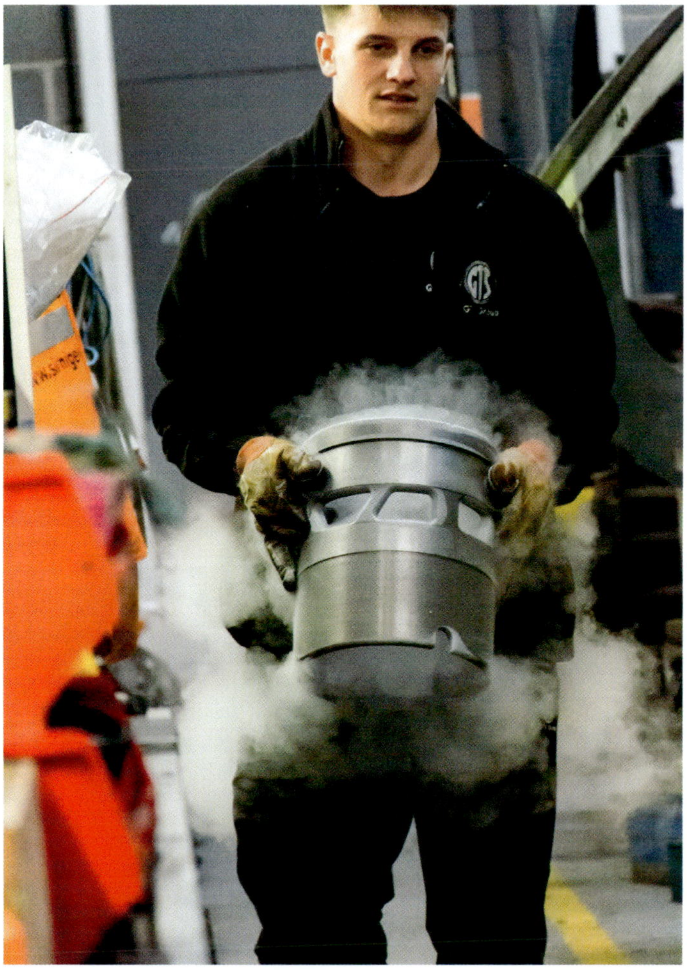

With time of the essence a shrink cooled valve liner is moved from the liquid nitrogen tank to be fitted back into No.7

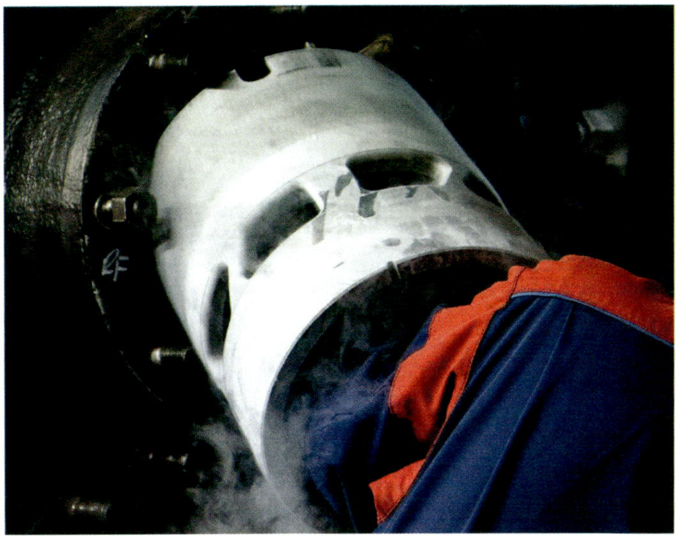

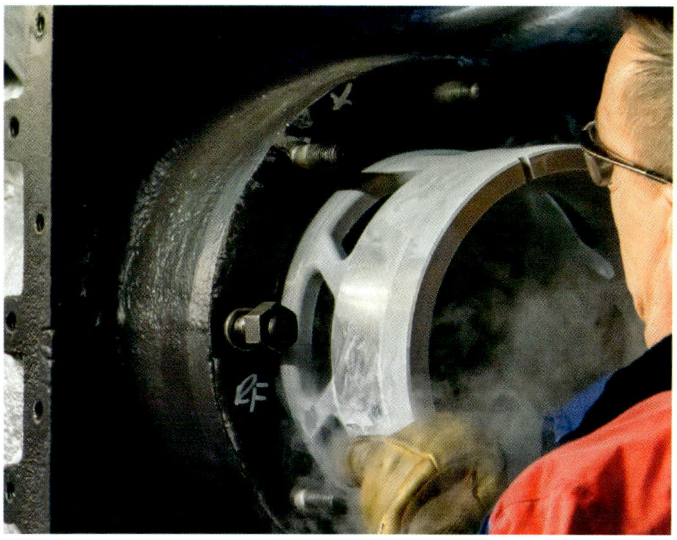

A cooled valve liner is placed into the valve housing on No.7's right hand side cylinder casing

The alignment of the liner is checked inside the valve housing

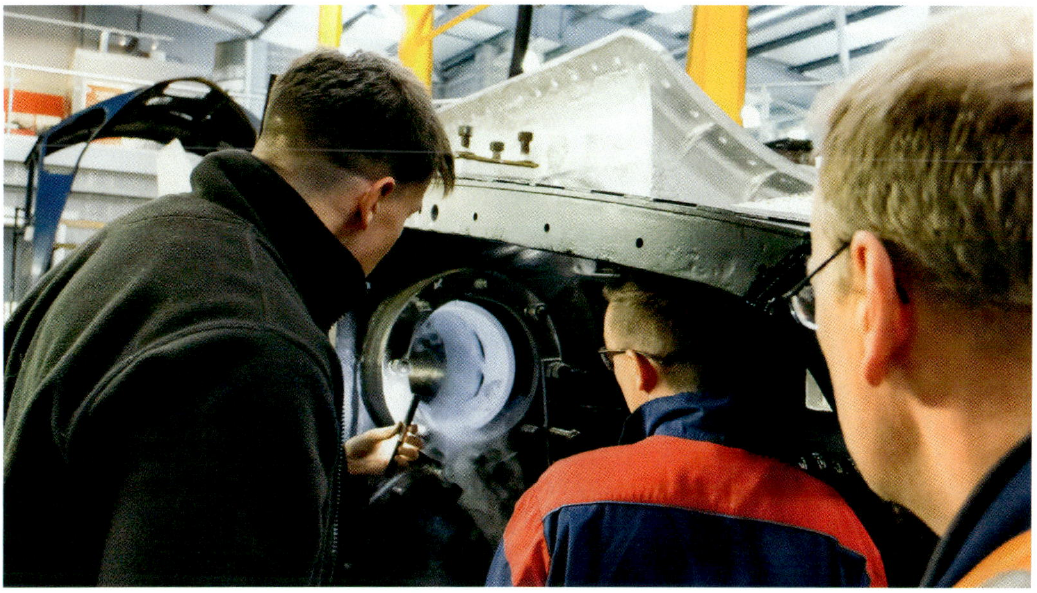

With the liner successfully installed the clearing liquid nitrogen is reminiscent of the more familiar steam surrounding the cylinder

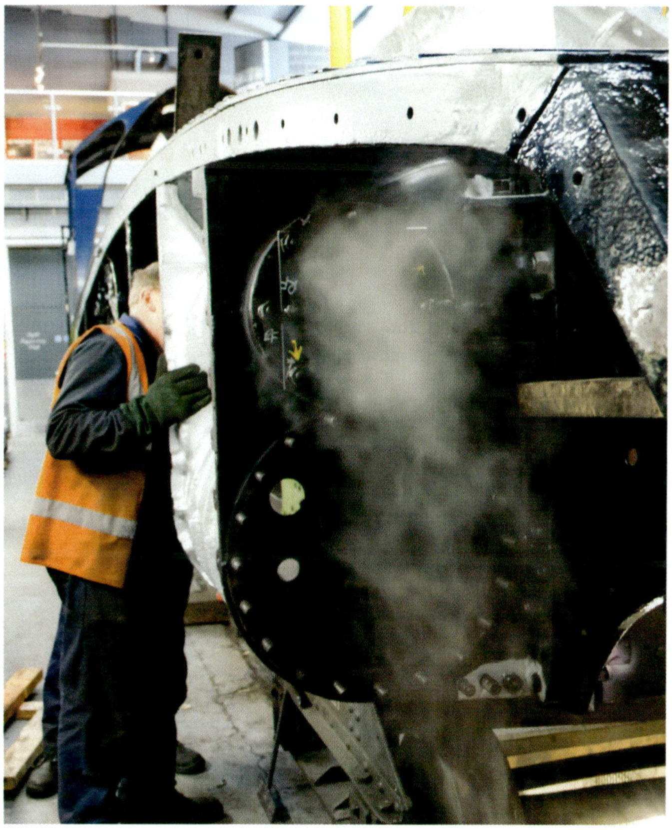

7

WHEELS ROLL BACK

September 2017

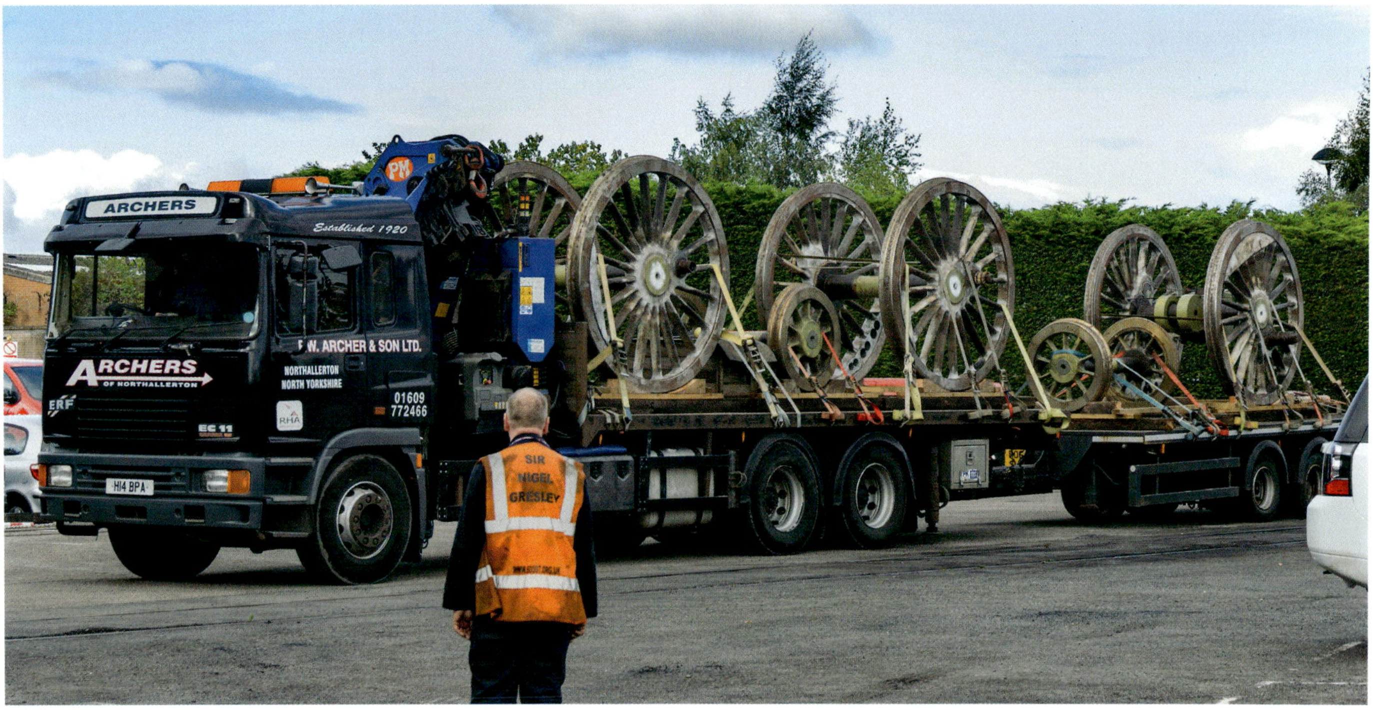

The wheels arrive back to the NRM North Yard in York

You would be forgiven for thinking you had already seen these images, as they are like the earlier set showing the wheels being despatched for refurbishment. However, the accompanying images are the wheels returning from Buckfastleigh after being refurbished and retyred at the workshops of the South Devon Railway.

It was all hands on deck as everybody rallied around to help transport driver Paul Archer unload when he arrived, and to everybody's relief, because of menacing dark clouds overhead, all the wheels were safely returned into the workshop in good time.

The team had already spent many hours preparing and repainting surfaces on the frame but now their attention was to turn to the wheels with preparation being the key yet again.

It wasn't just about painting of course, as there were a lot of more technical tasks being undertaken by the team.

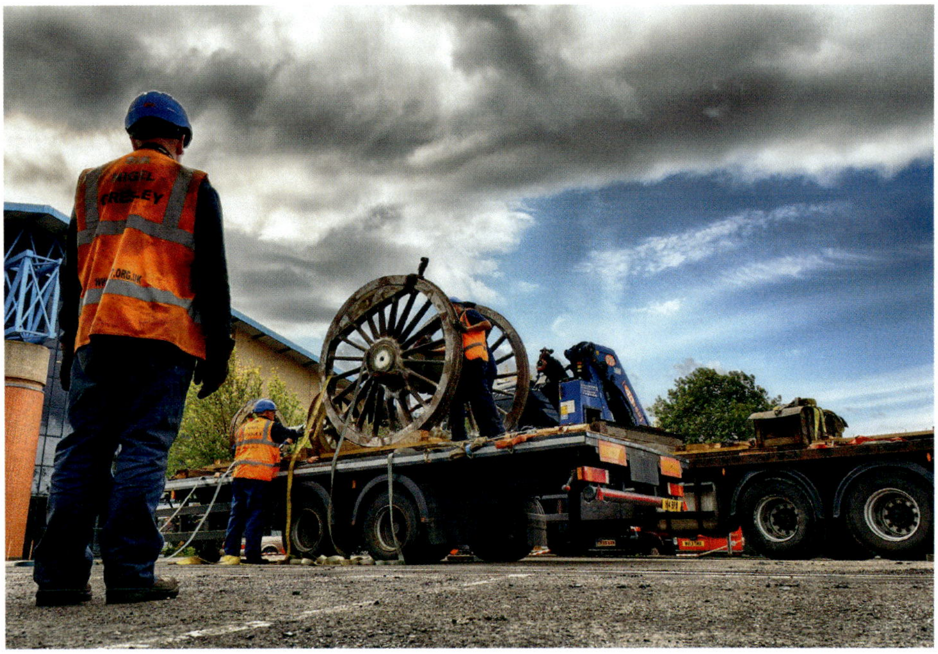

Darrin Crone supervises the unloading of the precious cargo in the North Yard at the NRM

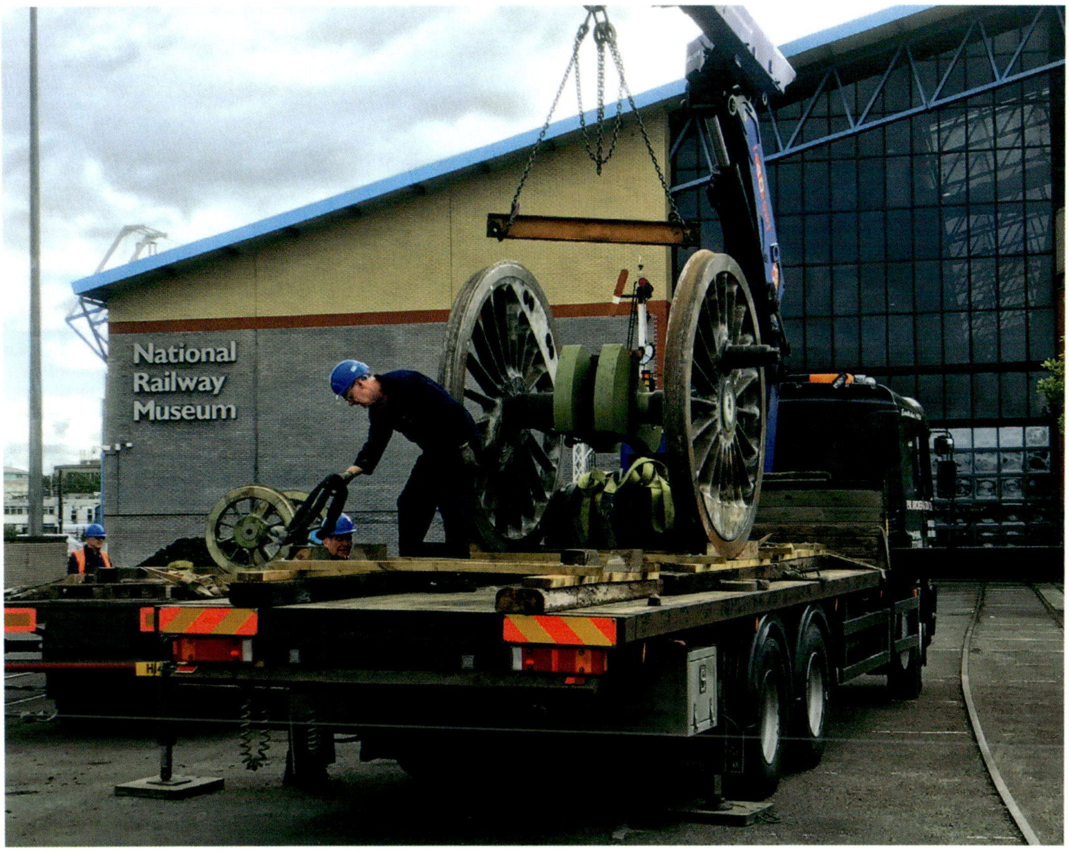

The main drive wheelset is prepared for lifting off the transport

Measurements could now be taken from the returned wheelsets to enable adjustments of the axle boxes and more rivets were being fitted. Preparations for lifting the tender's tank from its frame had started with the arrival of lifting equipment. I also came across a new term, to me, of a "clack box" which I learned was a ball valve to control the flow of steam. If you wish to understand more of the technical aspects the details are contained in Darrin's excellent technical reports on the Trust website (www.60007.org.uk).

On my next trip I found the left-hand side frame had been painted with a black topcoat but also had what looked like a washing line running down its full length from the cylinder block to the cartazzi.

All wheelsets safely back in the workshop

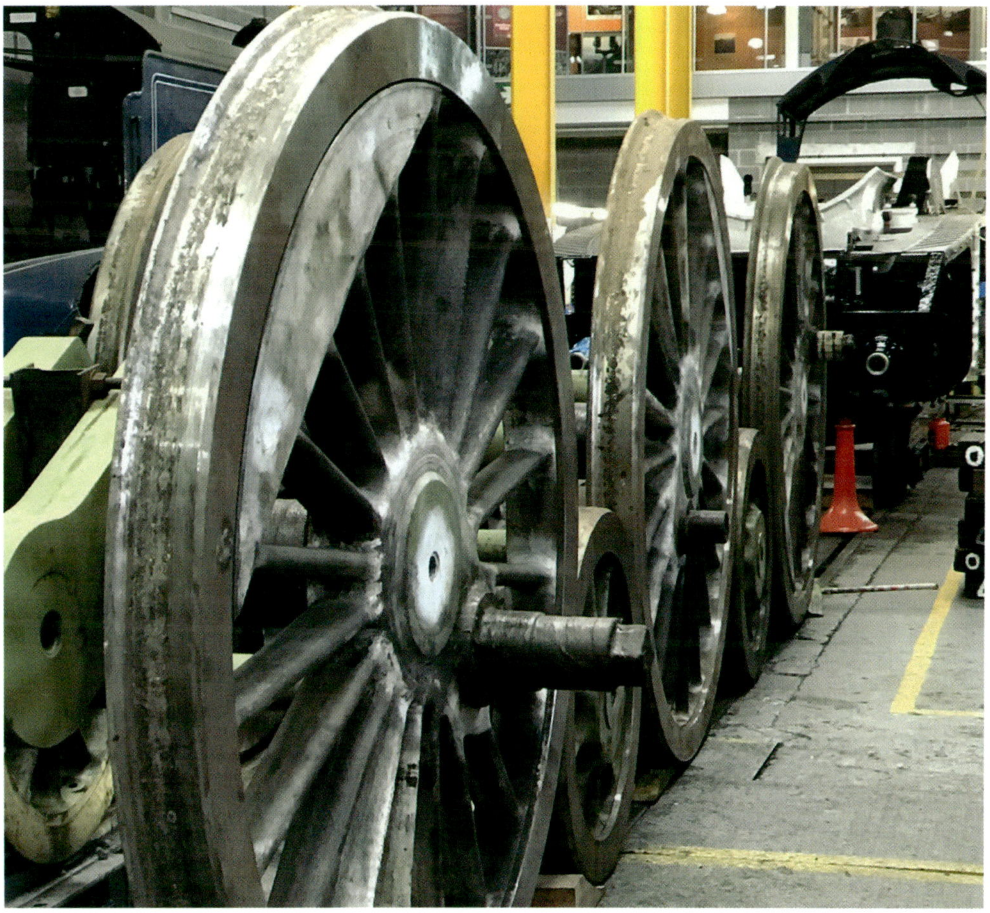

November 2017

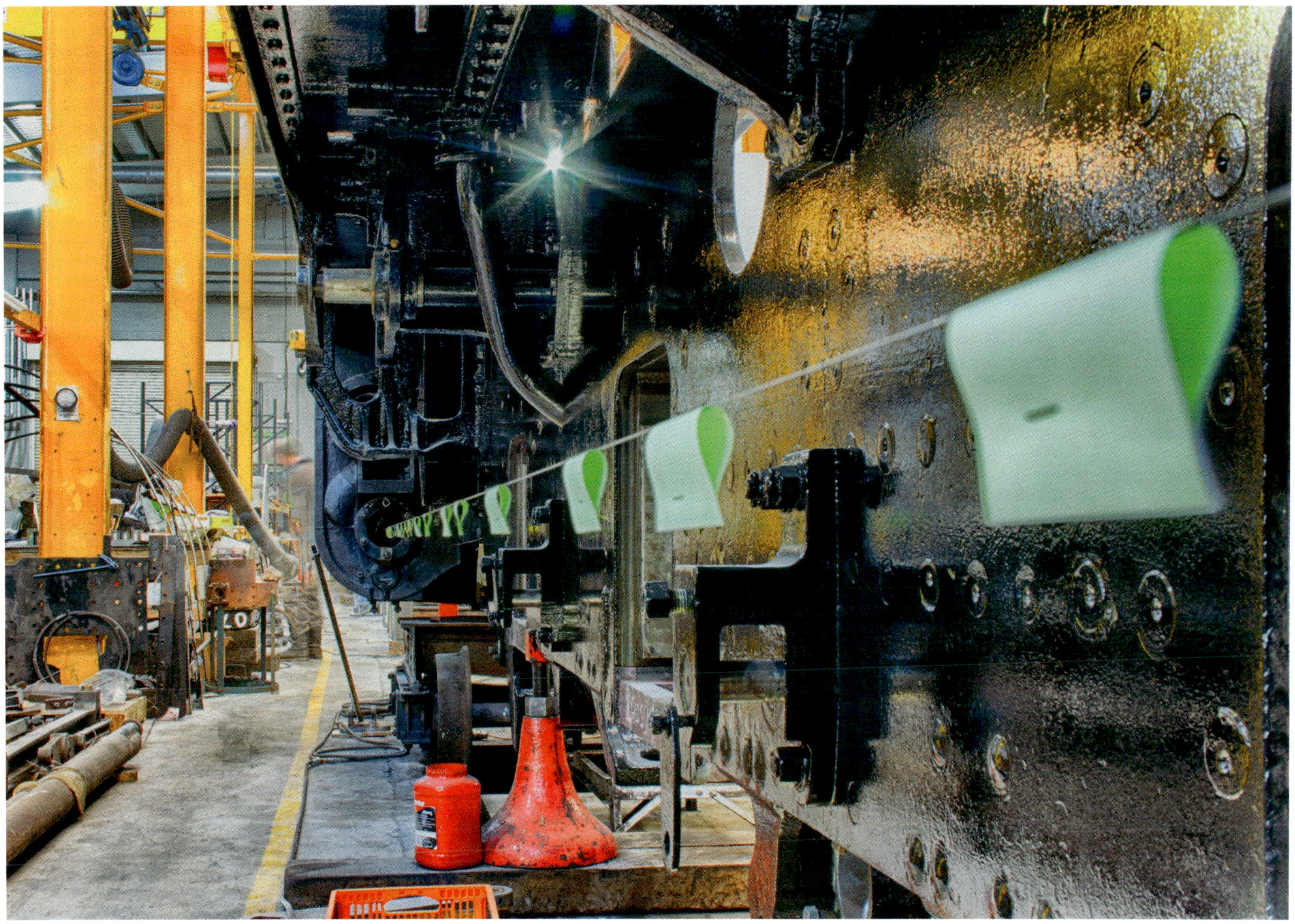

I wasn't quite sure where the last two years had gone since No.7 was moved from Grosmont to York for the overhaul, but two years it was. Back in LNER and British Railway days the job would have been completed a lot quicker, but then, of course, it would have been completed by a team of full-time engineers working around the clock with everything done all under one roof. No such luxuries for the Trust team of volunteers. Expertly skilled and dedicated, they gave up their time for free. Some were able to volunteer a day every week or more, some less often, but nonetheless dedicated to the cause, travelling from as far afield as London. Juggling the tasks allocated to them was skilfully managed by Darrin,

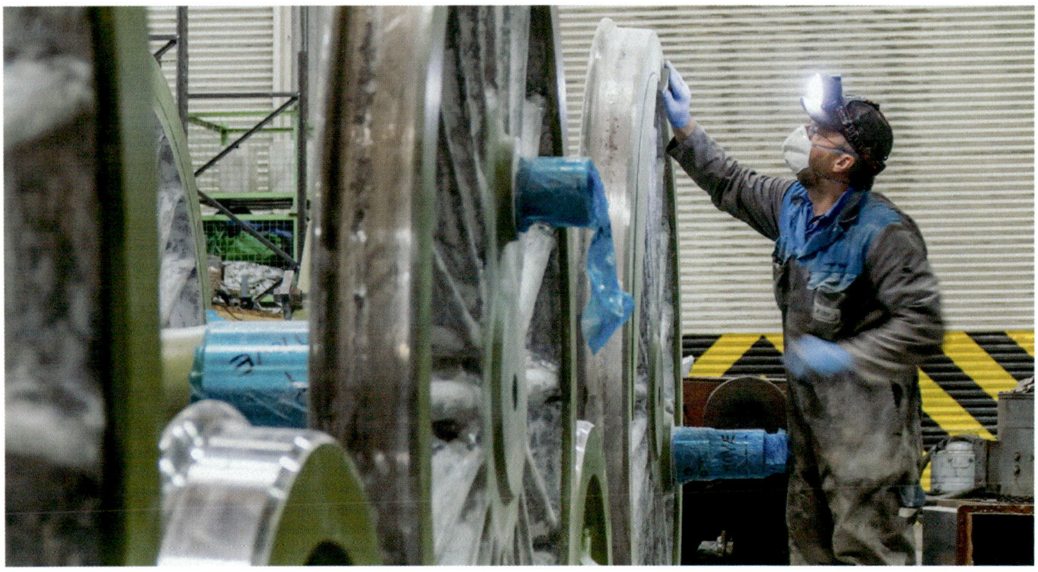

Roger Turnbull rubbing down the wheels in preparation for painting

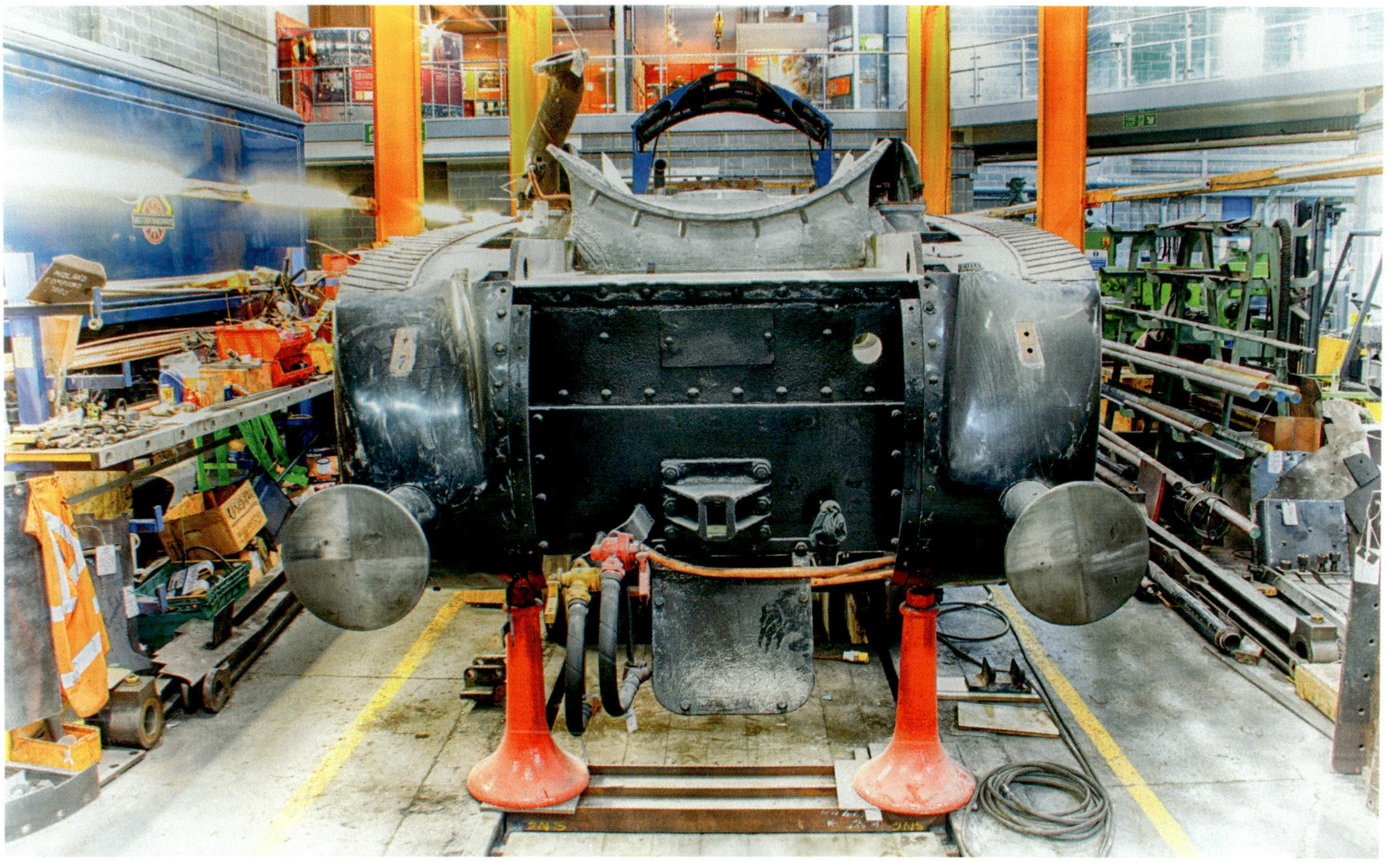

A taught line is stretched from the cylinder centre to check alignment with what would be the axle centres

who had built up a reliable and trustworthy team. With several jobs outsourced Darrin could not always be on-site, so he appointed a team of RO (Responsible Operatives) to lead the team in his absence, many of whom had experience of managing people in their career jobs. The camaraderie of this wonderful team never waned regardless of the tasks required of them with a sense of humour being a necessary requisite, particularly in tea and lunch breaks in the blue cabin. Unless there was a specific 'Milepost' event taking place, my days revolved around the morning tea-break and lunch time, arriving before the former and leaving after the latter and always managing to take some pictures in the hours in between. There was some method here, as arriving when I did Darrin had chance to set the team their tasks and thus had a little more time to give me if I required it. I arrived on one November morning and saw a washing

Andrew Jacques

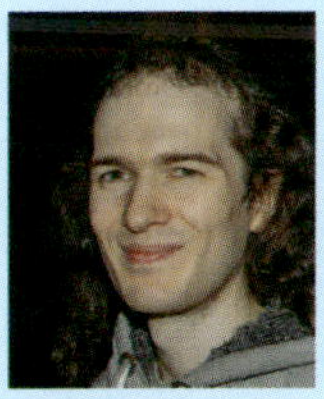

The first time I saw the Sir Nigel Gresley locomotive was at the North Yorkshire Moors Railway as a young child which sparked my interest in steam locomotives and led to my first venture in volunteering on a J27 overhaul in Darlington in my native north east England.

During my time at Swansea University studying aerospace engineering, my 'Year in Industry' was spent with the Sir Nigel Gresley Locomotive Company (SNGLC) working on the overhaul of No.7 and this contributed directly to me graduating top of my year in 2019. Following my graduation, I moved to Toulouse pursuing a masters degree at ISAE Superaero, and volunteering with a local charity on the restoration of a P9.

In October 2022 I will be going to Glasgow to do a PhD, where I hope to find a fourth steam engine to work on, although I doubt I will ever find anything as iconic as Sir Nigel Gresley.

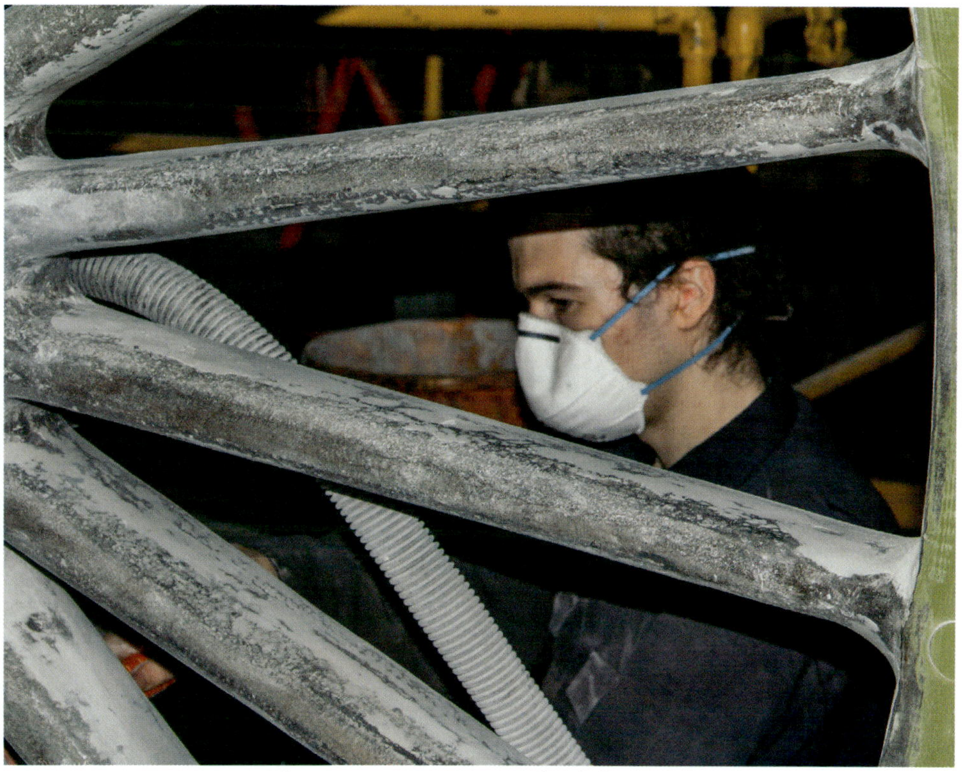

Andrew Jacques preparing wheels for painting

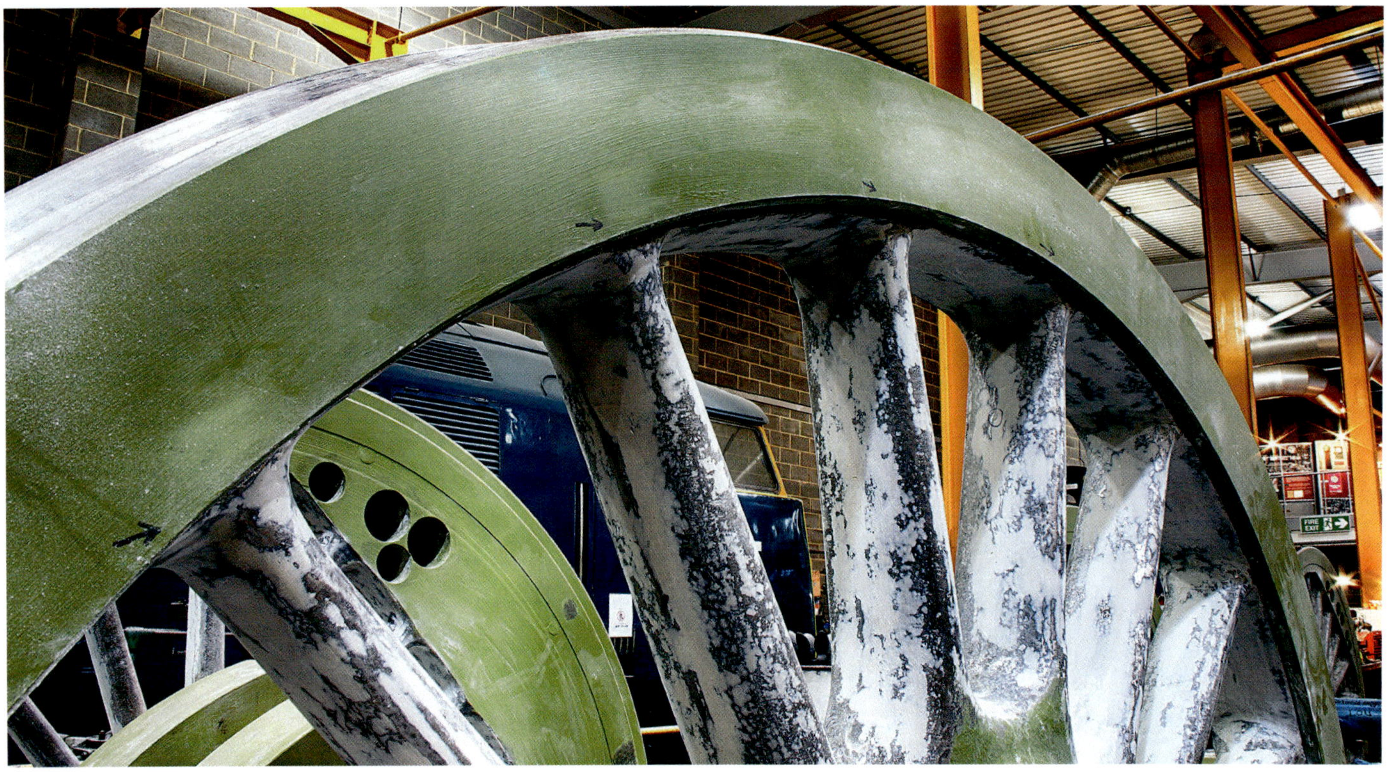

The outer wheels with primer applied and the spokes rubbed down awaiting primer. The inside of the far wheel showing drilled out holes which balance the wheel weight similar to adding weights to car wheels

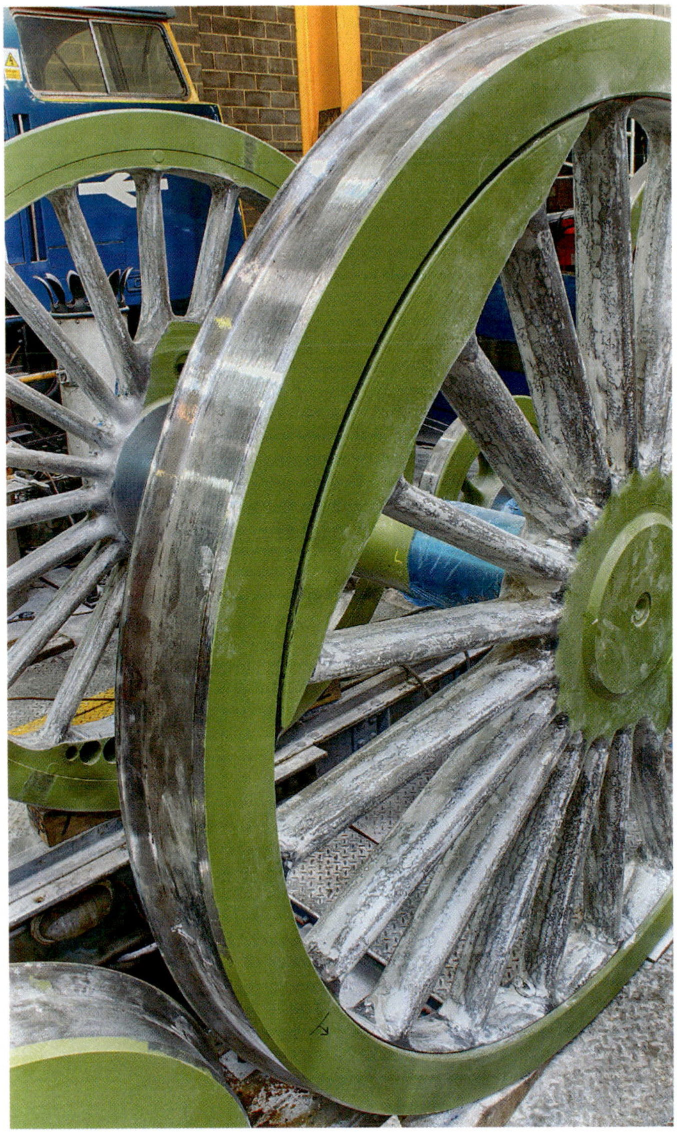

The shiny new tyres around the outer edge of the refurbished wheels

line strung down the side of No.7 with what at first glance appeared to be post it notes attached. It wasn't a washing line of course, as it was pulled straight and taught, more like a trip wire from (if anybody remembers that far back) the film 'Where Eagles Dare'. I discovered it was to check the alignment of the centre of the cylinders with the position of what would be the centre of the axles, and it was very important to have this correct with so many parts moving at high speed.

Preparing the wheels for painting was still taking place and desiring a superb smooth finish, it was decided to apply filler paste to the uneven surfaces of the cast wheel spokes. Anybody who has done this on a car will know it is easy enough to apply but a lot of hard work rubbing it back down to a smooth finish. Nothing to worry about there then, as equipped with suitable PPE (Protective Personal Equipment) the team, as usual, rose to the challenge. What is apparent in these latest images of the wheels is the shiny new tyres and the green primer applied to the outer wheel circumferences. The other thing that caught my eye were the large holes on the inside of the wheels' outer diameter, which I learned was the equivalent of a car tyre fitter adding weights to balance the wheels. Except in this case the balancing of the wheels is done by drilling out part of the wheel thus reducing the weight. This can be seen in a couple of images here and of course this is not a recent balancing adjustment; it dates back to when the wheels were originally cast.

There are clearly a lot of surfaces on wheels which you want to keep free from paint such as the axles and blue protective plastic was fitted around all these areas whilst this work was undertaken. When I returned just before the Christmas break the wheels had received a topcoat of black gloss and were now looking like brand new wheels. Looking like brand new was a term that was to become quite familiar.

LOOKING LIKE NEW

February 2018

The hard work in preparing the wheels for painting had progressed up to and beyond the Christmas break. Now in the second month of the new year there was a topcoat of black gloss on the wheelsets which was to be further improved by a finishing layer of lacquer.

With my privileged access on the floor of the workshop allowing me to get up close and personal with No.7, most of my time would be spent there. However, the gallery walkway, which overlooked the workshop floor, did offer a different camera viewpoint and from time to time I could

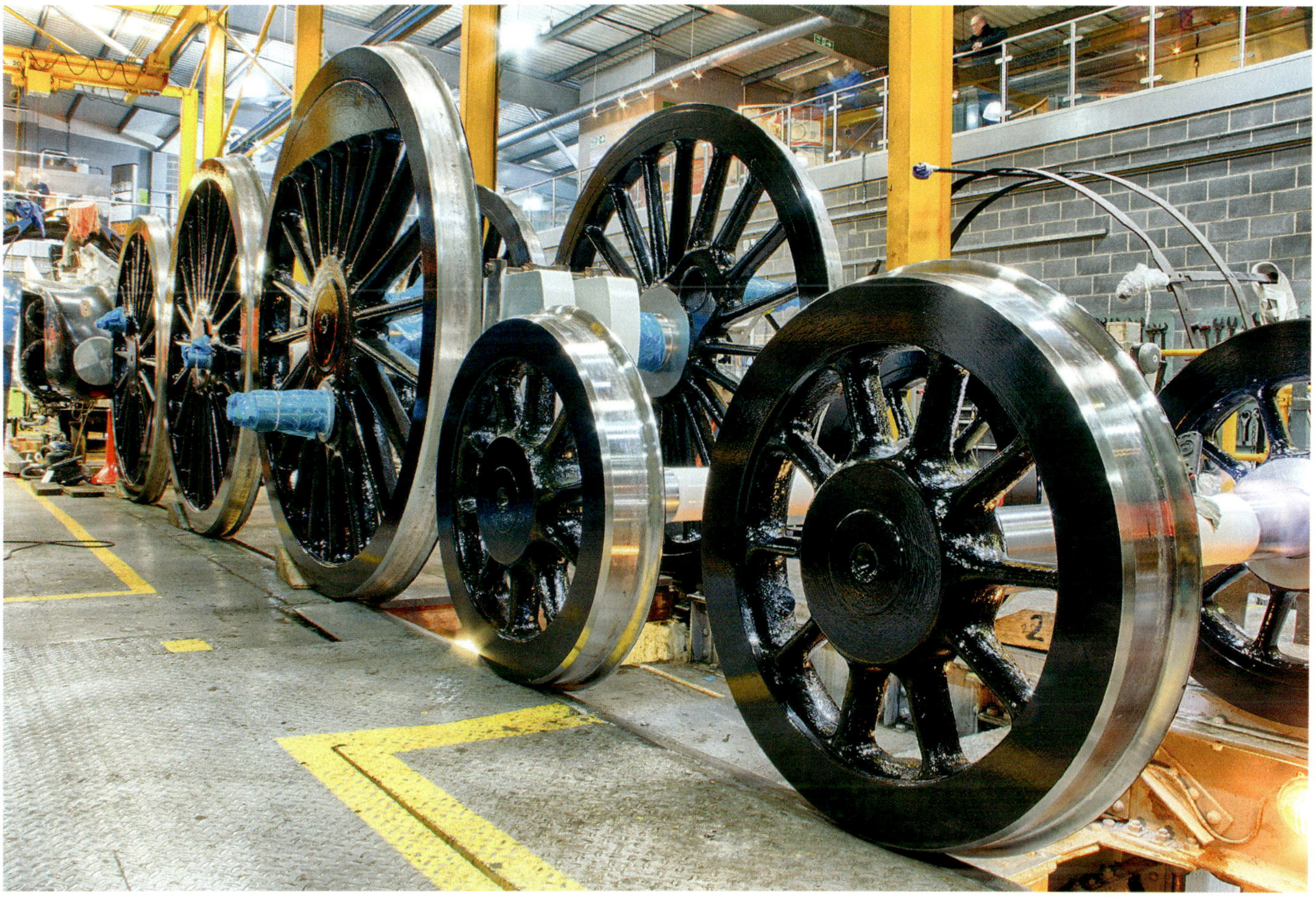

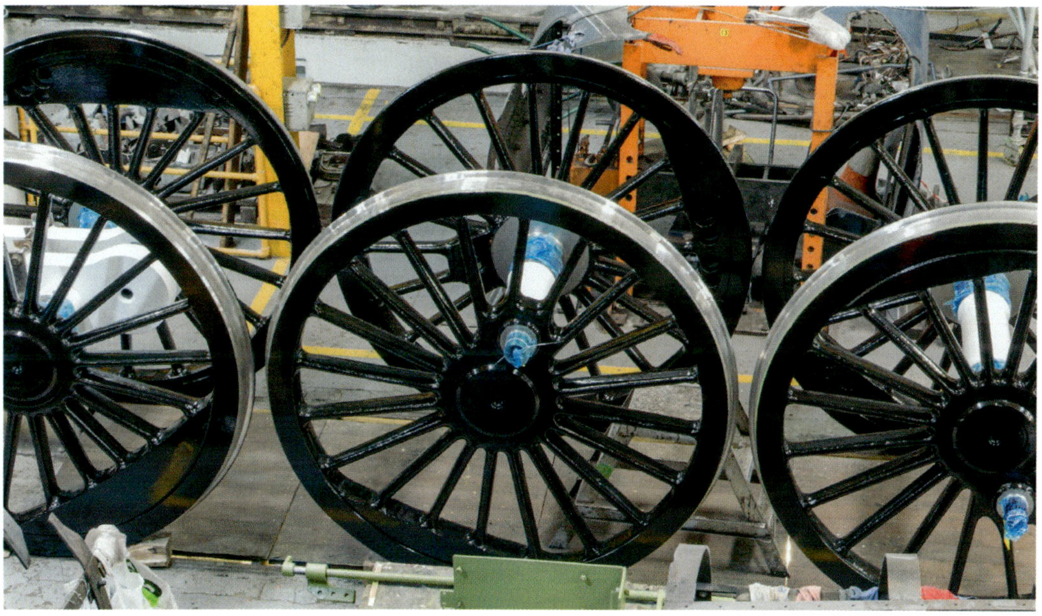

ABOVE & RIGHT - Looking like new, the refurbished wheels with a top coat of black paint applied

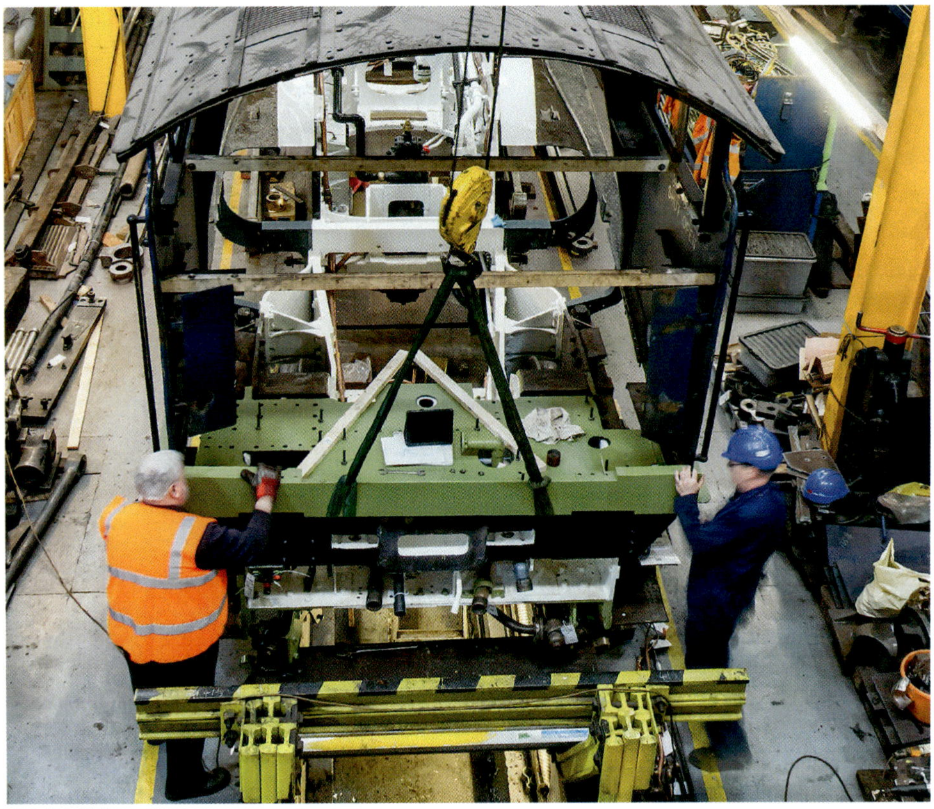

The viewing gallery at the NRM provides a location to capture Peter Ellis and Darrin Crone getting to grips with the cab floor aided by the overhead crane

be found wandering along the gallery with camera in hand. Thus, with my 'Sir Nigel Gresley' clothing on, I found I was to become very popular with the museum's visitors, and what was intended as a ten-minute diversion often turned into something far longer as I tried my best to answer all the questions. It was obvious that far from being out of traffic and out of popularity No.7 was still in the shop window, which was hugely important during this lengthy absence from traffic. I have included a couple of shots in this section taken from the gallery; one showing the newly painted wheelsets and another showing Peter Ellis and Darrin, with help from the overhead crane, starting to re-assemble the cab floor.

Work had started on the tender, work which was going to feature very prominently in the months to come and on my visit, Dave Lee was busy on top of the tender with his oxy-acetylene equipment.

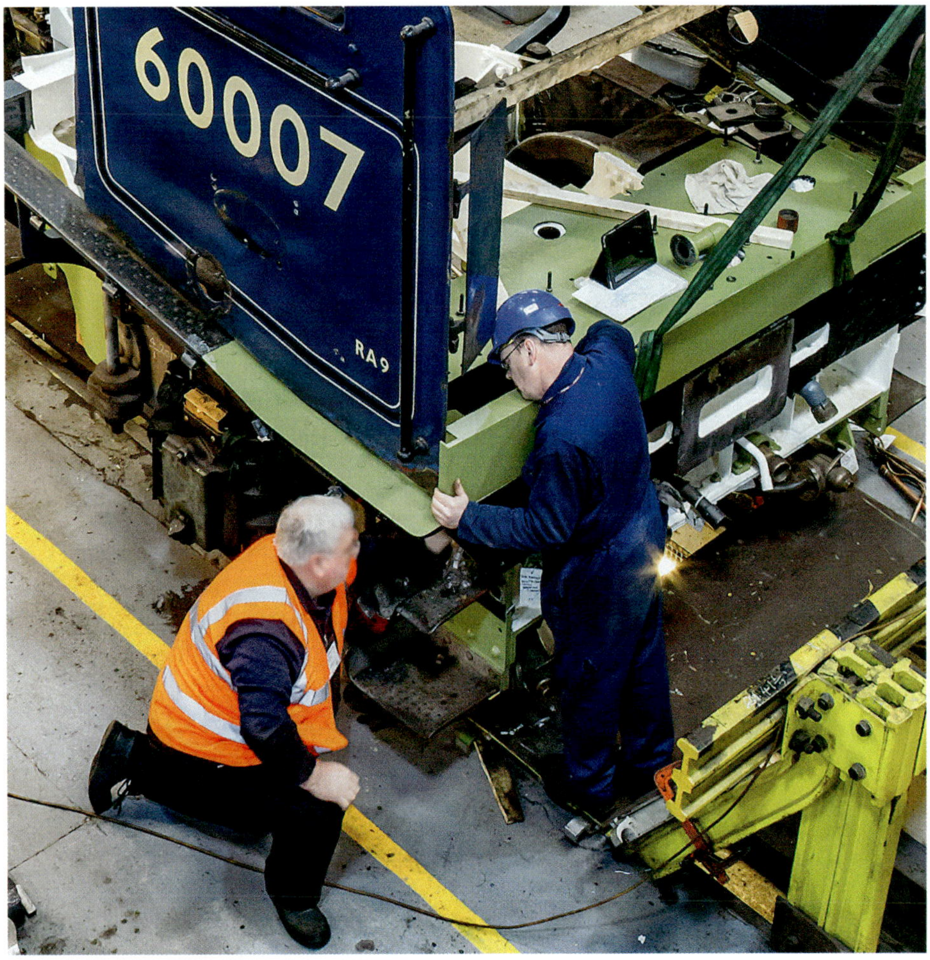

Another shot of Peter & Darrin working on the cab floor components

MARCHING ON

March 2018

A little bit of tender loving care was now being given to No.7's tender. This is one of the famous corridor tenders which allowed crew changes on the non-stop Flying Scotsman service, and it was the roof of this narrow corridor that now needed a lot of attention. The steel plate was badly corroded so it was decided it should be replaced. Oxy-acetylene burning expert, Dave Lee, was given this task and if you look closely at the image showing the removed roof of the corridor you can spot him. "How is it going?" I asked. "Very well," replied Dave and added jokingly "We are going to put a sunroof in." Capturing sparks in a photograph needs a long exposure, in this case 5 seconds, so the tripod was needed and as it was situated on the same scaffolding I was standing on, I needed to hold my breath during the exposure to avoid any unnecessary vibration. With more time to prepare the shot I could have left the camera behind on

"We are putting a sun roof in"

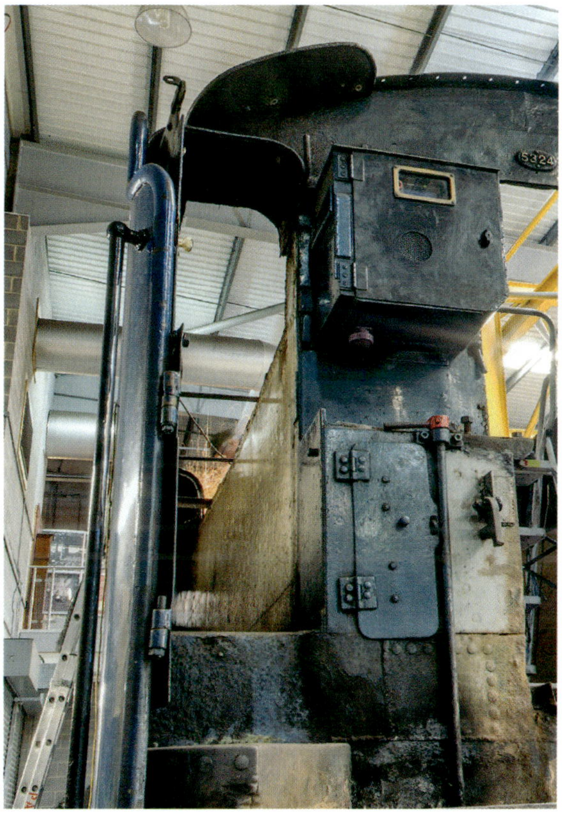

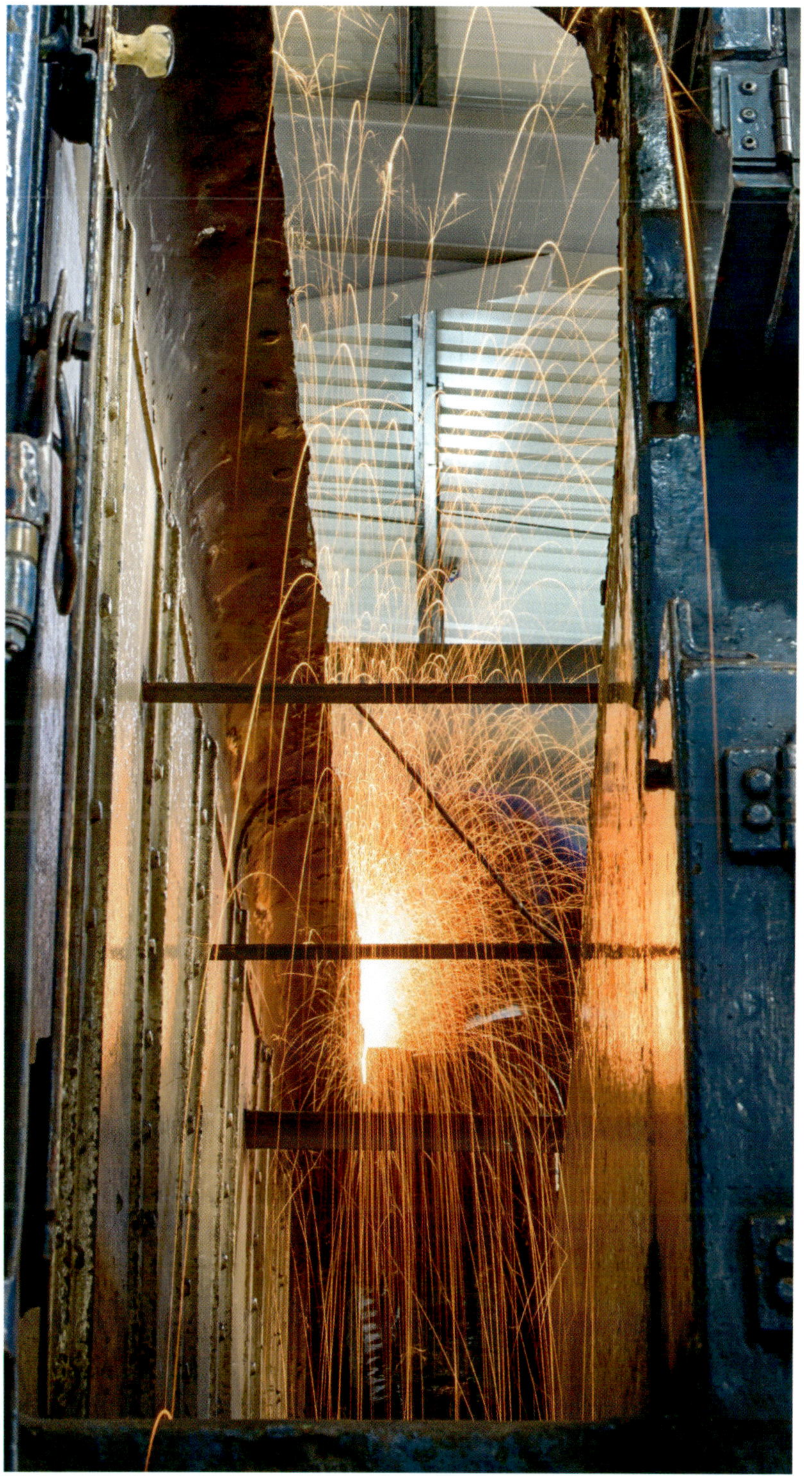

ABOVE & RIGHT - Somewhere above the shower of sparks is Dave Lee using oxy-acetylene burning equipment to remove corroded sections of the tender's corridor roof

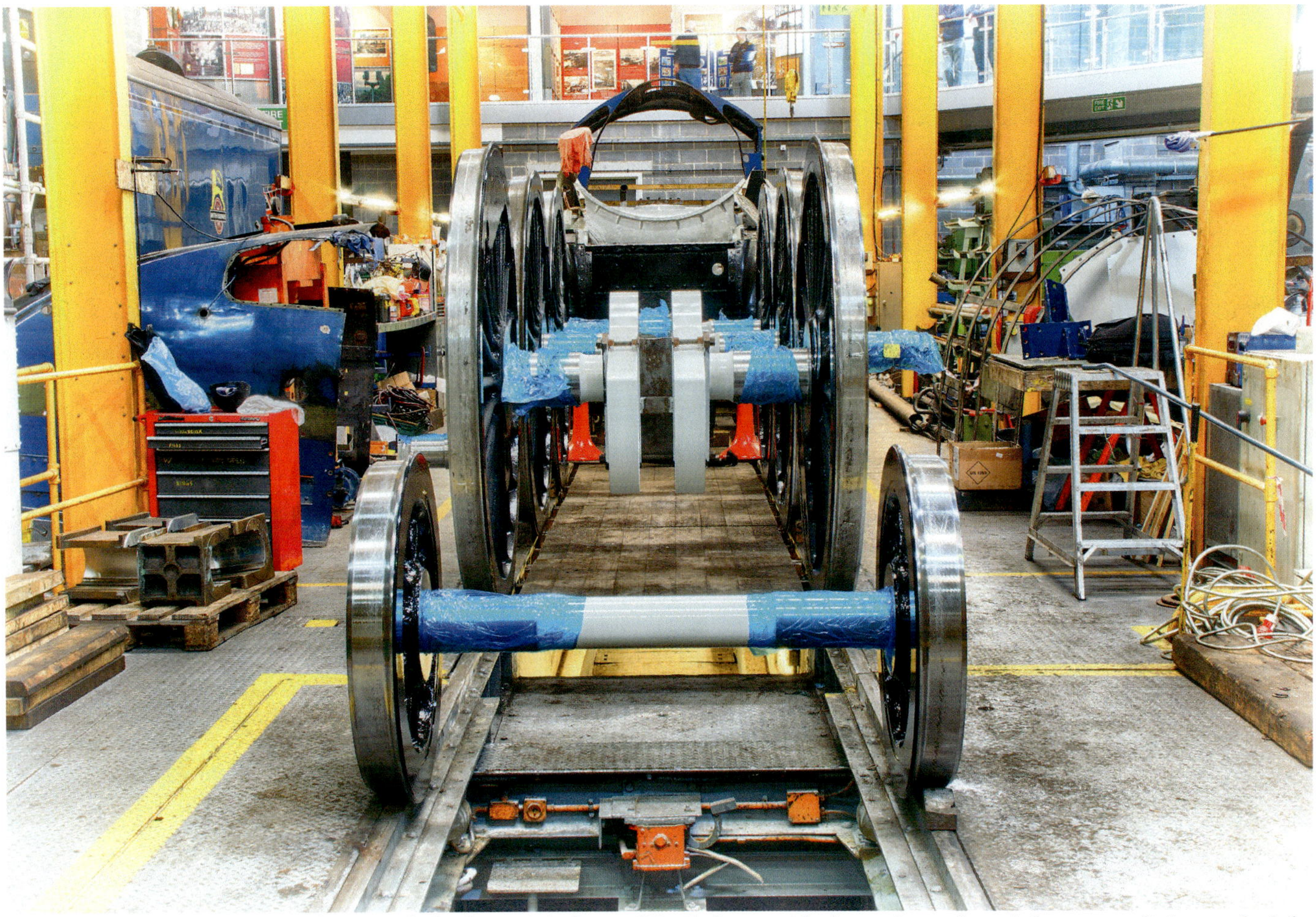

Fully refurbished, painted and lacquered wheelsets sit in front of No.7's frame with the tender and smokebox streamlining on the left and the streamlining's crinoline formers on the right

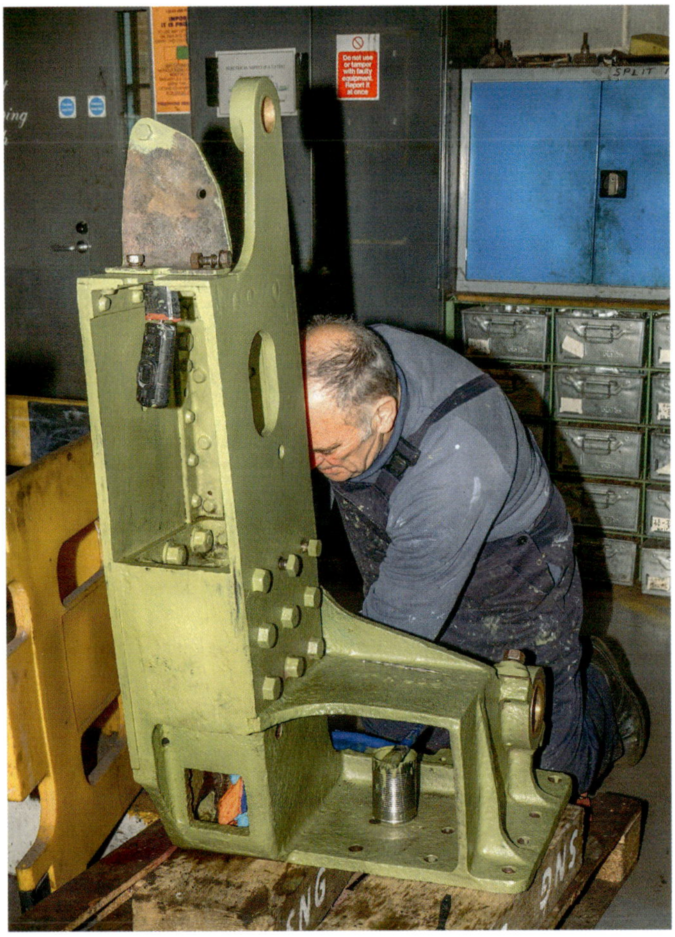

Peter Brackstone applies undercoat to the reverser

the scaffolding and used a remote shutter, but this was live action shooting and I had to hit the shutter button when the opportunity arose. I think the shot has worked out alright.

The wheelsets had now been painted and were standing in front of the frames on the same track, affording a good opportunity to get some shots with them all lined up together and better still, with a rare occasion of nothing being immediately in front, I was able to get a front on view. This shot was interesting as it showed just how much of No.7 was scattered around the workshop floor. On the left the tender stands on a parallel track with the streamlining from the smokebox next to it; in the centre are the wheels with the frames and cab behind and finally, on the right are the crinoline formers which fit around the boiler and create the

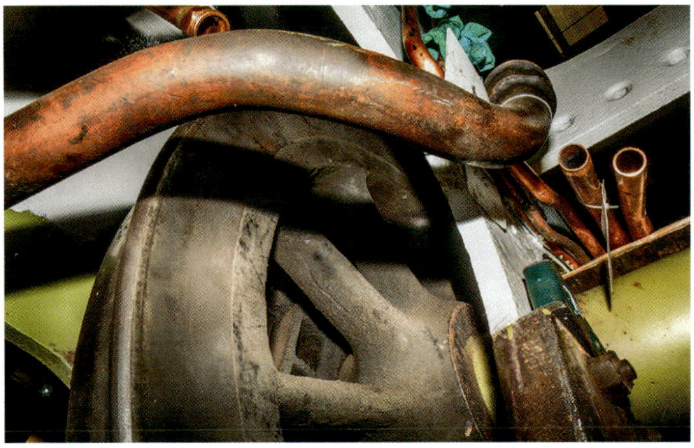

New pipework installed by Peter Ellis around a cartazzi wheel

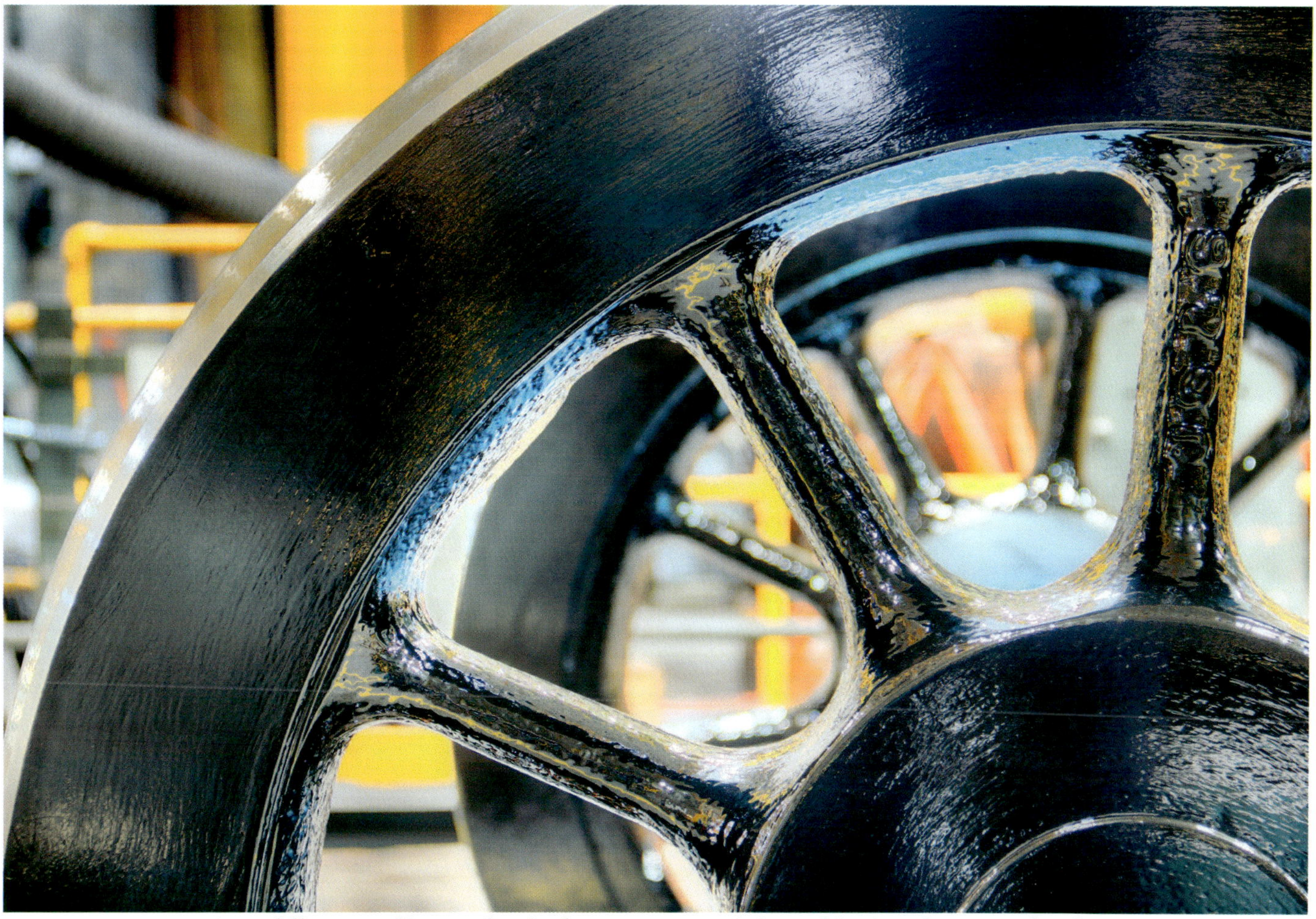

A carefully cropped image of a wheelset creates an artistic combination of curves and angles

framework for the streamlined panelling.

The reverser, which had been removed from the cab, had been cleaned and overhauled and was now in the capable hands of Peter Brackstone and his paint brush. Peter Ellis was making progress with new pipe runs under the cab and a multitude of other tasks were being undertaken. Andrew Jacques and Rod Thomas are captured on camera from the workshop gallery undertaking some of these tasks.

Returning onto the workshop floor, my attention was once more drawn to the wheels. There is something about spoked wheels that makes them very artistically appealing and, not for the first time, I found myself taking lots of shots from various angles but it wasn't until I returned home and started editing the shots that a particular favourite, shown here, jumped out. Most photographers will tell you that you sometimes see something in an image, but you can't put your finger on it and it only becomes apparent when you start cropping the image in different ways - this was one such occasion.

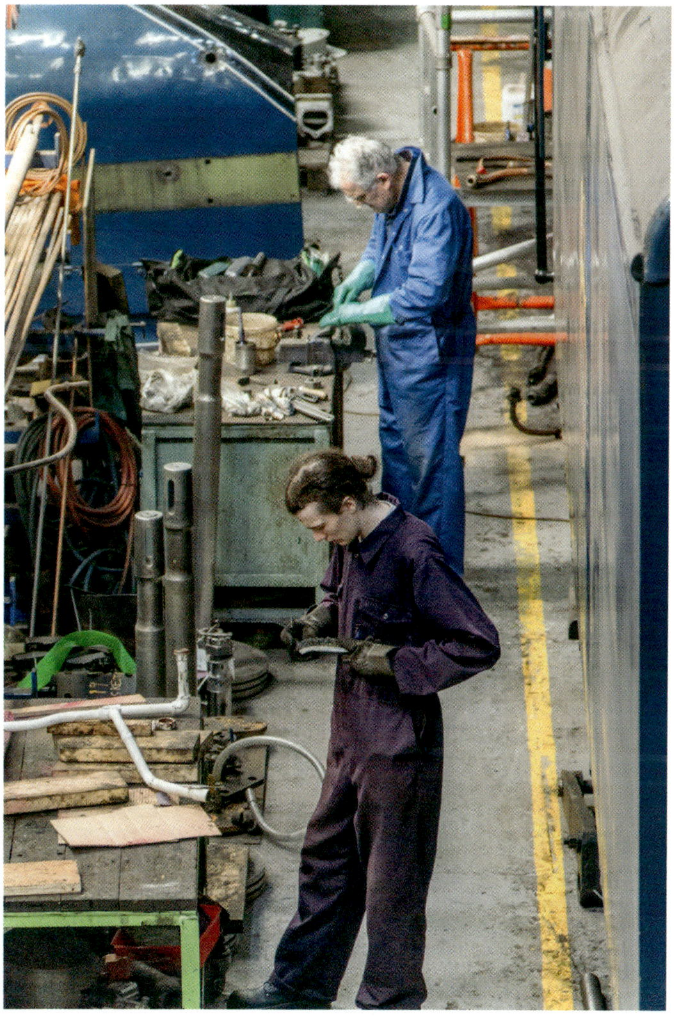

Rod Thomas (top) and Andrew Jacques hard at work on the workshop floor

DISCO LIGHTS

April 2018

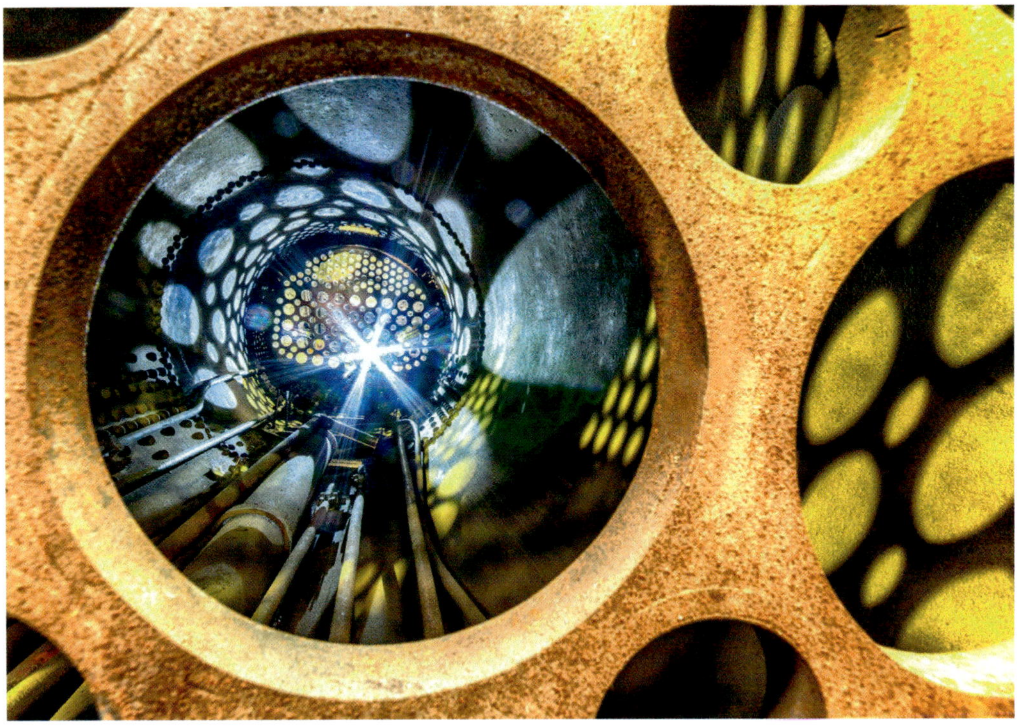

A point light source creates a disco lights effect on the inside of the boiler

"Oh, it's alright, the boiler is upside down, so the steam dome is on the bottom," was Darrin's reply when I asked how he expected me to get my XL physique inside the boiler, after he told me there were some great shots to be had inside. He had just returned from one of his routine update trips to the Llangollen Railway workshop, where the firebox copper tubeplate had been put in position, and said I should go as he had noticed how dramatic an effect the light passing through the tube plates was having on the inside. "Ah, well that's ok then," I replied and promptly booked an overnight stop at one of Llangollen's many hostelries.

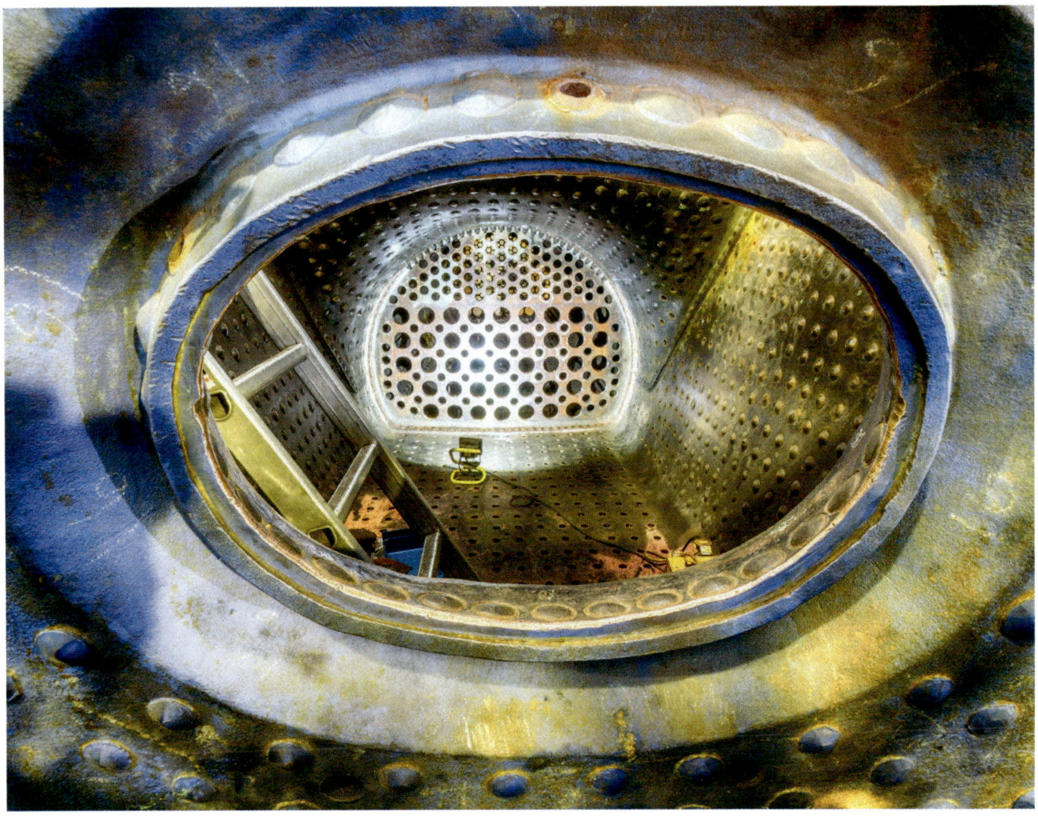

The new tube plate in position at the back of the firebox viewed through the firebox door opening

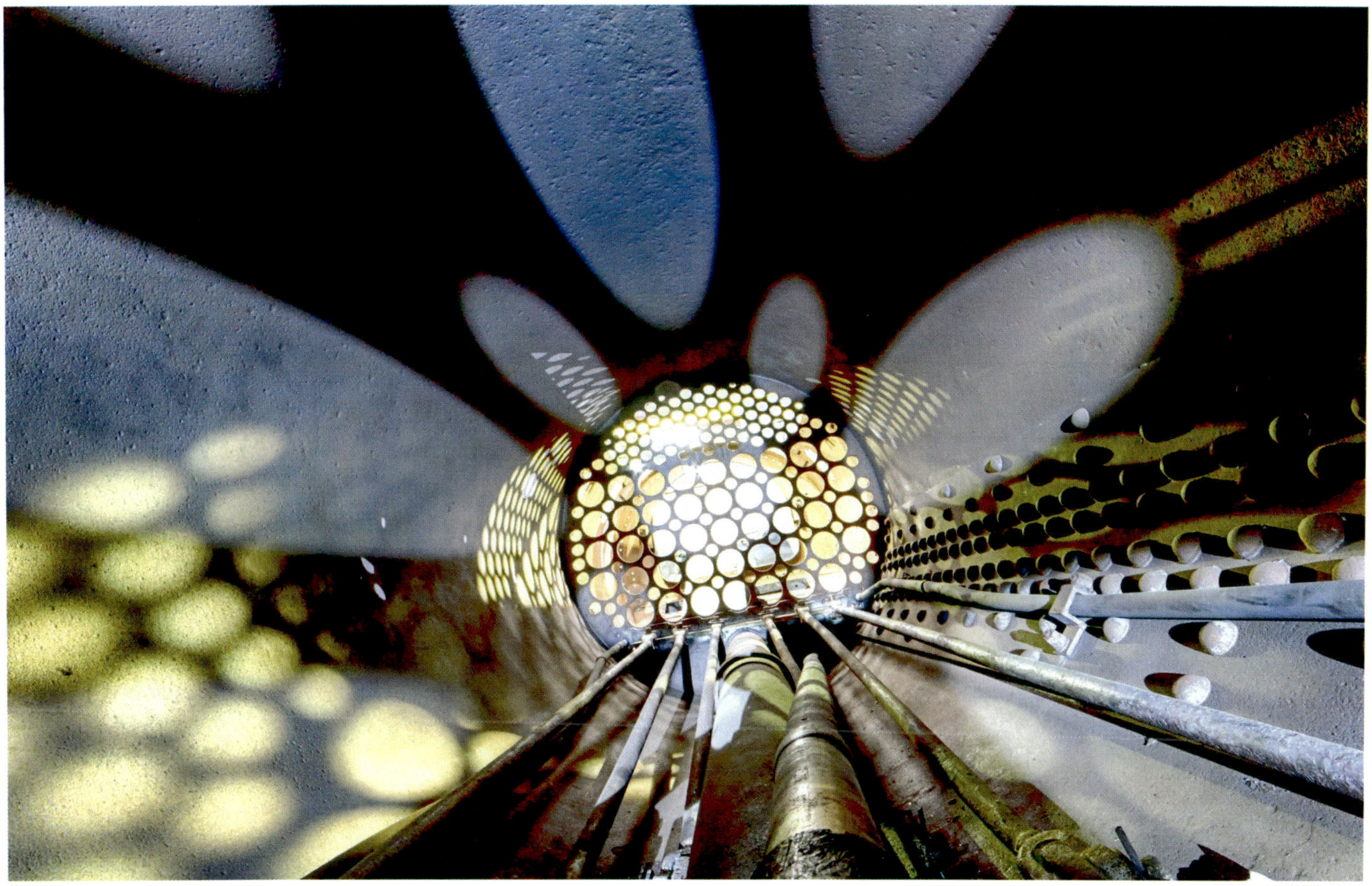

The disco lights taken after I climbed through the steam dome with my camera and tripod from below as the boiler was currently upside down

Malc Bateman, who had been staying in the area, diverted his return home to meet up with me and we joined up with Paul Aston, our resident representative in Llangollen. We promptly reported to the workshop where I knew something was special on this my third trip to Llangollen. It wasn't raining or snowing as on the two previous visits - surely a good omen.

The boiler was upside down as Darrin had said, which did mean I didn't have to climb on top to gain access through the steam dome - the only way in and out. However, I was expecting it to be higher off the ground with a ladder to

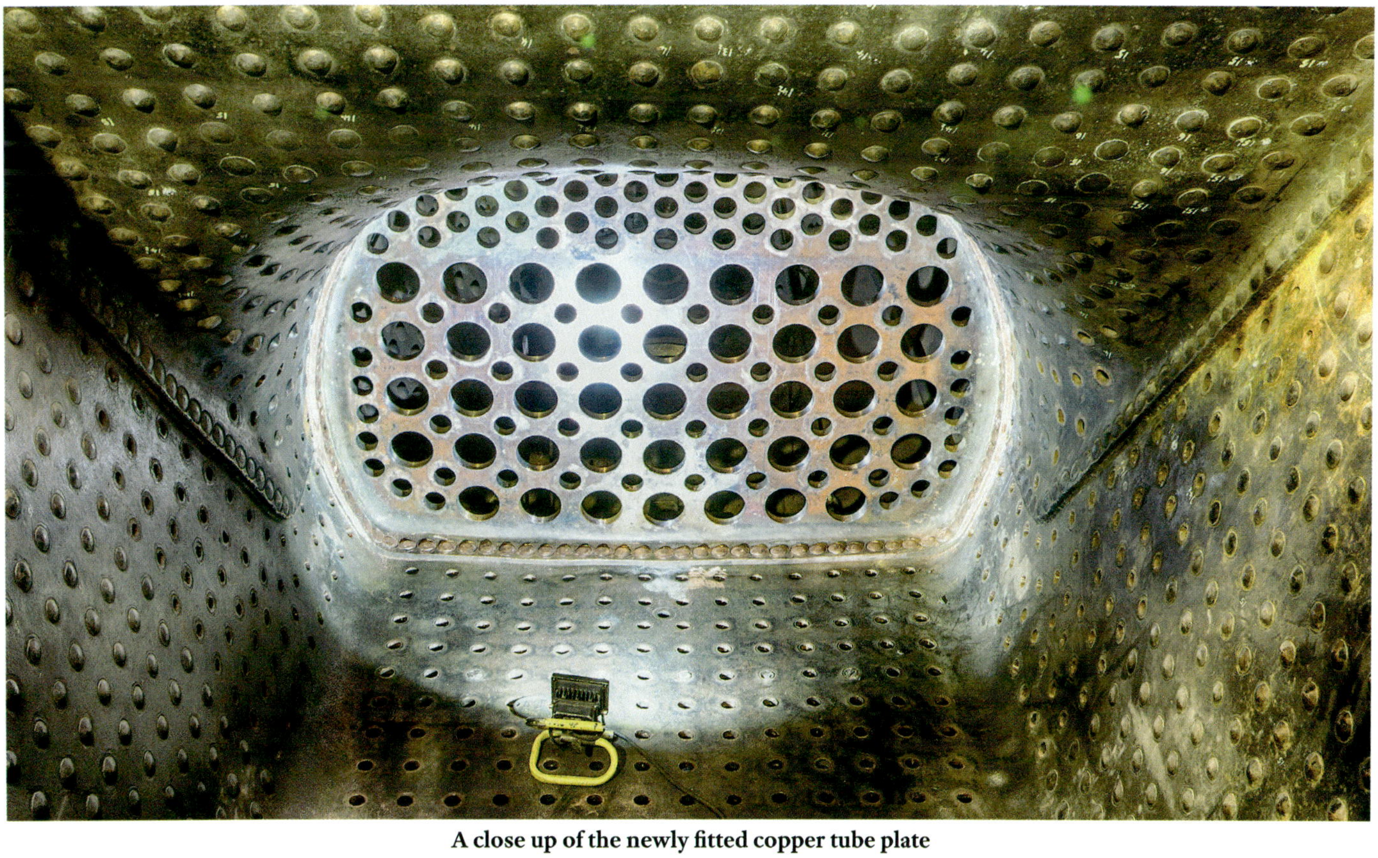

A close up of the newly fitted copper tube plate

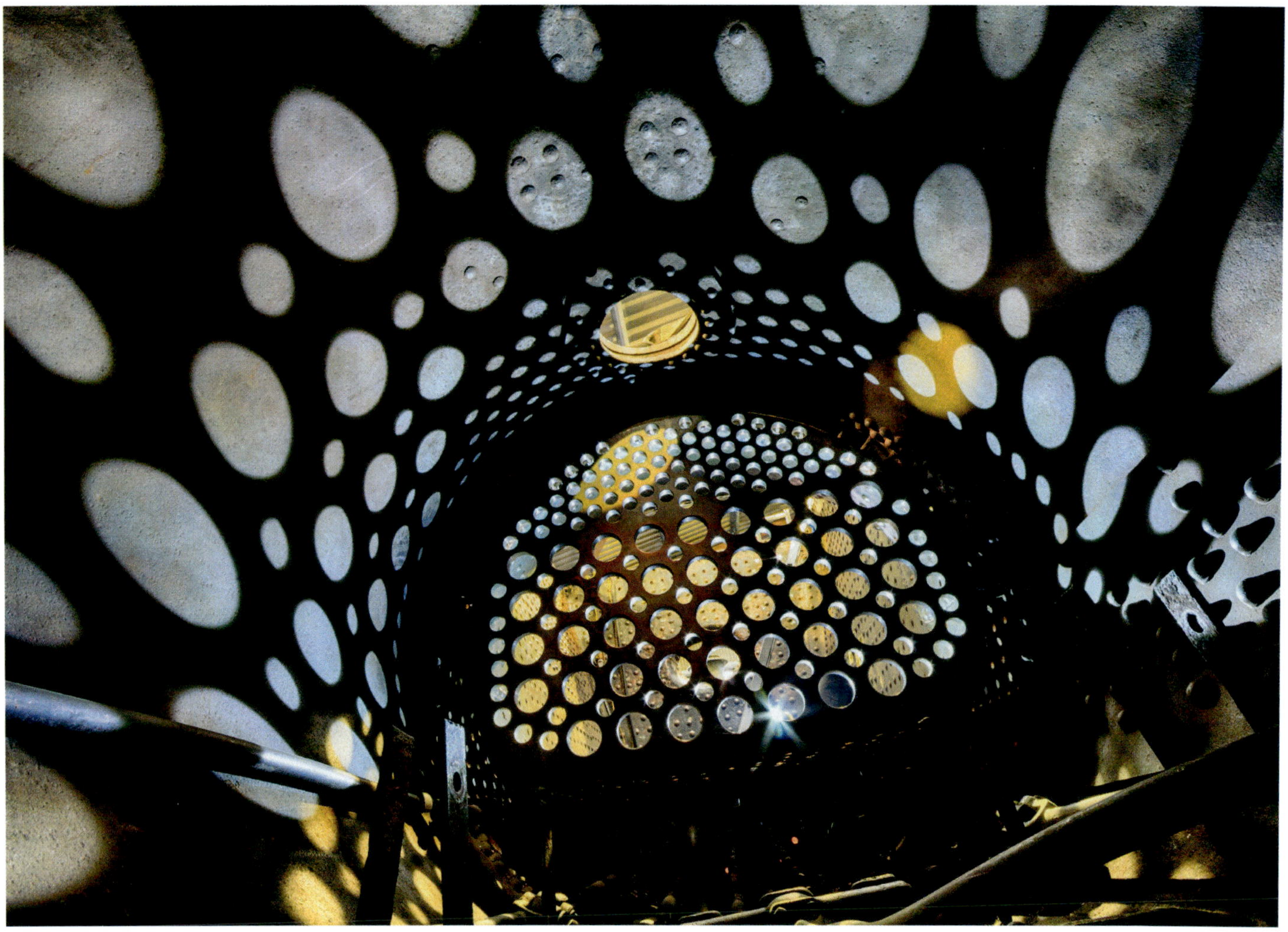

Another shot of the dramatic lighting effect inside the boiler

climb up and through for entry, but I was wrong. It was about 18" off the ground. Having come this far I wasn't giving in that easily, surely it was no more difficult than what pot holing required to be done routinely. Eventually I was through; it was dull and monotoned, but as my eyes readjusted, I could see faint patterns forming on the boiler walls. Shouting out to Paul and Malc I requested a single lamp to be randomly positioned in the firebox and, eureka, I was transported back 40 or more years to the last time I regularly spent Saturday nights in a discotheque.

Reaching back outside I lifted my tripod and camera into the boiler and set about collecting my shots. It was now that I realized I didn't actually need to be inside the boiler at all, it was the equipment that did. Still I had overcome a phobia of climbing through tight squeezes into a confined space that only had one way in and out.

Job done - well half done. The images still needed editing but that could wait until I was back home. Tonight was for a nice meal and a couple of beers. Tomorrow Malc and I would take a ride on the Llangollen Railway, which was a sheer delight, as this has to be one of the most scenic railway lines in the UK.

BOGIE ON TRACK

8

May 2018

"We need a volunteer to go in the pit Peter"

Rebuilding No.7 started some 18 months ago in May 2016 with the refitting of a sandbox and had continued steadily since then, but I think there was a step change now when the bogie frame was reunited with its wheels. There were many days when the team members were working as individuals on the various tasks allocated to them, but the milepost events pulled everybody into one big team. Today was one of those days.

Today, to me, contained a 'wow' factor as I had witnessed the refurbishment of many components, and seeing so many of these refurbished components coming together with the bogie reassembly was immensely impressive. If I hadn't have known better, I would have thought I had been transported back in time to the Doncaster works of the 1930s and I was watching brand new components being assembled.

With the many heavy components involved and the overhead crane in use, it was a hard hat day for the team and a long lens day for me as getting the close-up shots and staying out of the way at the same time required that. Darrin was in fine form directing operations by starting with "We need a volunteer to go in the pit Peter," and with a glance at the sky, (well workshop roof), Peter Brackstone duly 'volunteered'. There was no real choice here as the exceptionally heavy axleboxes needed positioning on the wheelset axles and it made sense to have your strongest man in the pit for that. Peter B was that man. With Darrin on overhead crane duties, Richard aligned the axleboxes before they were lowered into position on the well oiled axles. To stop the top heavy axleboxes slipping around the axles and falling into the pit,

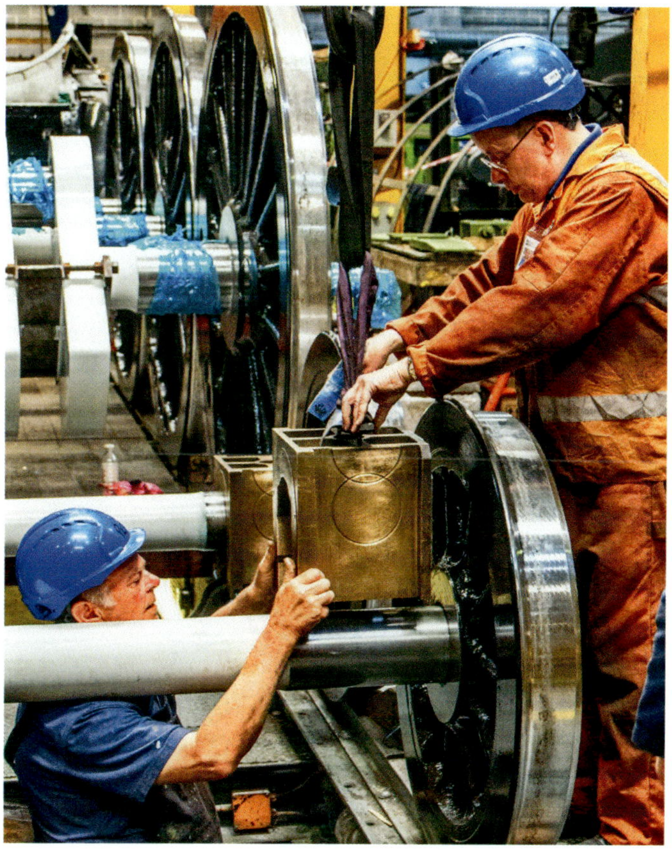

ABOVE & BELOW - With the overhead crane taking the weight, Peter Brackstone (in the pit) and Richard Swales guide an axlebox back onto an axle, Darrin Crone is at the controls of the crane

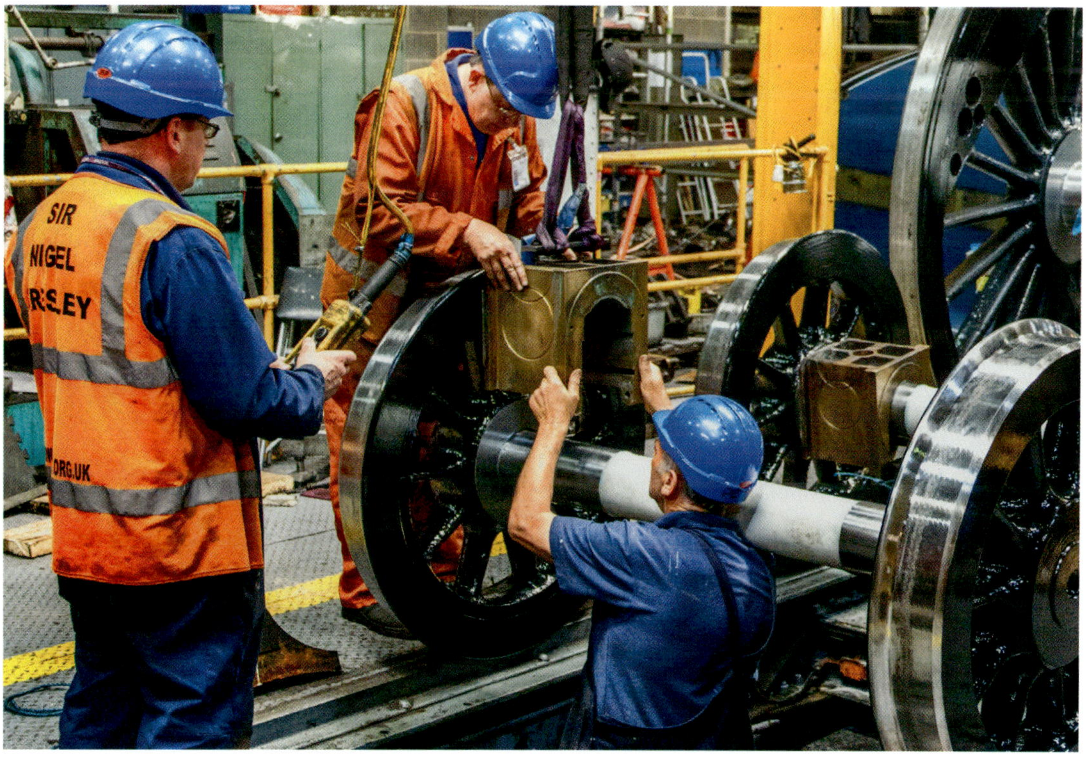

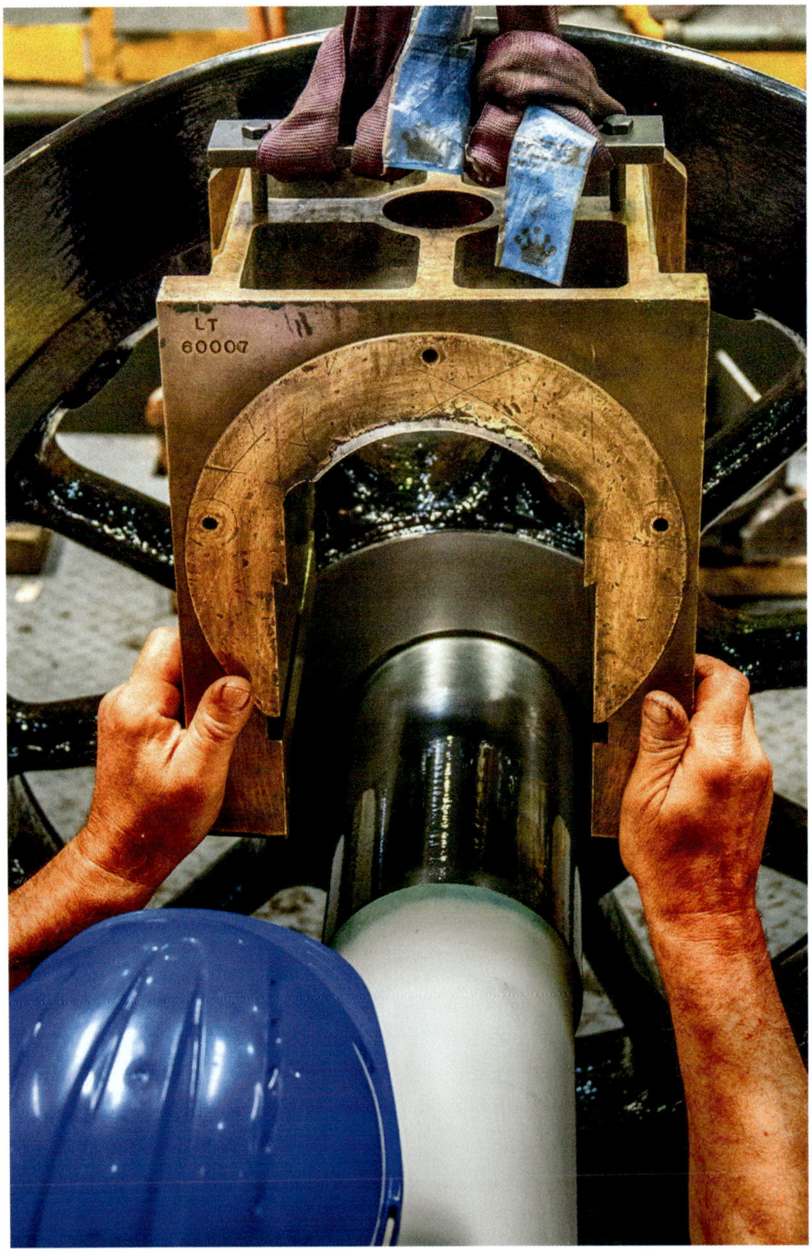

Peter Brackstone carefully aligns an axlebox

counter weights in the form of old brake shoes were temporarily fitted until after the frame was lowered and Richard joined Peter B in the pit to complete the operation by fitting the axlebox sliding trays and securing the fully assembled boxes to the axles and frame with the horn stay brackets.

So far only Richard, Darrin and Peter B had been involved but now it was time for more hands-on deck. Tony and Andrew formed the cavalry for the next operation which was the mating of the bogie frames with the wheelsets. This was another NASA style docking manoeuvre operation, and with Richard on overhead

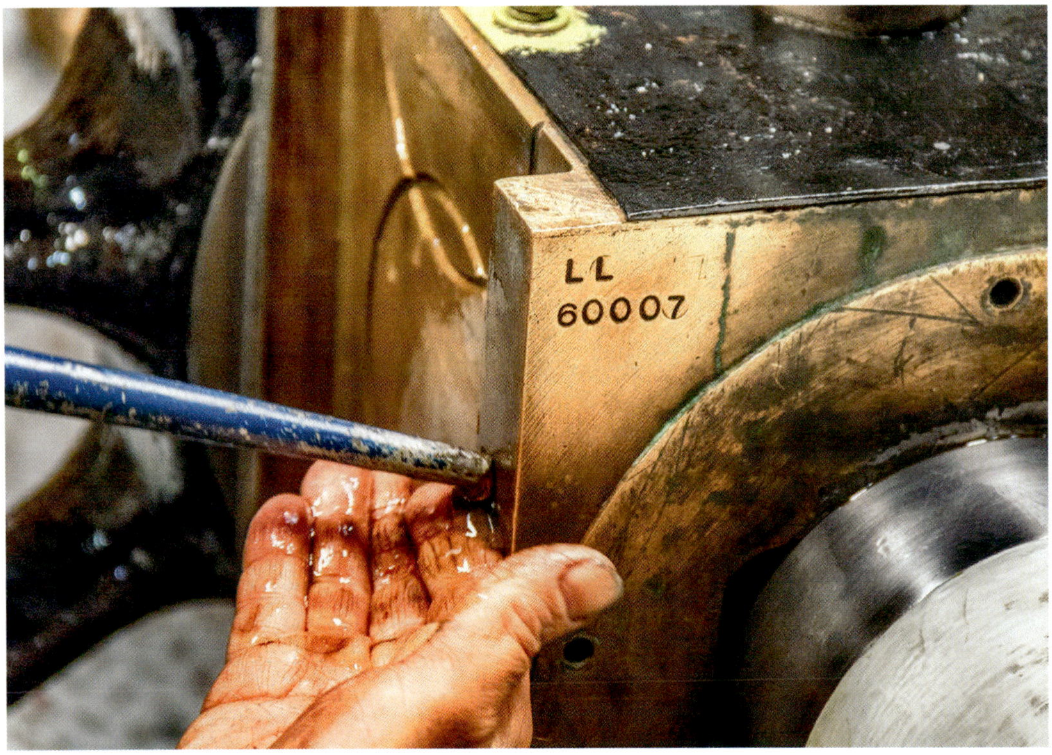

Plenty of oil is used to help the task of refitting the axleboxes run smoothly

TOP to BOTTOM - With axles in position the frame is lifted by the overhead crane, moved over the wheelsets and lowered carefully. Richard Swales operates the crane whilst Andrew Jacques, Tony Bickerstaffe, Peter Brackstone and Darrin Crone guide the frames into position on the wheelset axles.

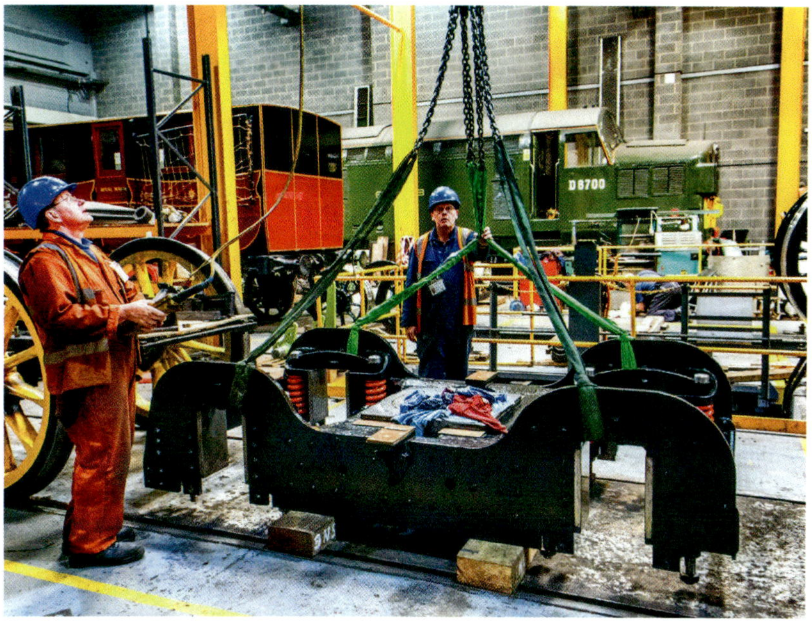

crane operation a final check was made on the wheelsets positions before he swung the frame carefully into position above the wheelsets. Andrew, Tony, Peter B and Darrin each took a corner of the frame to steady the lowering onto the axles. Keeping everything square and level was crucially important for the successful outcome and millimetre by millimetre the frame was lowered onto the axles until the team's precision work was perfectly completed at the first time of asking, as usual. It was just as well as it was one o'clock and Peter B had more important tasks to complete with the kettle in the blue cabin, as lunch time had arrived once more.

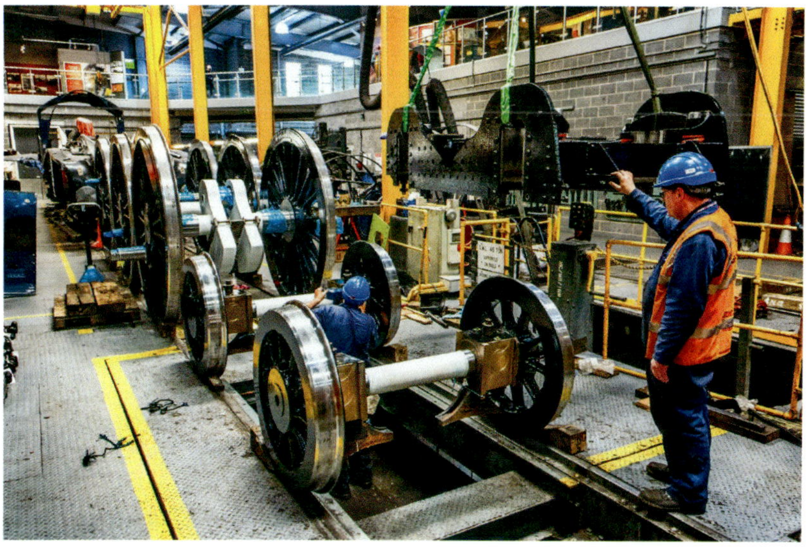

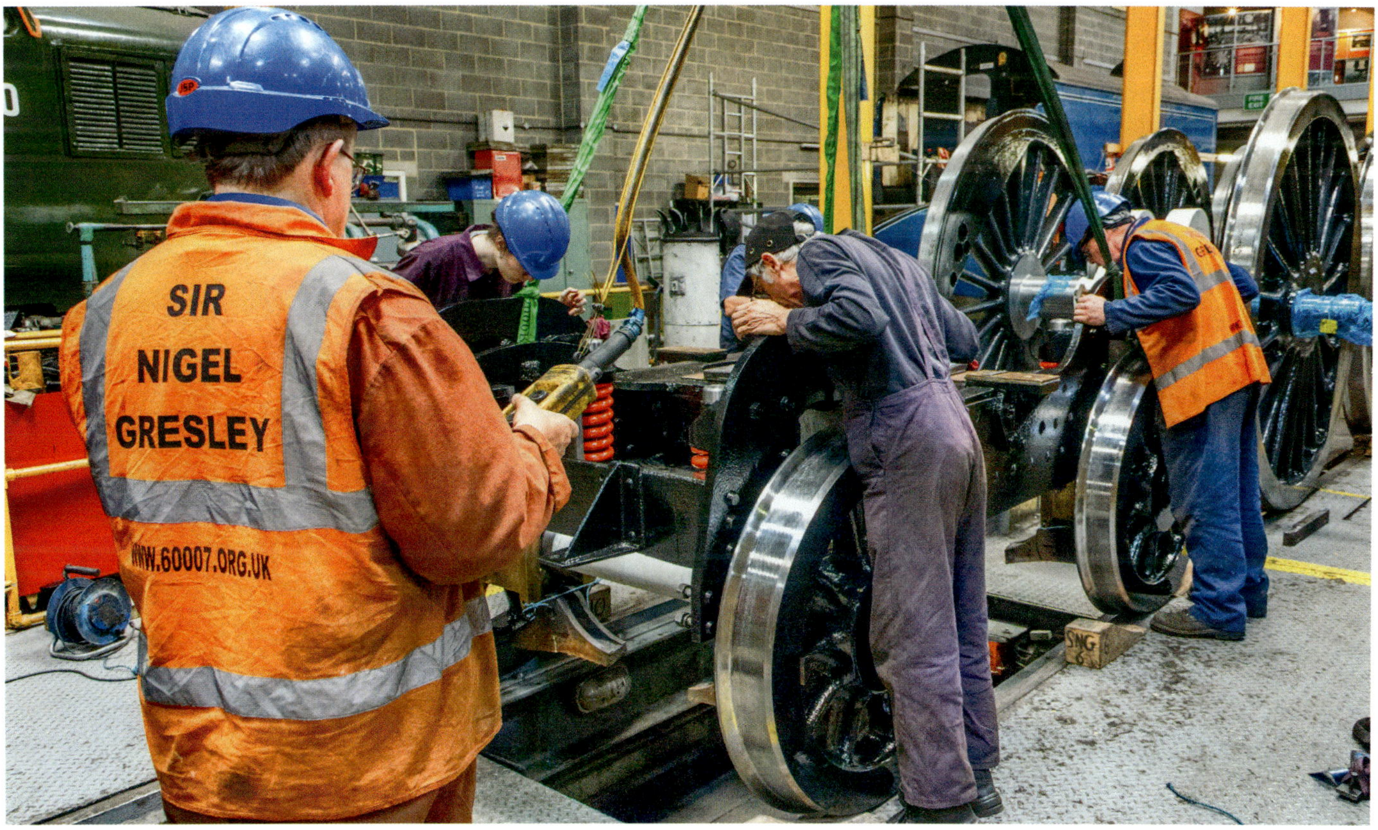

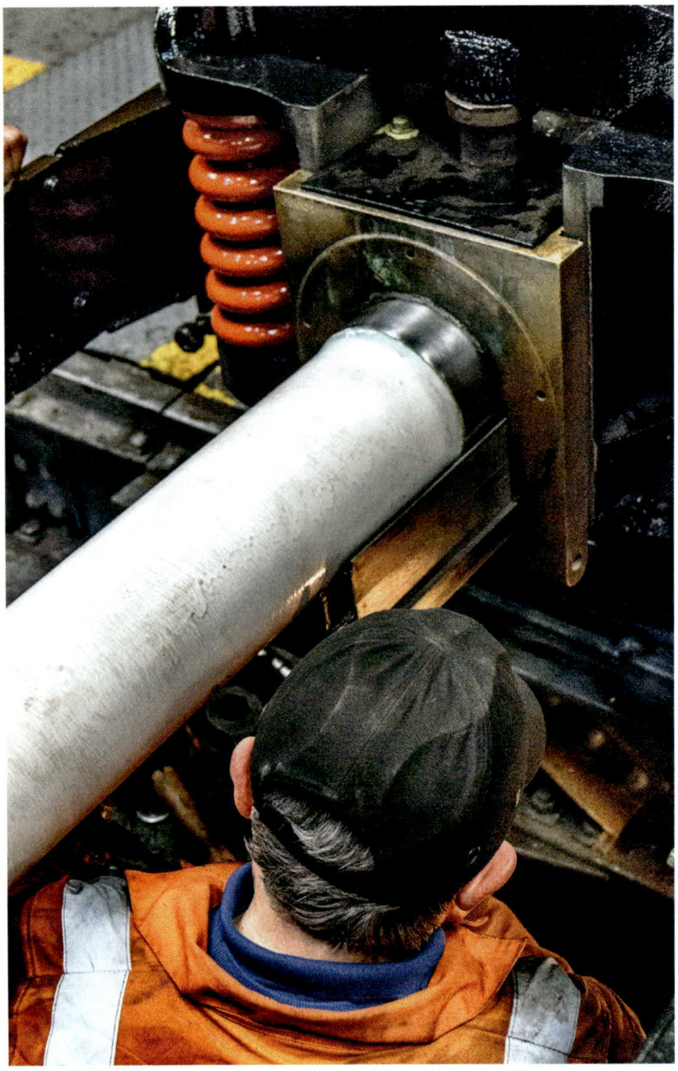

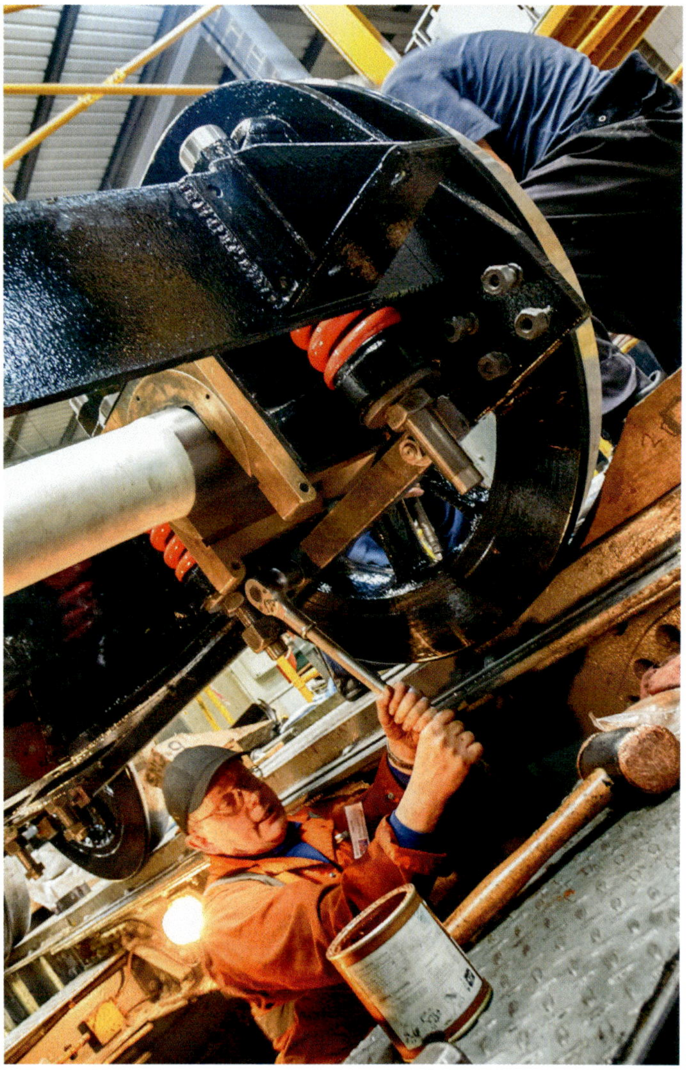

Richard Swales completes the reassembly by fitting the axlebox sliding trays (left) and tightening the horn stay brackets (right)

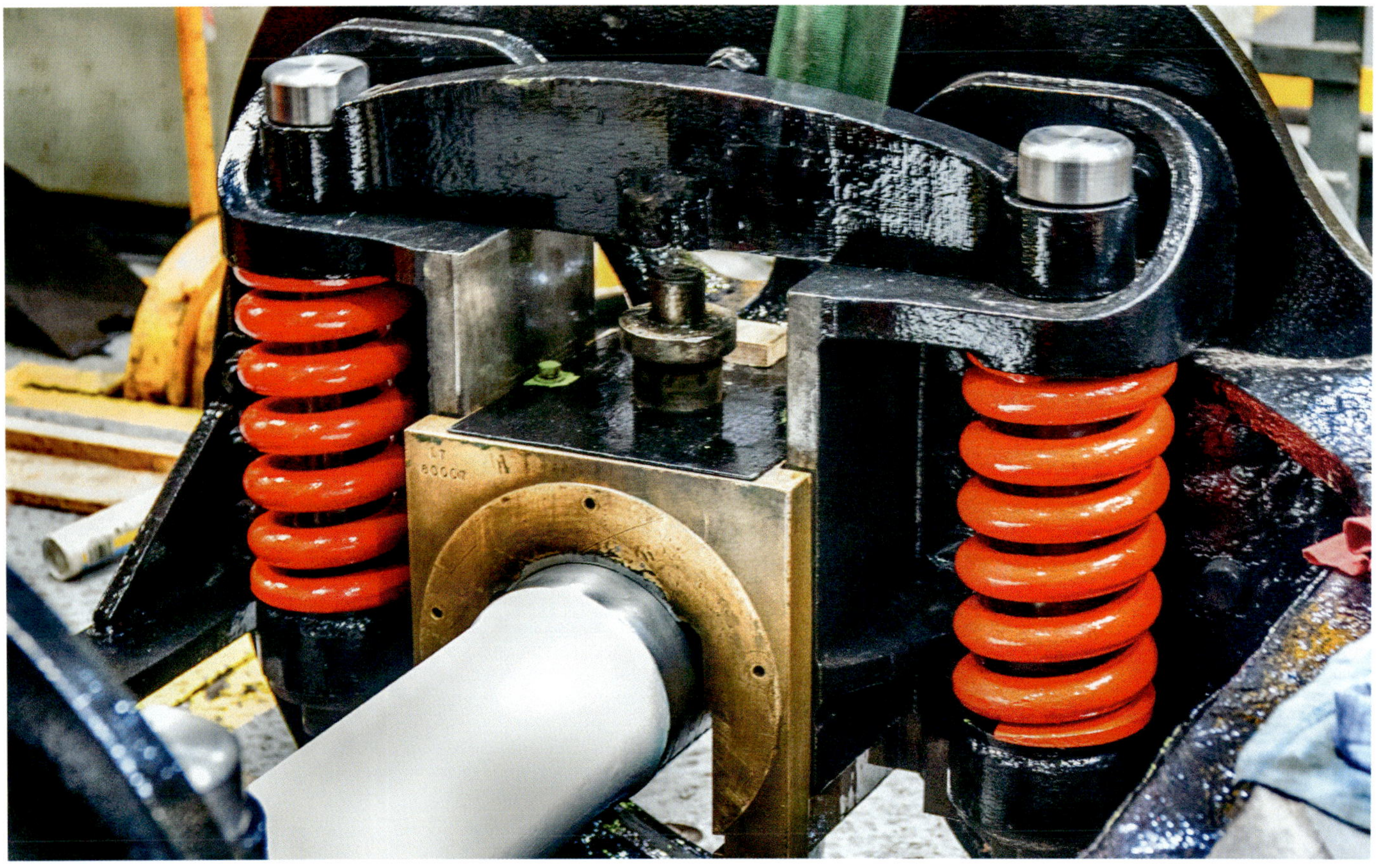

Wheelset axle with axle box and frame completely reassembled and looking like new after refurbishment

TENDER LIFT

9

July 2018

I have to be honest, I had not given much thought to No.7's tender, but that is probably because I was focussed, (no pun intended) on No.7, which was getting all the attention. That was about to change. There was work, a lot of work, to do on tender 5324 - which I was to discover is quite famous itself.

Tender 5324 is one of the ten corridor tenders built in 1928 for the non-stop Flying Scotsman train service and was allocated to the locomotive of the same name when the famous A1/A3 achieved the first officially recorded 100mph run hauling a Kings Cross to Leeds train on November 30th 1934. It was transferred to No.7 on August 8th, 1943 and was to become a speedster again when No.7 set the post War steam speed record of 112mph on May 23rd, 1959.

Darrin and Richard had identified several areas which needed attention, more of which will be revealed in later pages, so a plan was drawn up to address them all. The first major task was to separate the tank (top bit that holds the coal and the water) from the frames and wheels. Special lifting equipment would be needed, and this was provided by Durham Lifting. Darrin also appointed Trust volunteer Roger Turnbull as Lift Team Leader. Roger is a retired power station engineer with good experience of heavy lifting procedure so was a logical choice to lead the team.

Having no idea what the lifting gear would look like other than some form of crane, I was to discover that it consisted of hydraulic jacks. After the tank was lifted, it would not be

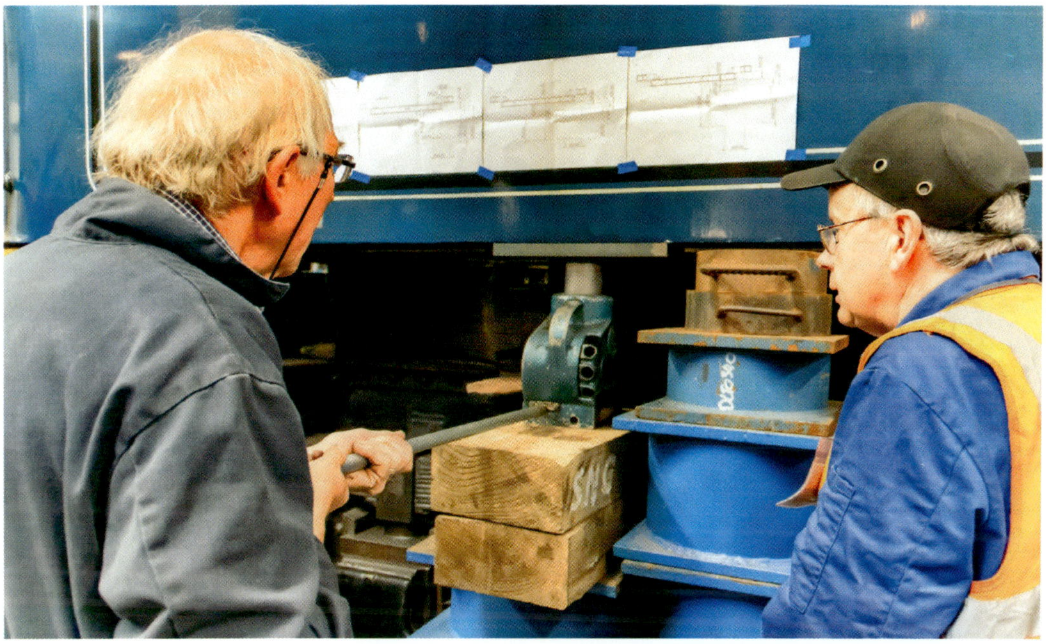

Bob Shearman operates a tender lifting jack as Chief Mechanical Engineer Richard Swales carefully monitors the lift

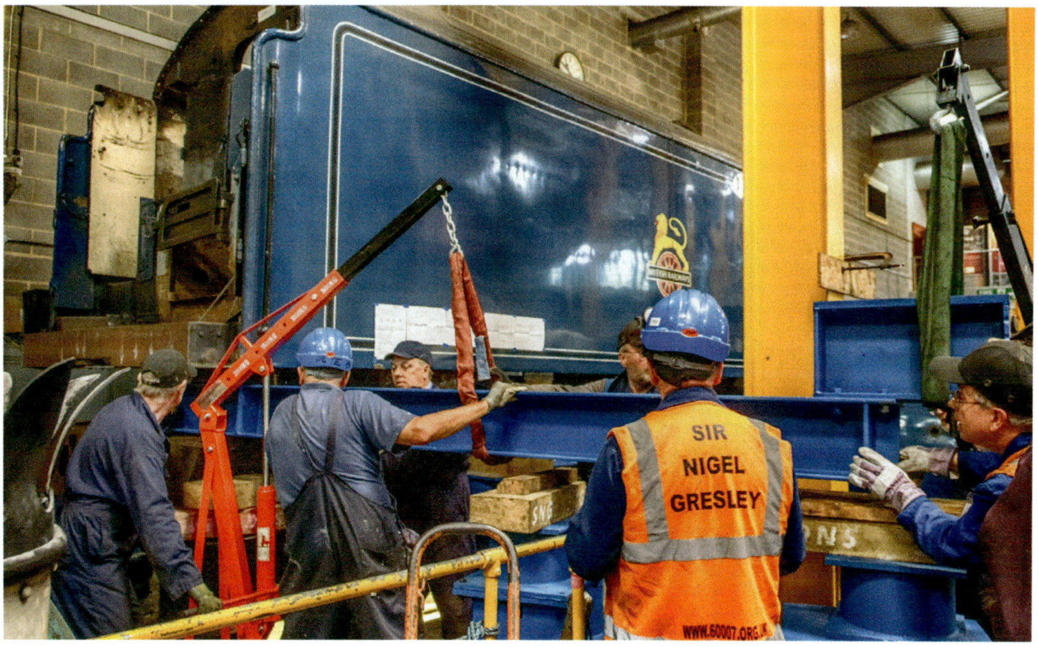

With the tank lifted by jacks a cross beam is placed across the lifting stools which the tank will be lowered back onto

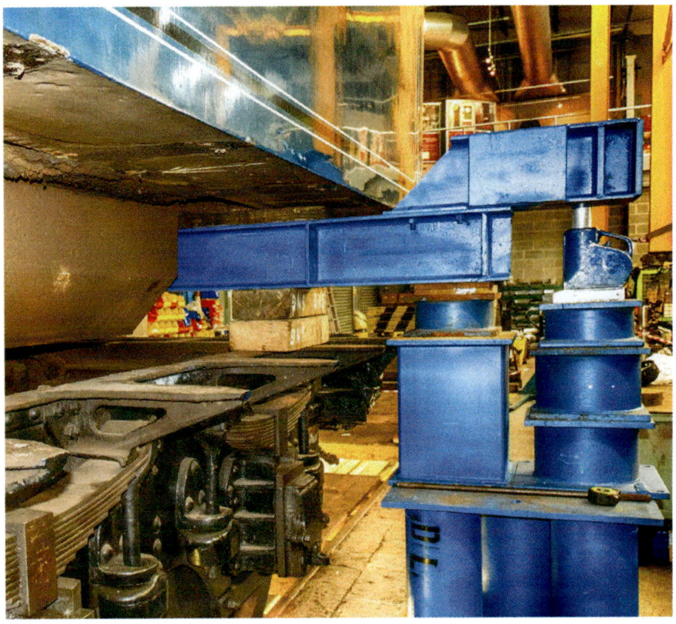

Close up of the lifting stools and cross beam supporting the tender tank

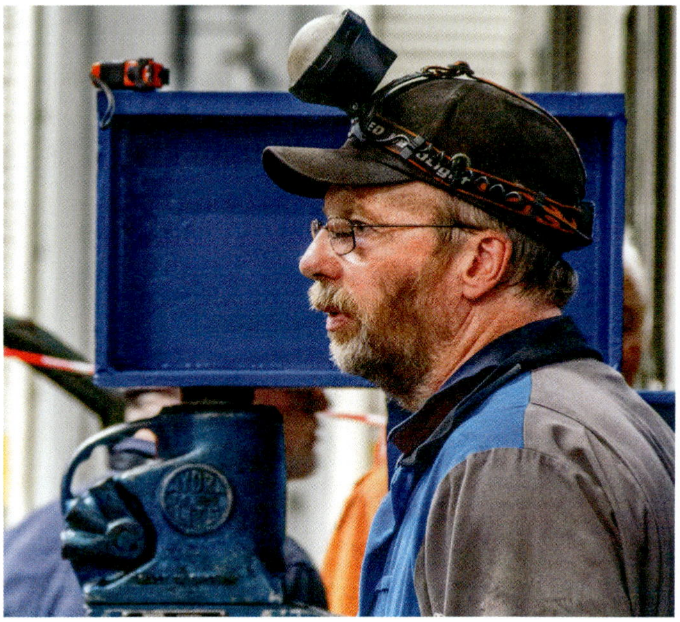

Lift Team Leader Roger Turnbull

sitting on the workshop floor as a lot of the work needed to be done underneath the tank, so a number of various sized stools would be used to sit the tank on and keep it at a workable height. The tank would be jacked up at all four corners until it was clear of the frames and wheels, which would then be hauled clear before the tank was lowered back onto the stools. Due to the weight of the stools and lifting equipment it was necessary to have three team members at each corner and Roger free to oversee the operation, as keeping everything level at all times was incredibly important.

The tentative first lifts commenced until the tank was clear enough to enable lifting beams to be fitted across the top of the stools so the jacks could then lower the tank onto the stools, and the jacks could be removed. This was a complex and timely operation, which I have very much simplified for

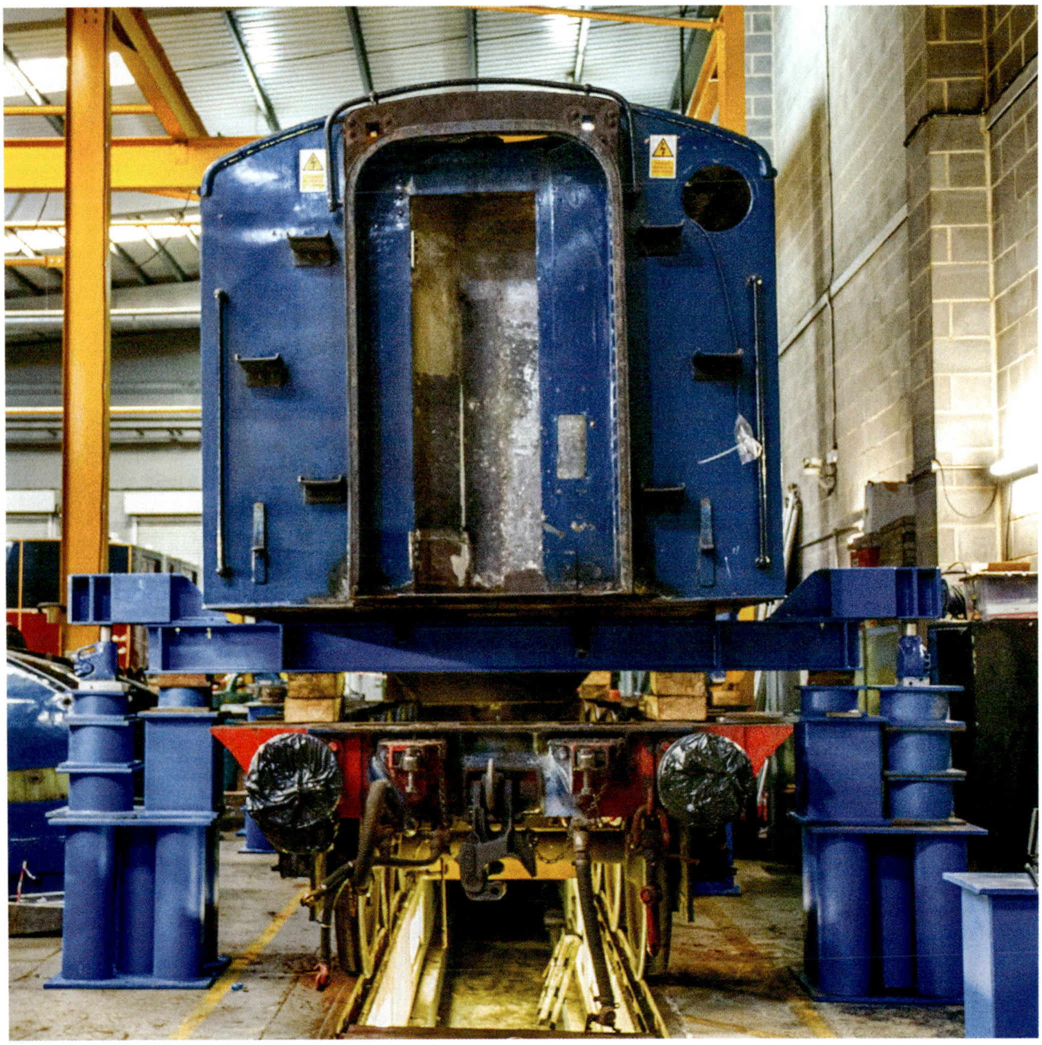

The tank sits on the cross beams clear of the frames and wheels

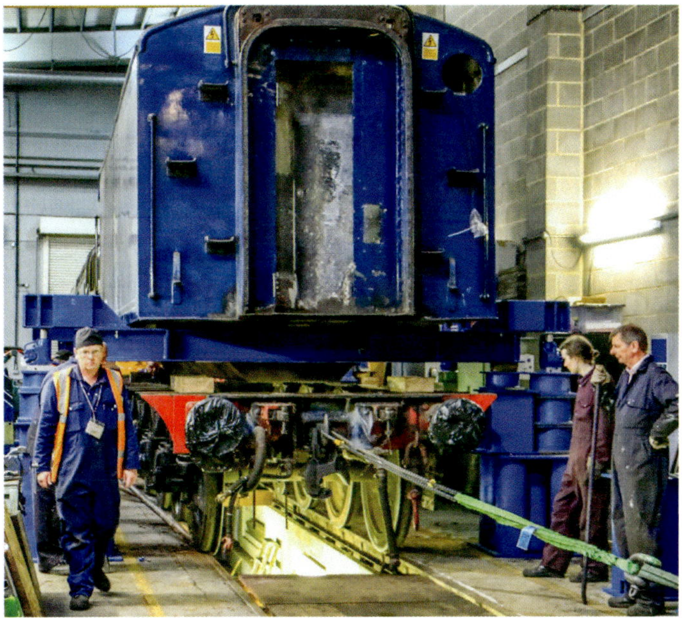

The frames and wheels are hauled clear from underneath the tank

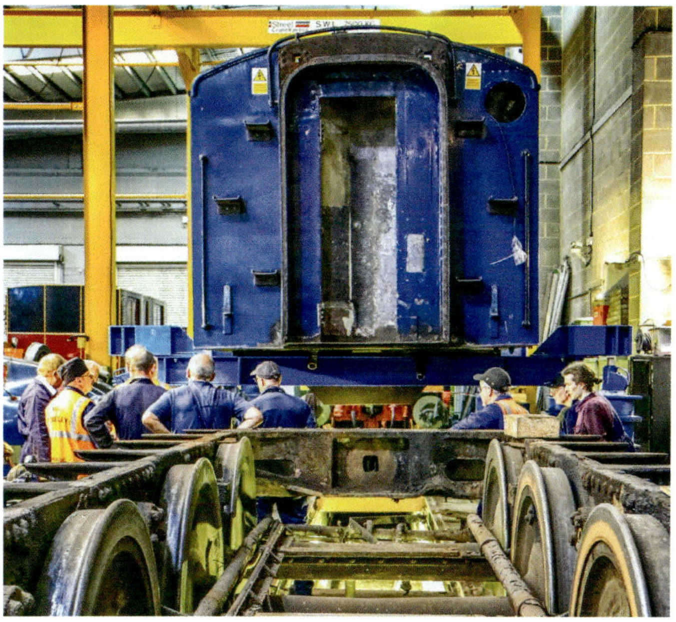

Frames and wheels are clear from underneath the tank

conciseness here, but as usual, completed with professional due diligence from the assembled team of Trust volunteers. The final job to complete the task was when the frames and wheels were winched clear from underneath the tank. The proud team then assembled for a memorable group picture standing around the frames and wheels with the tank in the background.

Now the team's focus really was on the tender. Painting, pipework, platework and numerous other jobs were to be undertaken in the weeks and months ahead. Of course, the tank had been lifted clear and separated from the frames and wheels, but the frames were still sitting on the wheels and they too would need separating, but that was for another day. The team could be rightly proud of today's accomplishment.

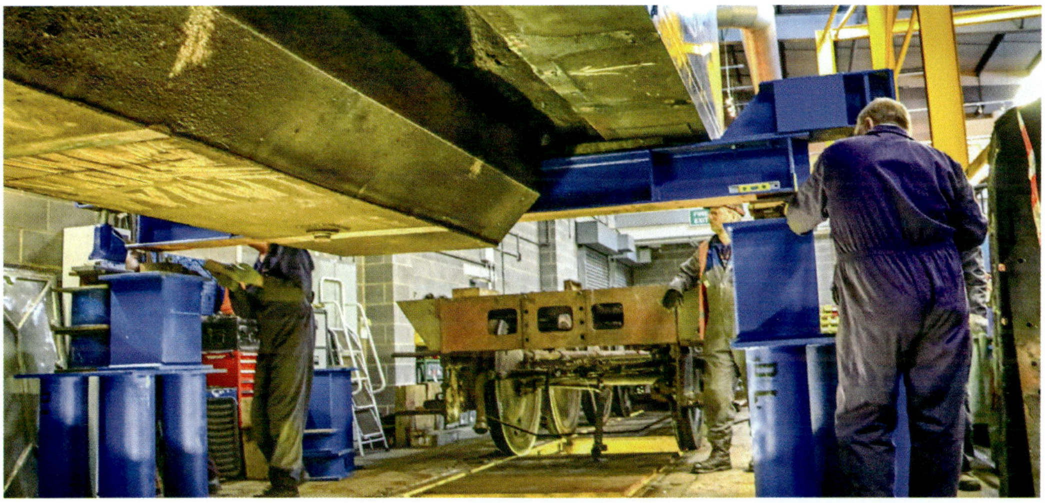

With the tank supported on the lifting stools and cross beams the frames and wheels are clear from underneath

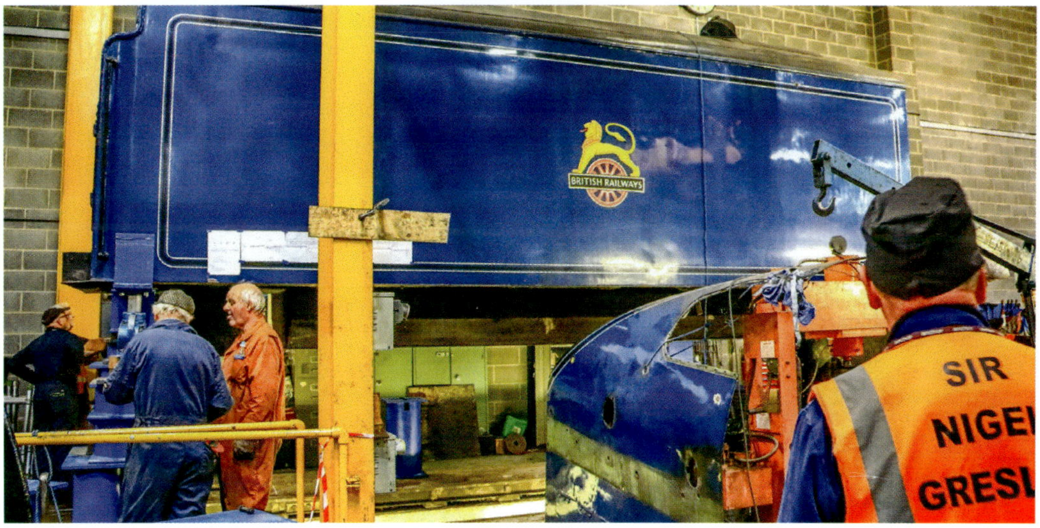

With the frames and wheels removed the tender tank appears to be floating

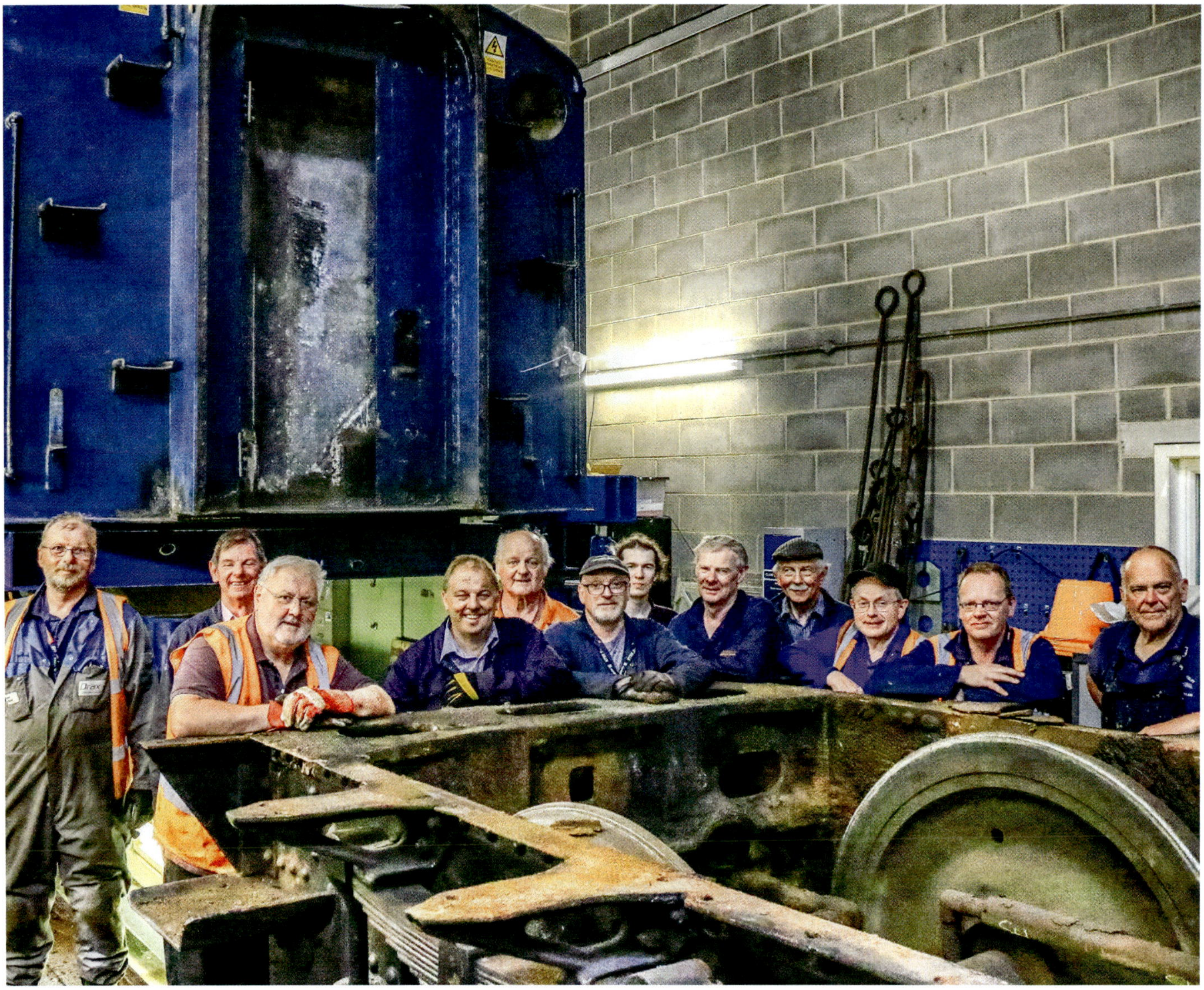

Tender Lift Team pause for a team picture after completing the tank lift and hauling the frames and wheels clear from underneath. Left to right - Roger Turnbull (Lift Team Leader), Malc Bateman, Peter Ellis, Richard Hawkhead, Trevor Wilford, Andy Barwick, Andrew Jacques, Phil Slater, Bob Shearman, Richard Swales (Chief Mechanical Engineer), Darrin Crone (Locomotive Engineer), Peter Brackstone

Richard Hawkhead

I joined the engineering team in 2018, having been an electronics engineer repairing valve TV sets before the "solid state" revolution took over, and I eventually spent the last 30 years in the IT industry. I also have a mechanical engineering background, albeit an agricultural one due to my farming activities at home, so I brought some odd, sometimes non-standard resolutions to issues during the rebuild, as well as becoming the heavy haulage member of the team thanks to my access to large trailers. Within the NRM workshop I worked mainly on the tender restoration, which involved separating the tank from its frame and then replacing large steel pieces in the rear section and the tender front. I never realised a corridor connection frame was so complex! I am a fireman at NYMR, and a support crew member for the Thompson B1 61264, which has allowed me to gain mainline experience, and by the time this book gets published we should be out on the iron roads of the UK, bringing this iconic loco to the enthusiasts and supporters of historic British engineering.

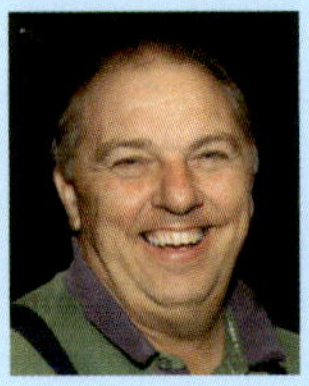

Phil Slater

A mate of mine once described me as a frustrated mechanical fitter and I have to agree with him. Unfortunately my working career in the power stations took me away from my love of using my hands working on mechanical bits and pieces. When I retired and got the opportunity to get my hands dirty on Sir Nigel Gresley, it was a no brainer.

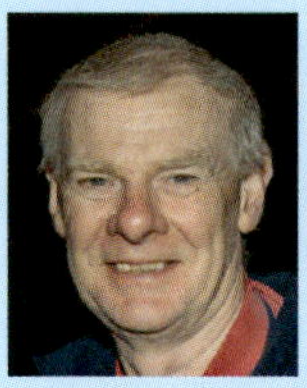

I have to say most of what I've done at York during the overhaul has been satisfying in many ways. The pride of being involved in such an iconic steam locomotive will be something I'll remember and the banter with the rest of the volunteers has been great. Highlights for me have been getting the tender tank off and the boiler on, but without a doubt the biggest highlight will be seeing Sir Nigel move under her own power. I've no doubt it will be a very emotional time.

Trevor Wilford

I served my time at Cargo Fleet Iron Works, as an apprentice boilersmith in the loco sheds, working on steam locomotives. This involved renewing boiler tubes, boiler stays, riveting and general repairs on the locos. I did plating burning on all the plants in a steelworks, followed by a spell at ICI, then work in shipyards and fabrication yards in the UK, Germany and Holland. After retiring, I took a trip across America, Hawaii, Fiji, New Zealand, and Australia. I teamed up with Darrin in 2016 on the SNG overhaul, assisting with all the general repairs.

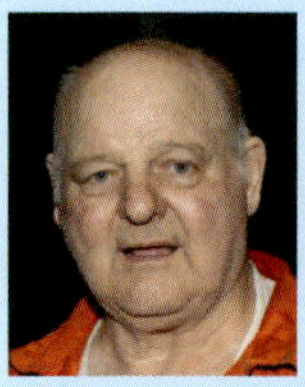

TENDER LOVING CARE

July 2018

"Who has eaten all my Jammy Dodgers?"

A week after the tender lift had been completed, the axle boxes had been removed from the wheels and the frames had been lifted clear of the wheelsets. The springs and hangers had been removed and the cleaning of many years of accumulated grime was taking place before needle gunning commenced to get the frames back to base metal.

After a lot of handheld shots on my previous visits to capture the 'as it happens' action images, I was back to the more relaxed, composed shots with my camera on a tripod. This allows longer exposures at small apertures and low iso to compensate for the low lighting levels in the workshop and to give a higher quality image. As I have mentioned earlier these longer exposures will result in any movement being blurred, and people movement giving the effect of high-speed action which can be quite amusing. Bob Shearman, not for the first time, was caught on one of these shots.

There were a lot of different tasks underway, so I took the opportunity, after the team had left for lunch, to capture a general shot from a corner of the workshops which shows several of them. My image shows, below the cab, the superheater header and brake cylinder components. In the centre are sections of the streamlining and on the right are

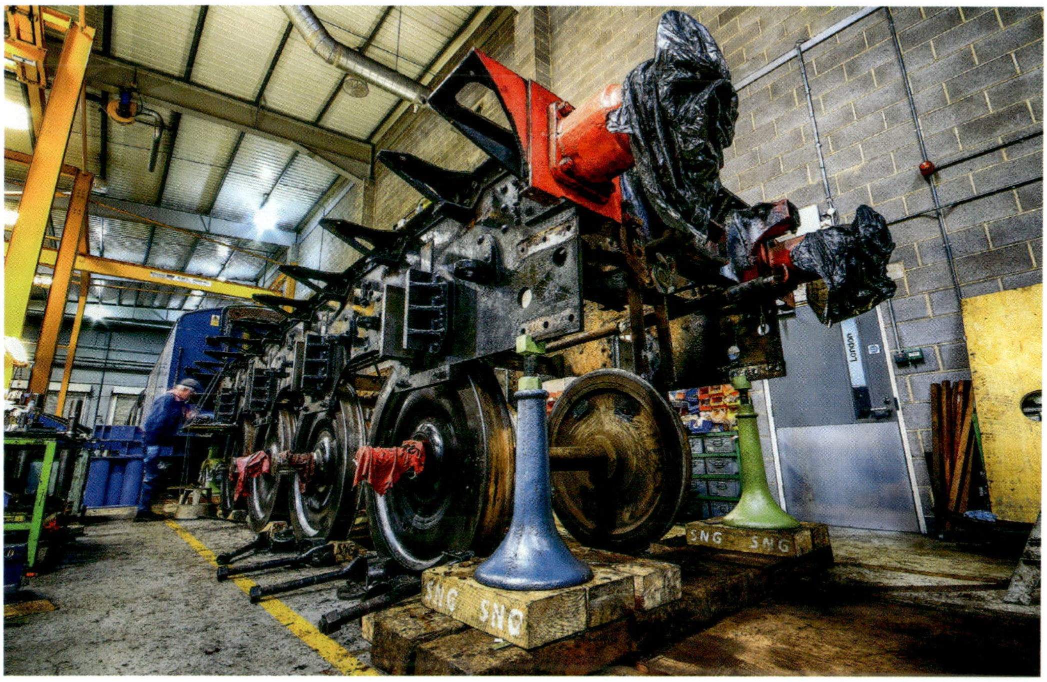

LEFT & BELOW - The tender frames raised above the wheelsets. NOTE - A blurred Bob Shearman, a result of a long camera exposure

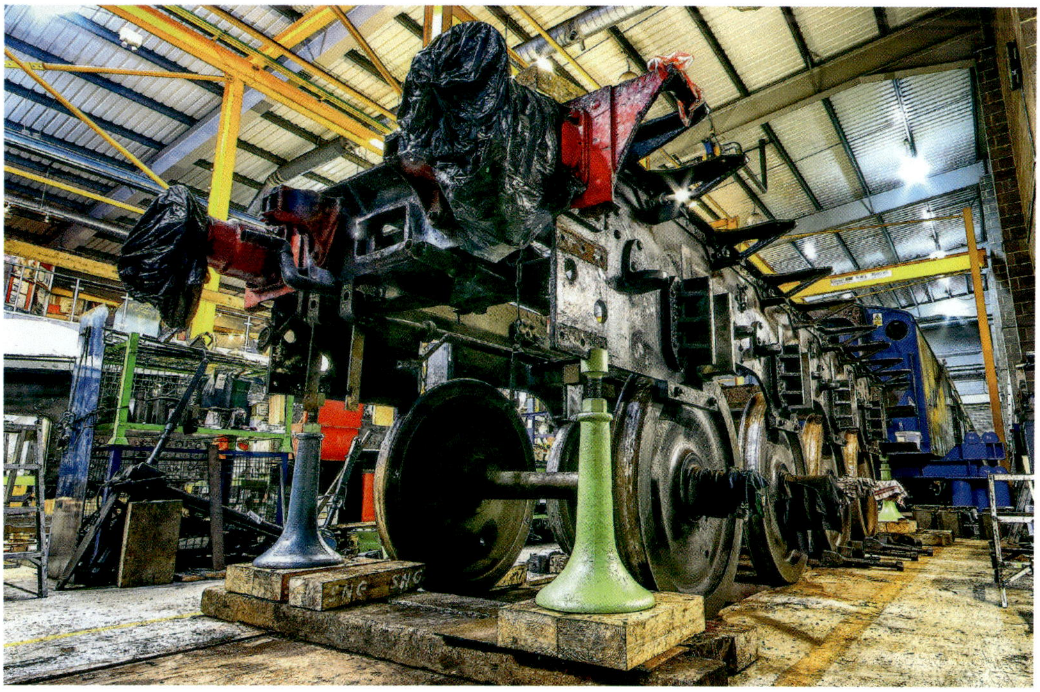

the tender frames sitting on raised jacks clear of the wheels. Looking carefully through the round window on the right of the tender tank a section of daylight can be seen where part of the corroded corridor roof had been removed.

I had a routine on my visits. Unless it was a 'Milepost' event I could generally capture my update images in a couple of hours, so I always arrived around the 11am tea break and left after the lunch break at 1pm. Contrary to what you may be thinking there was more to this than just having a passion for enjoying a cup of tea, as the banter in the blue cabin made these timings well worth the effort. On this occasion Darrin was the target of a joke. Darrin had a love for jammy dodger biscuits and was quite keen to ensure he got his fair share from the packets, but Peter had carefully kept an empty packet in good condition so when Darrin requested they were passed over to him he was quite gobsmacked and momentarily speechless to discover the packet empty. After a glance around the cabin at all the innocent faces Darrin eventually broke the silence with "Who has eaten all my jammy dodgers?". More innocent faces for a second or so until Peter produced another packet that only had a couple missing and a relieved look washed over Darrin's face to the amusement of the rest of the team. Of course, capturing the moment on camera would have been the cherry on the cake, or jammy dodger, but somehow I think sitting with my camera at the ready would have given the game away.

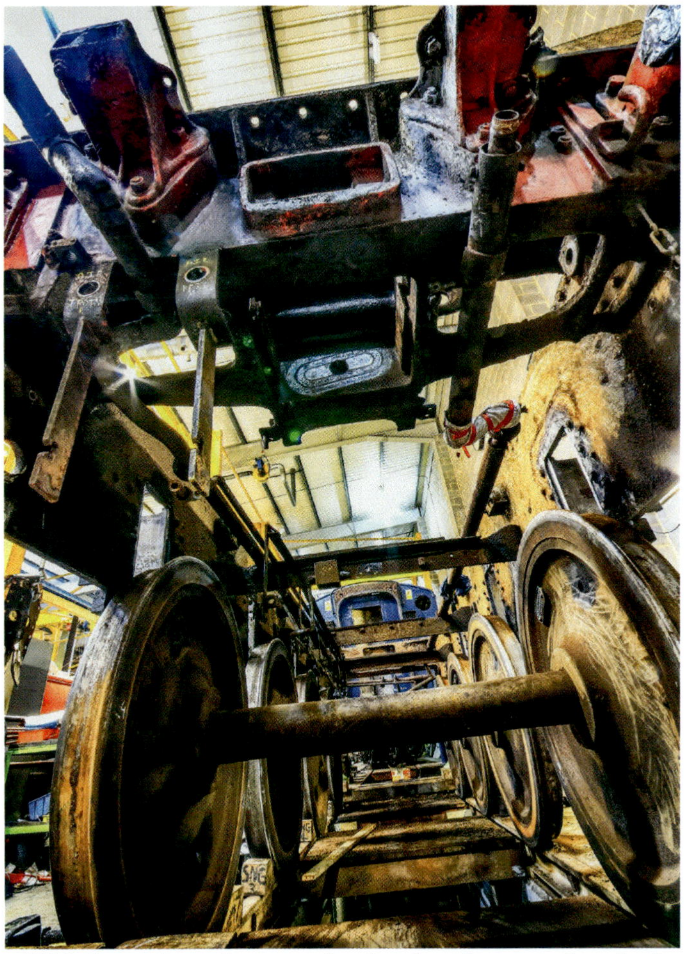

With the tender tank removed a view from between the raised frames

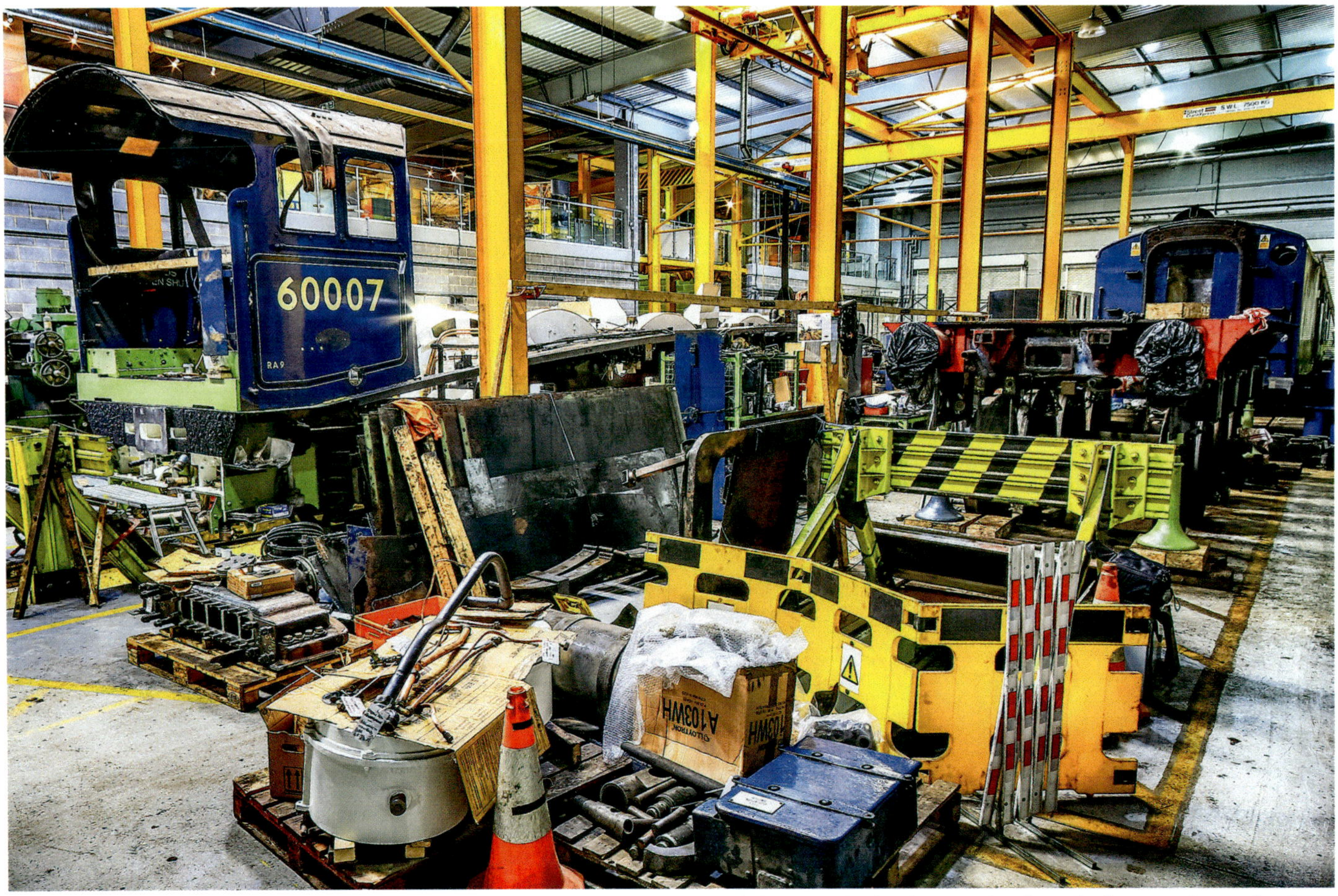

The work at this stage was taking up much of the available space in the workshop. In the background is the cab, frames, drive wheelsets (out of view) tender frames, wheels and tender tank. In the foreground left to right superheater header, streamlining panels and a host of various brake components amongst many others too numerous to mention

WET AND DRY

August 2018

The first time I joined the team back in January 2016 I asked what this large block of metal, which had already been removed from No.7 was, and was told "Oh that's the superheater header". Well, I had come across the term before so I did realise it was a little more than something you got for those very cold winter nights, but I didn't ask for a more detailed explanation for more than two years as nobody was paying much attention to it. Now that the focus had finally switched on to it, I started asking questions and began to realise just what an important piece of the boiler the superheater was. Bear with me here as I give you my photographer's technical description. On a steam locomotive the steam that pushes the pistons which drive the wheels is created by the hot gases generated from the fireman's fire passing through boiler tubes which heat up the water surrounding them until it boils into steam. I hope you are still following me as it is a little more complicated than that. This is where the superheater comes in. The steam in the boiler is now classed as "saturated steam", i.e. it still has water vapour present, so to turn it into more efficient, drier steam it is sent to the superheater which then passes it back into the boiler tubes via smaller tubes (called the superheater elements) inside the bigger ones which are known as the large flues, where it is heated to an even higher temperature

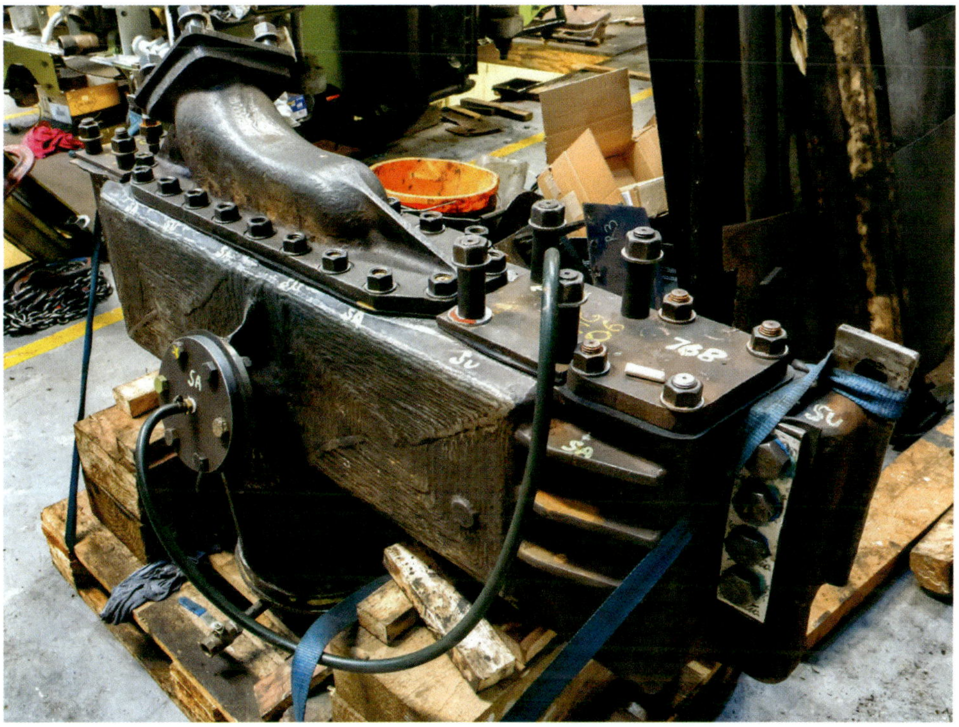

The superheater header viewed from the smokebox side sits on a pallet ready for refurbishment

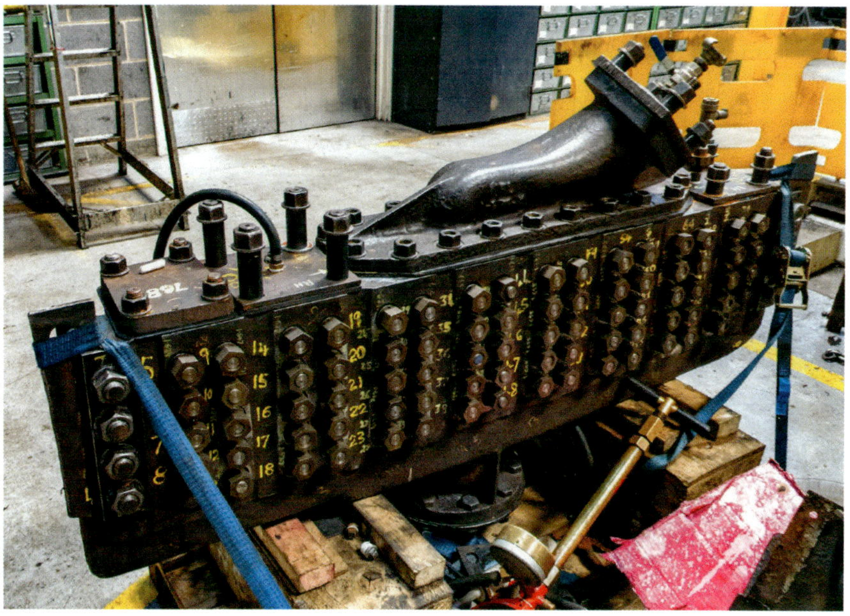

Blanking plugs replace the tubes during refurbishment work on the superheater header

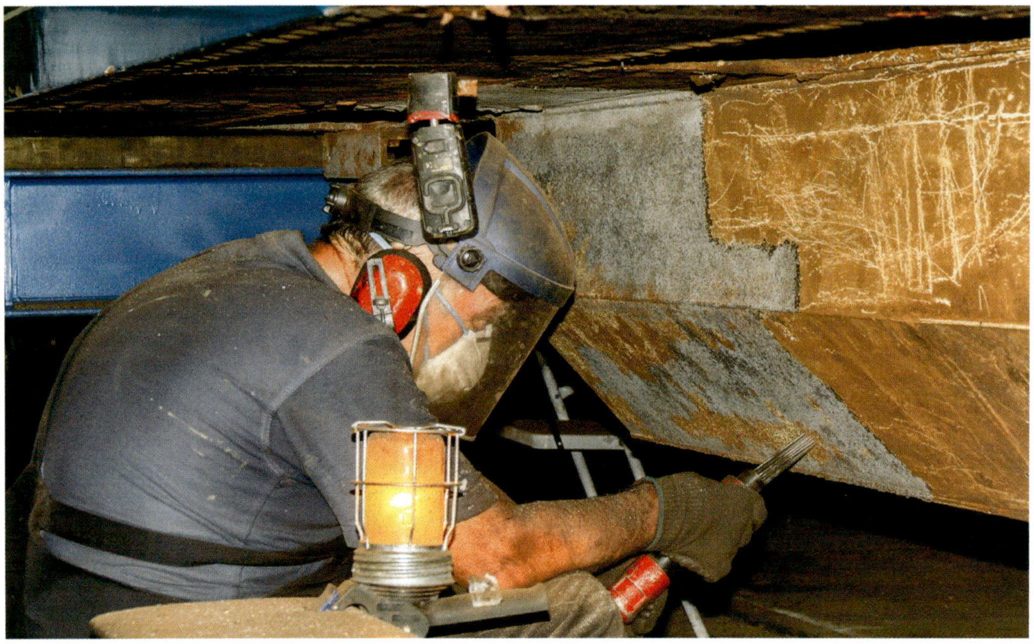

Peter Brackstone works with the needle gun on the underside of the tender tank

turning it into highly efficient dry steam. The dry, hotter steam returns to another section of the superheater to be distributed to the cylinders and thus we have motion. Darrin had delegated the refurbishment of the superheater into the capable hands of former oilman, Andy Barwick, who had patiently answered my questions to give me a better understanding. Thank you to Andy for that but let me just add that if I have got it wrong it is nothing to do with Andy's explanation but more my lack of understanding.

Back with the tender, Peter Brackstone is seen here working on the underside of the tender tank with the needle gun, chiselling off old paintwork, and another picture shows the frames successfully needle gunned down to bare metal.

Peter Ellis was hard at work back on No.7's frames and cab pipework and the results of his efforts were becoming very evident as they stood out against the freshly painted frames. Whilst Peter was working with the angle grinder on old pipe brackets below the cab, I got the chance to set up my tripod and capture a long 15 second iso 100 f16 exposure looking through the cab and down between the frames, which captured Peter's endeavours with a few spark trails looping through the image. Unlike a video, a photograph can capture a single moment in time, but I thought the light trails here offered a compromise of adding a little live action to a still shot.

My last picture in this sequence was taken from the back of the cab and shows, with the cab floor removed, some of the refurbished pipework that runs underneath the floor. Since being a boy, I had always wanted to stand in the cab of No.7 and now I was doing just that but maybe not in the way I had first envisaged.

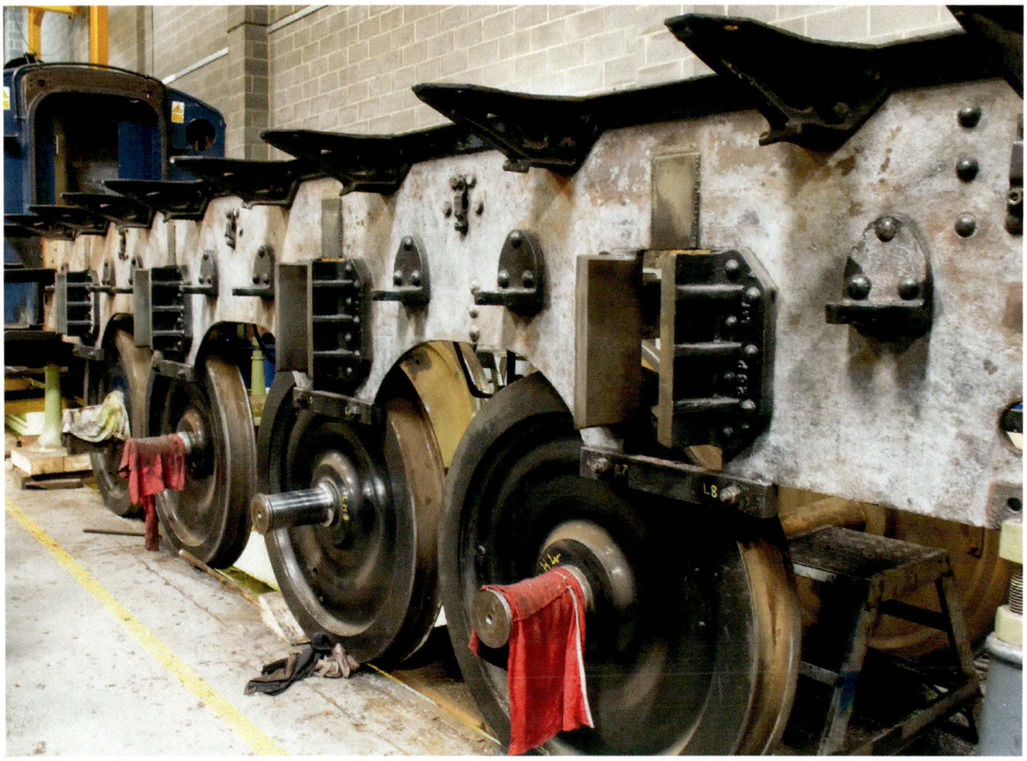

The tender frames after old paint has been removed to reveal the base metal

Peter Ellis uses the angle grinder to remove old pipe clips

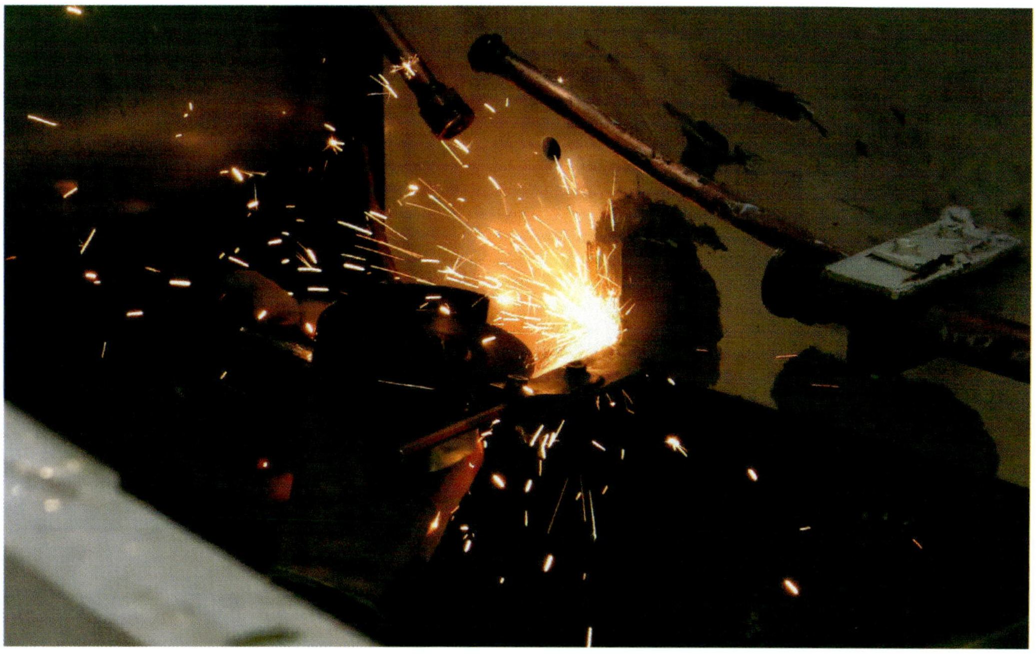

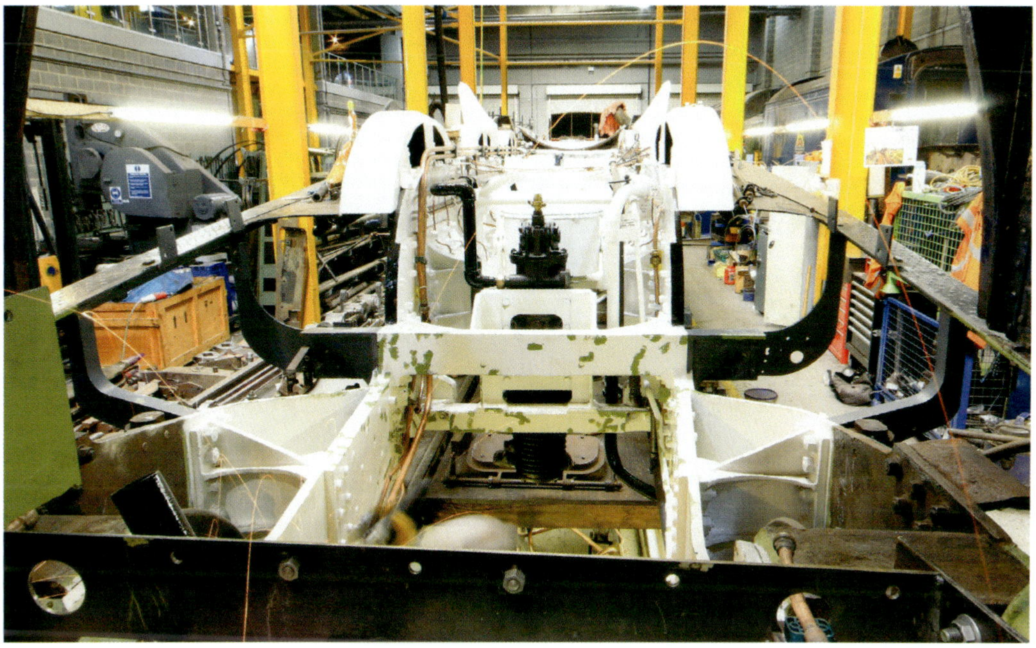

Sparks loop through the air as Peter Ellis uses the grinder below the cab floor

The removed cab floor allows access to refurbish pipework

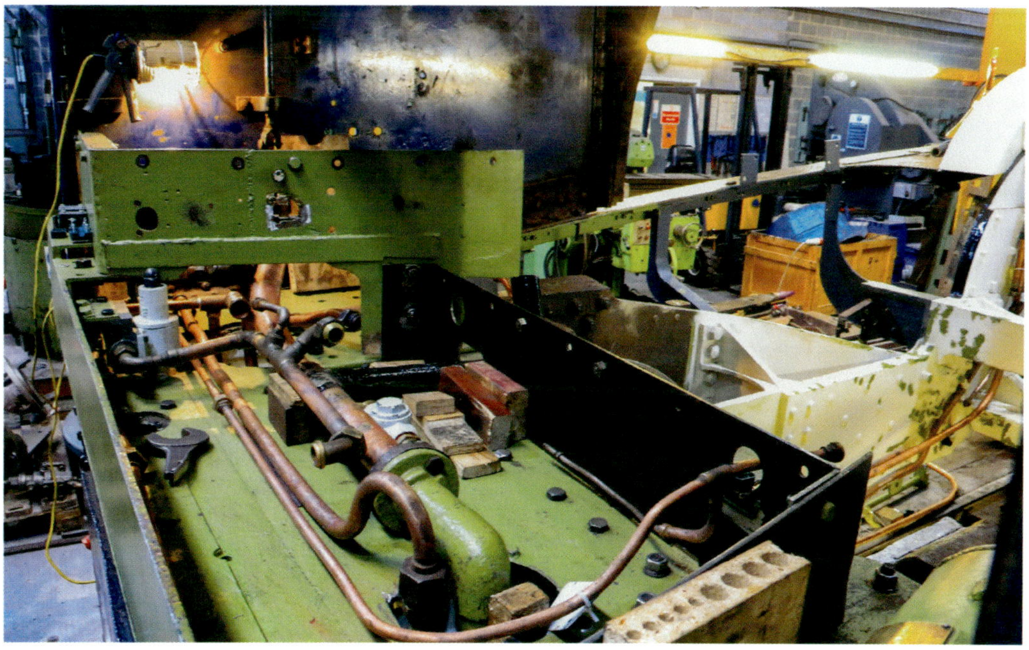

IN THE FRAME

October 2018

No.7 has always been a bit special amongst the famous steam locomotives, being the 100th Gresley pacific built at Doncaster works, and to honour the fact, No.7 was named after the great man in a ceremony at Marylebone station on the 27th November 1937. So, it was no surprise that, with the NRM workshop having a substantial viewing gallery, No.7 was becoming the subject of many gallery photographers' camera clicks. I noticed these on several occasions. I couldn't help thinking how lucky I was to be so up close and personal with my camera down on the workshop floor.

My first shot here is taken from the pit between the frames looking back through the cab towards the gallery. Nobody was pointing a camera back at me today, so I guess all the gallery visitors were camera shy.

The shots taken between the frames all show the considerable amount of work that had gone into refurbishing the pipework and, particularly, the progress made on the pipes below the cab floor when compared with the image on page 84. I was grateful to Peter Ellis, who took time out to

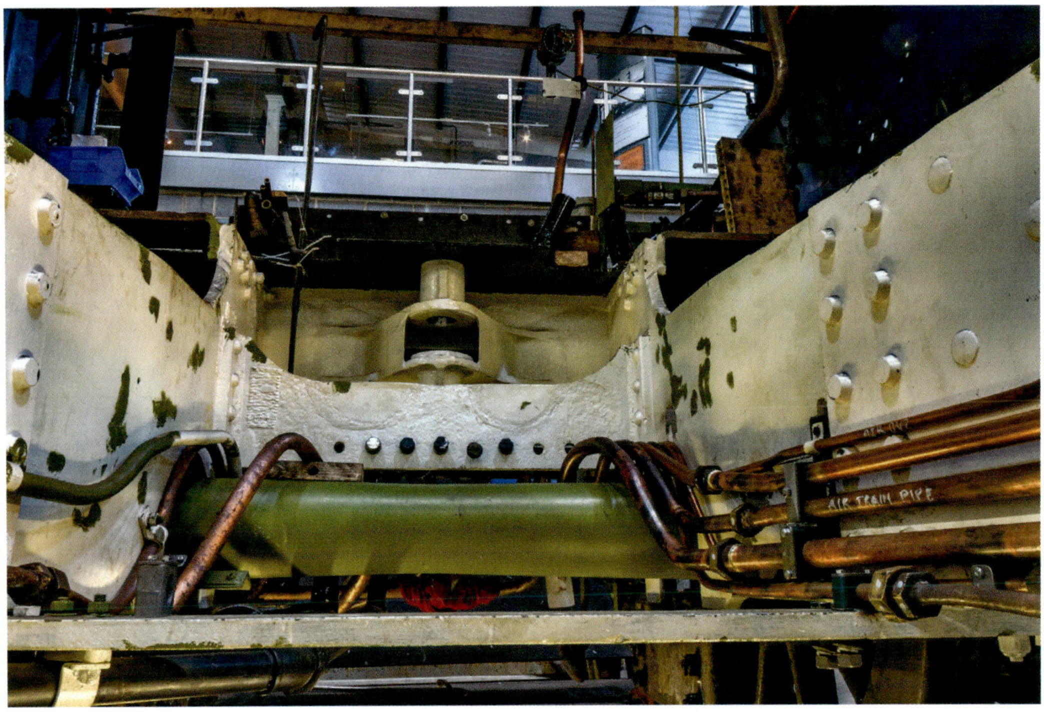

The view from the pit below the frames looking back up through the cab towards the gallery where there would often be plenty of cameras clicking, but not today it seems

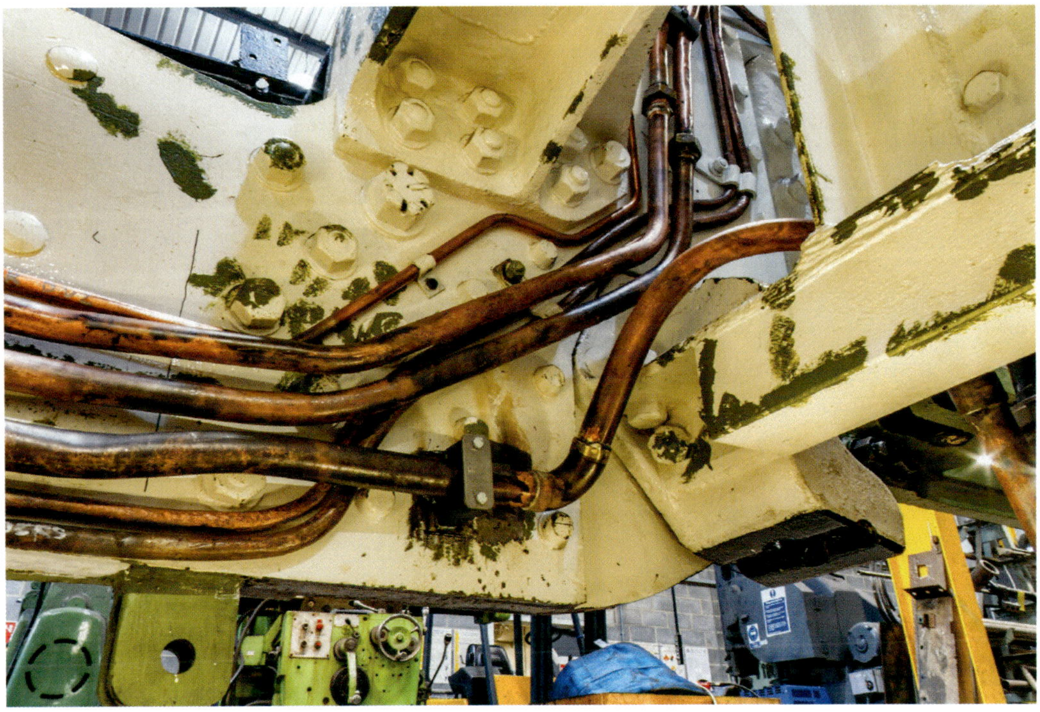

Refurbished pipework underneath the cab floor. The fitting of the pipes has left a bit of retouching to be required on the recently repainted frames

Peter Brackstone using a tender axle as a seat gets to work with a wire brush attachment refurbishing the inner side of a tender wheel

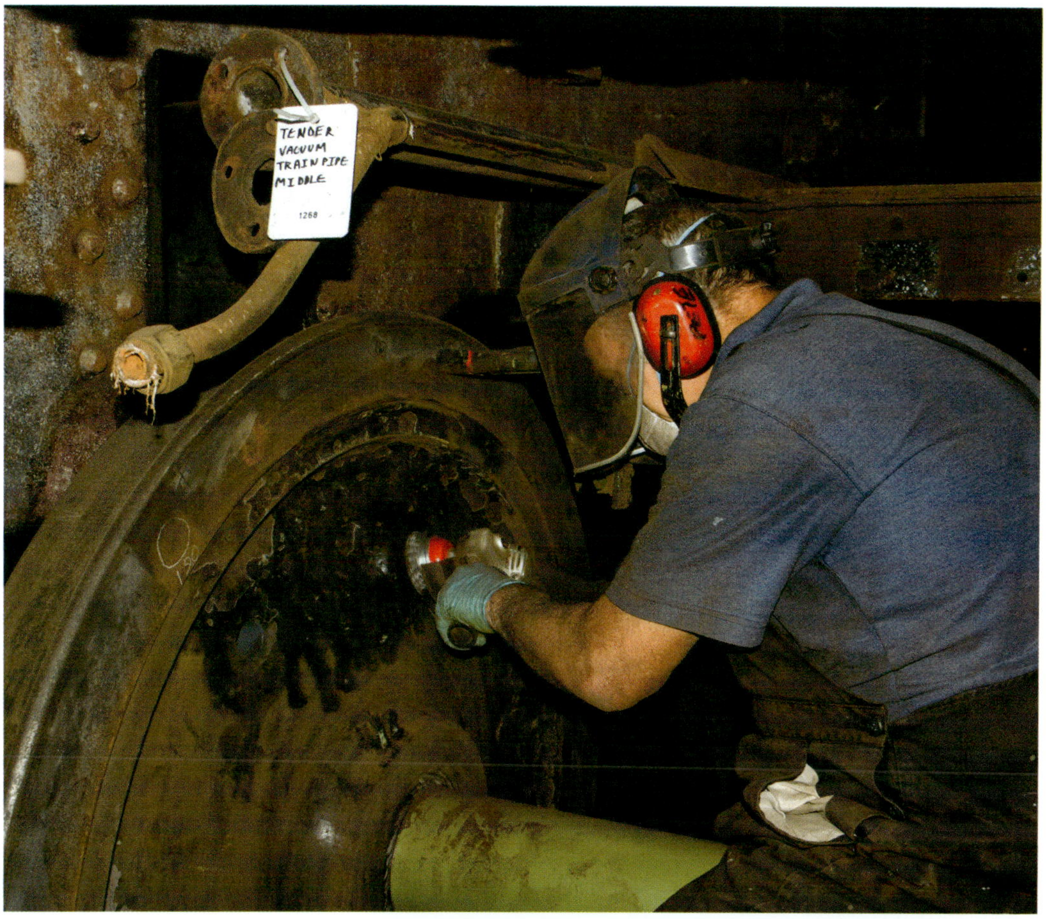

show me in more detail some of the work he had done with the pipes, several of which, thanks to Peter, had improved the pipe runs.

Away from the frames work was progressing on the tender's tank, frames and wheels. Malc Bateman is using the needle gun to remove old paint from the buffer beam and Peter Brackstone uses an axle as a seat whilst working on the inside face of a tender wheel. Another image shows the freshly painted underside of the tender tank. Two more images show the tender wheels after undercoat has been applied to their freshly refurbished surfaces.

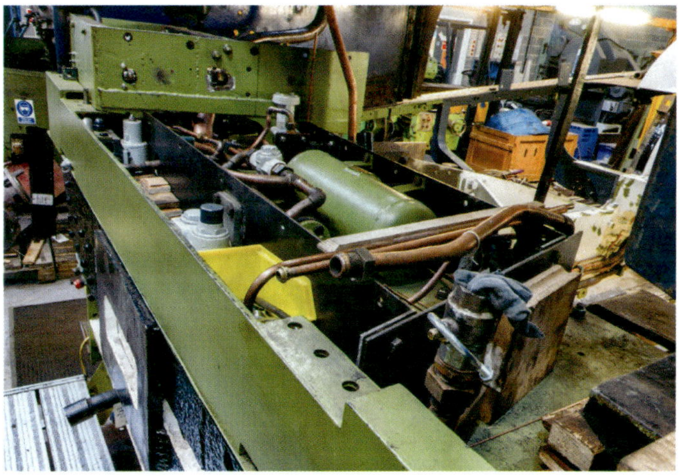

With the cab floor removed more components and pipework are reinstated after being refurbished

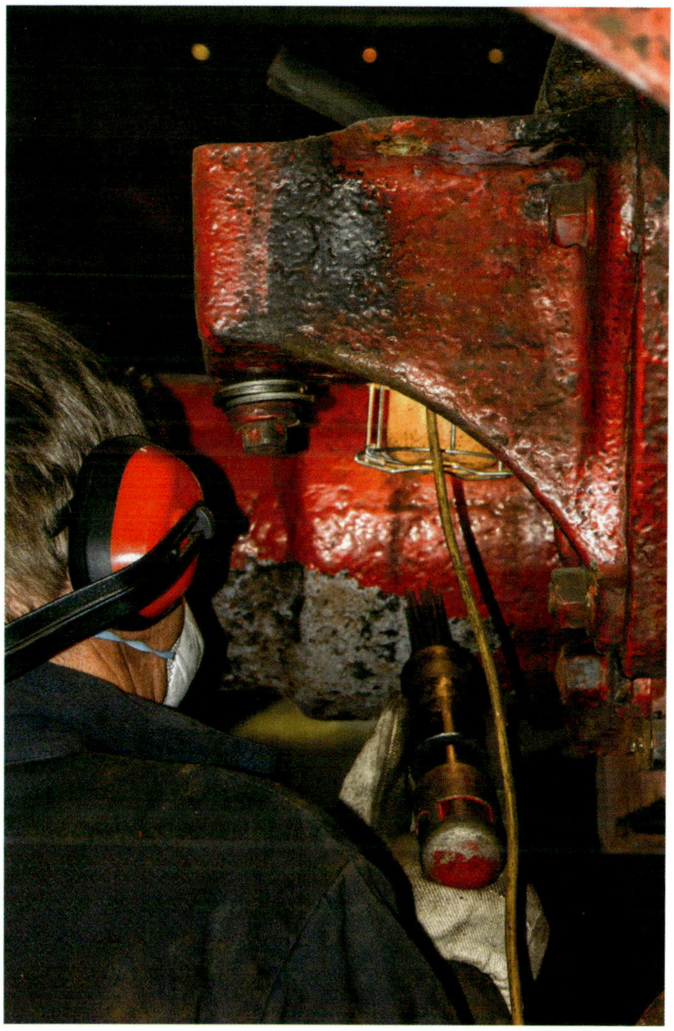

Malc Bateman uses the needle gun on the tender buffers

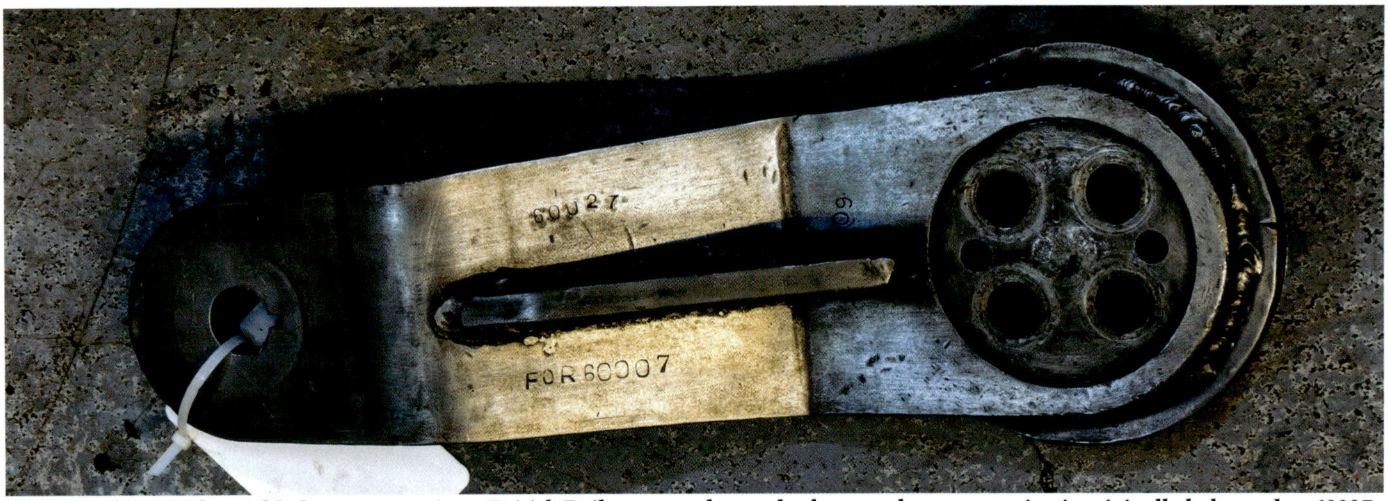

The regulator with a welded repair covering a British Railways number and other numbers suggesting it originally belonged to 60027 Merlin but was transferred to No.7

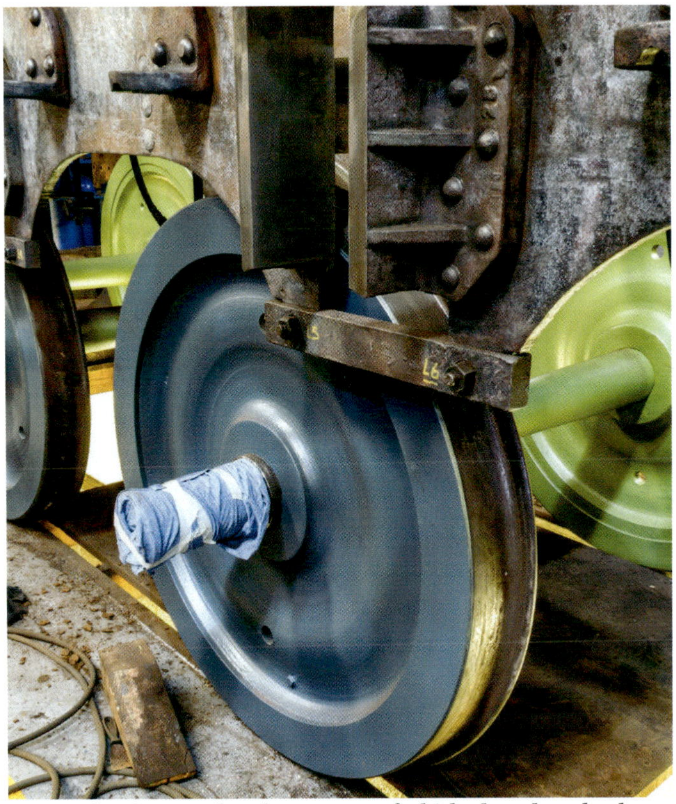

Fresh primer and undercoat on refurbished tender wheels

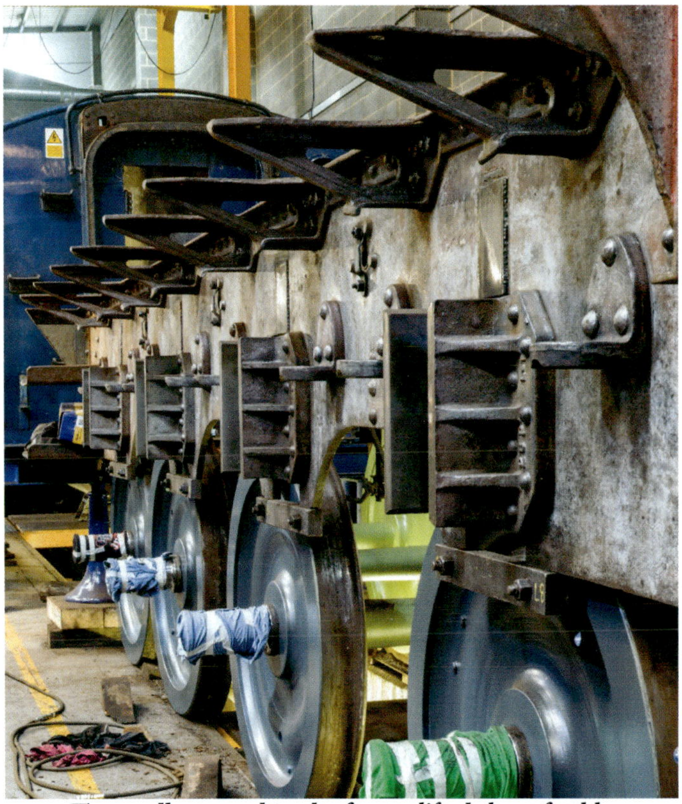

The needle gunned tender frames lifted above freshly undercoated wheels

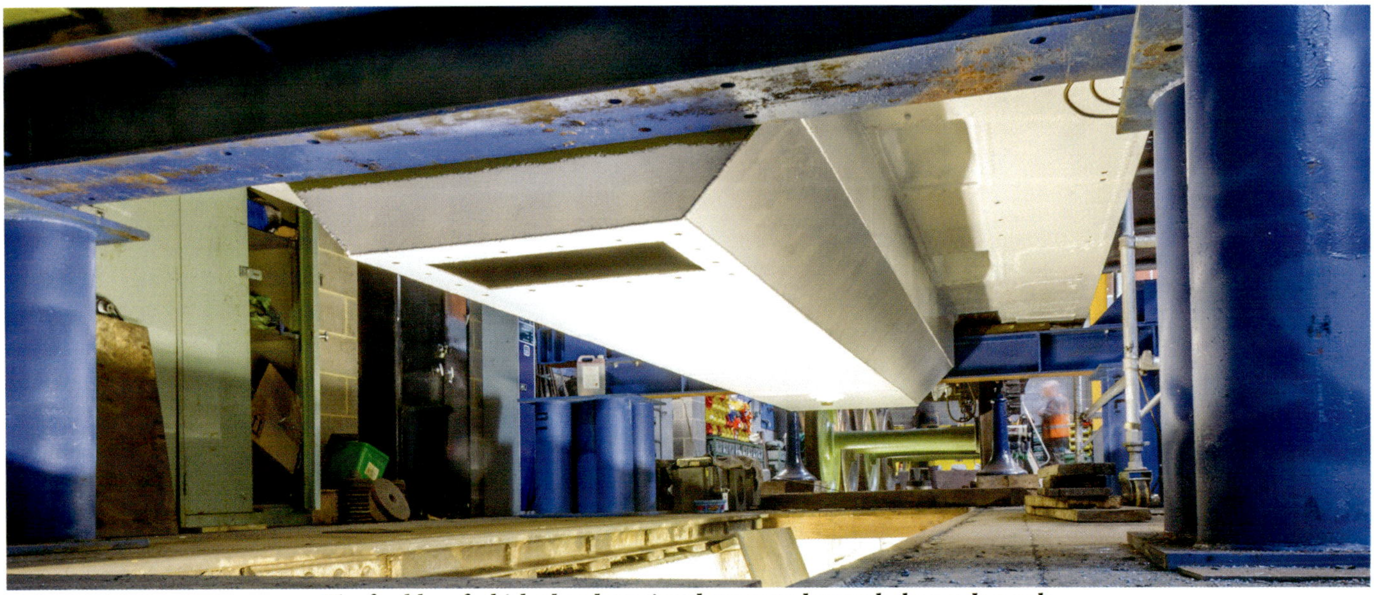

The freshly refurbished and repainted sump underneath the tender tank

The overhaul was a bit like an archaeological dig at times, as on a regular basis it was revealing stampings and other marks from the past. One of these was on the speed regulator, where there already appeared to have been a repair which was hiding the British Railways number. The new numbers seem to indicate that the regulator previously belonged to 60027 Merlin but was transferred to No.7 or 60007 at some point. None of these numbers belong to LNER so it looks as though the regulator was not the original.

Travelling from his London home to volunteer, Tom Bell was photographed cleaning a middle spindle extension rod side rod. Another, in good Blue Peter terms, which he had prepared earlier is below the right-hand cylinder alongside the accommodation bogie, this one a connecting rod.

Tom Bell

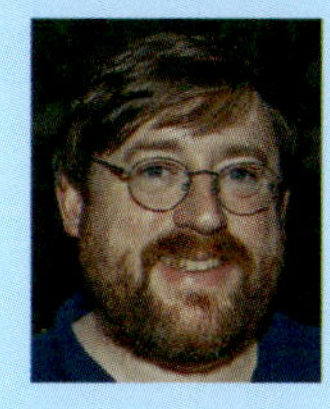

Being a musician by trade, a railway workshop is fairly unusual territory for me! I play the pipe organ, and my concert work and teaching can take me all over the world. Now and again I combine business with pleasure, and find a railway museum somewhere. Amongst other things, I have enjoyed some of the 'archeological' elements of Sir Nigel's latest overhaul, which came as a consequence of my championship needle-gunning. Who knew that some of the brake gear came from the W1?! Alas, work and family have limited the amount of time I have been able to devote to Sir Nigel, but nevertheless I have been thrilled to make the contribution I have. I've really appreciated the generosity of several vastly more experienced engineering team colleagues in sharing their wisdom and knowledge when I was tackling something new. I have made new friends, developed new skills, and learned new words. Most of the latter were names Bracko gave me after an incident involving the microwave and a fish curry... Sorry, Pete!

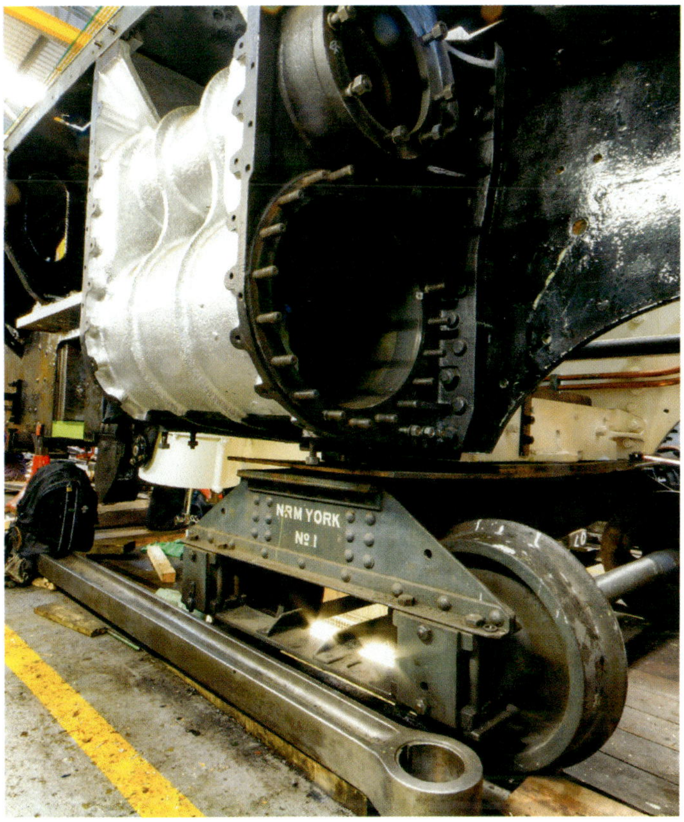

Cleaned connecting rod placed alongside No.7 next to the accommodation bogie

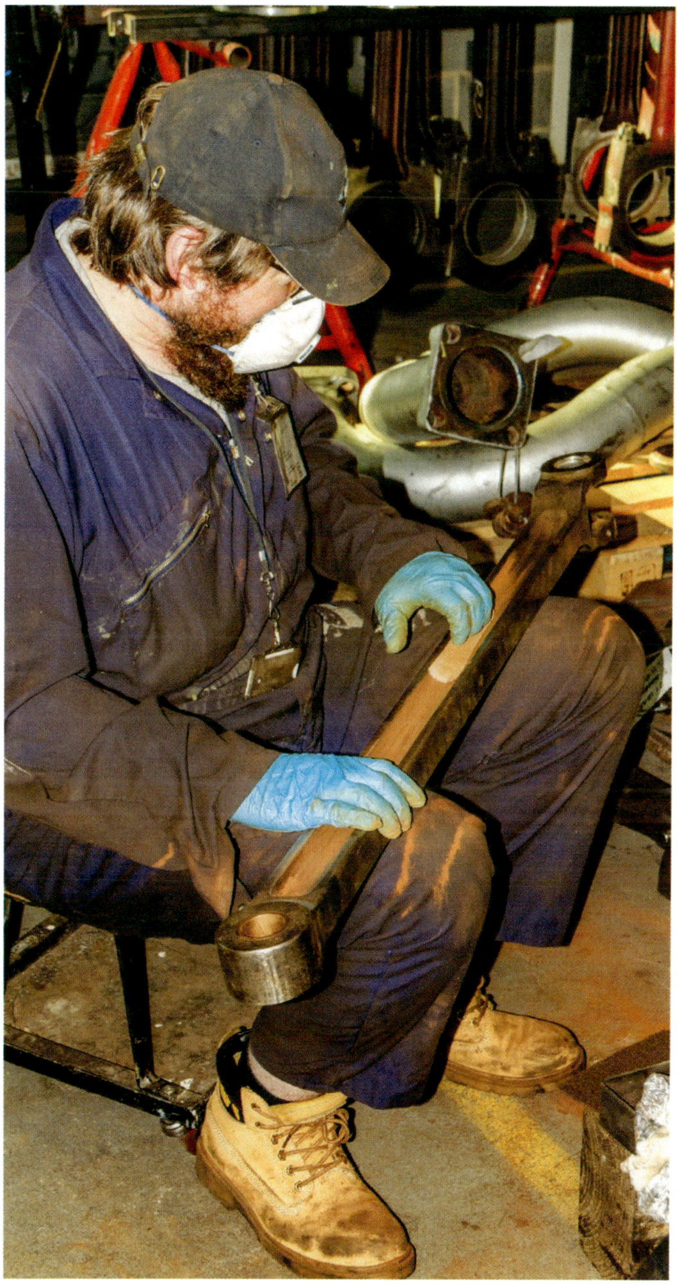

Tom Bell cleaning a valve gear middle spindle extension rod

THE HORNS

November 2018

I had come across the term of 'horns' (when the wheels had been removed from the frames) back in May 2016, so I knew they were just like their name suggests - a pair of horns on either side of an axle box between the frame and the axle box. I guess they could equally have been called ears. Today, CME Richard Swales, assisted by Richard Hawkhead, had measuring equipment out to adjust the gap between the horns in preparation for the refitting of the driving wheels. This adjustment was needed to ensure the axle boxes fitted properly in the frames as either too tight or too slack would cause damage to the axle boxes. The horn on one side of the axle box recess is split into two pieces by a slight diagonal, therefore moving one against the other will increase or decrease the overall thickness thus closing or opening the gap that the axle box sits in. When the wheels were first dropped from the frames and before the horns were removed, a scribe mark was placed across the two sections of each horn. By lining up these marks when the horns were refitted it enabled the gap to be approximately set before the axle box was refitted. More accurate measurements taken across the actual axle box and then applied to the horns would fine tune the required gap.

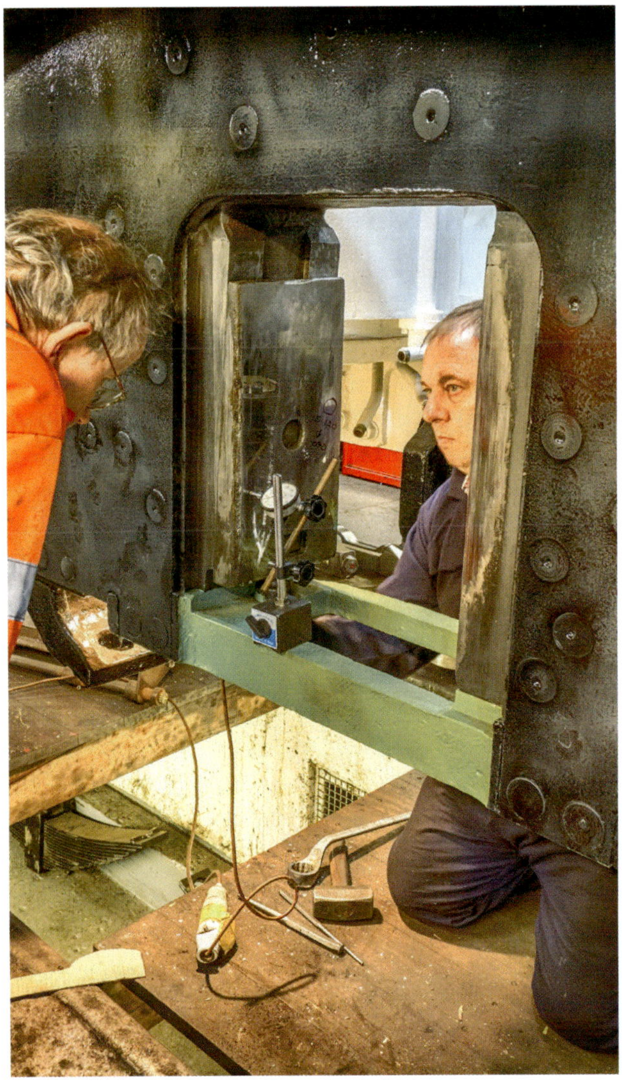

Richard Swales (left) and Richard Hawkhead setting up measuring equipment on the leading wheelset(RS) horns

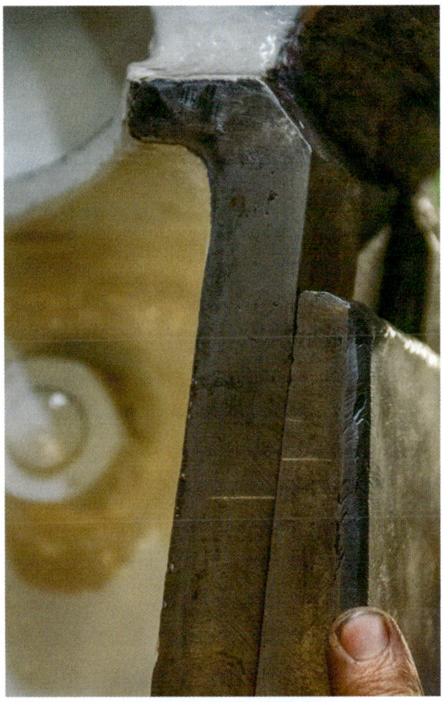

The horns slide against each other to line up with the scribed mark. This one being the left hand side

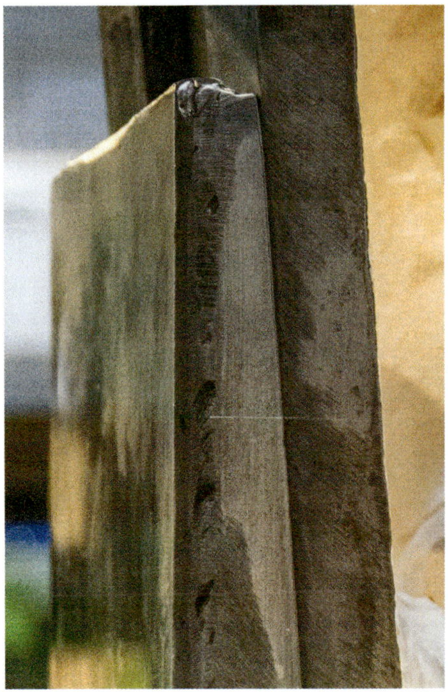

The scribe mark is aligned. This one on the right hand side of the horns

WHEELS UP

10

December 2018

It was nearly time for the Christmas break and one big push, literally, was taking place for a well-earned Christmas present of reuniting wheels with frames. The leading wheelset and the bogie were to be reunited with the frames and that would involve pushing the frames into position over the wheeldrop. Pushing is aided with the help of pinch bars under the wheels and the team were now well practiced in the procedures having done it before when the wheels were removed in May 2016. However, moving one of the huge driving wheelsets wasn't to be done without a degree of anxiety amongst the team which was to be expected for such heavy components. Word had spread around the museum's visitors and a good audience had gathered on the gallery to watch the show, which was not going to be a Christmas pantomime as the team, yet again, put on another first class and faultless performance.

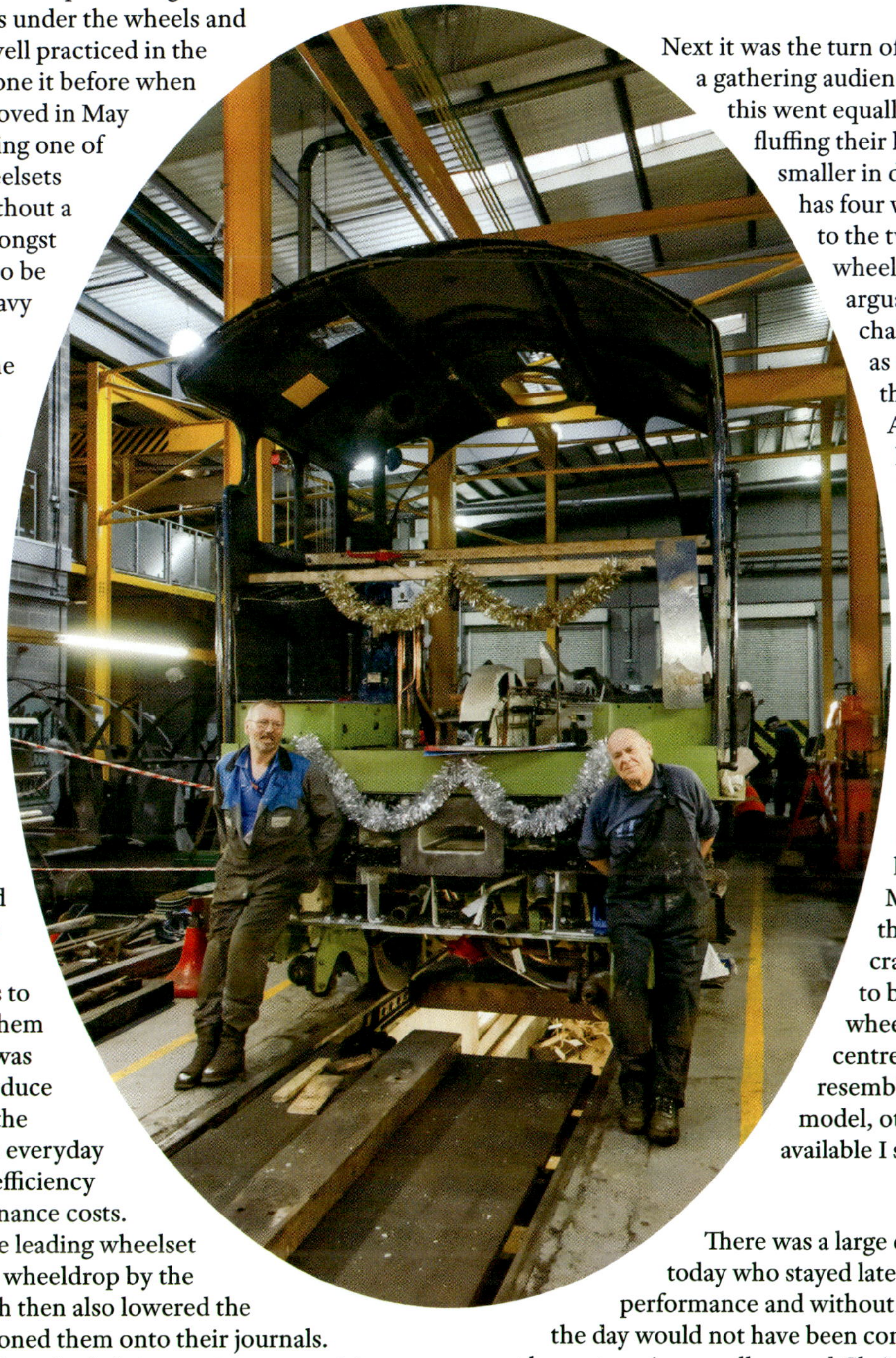

Engineers blue was applied to axle box surfaces which would reveal any high spots when test fitted and allow Richard Swales to set about removing them very diligently. This was important work to reduce the resistance when the wheels are turning in everyday use, thus increasing efficiency and reducing maintenance costs. When completed, the leading wheelset was lowered into the wheeldrop by the overhead crane which then also lowered the axle boxes and positioned them onto their journals. The image of Richard Swales helping position one of the axle boxes on the journals emphasises the huge size of these axle boxes. With the axle boxes in position the wheeldrop traversed across to be in alignment below the frames before it then slowly lifted the wheelset up inch by inch towards the frames. Groups of engineers on both sides of the frames watched closely as the wheelset completed its journey successfully. The hornstays were then refitted below the axle boxes to secure the wheelset in place.

Next it was the turn of the bogie and with a gathering audience on the gallery, this went equally well, with nobody fluffing their lines. Although smaller in diameter, the bogie has four wheels as opposed to the two on the drive wheelset and is therefore, arguably, a more challenging operation, as at 3' 2" diameter they are not small. As the bogie was lifted into position by the wheeldrop I couldn't help but notice that it was connected to the frame in a similar way to a model engine, a fact that was confirmed by Richard Swales when I commented. In simplistic terms a big centre pin held by a big nut. My image shows the bogie in the crane harness ready to be lowered into the wheeldrop and the large centre bearing clearly resembles the Hornby model, other models are available I should add.

There was a large cast of engineers today who stayed late to put in an extra performance and without whom the tasks of the day would not have been completed. Everybody went home to enjoy a well-earned Christmas break. The trailing wheelset would be the next set to be fitted early in the new year.

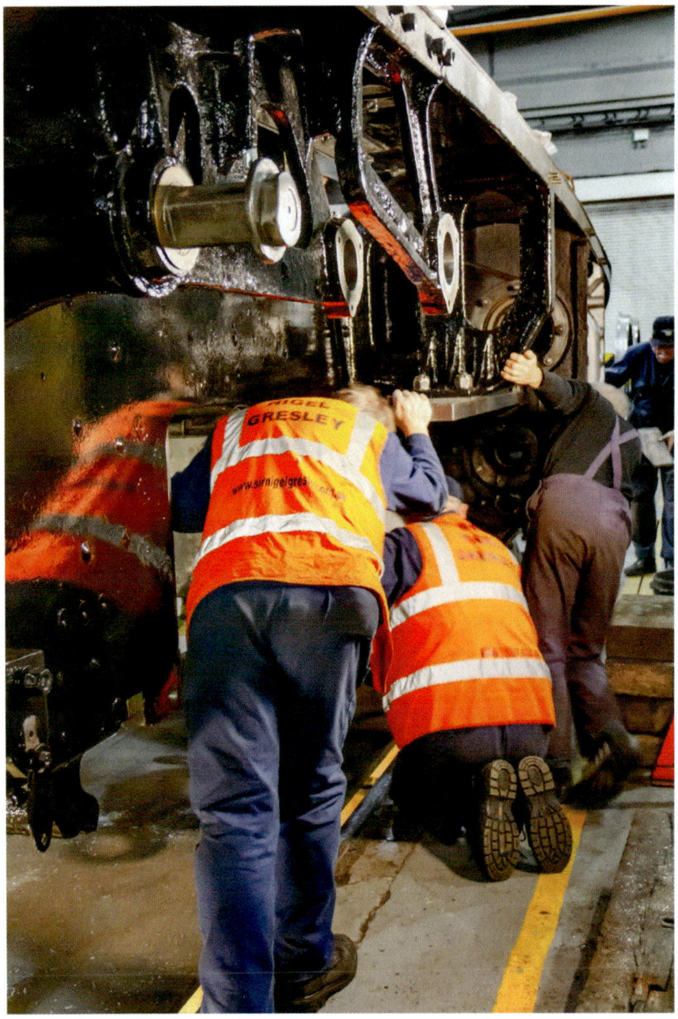

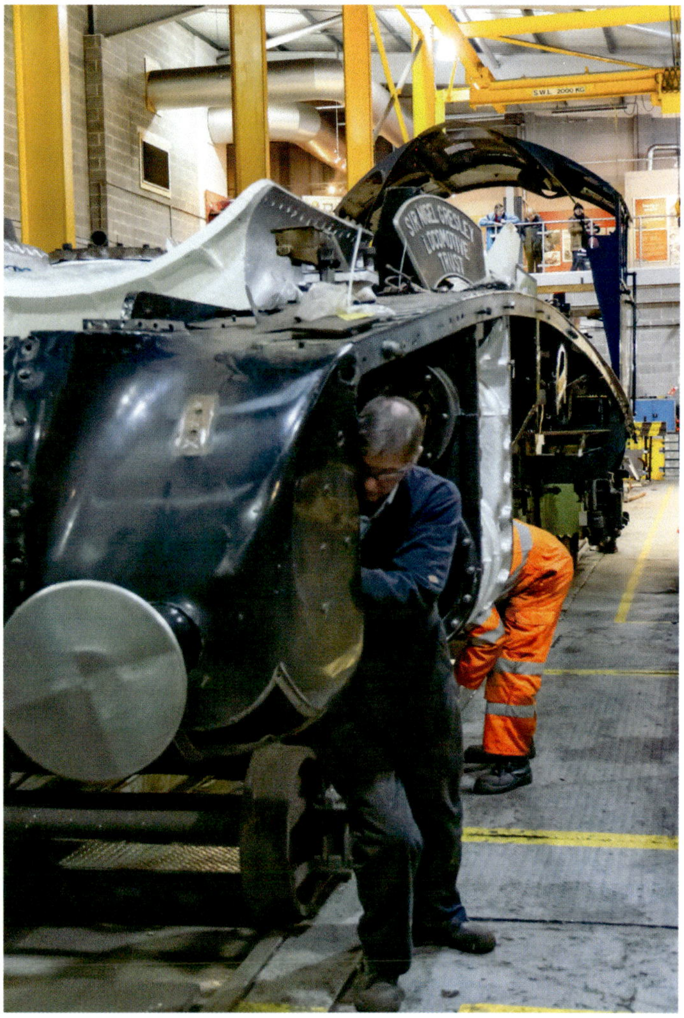

The team get to grips with pushing No.7 into position over the wheeldrop

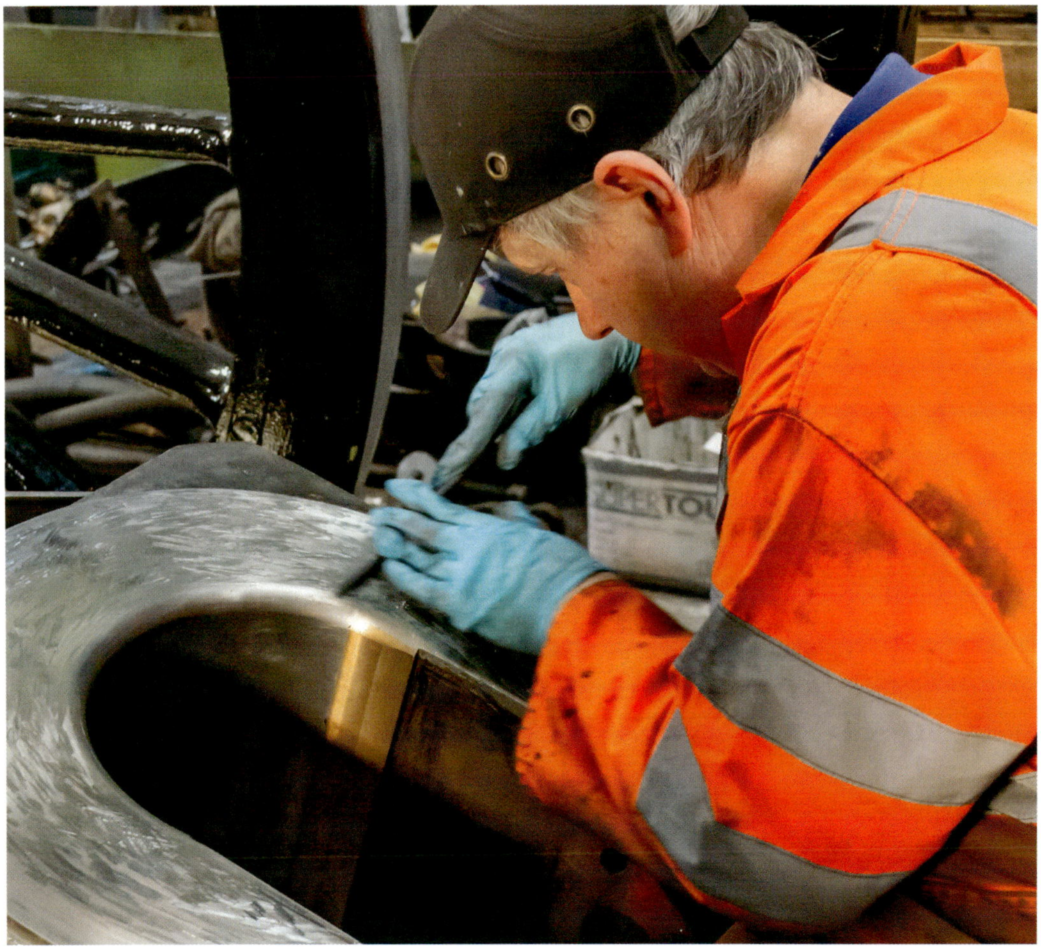

Richard Swales removing high spots on the axlebox surfaces

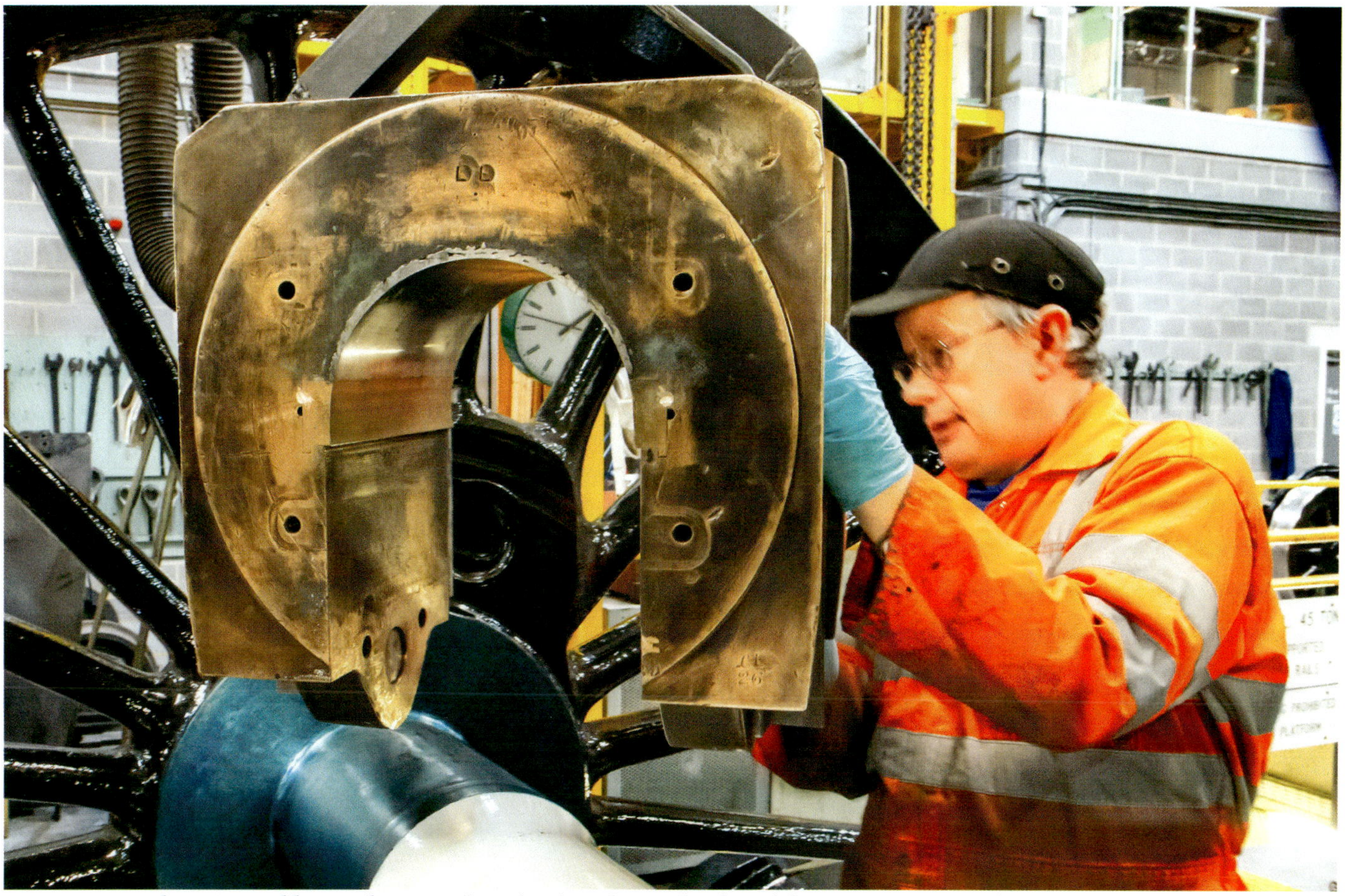

With engineers blue on the journals Richard Swales guides the axlebox into position in the journal to check for high spots on the surfaces

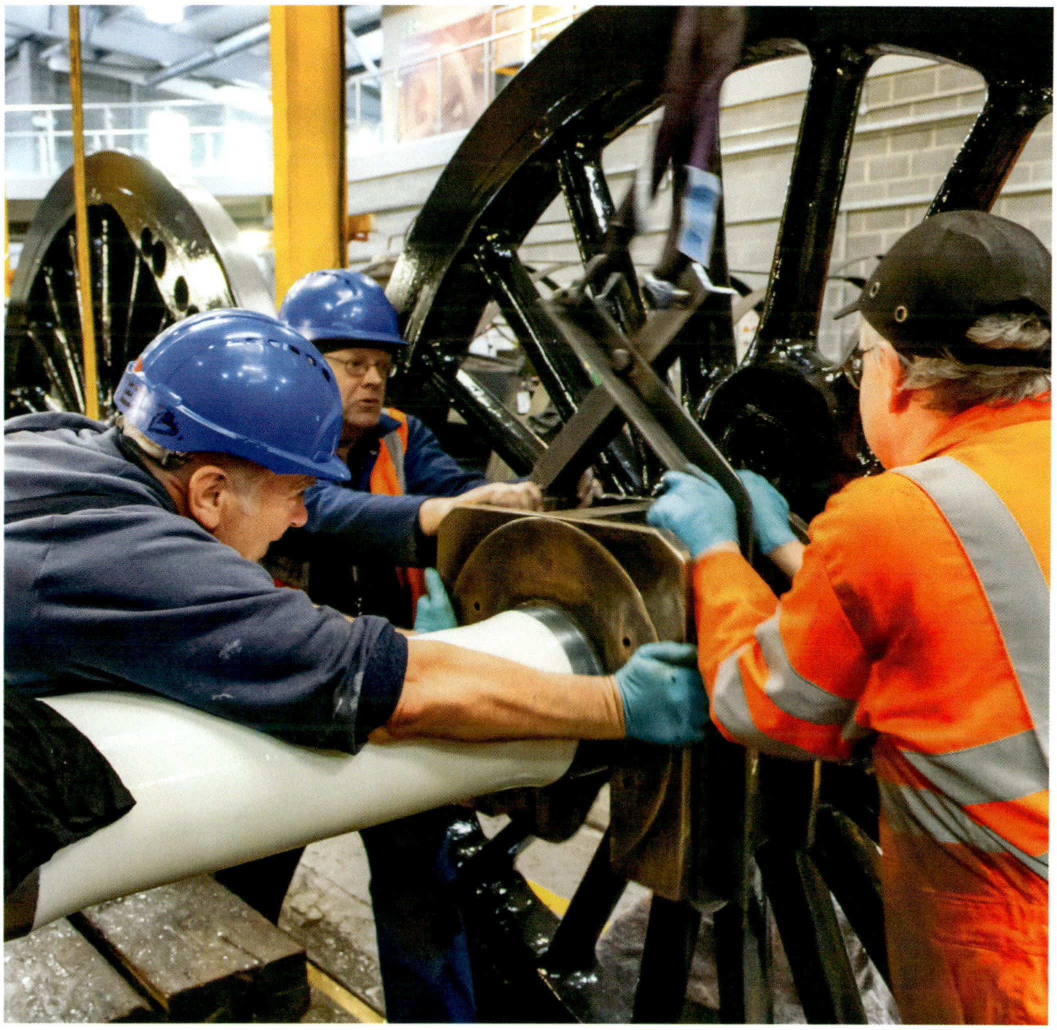

Peter Brackstone and Darrin Crone assist Richard in positioning the axlebox on the journal

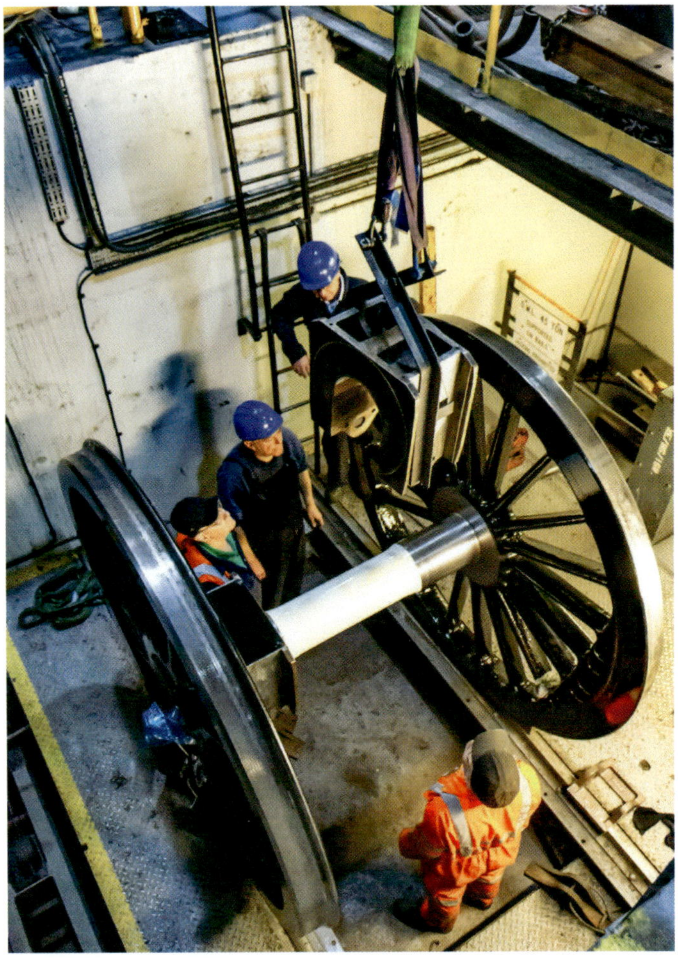

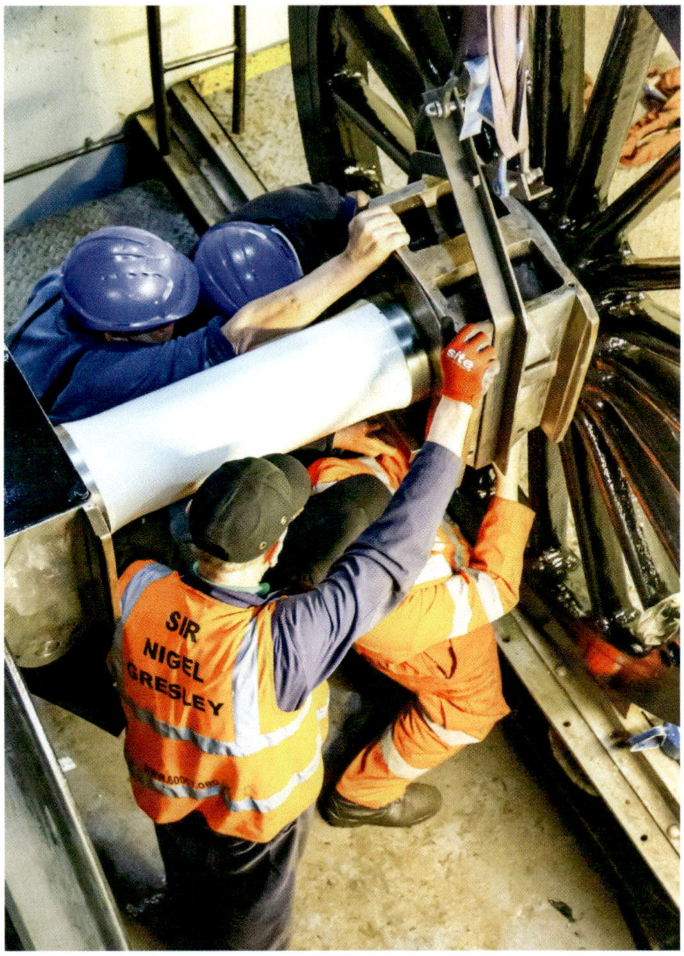

With the wheelset sitting on the track at the bottom of the wheeldrop an axlebox is lowered into position on the axle journals

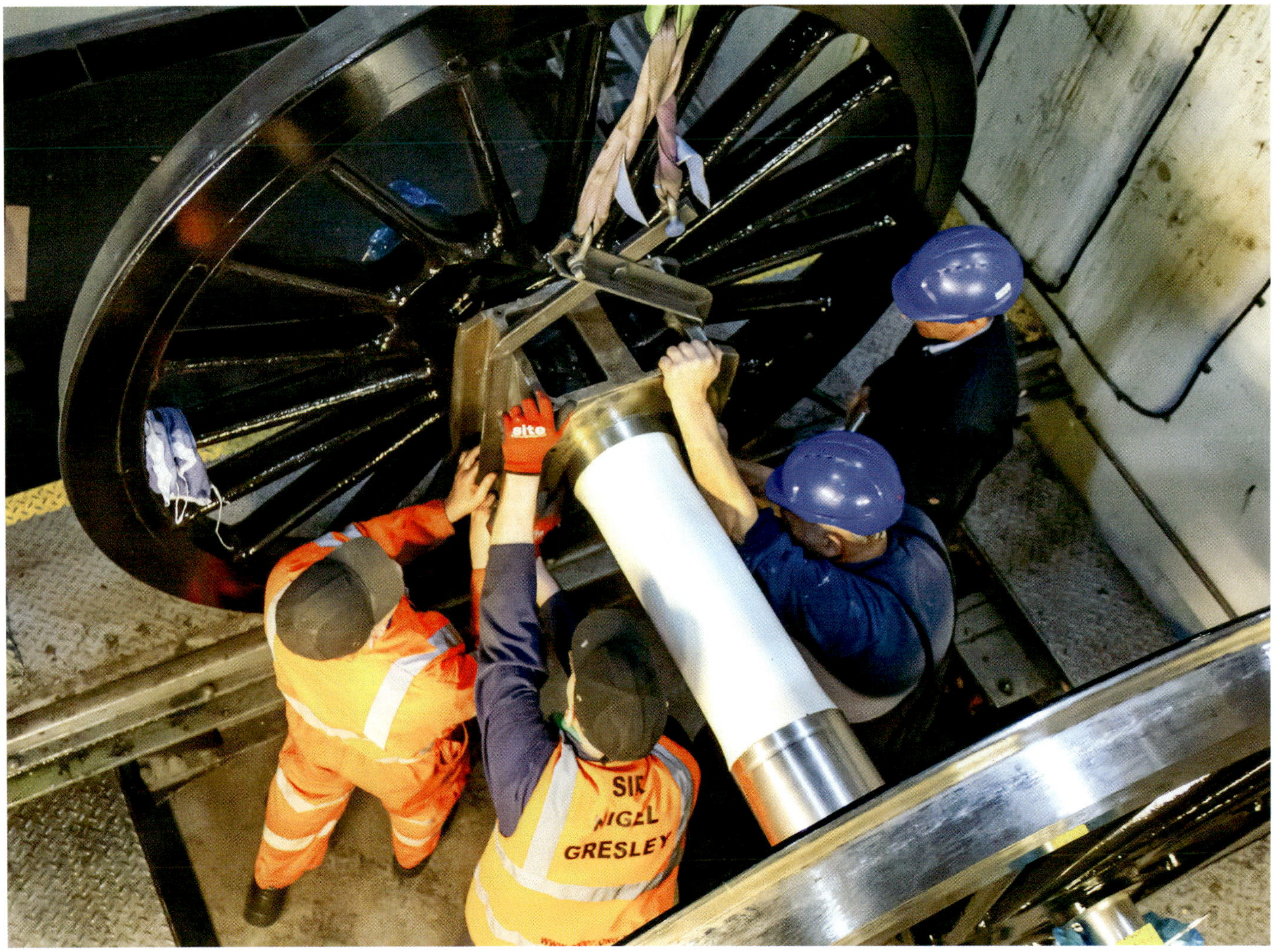

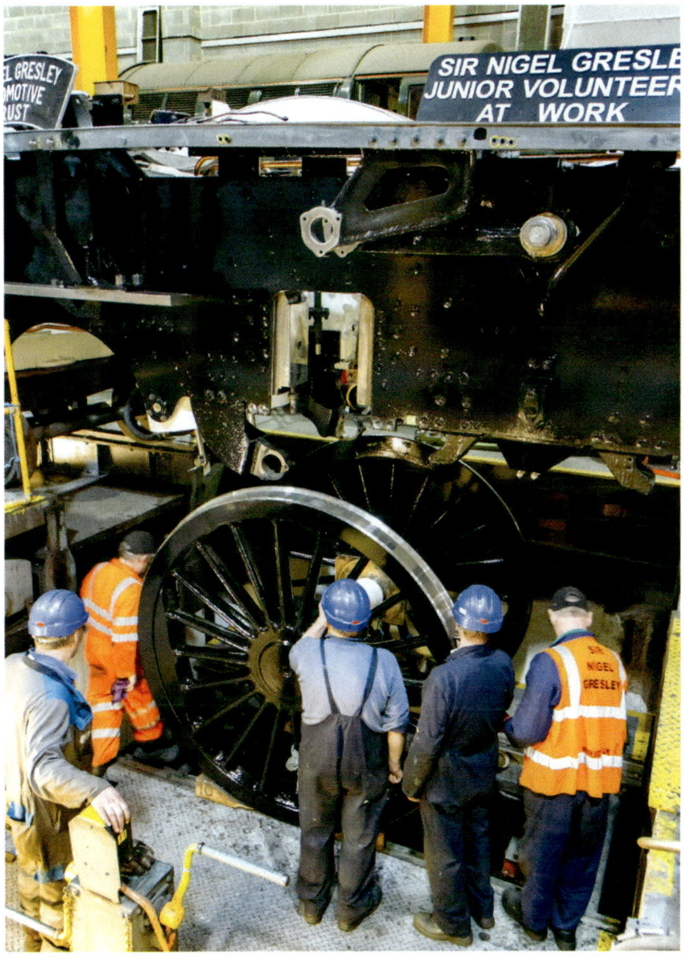

The wheeldrop is slowly raised back up to meet No.7's frame

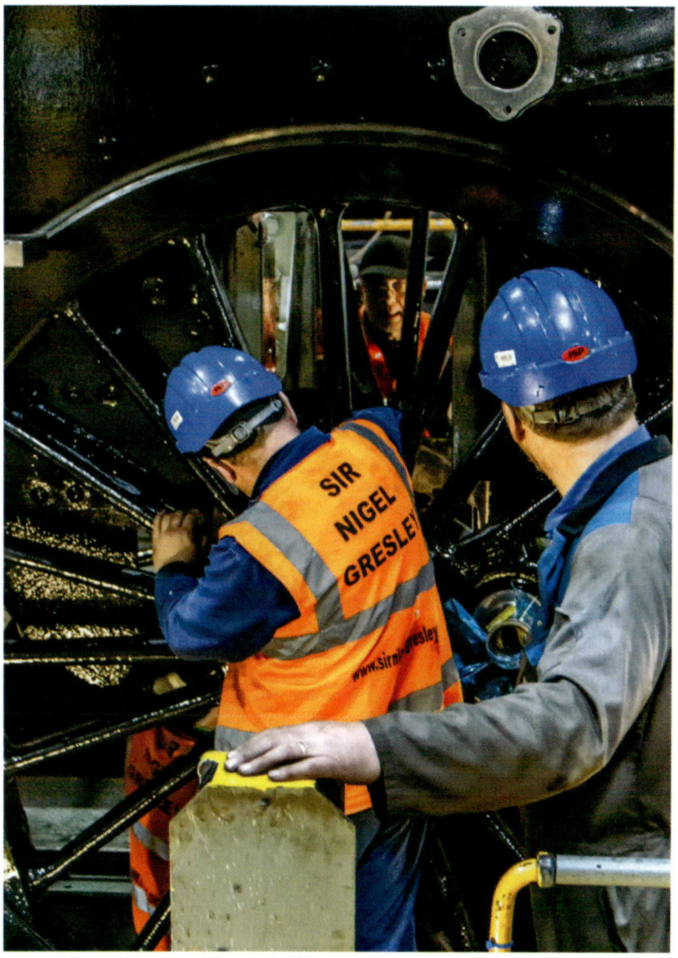

Richard Swales (far side) Darrin Crone and Roger Turnbull (near side) ensure the axleboxes enter the horns squarely as the lift continues

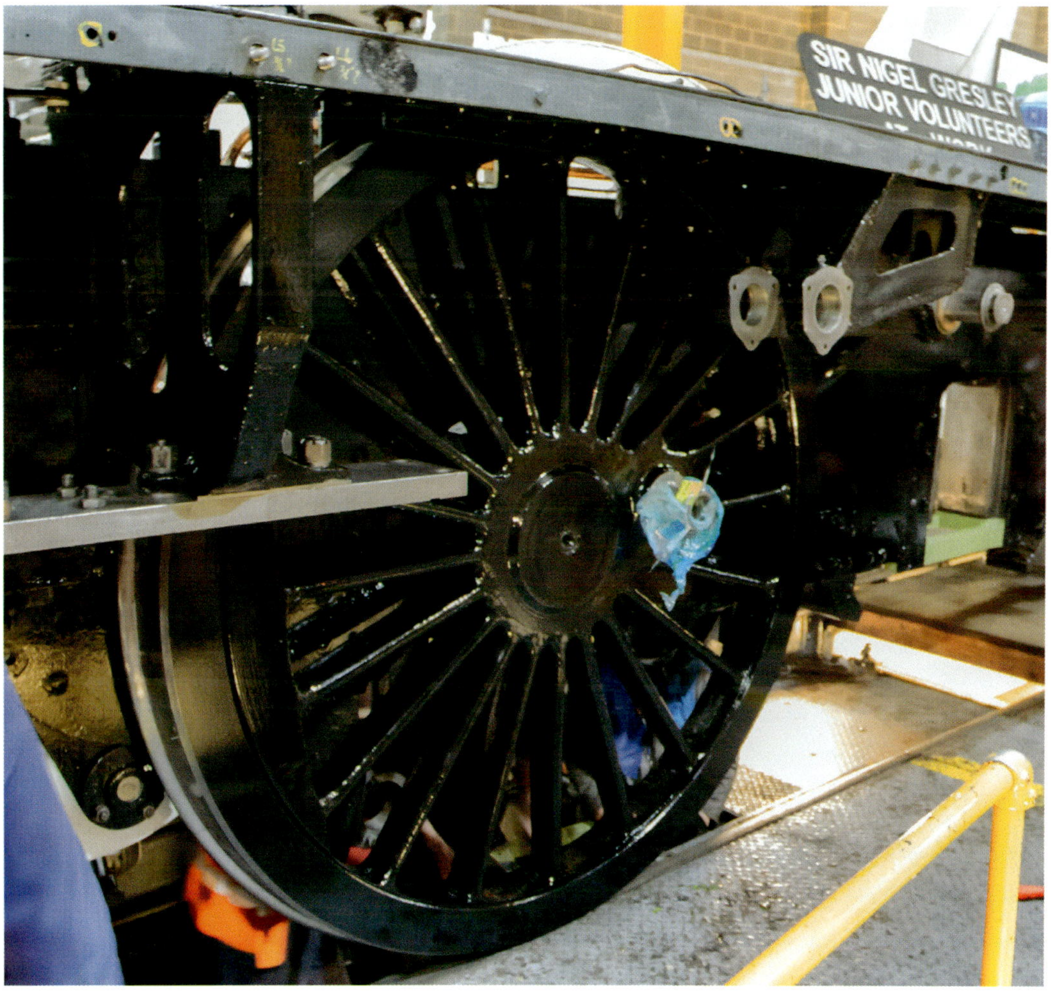

The wheelset is fully lifted into position

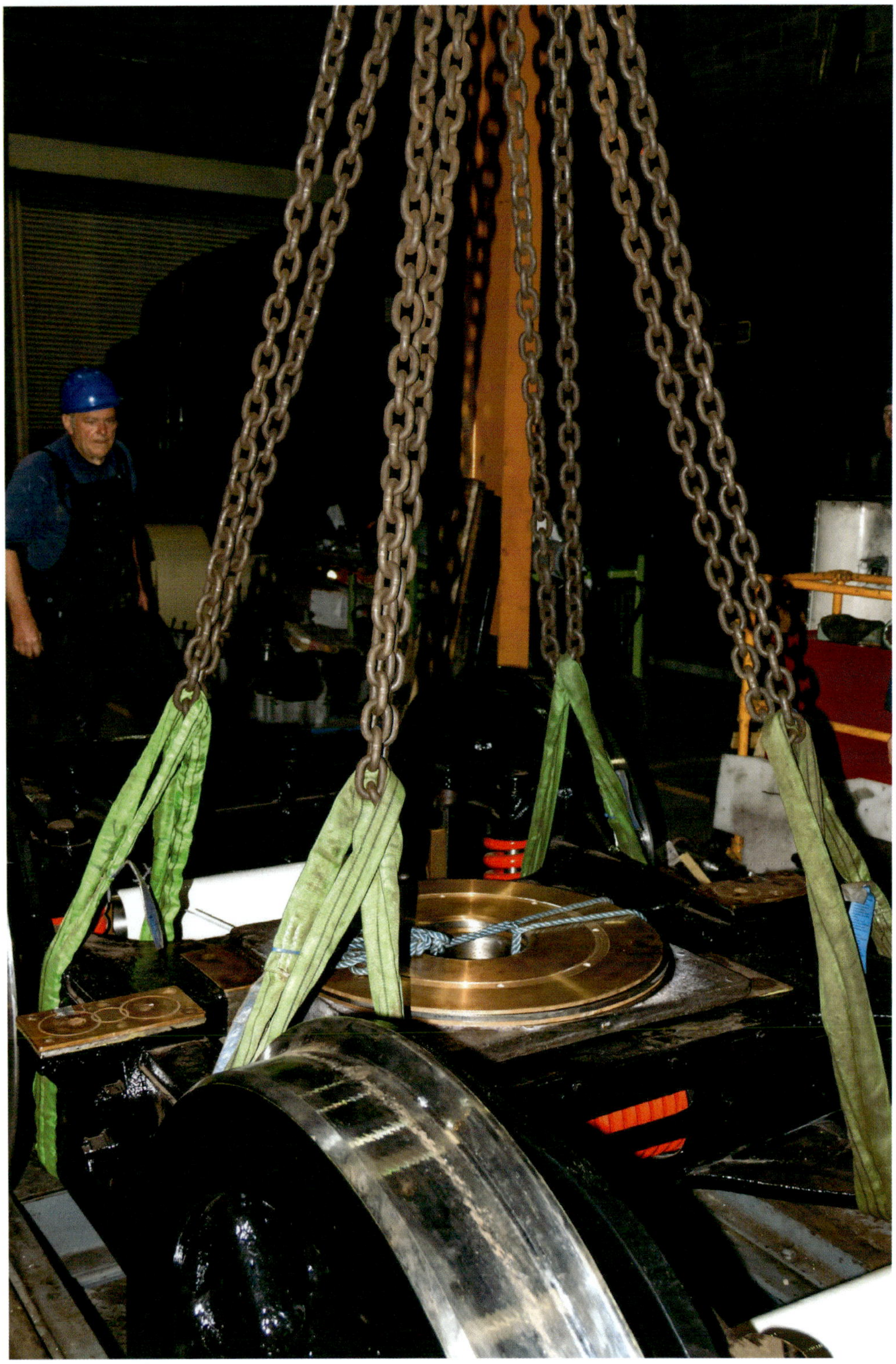

The overhead crane takes the weight as the bogie starts its journey into the wheeldrop where it will then be elevated back into position between the frames

WHEELS UP 2

February 2019

We were now two months into the 4th year of the overhaul and things were progressing nicely. In Llangollen progress was being made with the boiler and back in York the frames and the wheels were getting reacquainted. Like the engineers themselves I had a preferred day for my visits to York - unless it was a Milepost event it was more than likely a Thursday. This, however, meant that I tended to meet the same people and I therefore made a conscious decision to vary my days for the rest of the overhaul in order to catch a greater number of the team. Today was a Friday and I was doubly rewarded as new faces to me were Scott Middlemiss and Tom Readman, both were junior engineers who would normally only be able to attend on Saturdays due to school commitments, but it was half-term and they were putting in extra days. These lads are vitally important to the future of No.7 and I so much hope that in years to come they are senior members of the team and reread these words saying, "Do you remember when.........." and "How young did we look there?". Today, they would be key players on the major task of refitting the trailing wheelset.

Working on a Friday was an unusual event for the team, but Darrin was keen to get this second drive wheelset back into the frames before the end of the week and utilising the juniors was all part of the plan. No sitting on the sideline watching for the juniors, the policy was that you learn best by being in amongst it with experienced eyes looking over you and the next time you do the job in the future you would be experienced in the task.

One of the most difficult tasks with fitting the drive wheelsets is removing the hornstays from the frames and refitting them after the wheelset is in position. This is on account of their heavy weight, but the team had devised a way of utilising one of the workshop's overhead cranes to take the weight. The image with Roger Turnbull and Richard Hawkhead removing one of them shows a purple crane strap passing through it and taking the weight off the engineers' arms and shoulders.

This trailing wheelset was to go into frames without too much difficulty and I would be able to leave for home perhaps a little earlier than normal which is always good on a Friday. But before leaving for the day a lot of the freshly topcoat painted components caught my eye, and being refitted to the newly refurbished pipework, they certainly looked smart.

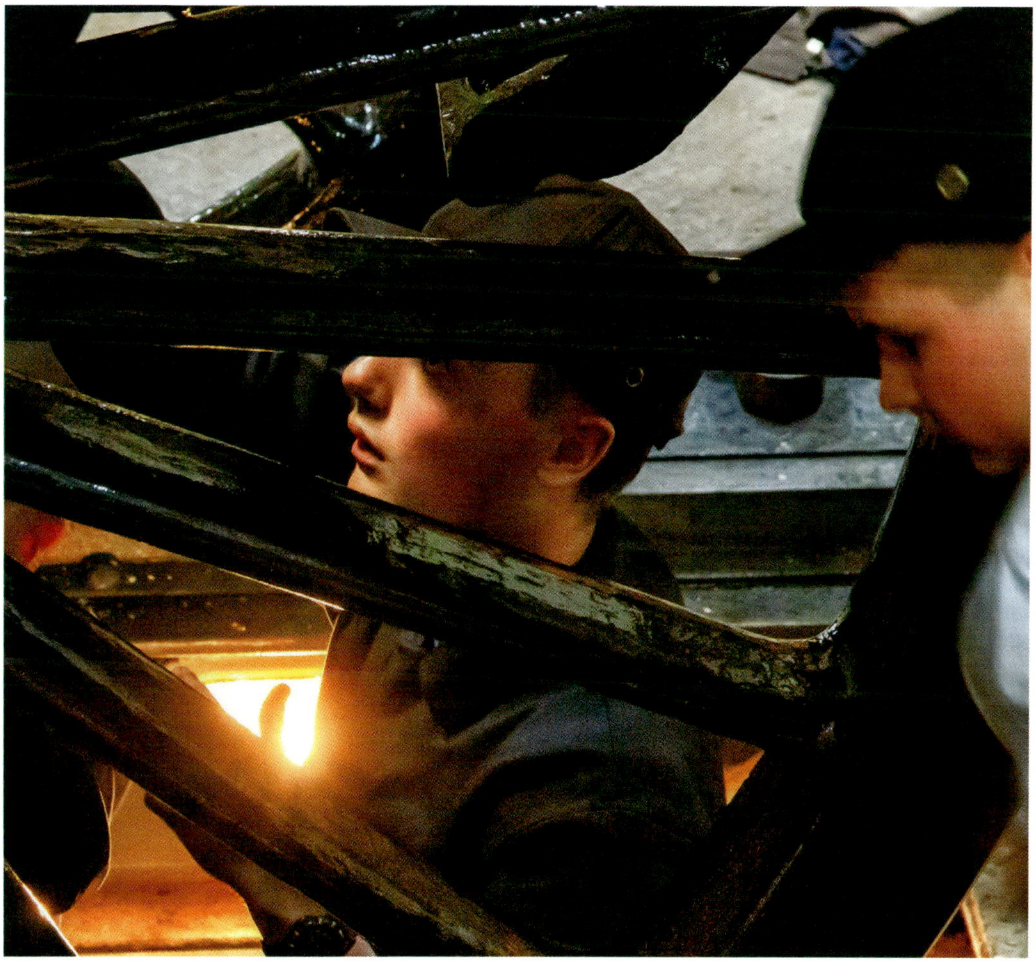

Juniors engineers Scott Middlemiss and Tom Readman getting hands on experience refitting the trailing wheelset

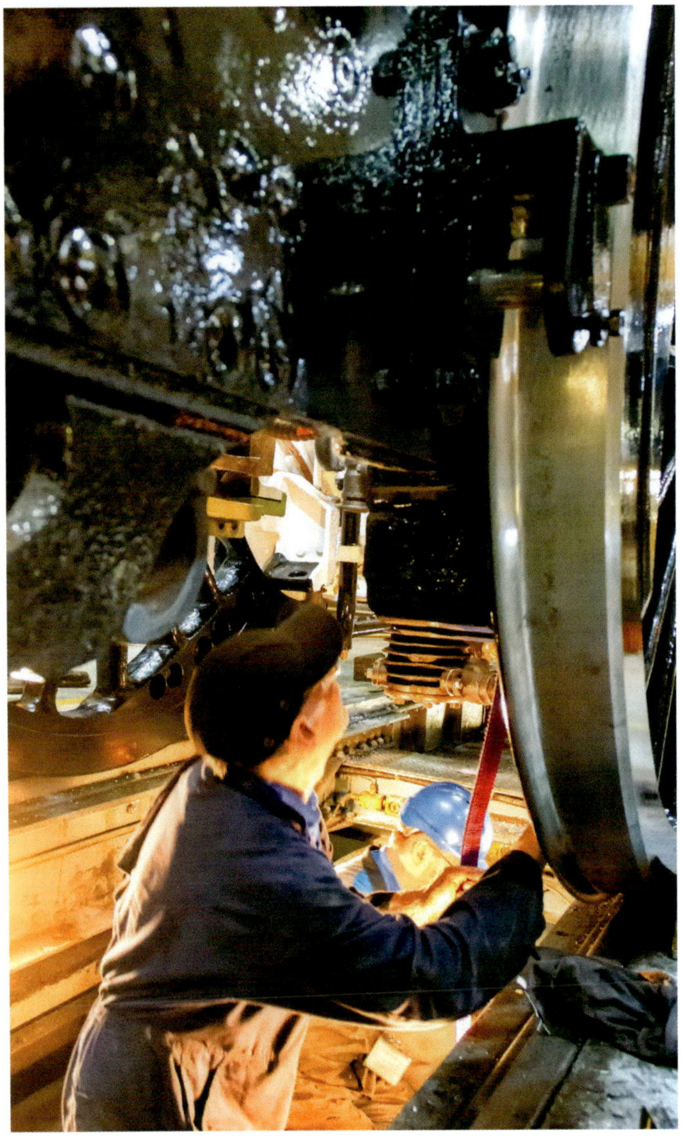

Phil Slater watches proceedings closely from between the frames

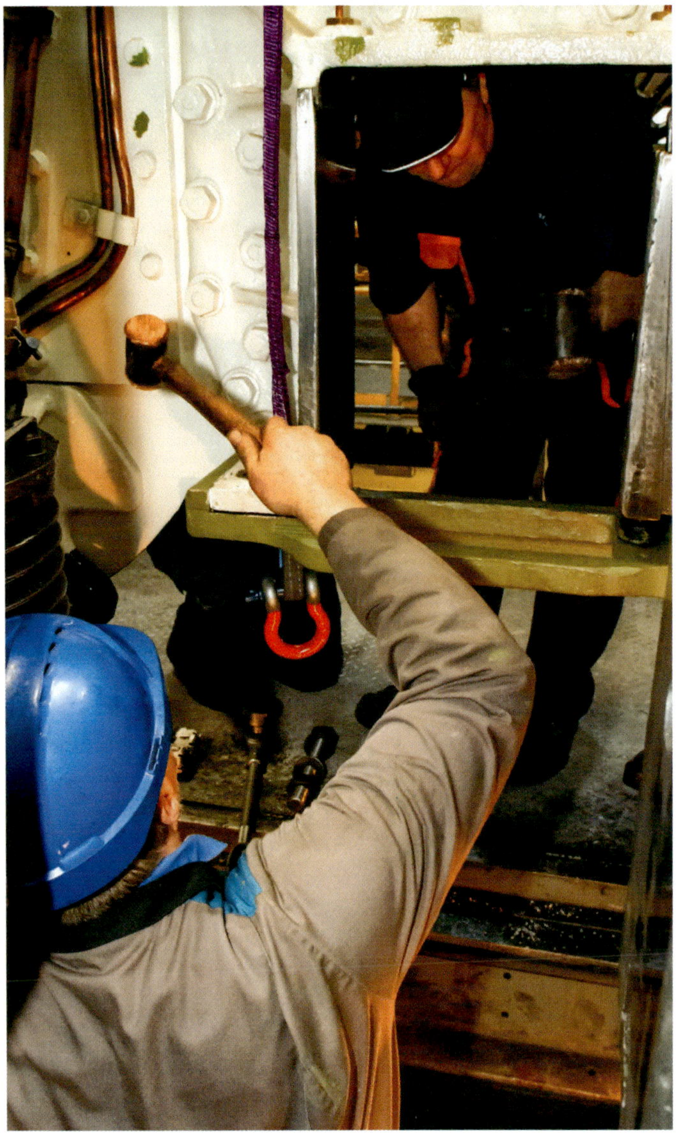

With the help of the overhead crane Roger Turnbull (left) and Richard Hawkhead remove one of the hornstays which will then allow the wheelset and axle box entry into the frames

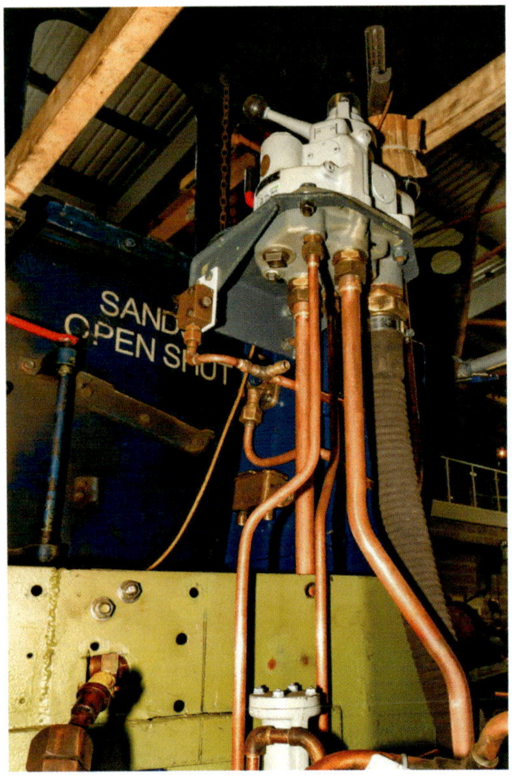

The Reverser and Air brake sporting shiny new paintwork and pipework courtesy of team members led by Peter Brackstone and Peter Ellis

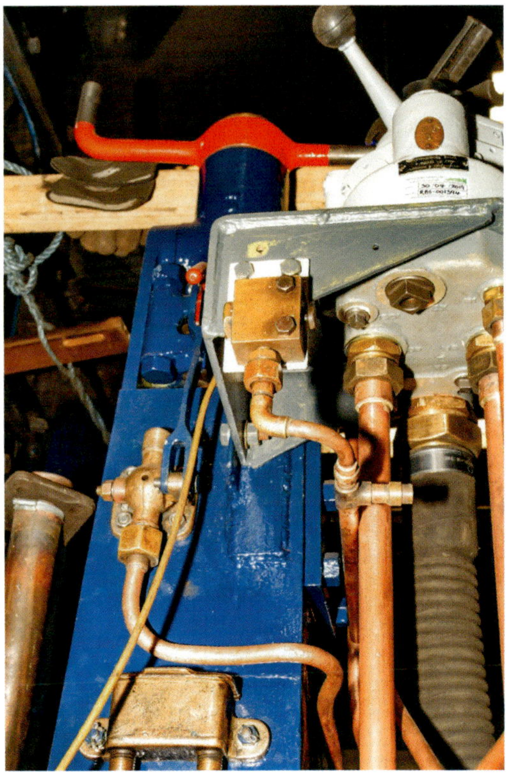

JUNIOR ENGINEERS

May 2019

Three months had passed since my last visit to the workshop; a disgraceful neglect of my duties you may think (and so did I for that matter), but there were mitigating circumstances. There was plenty happening in the workshop, but photographically much of it was similar to what I had already covered, so I was really waiting for the next Milepost event. However, I had made a personal commitment to vary my attendance days to feature different members of the engineering team in my images and, in particular, the Juniors. So, on this Saturday in May I set off for York to do exactly that.

Junior engineers Tom Readman, Oliver Barker, Luke Perry and Seb Lorch were all in attendance this morning as representatives of the 007 Gang and watching the lads at work took me back many years to my

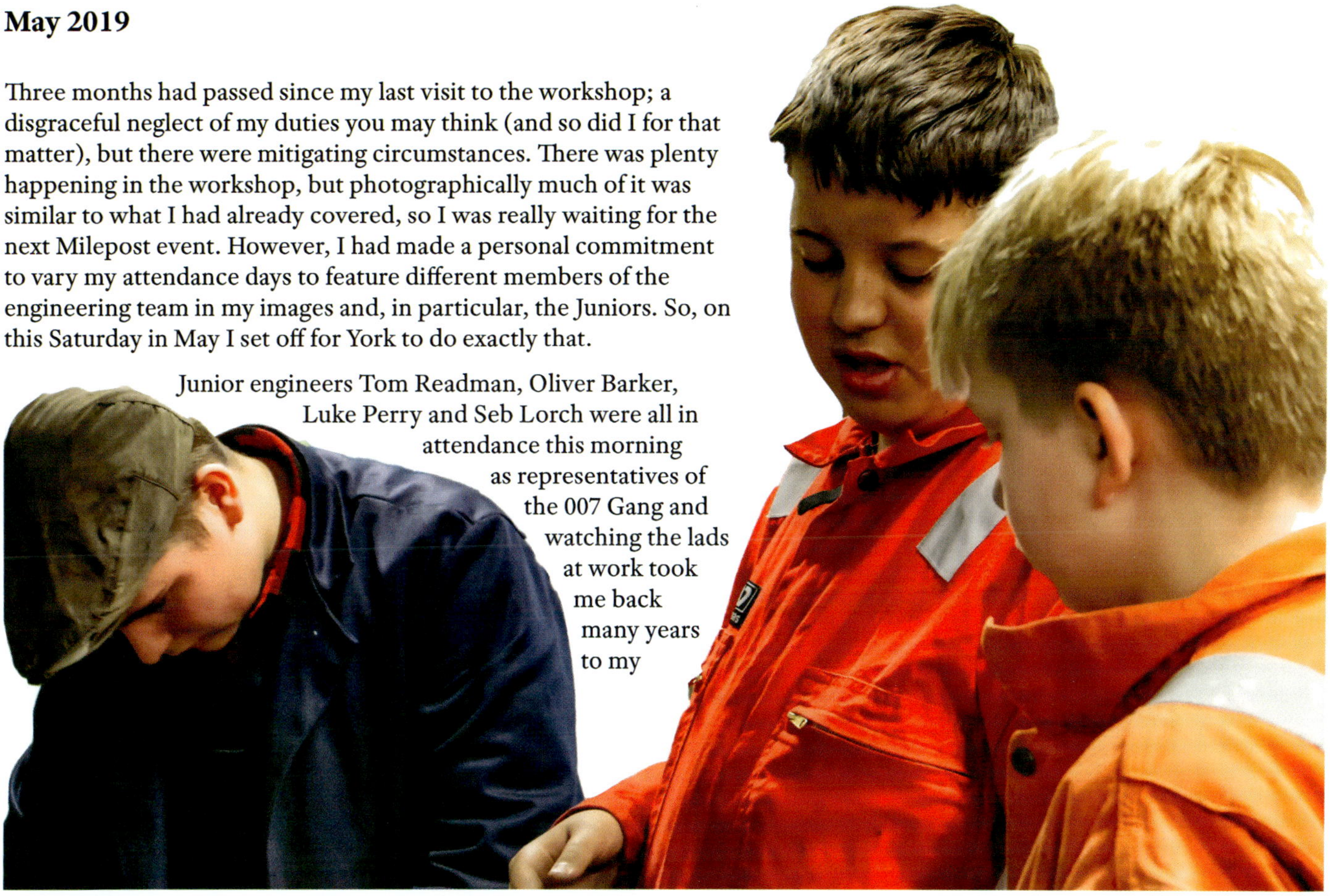

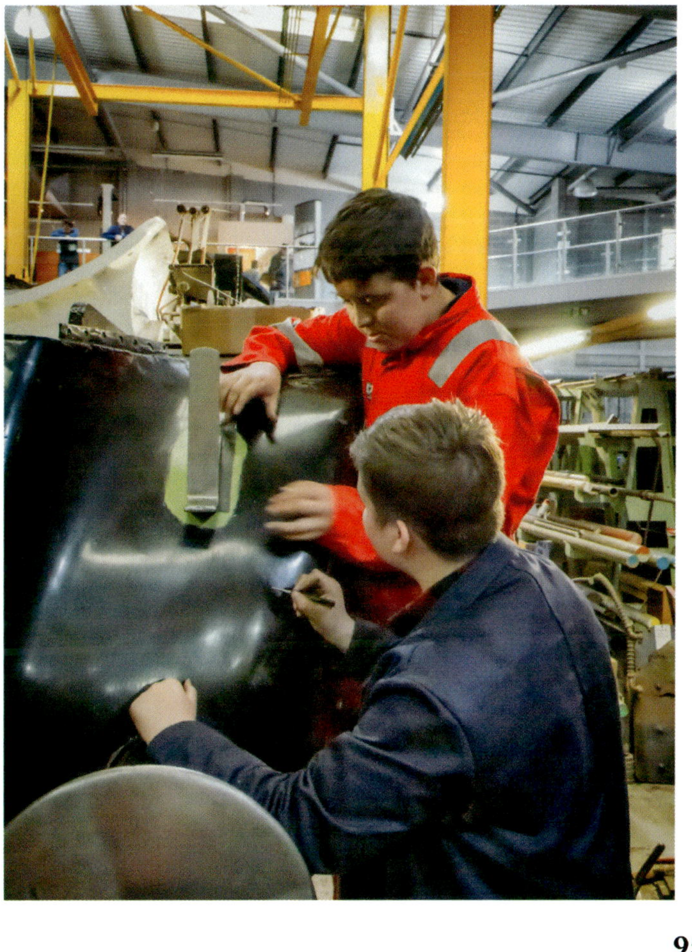

ABOVE - Member of the '007 Gang' of juniors volunteers Tom Readman, Luke Perry and Oliver Barker discuss the work allocated to them by Darrin Crone

LEFT - Luke Perry and Tom Readman at work on No.7 fitting a new lamp bracket to No.7

BELOW - The lamp sits on its new bracket made by Luke Perry and Tom Readman

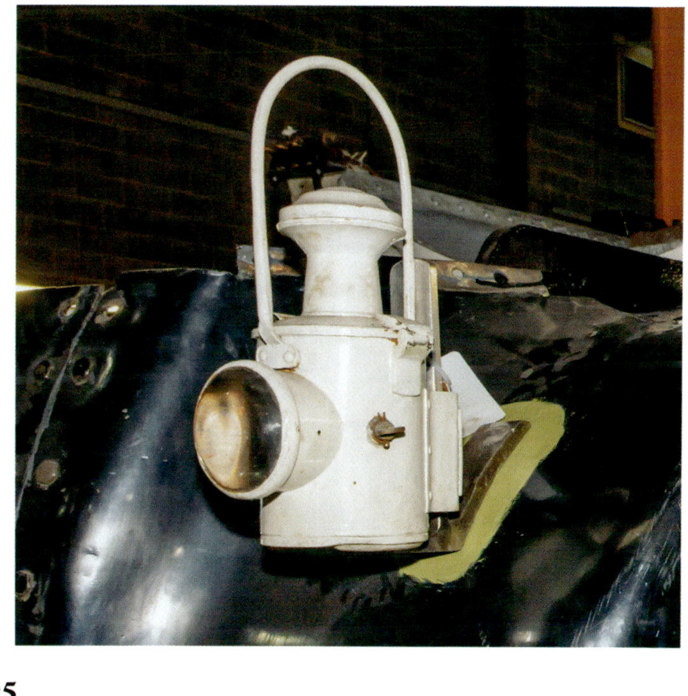

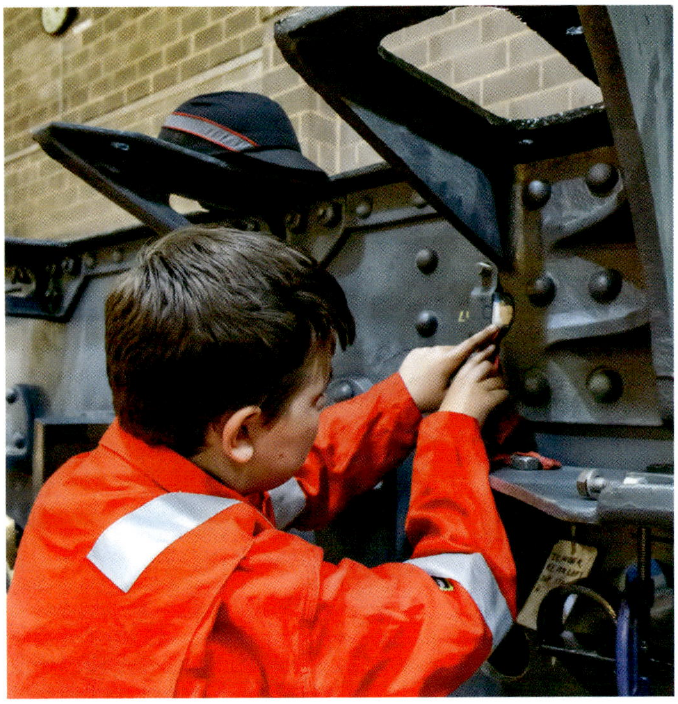

Luke Perry working on tender frames

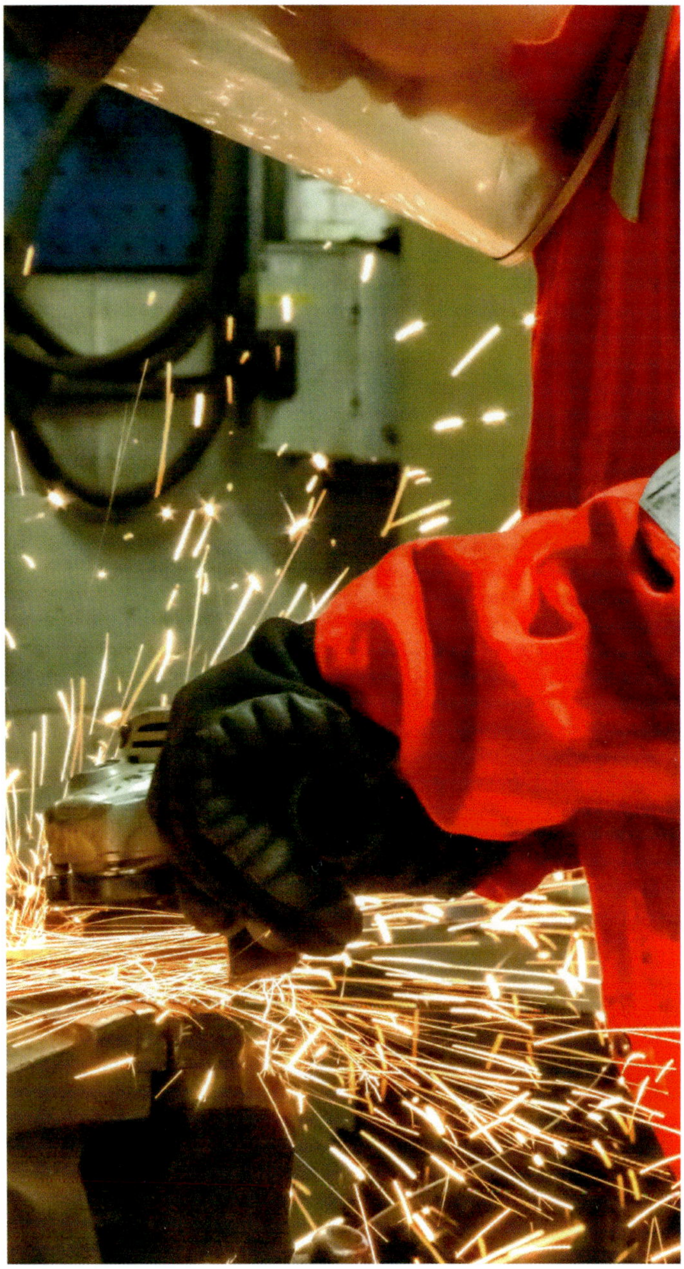

Luke Perry working with the angle grinder on a bracket for the tender frame

metalwork classes as a schoolboy. I couldn't help but think how well they were doing compared to my average efforts. I wasn't great but I am sure that with the right mentoring, such as the lads were benefitting from, I may have managed better. I was particularly impressed with the grinding skills I witnessed from Luke Perry assisted by Tom Readman and the ability of Oliver Barker to cut carefully and accurately with a hacksaw, i.e. a straight line, something I failed to master. I guess they could have taught me a few things as the most I had contributed so far was a brief stint with a paintbrush and some needle gunning, which I think I have mentioned already.

This was, for me, a shorter visit than most and I left immediately after having lunch but not before taking a few snaps of the team during their break in the blue cabin.

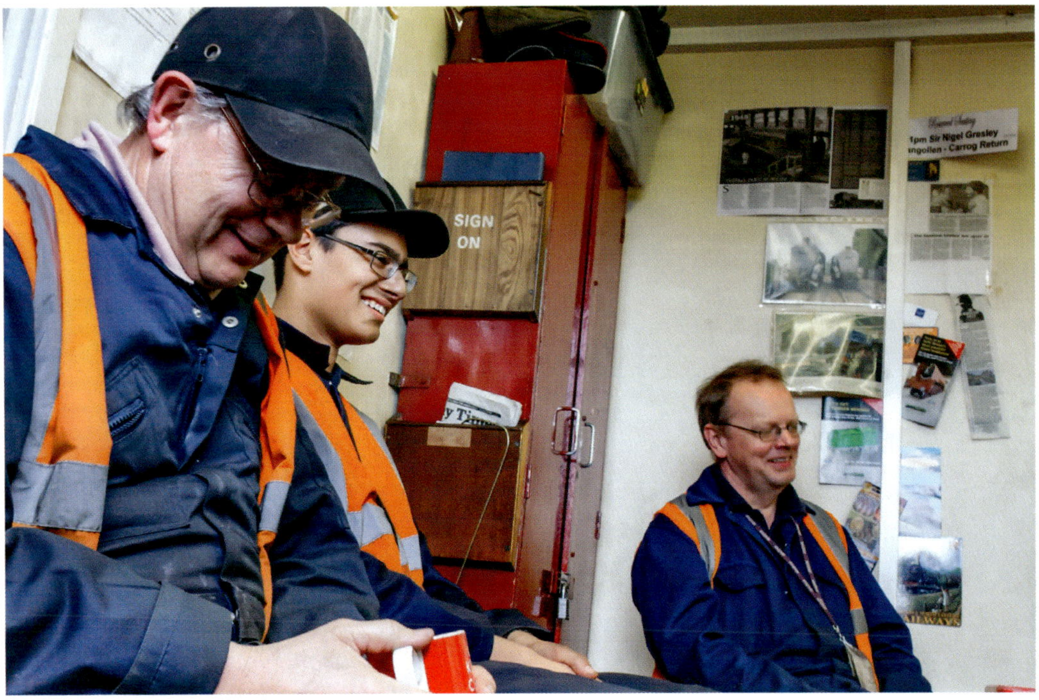

Richard Swales, Seb Lorch and Darrin Crone share a joke

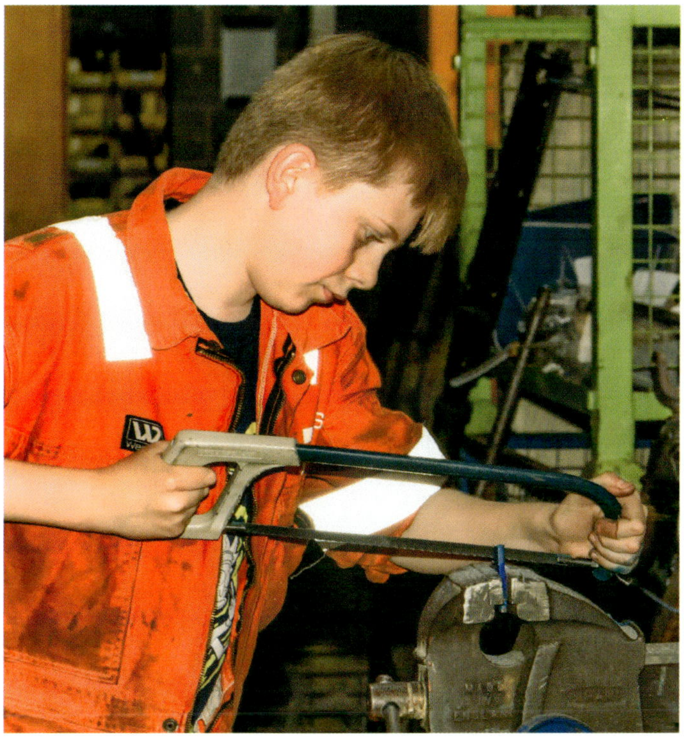

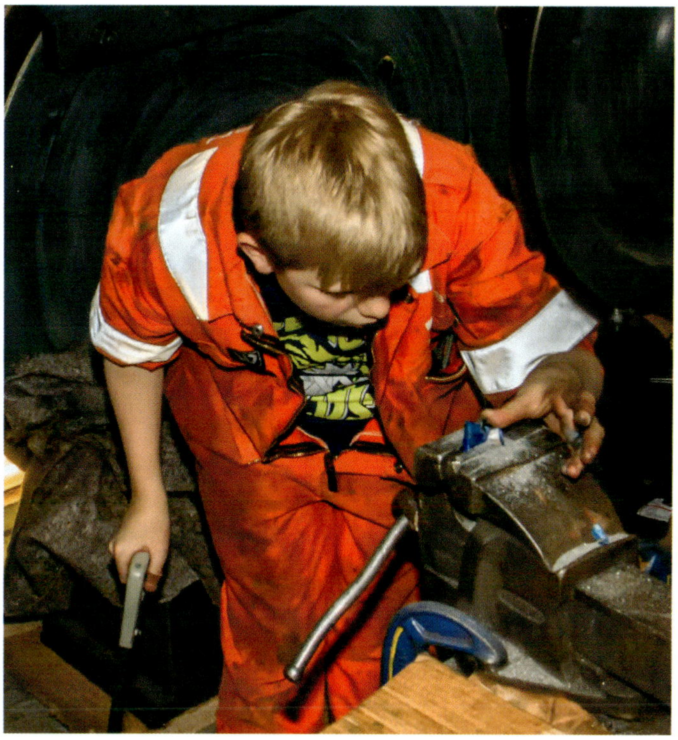

Oliver Barker works carefully and accurately with the hacksaw to repair a bracket for No.7

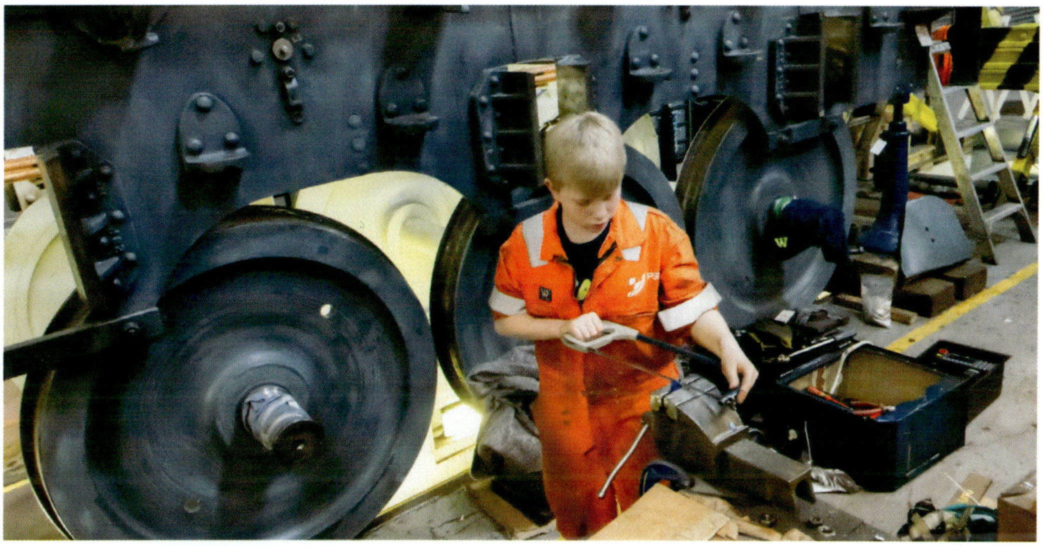

Bryan Orange, Mike Page, Luke Perry and Oliver Barker enjoying their break

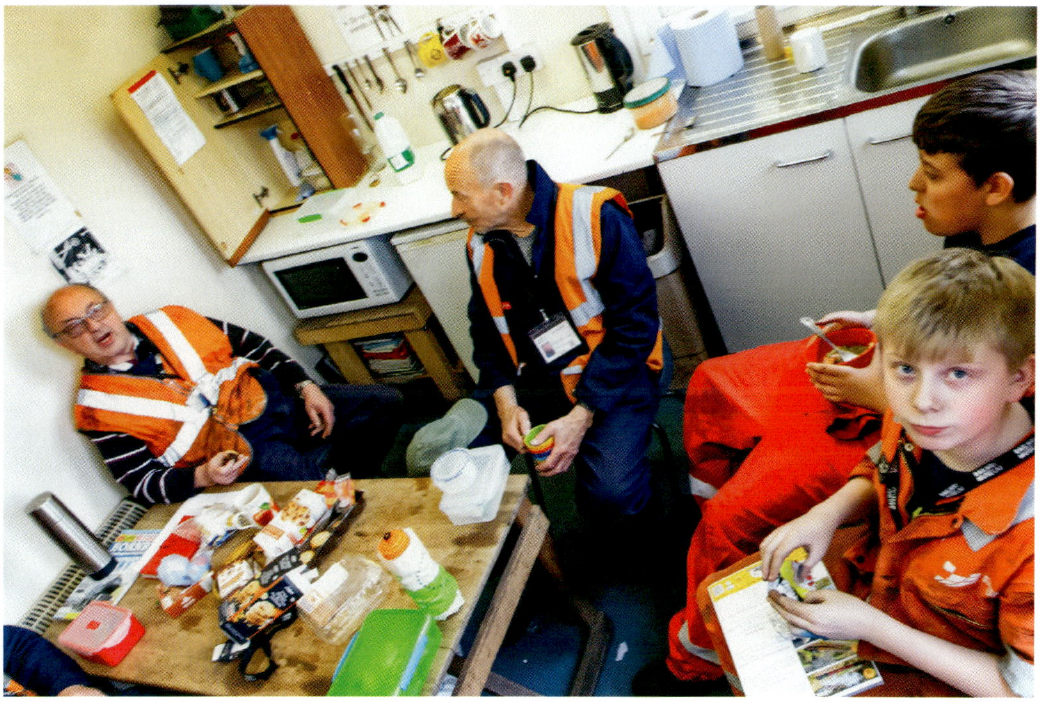

June 2019

Over in Llangollen progress was being made with the boiler. Successful hydraulic tests of the main steam pipe had been carried out and the refitting of the superheater was completed. The refurbished ashpan had been returned to York and refitted back into the frames. This required the removal of the cab and lots of other refurbished components, giving a sense of déjà vu but assembling them in the knowledge that they would have to be removed again was necessary as it was important to know everything fitted correctly before the boiler arrived. The cab, now positioned behind the frames, would not be refitted until the boiler was back in the frames. With the cab removed I took the opportunity to take an unobstructed shot from the gallery looking down the frames. Once the boiler and cab were refitted the opportunity would no longer be available until the next overhaul in ten years' time. After the ashpan was fitted painting of the cab floor commenced and again this required removal of newly fitted components, primarily pipework, so access could be obtained. Over on the tender new plating was on order to replace badly corroded areas around the coal door and cosmetic work was taking place at the back with sanding block and paint brush.

With it not being a milepost event day, I had a free hand to wander around looking for shots and with nobody working on the frames it was safe to venture once more into the pit. On my very first visit Darrin had taken me down there and I had taken shots which I thought would be good to replicate here for comparison purposes even though, this time, there was no boiler above me.

With the news from Llangollen that the boiler was almost ready there was a new sense of enthusiasm around as everybody was eagerly awaiting its return. The light at the end of the tunnel was just starting to flicker, bringing a smile to all. The light-hearted mood was reflected in the blue cabin at lunch time where the conversation today focussed on Malc's scotch eggs and to how do they cook the outside without overboiling the eggs inside! It seems that the combined engineering brain power of the assembled team could not equal that of a single cook as no acceptable to all conclusion could be reached. We settled the debate by agreeing our resident meteorologist Bob would ask his wife Jackie, our resident baker of fine cakes, for an answer.

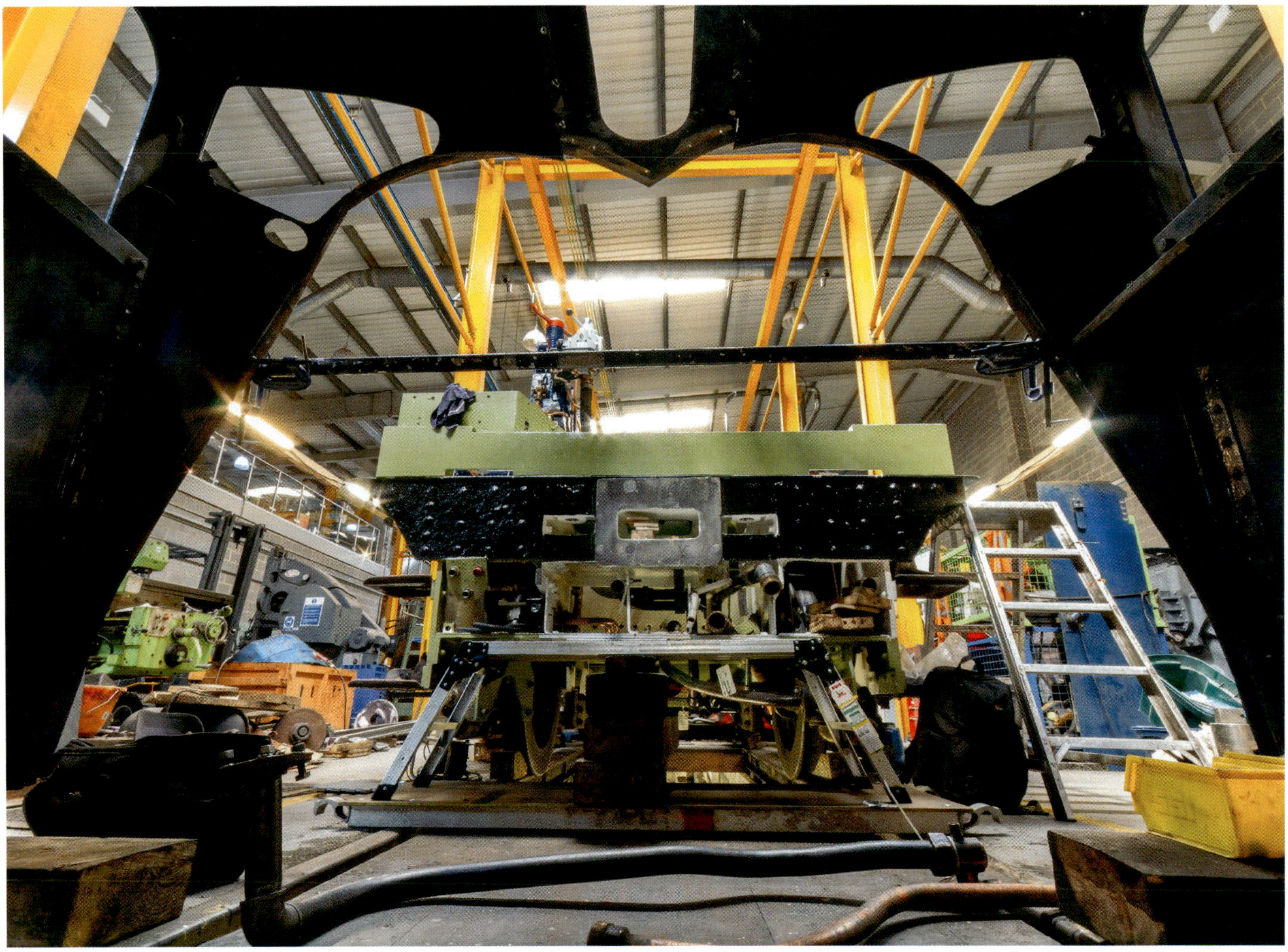

The view back through the cab after it had been removed in readiness for the return of the boiler

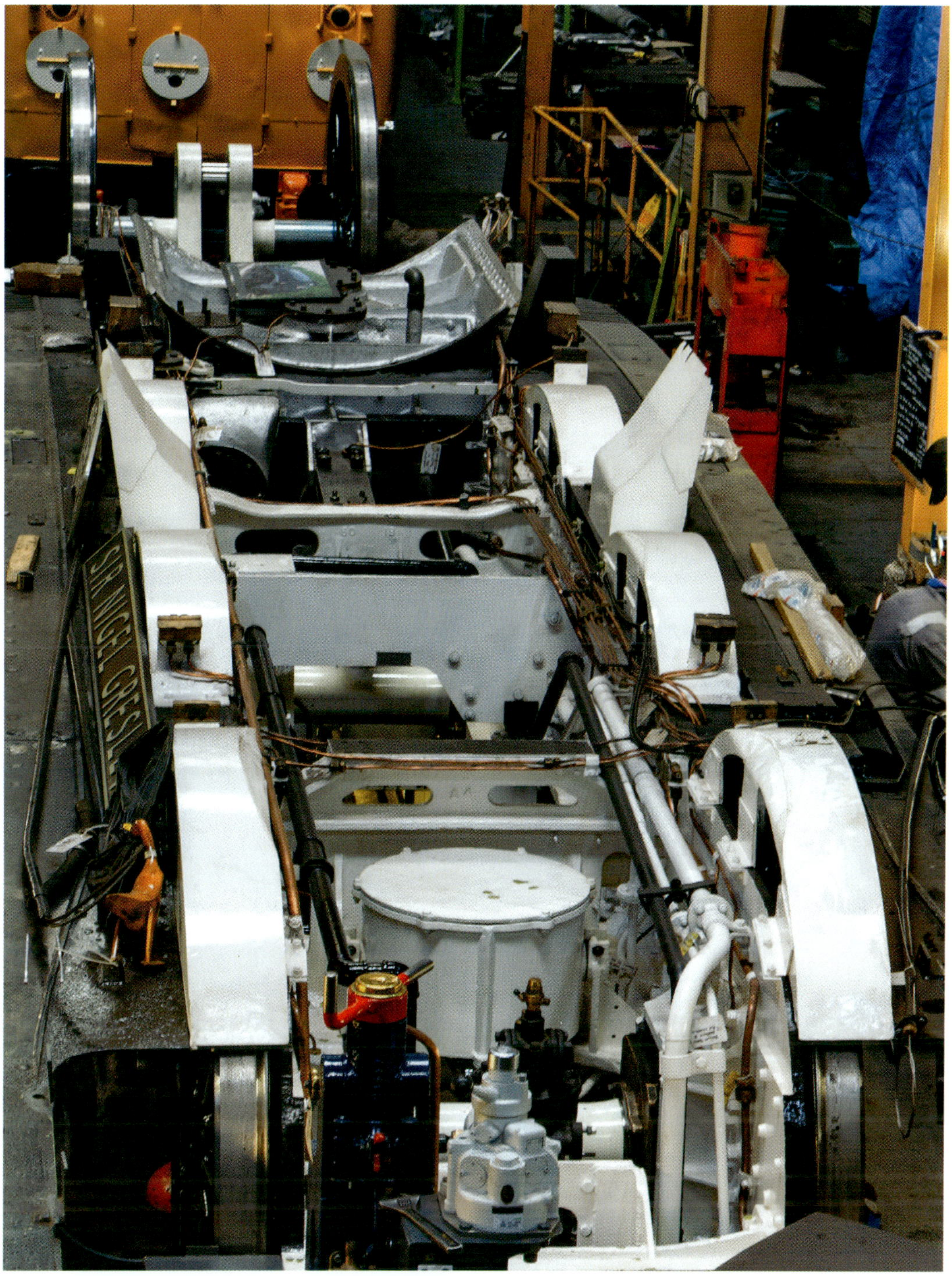

With the cab removed the view from the workshop gallery provided an uninterrupted view of the refurbished frames sitting waiting for the return of the boiler

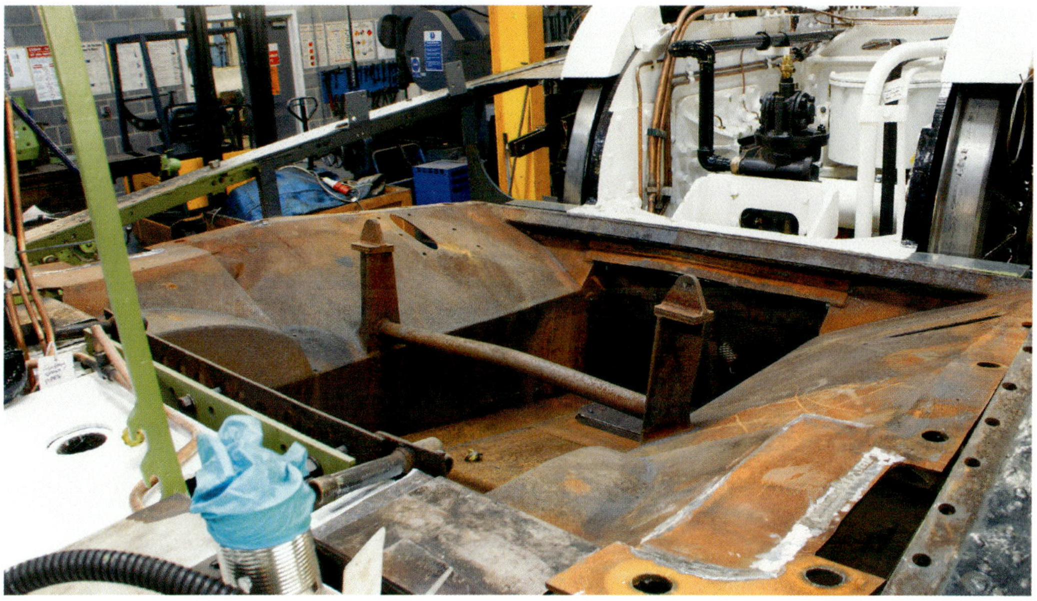

Recently returned from Llangollen after being refurbished the ashpan is back in place between the frames

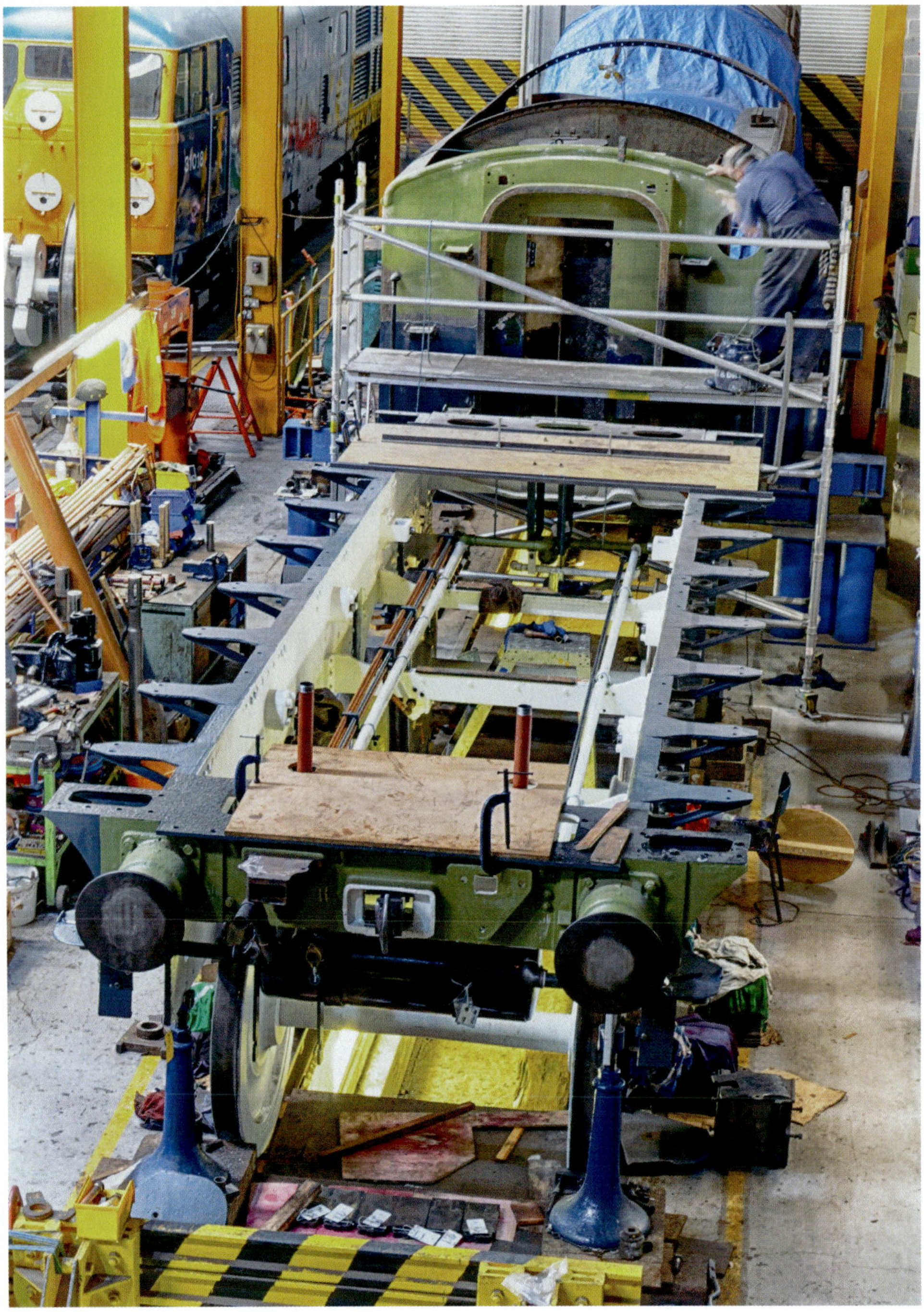

The gallery view showing the tender frame raised above its wheels and painting work proceeding on the tender tank

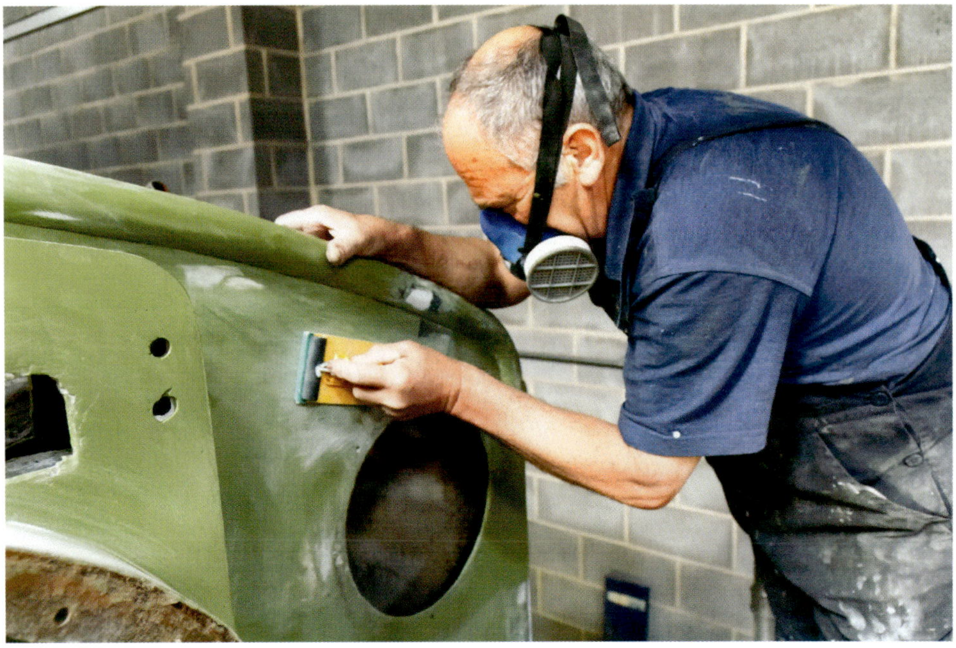

For a perfect finish Darrin has asked Peter to carry out some more rubbing down on the tender tank

The footplate view of the tender tank which is awaiting new steel plating to replace removed corroded plating

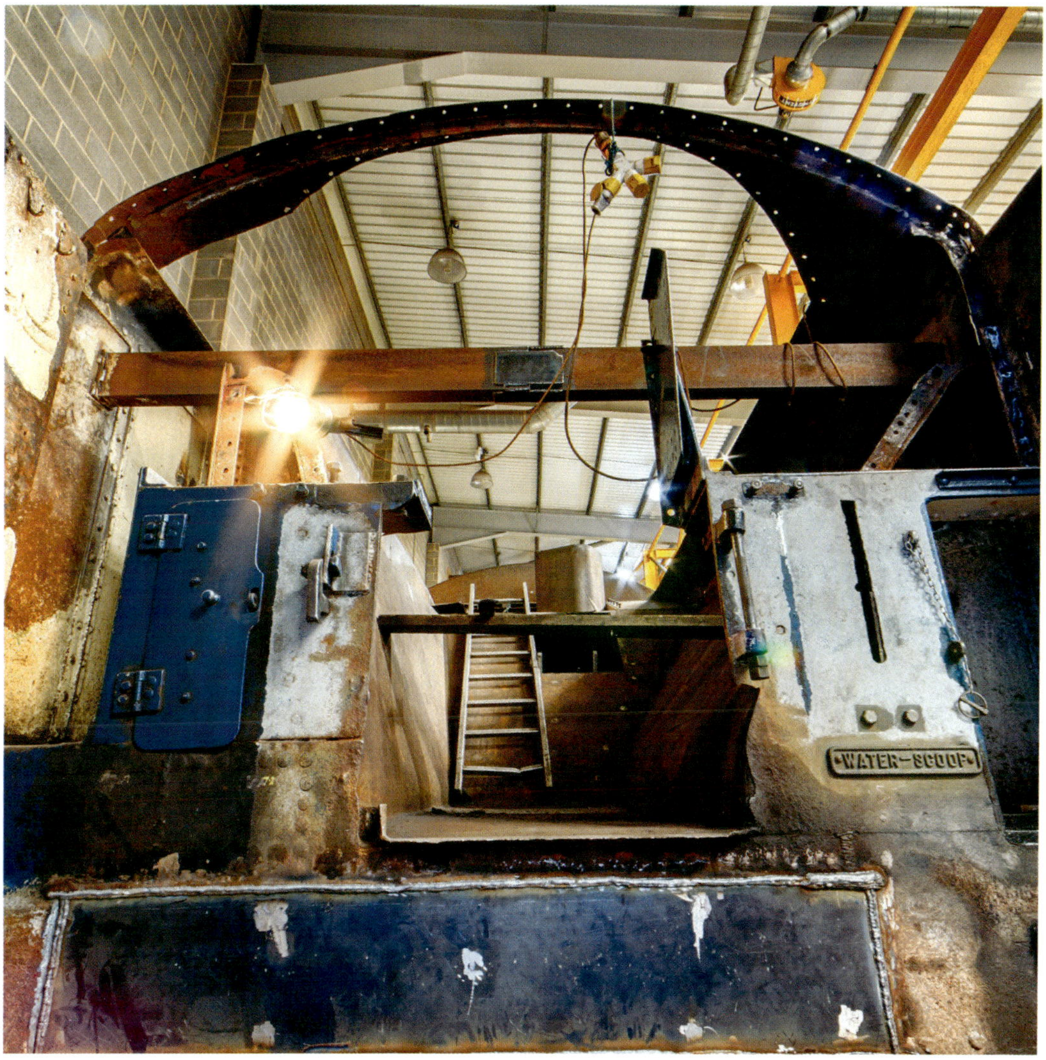

The view from the pit below the frames showing all the clean and freshly painted refurbished components

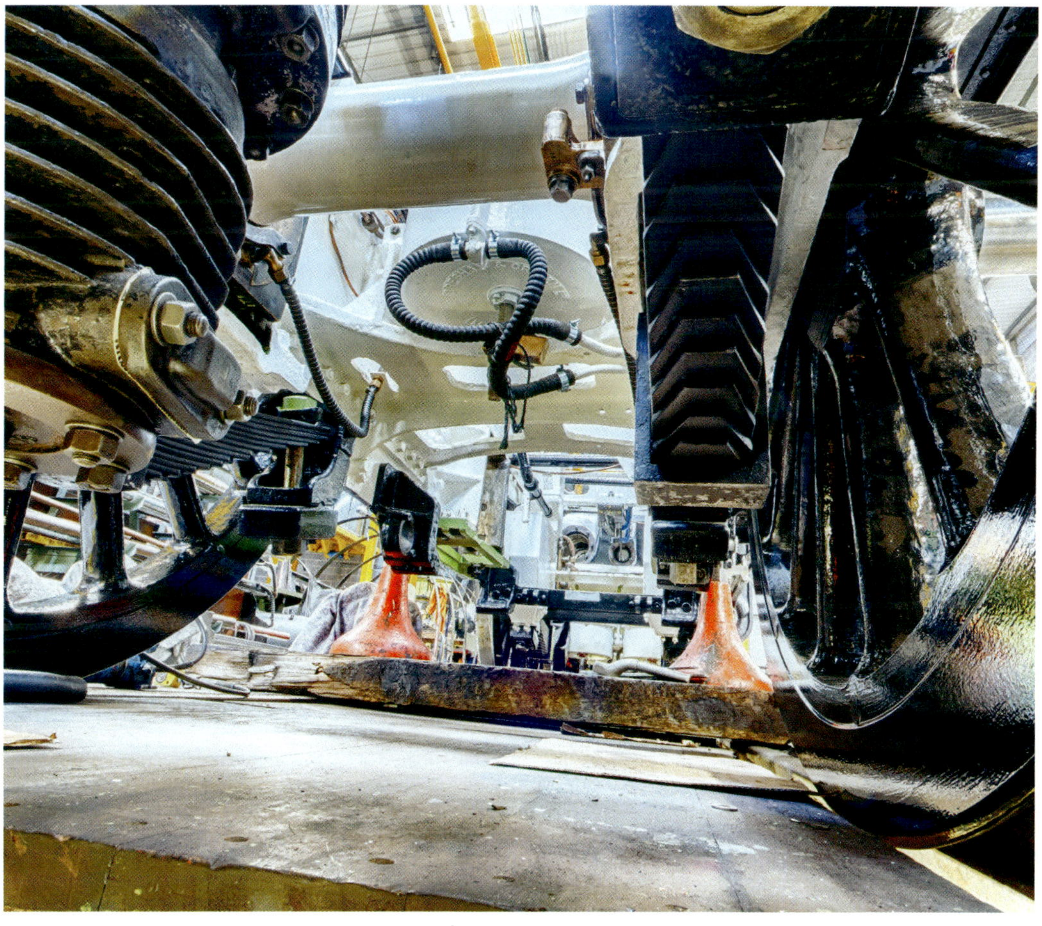

MIXED TRAFFIC

August 2019

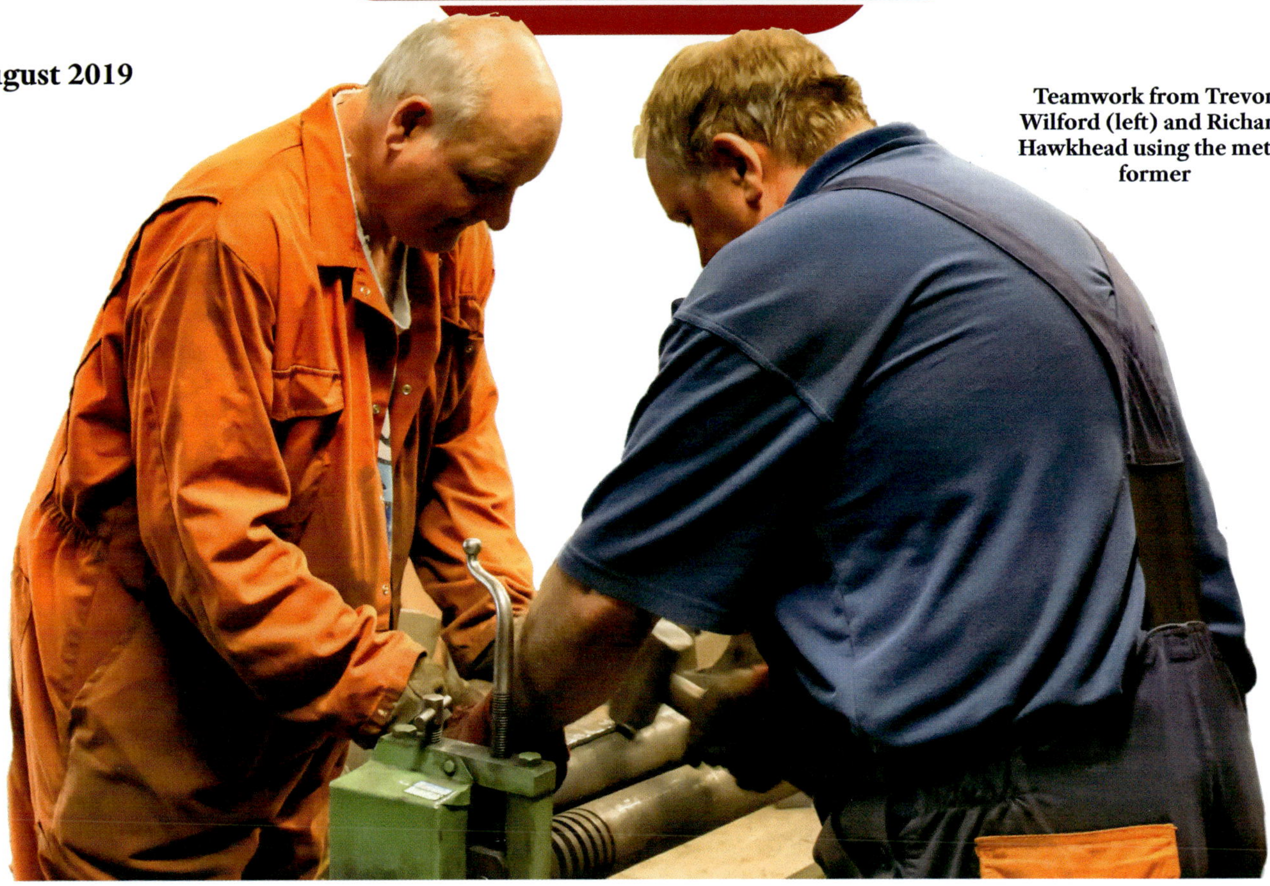

Teamwork from Trevor Wilford (left) and Richard Hawkhead using the metal former

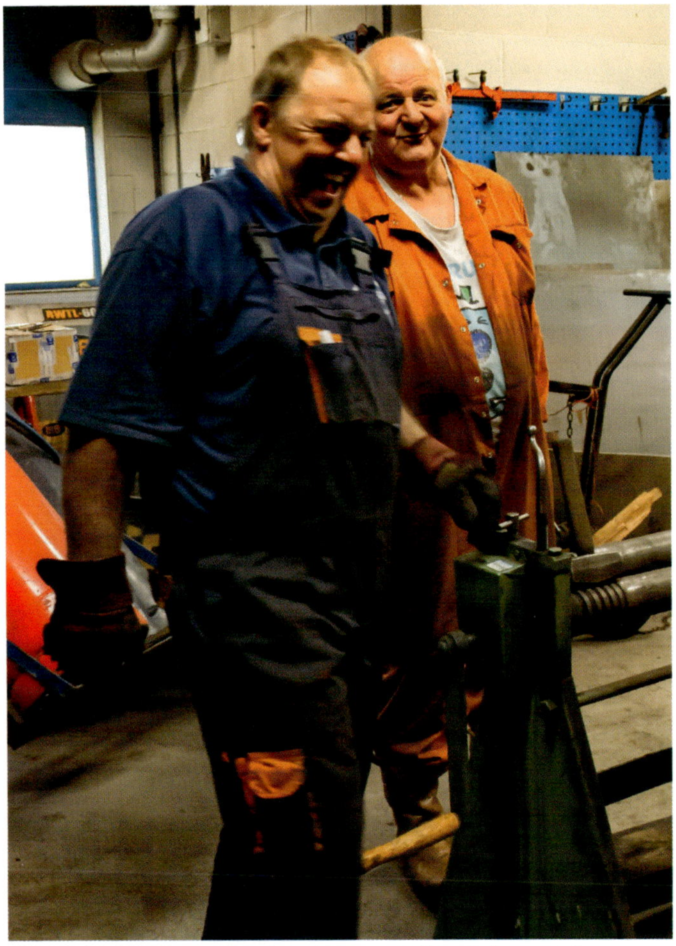

Richard Hawkhead and Trevor Wilford share a joke

I made a couple of visits to the workshop during August, they were, to use a railway term, mixed traffic events with a wide variety of tasks taking place. Rod Thomas was reshaping pipes, Barry Horton and Peter Ellis were working on tender tank air cylinders and Dave Lee, with Trevor Wilford and Richard Hawkhead continued with the roof on the tender corridor. On my second visit I found Andrew Jacques checking for cracks in valve gear lifting arms with 007 gang member Scott Middlemiss. Peter Brackstone was kept busy making tea and of course painting. Busy and bustling was very much the case this month. What did come over in these visits was the harmony and comradeship of the team members when working in smaller groups as opposed to the milepost event days when everybody was working as one team.

News had been received of more progress with the boiler in Llangollen with the superheater elements now expanded into place. The long awaited return of the boiler to be reunited with the frames and wheels was getting nearer.

With the completion of work on tender mechanical and electrical components, there was pressure on the painting team to complete the painting of areas of the tender that would be inaccessible to a paint brush when the wheels were reunited with the frames. The reassembly of the tender was being scheduled for September but would require coordinating with lifting specialists. I agreed with Darrin that that would be my next visit to the workshop, but little did I know at that moment there was to be an unexpected break in my attendances.

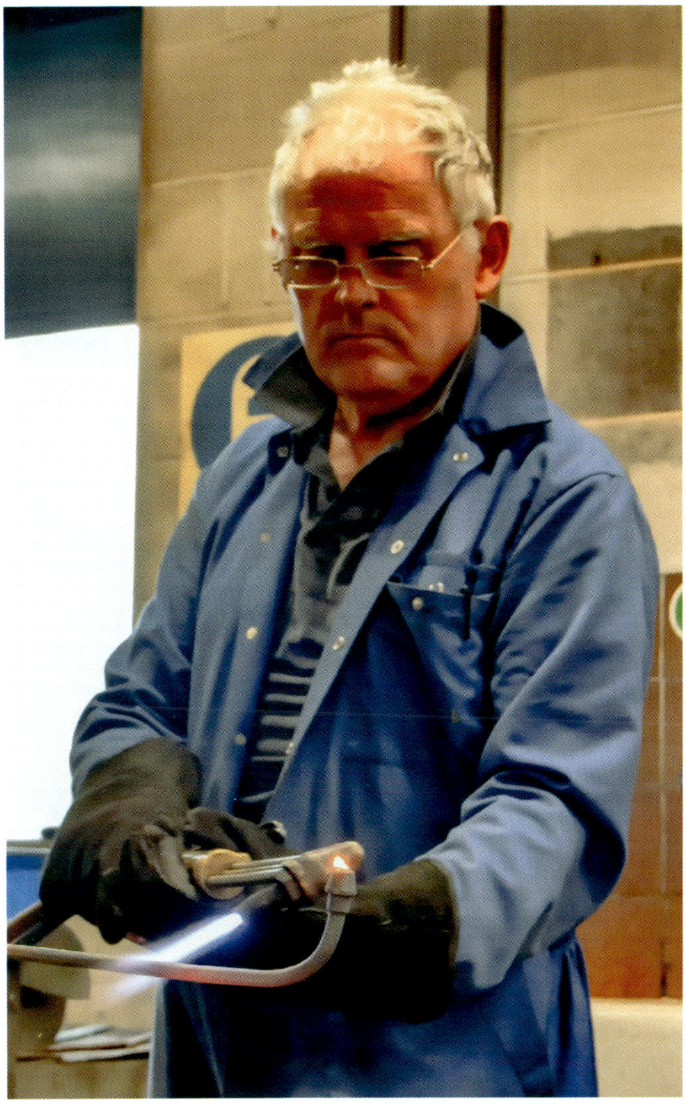

Rod Thomas using an oxy-acetylene burner to reshape pipework

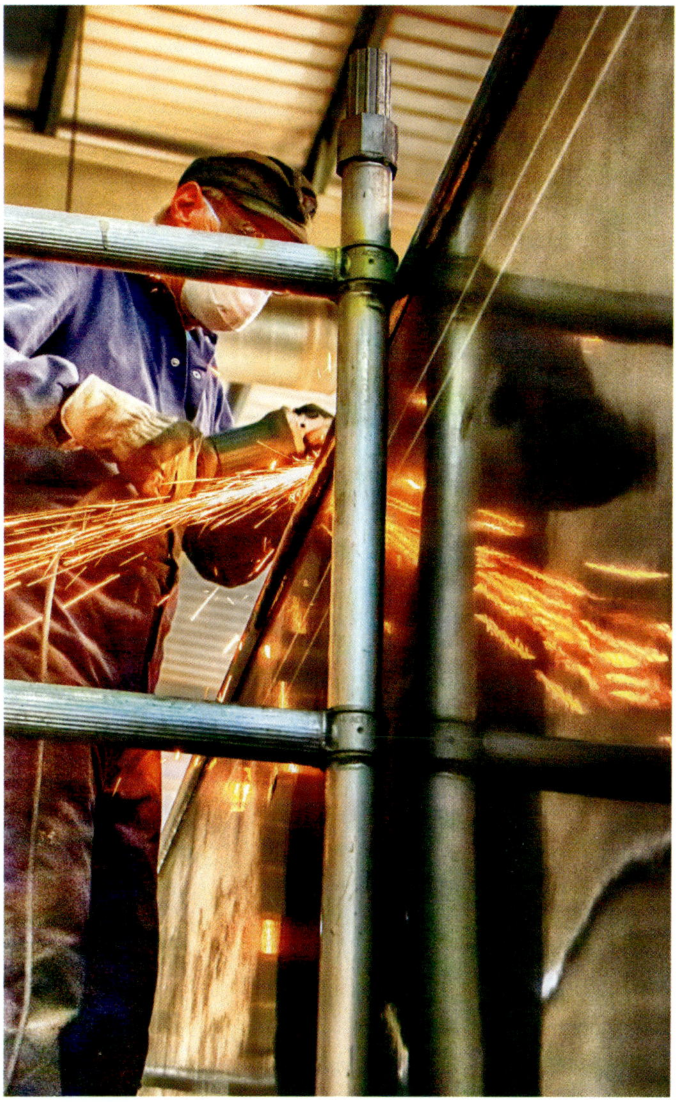

Dave Lee works on the newly installed platework in the tender's corridor roof

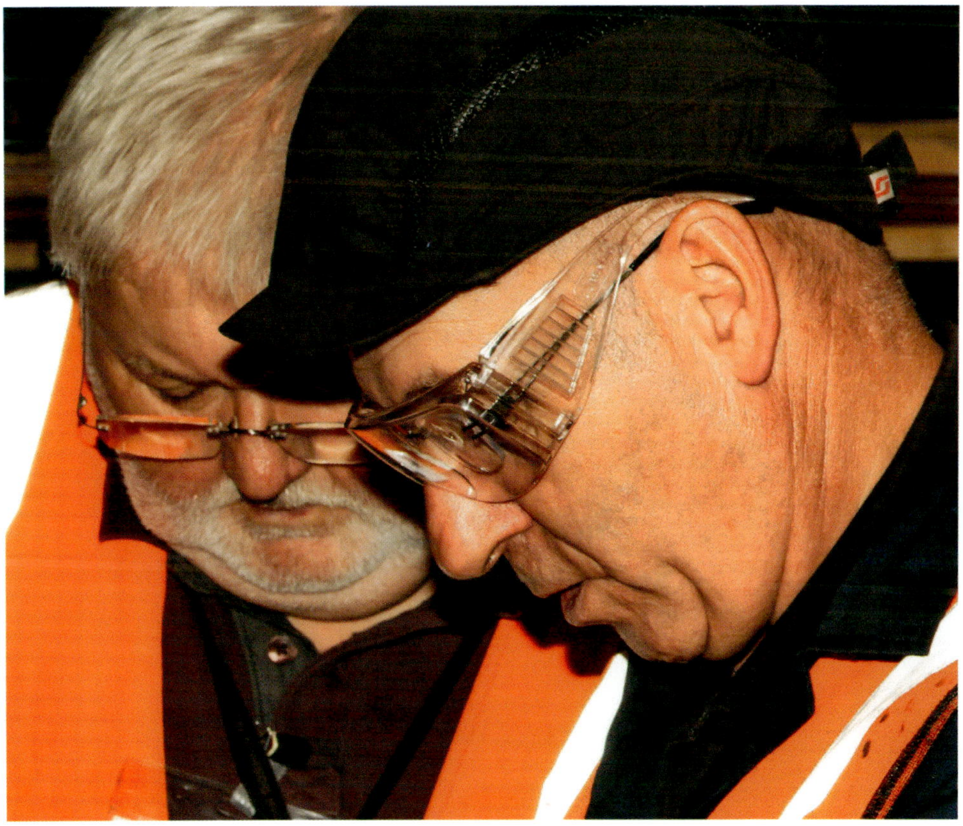

Concentration as Peter Ellis and Barry Horton work together to solve a problem

Barry Horton

A railwayman with a 40 year career at BREL latterly Wabtec Rail Limited at Doncaster where he spent 14 years as Project Manager overhauling, refurbishing and modifying DMUs and EMUs.

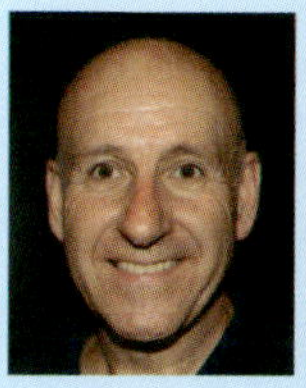

"I took early retirement in March 2019, I looked around for something that would be interesting to do and spotted SNG at the NRM, which offered me the perfect opportunity to get involved with something worthwhile, that interested me and allowed me to use the skills and knowledge I'd learnt over the years. To be involved with such an iconic locomotive is an absolute pleasure and something which I feel extremely proud of."

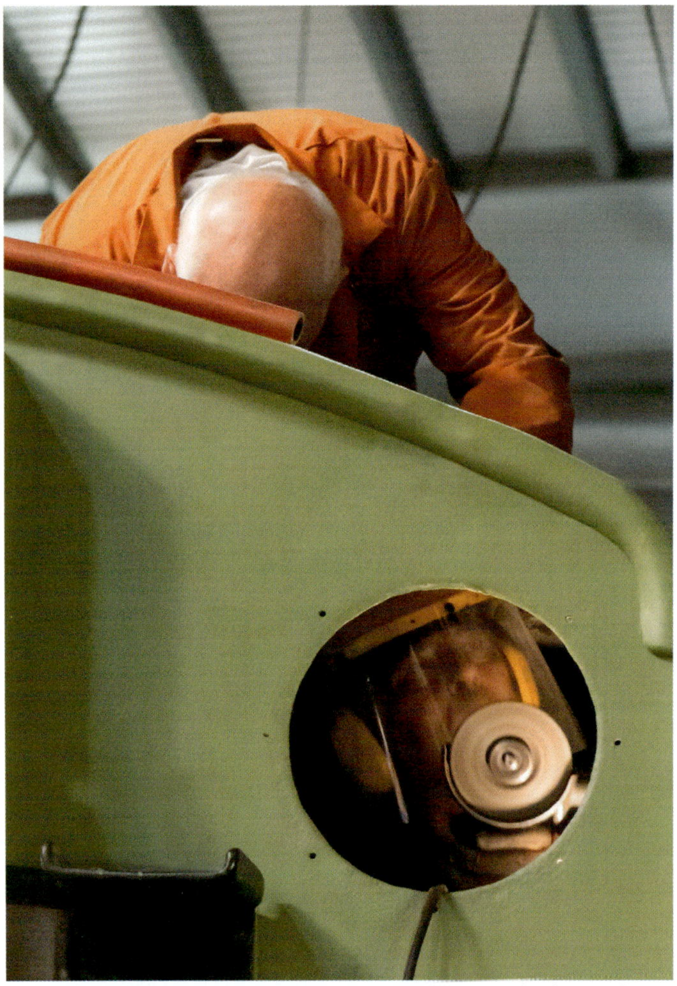

Looking like a spaceman Dave Lee working inside the tender corridor with the help of Trevor Wilford above him

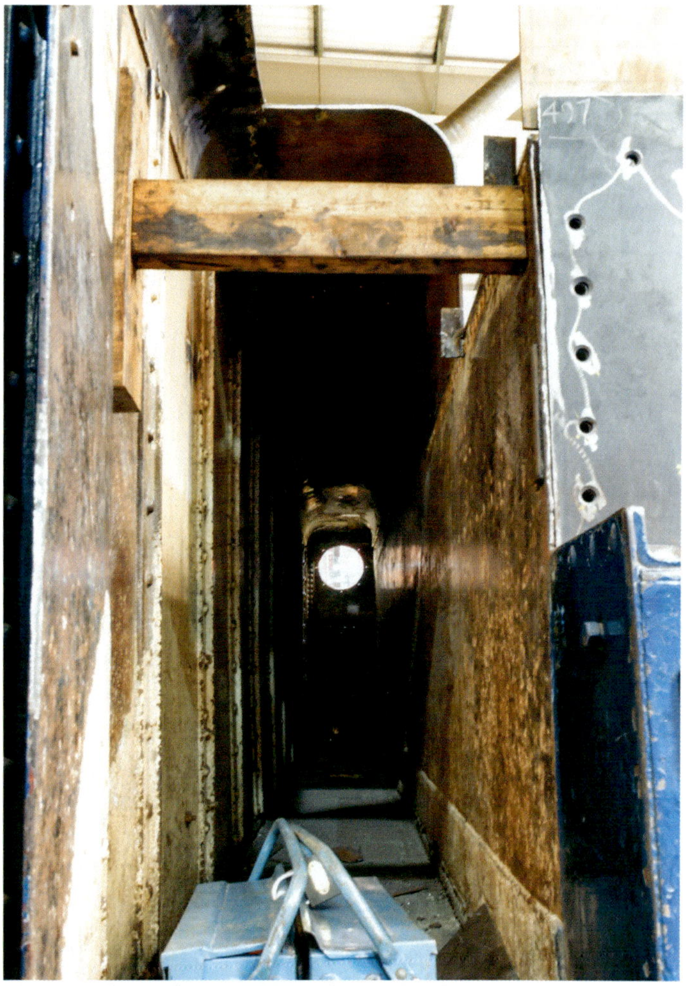

The view down the corridor in the tender showing newly installed steel roof plating which was replacing badly corroded sections

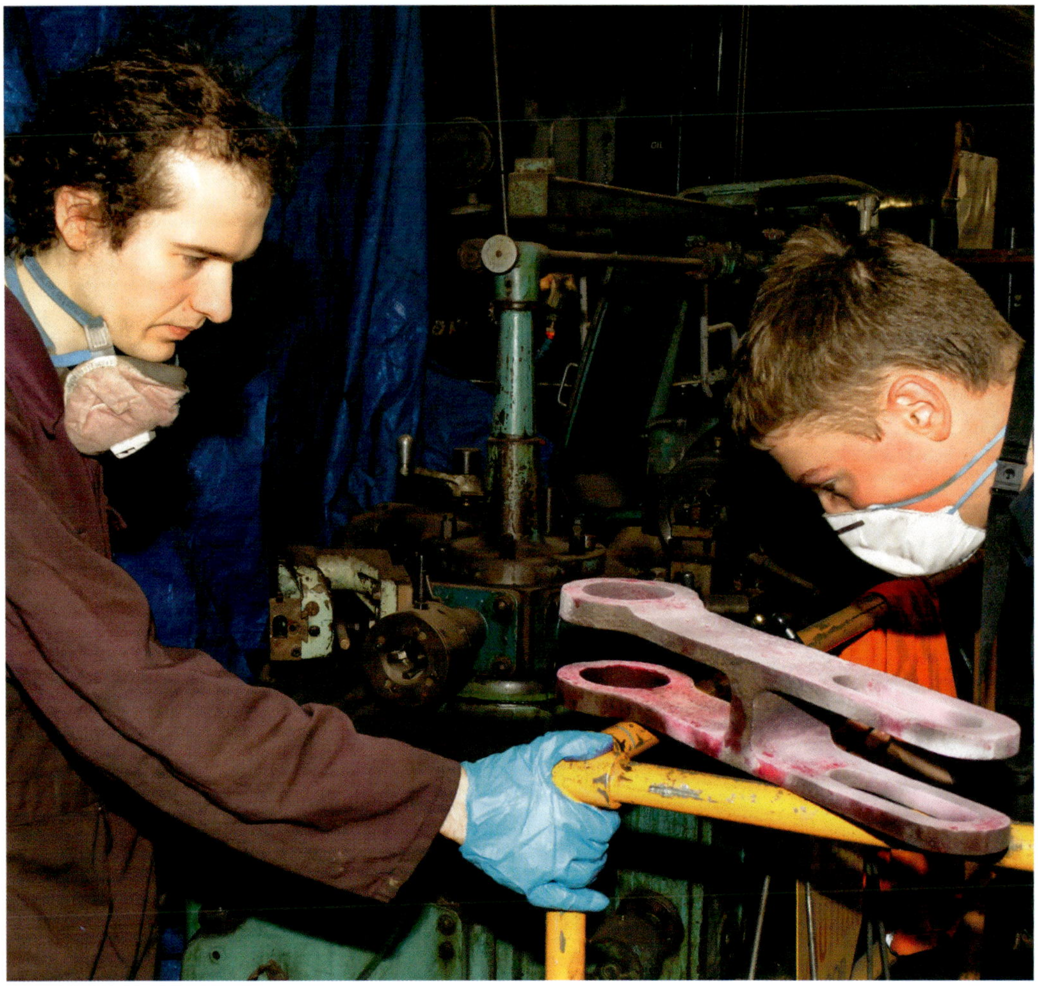

Andrew Jacques watches on as Scott Middlemiss checks for cracks using dye penetrant in a linkage

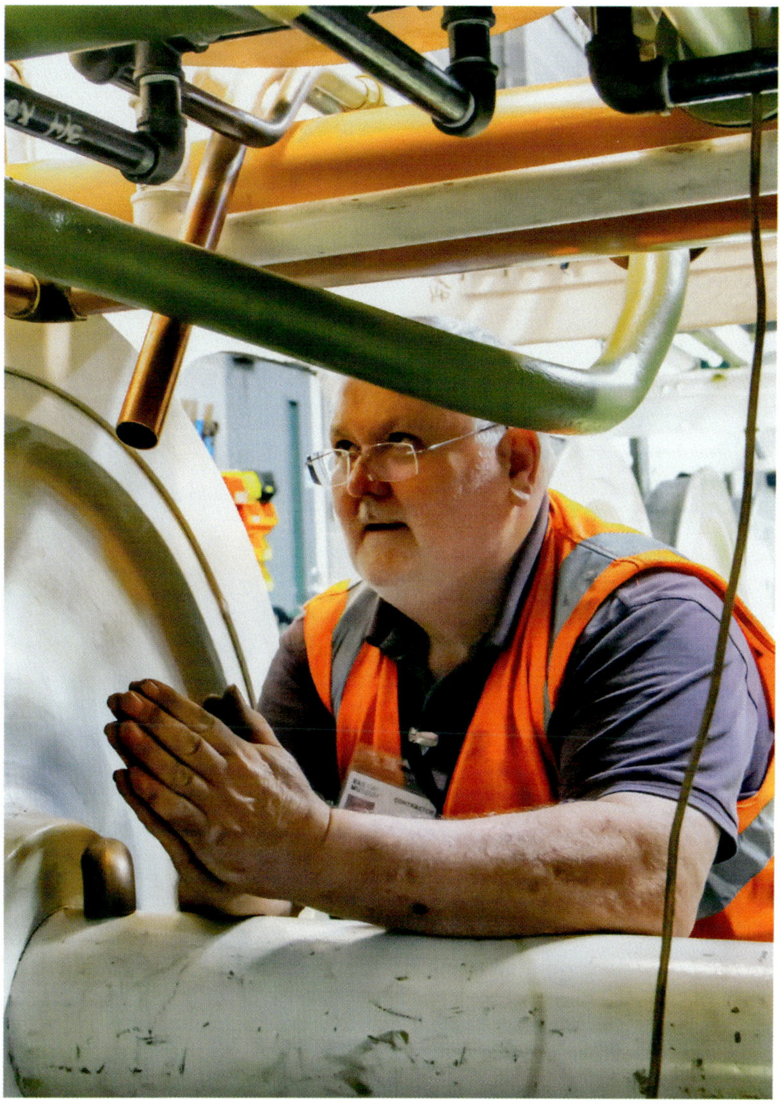

Peter Ellis is praying for devine intervention that his modified pipework under the tender fits well

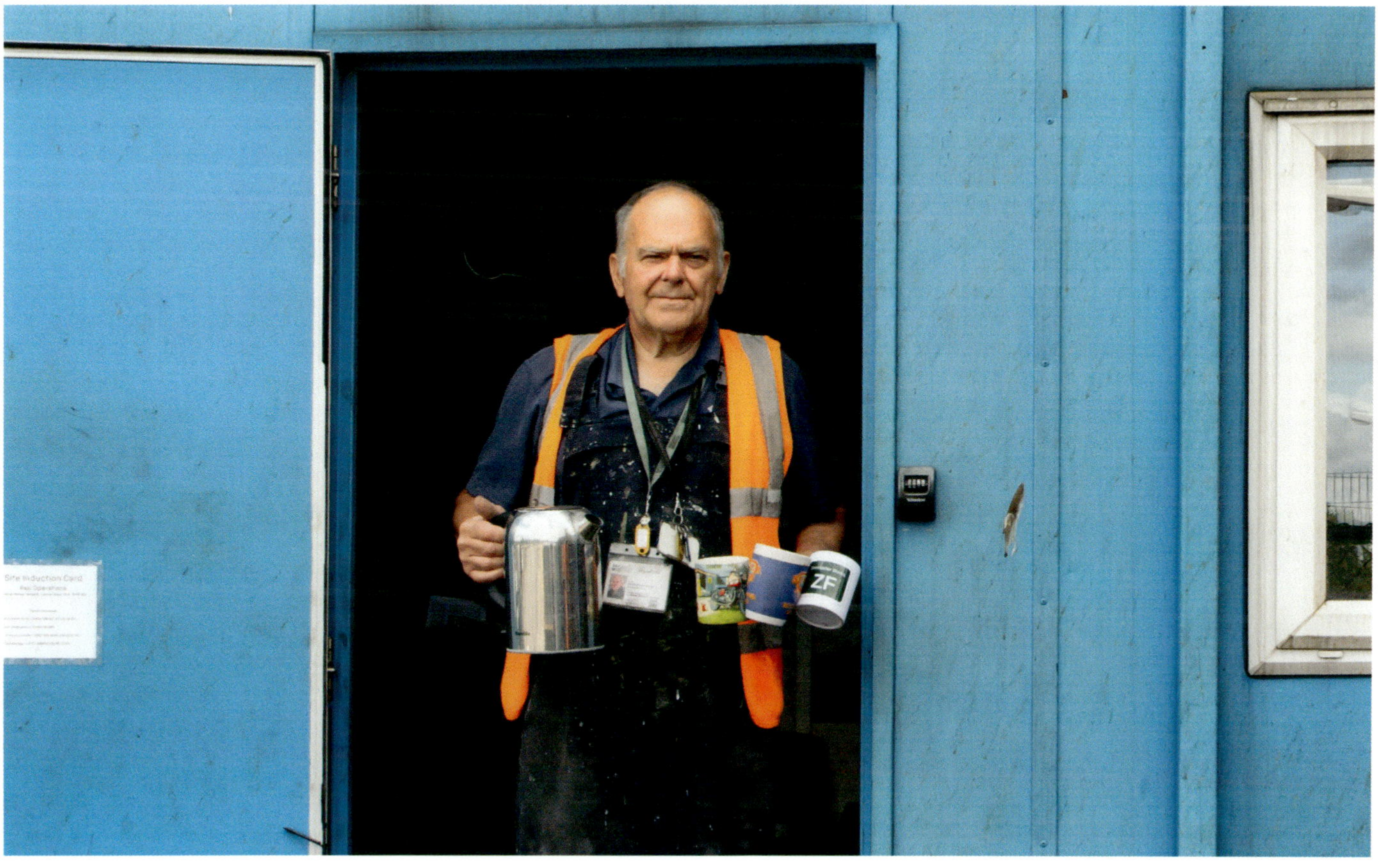

Peter Brackstone emerges from the blue cabin with more cups of tea for the team

TENDER COMES TOGETHER

September 2019

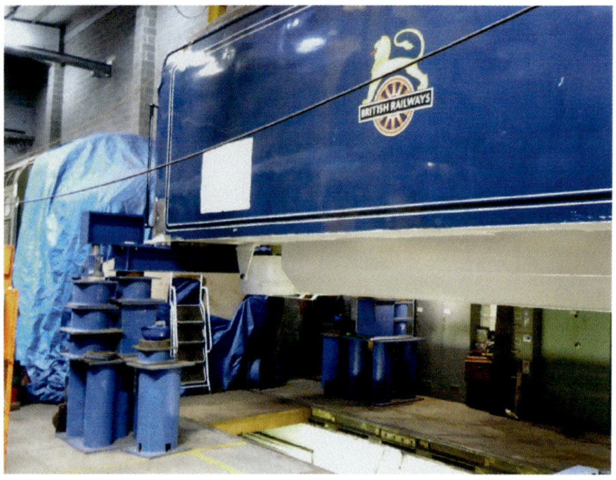

The tender is raised on stools to allow access for the frames and wheels to be pushed back into position

I couldn't attend the workshops to photograph the reassembly of the tender wheels with the frames and subsequently the tank due to an unexpected break which I will explain. So, I am grateful that Malc Bateman sent me a few of his images of the tank being reunited with the frames and wheels.

My photography in recent years was very much concentrated on landscapes and for practical reasons carrying my kit around was best accomplished with my photographers' backpack. I had never considered a handheld bag or case as they were just not suitable when hiking along countryside pathways or climbing up and down moors and fells. But, things had changed and now I was spending most of my shooting days at the NRM, which didn't really require a backpack, so I thought I will take the burden off my ageing back and get myself a nice new wheeled photographers' trolley case, preferably one that converts to a backpack for when I do go hiking for my shots. Trouble is I just never got around to it. With my next day at the museum approaching and again forgetting to get around to buying the new case, I had a brilliant idea for a temporary fix. These brilliant ideas can be dangerous and so this one proved to be. Somewhere in my loft I had a small wheeled case, the sort that can go onto an airliner as carry on luggage, so, if I used that and put my backpack inside it would do for my next trip and I would definitely go and buy a new trolley case afterwards.

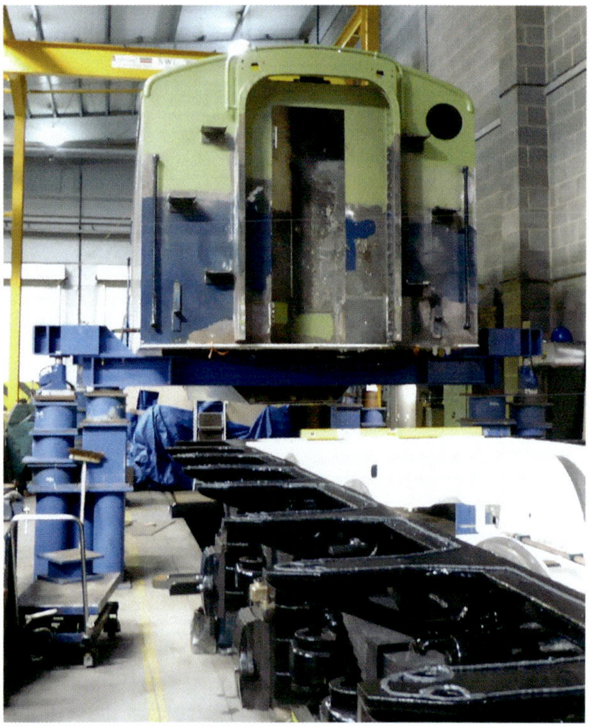

The frames and wheels ready to be pushed back under the tank

Off I went to my loft. Now where is it? Oh yes it would be over there where I haven't got any loft flooring. Never mind, just hang onto this roof beam and be careful where you put your foot. Thing is I wasn't careful enough. The last thing I remember is seeing daylight coming up through the hole in the bedroom ceiling I had just created................. after that nothing until I came round, thankfully, on the floored section of the loft with exceptional pain in my right arm. Looking on the bright side holding onto the beam meant I was still in the loft and not on the bedroom floor. The downside was my right arm's humerus had snapped under the strain of hanging on.

After a trip to A&E it would be many weeks before I got back to the workshop in York, so I thank you Malc Bateman for supplying pictures of the tender reassembly which you see here. This was an unexpected break in proceedings in more ways than one.

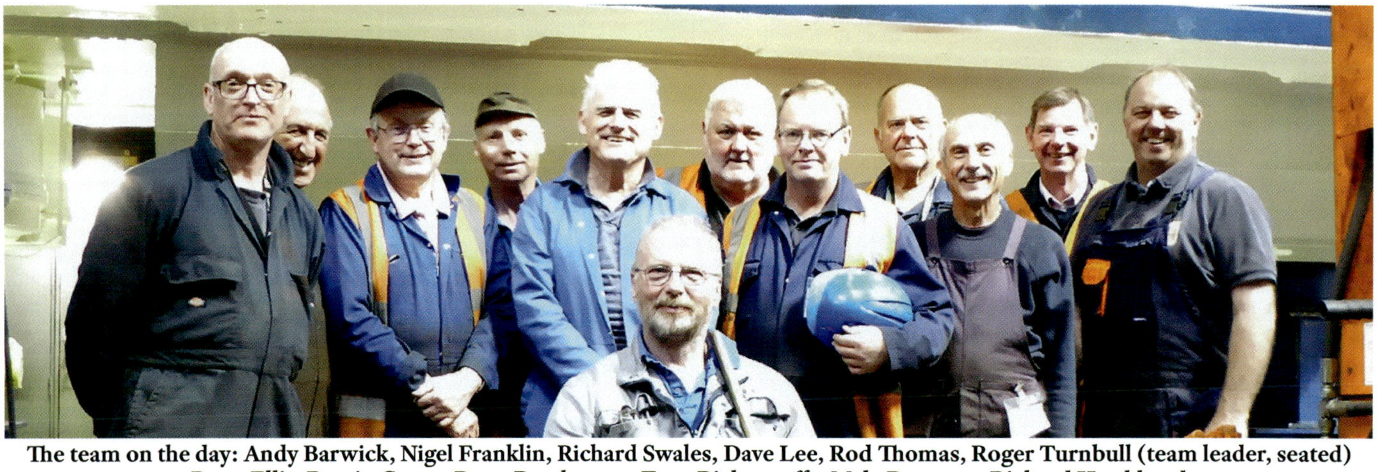

The team on the day: Andy Barwick, Nigel Franklin, Richard Swales, Dave Lee, Rod Thomas, Roger Turnbull (team leader, seated) Peter Ellis, Darrin Crone, Peter Brackstone, Tony Bickerstaffe, Malc Bateman, Richard Hawkhead

BOILER RETURNS

11

November 2019

"The 9:08 will be fine"

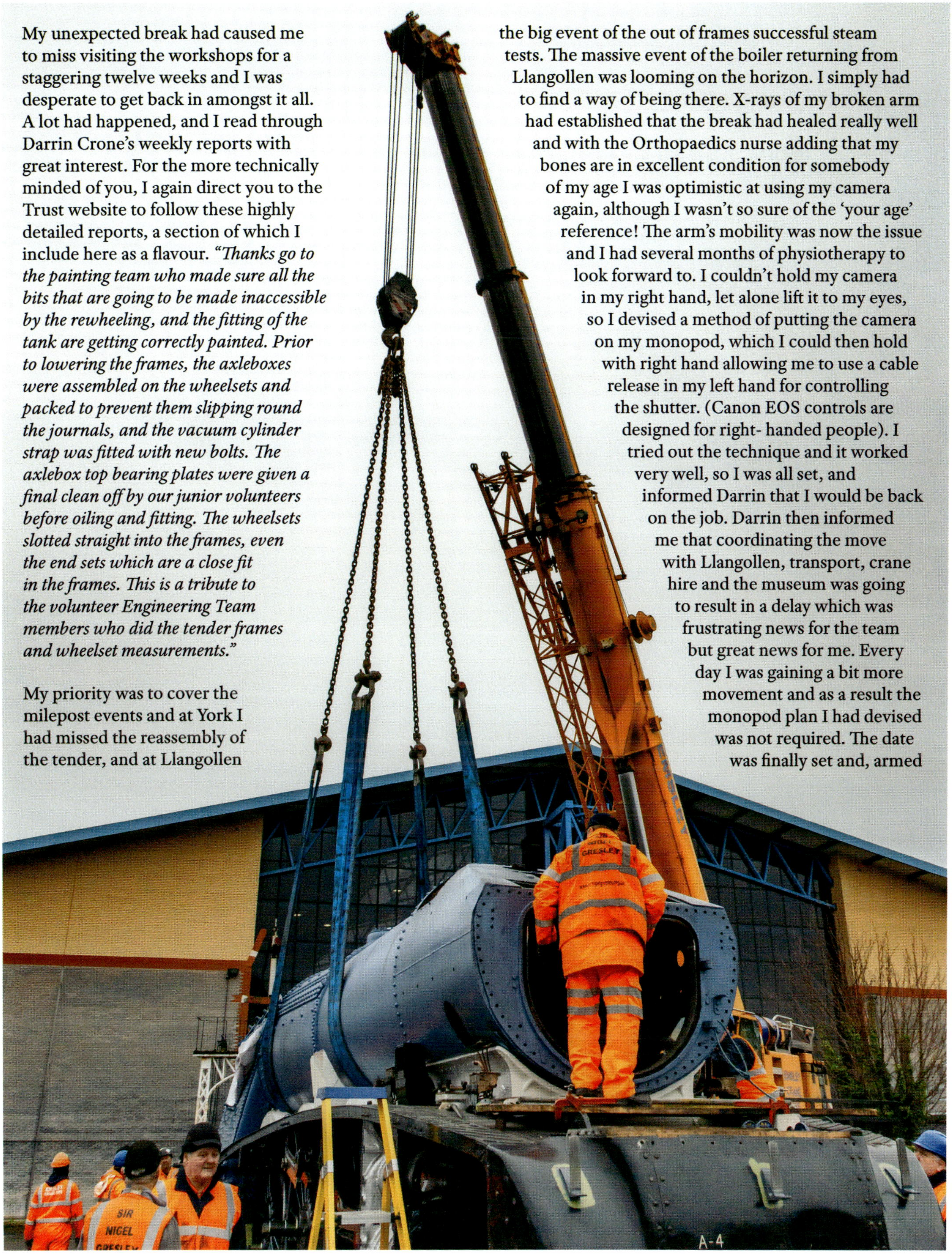

My unexpected break had caused me to miss visiting the workshops for a staggering twelve weeks and I was desperate to get back in amongst it all. A lot had happened, and I read through Darrin Crone's weekly reports with great interest. For the more technically minded of you, I again direct you to the Trust website to follow these highly detailed reports, a section of which I include here as a flavour. *"Thanks go to the painting team who made sure all the bits that are going to be made inaccessible by the rewheeling, and the fitting of the tank are getting correctly painted. Prior to lowering the frames, the axleboxes were assembled on the wheelsets and packed to prevent them slipping round the journals, and the vacuum cylinder strap was fitted with new bolts. The axlebox top bearing plates were given a final clean off by our junior volunteers before oiling and fitting. The wheelsets slotted straight into the frames, even the end sets which are a close fit in the frames. This is a tribute to the volunteer Engineering Team members who did the tender frames and wheelset measurements."*

My priority was to cover the milepost events and at York I had missed the reassembly of the tender, and at Llangollen the big event of the out of frames successful steam tests. The massive event of the boiler returning from Llangollen was looming on the horizon. I simply had to find a way of being there. X-rays of my broken arm had established that the break had healed really well and with the Orthopaedics nurse adding that my bones are in excellent condition for somebody of my age I was optimistic at using my camera again, although I wasn't so sure of the 'your age' reference! The arm's mobility was now the issue and I had several months of physiotherapy to look forward to. I couldn't hold my camera in my right hand, let alone lift it to my eyes, so I devised a method of putting the camera on my monopod, which I could then hold with right hand allowing me to use a cable release in my left hand for controlling the shutter. (Canon EOS controls are designed for right- handed people). I tried out the technique and it worked very well, so I was all set, and informed Darrin that I would be back on the job. Darrin then informed me that coordinating the move with Llangollen, transport, crane hire and the museum was going to result in a delay which was frustrating news for the team but great news for me. Every day I was gaining a bit more movement and as a result the monopod plan I had devised was not required. The date was finally set and, armed

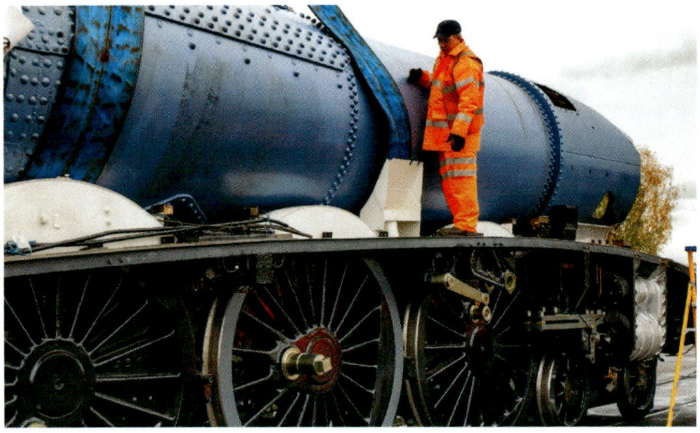

Adjusting the crane strops

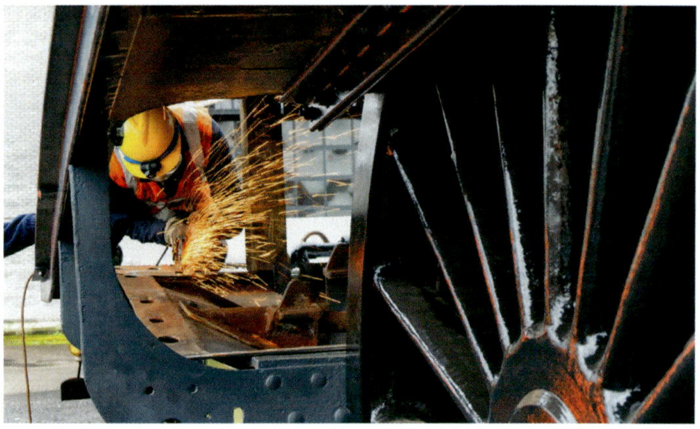

Andy Barwick grinding the ashpan to create a good fit with the boiler

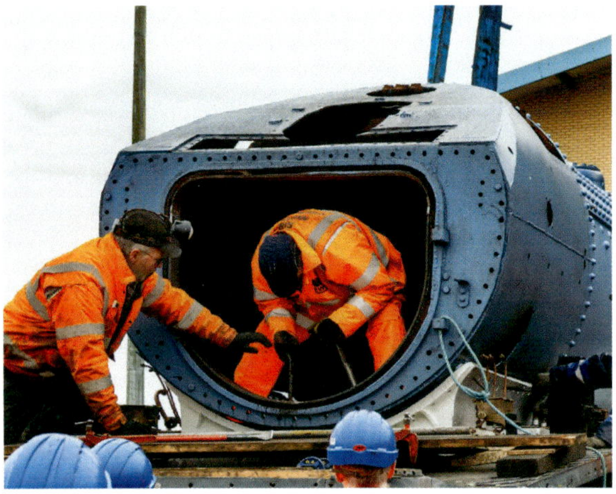

Lining up the boiler's bolt holes with those on the saddle

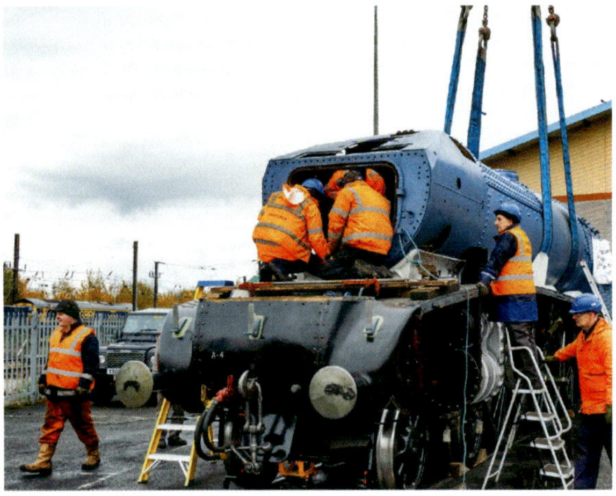

Checking hole alignments

with my new camera trolley bag, (yes I did finally buy one), I booked my Grand Central rail tickets as I was making progress but not enough to consider driving, but at least I was back in the game.

"The 09:08 will be fine," Darrin said, adding "The crane isn't booked until 08:30 and by the time everybody has had a cuppa and a chat nothing much will have happened until after your train arrives." My train arrives by passing the Railway Museum on the approach to York station, so as I was making my way to the carriage door, I was astonished to see the boiler hovering over the top of No.7's frames in the North Yard. Maybe I should have got the 07:08, I muttered under my breath. Now having to rush from the platform over the footbridge and up Leeman Road I was so pleased I had bought that new trolley case.

Except for me, everything had come together on time, and with a Yellow weather warning in place for prolonged heavy rain, it had been decided to lift the boiler directly from the road transport onto No.7's frames instead of to a resting place on the ground first whilst all took stock. I had missed the lift from the huge lorry but there were still plenty of picture opportunities to be had and I started testing my aching arm pretty much to the limit as I fired off just over 300 shots on the day.

For the engineering team this was a two-day event, as the night before No.7's frames had to be shunted from the workshop and into the North Yard. This involved passing through the museum's Great Hall, moving the Stirling Single, No.1, steam loco, deltic D9002 and the TPO carriage, none of which could be done whilst the museum was open, so it was a late night for all. The shunting was undertaken by the museum's staffed shunter who also shunted the Class 31 diesel from the adjacent track in the workshop to allow the tender to be

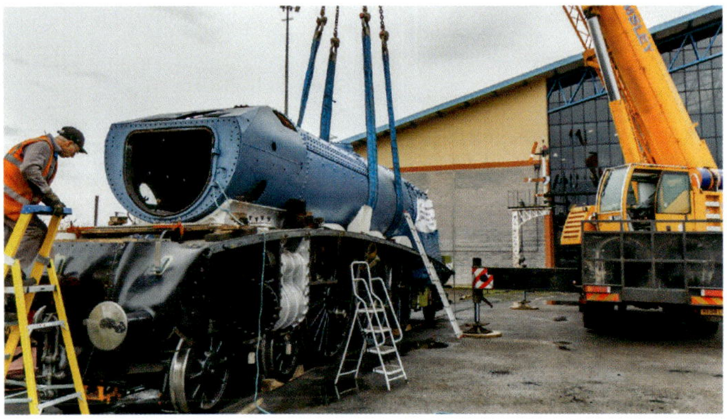

The boiler is lifted slightly to allow access for sealant and ashpan grinding

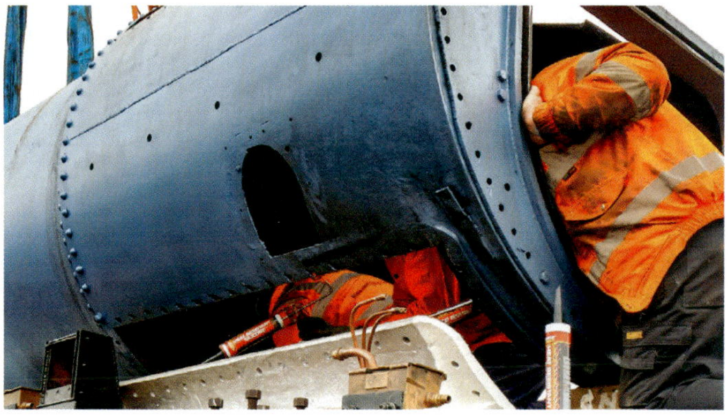

Sealant is applied to the saddle before the boiler is lowered fully into position

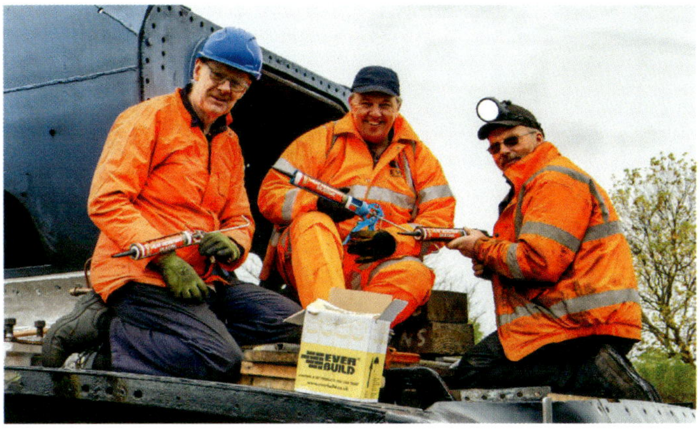

Phil Slater, Richard Hawkhead and Roger Turnbull with tubes of heat resistant sealant to bond between the boiler and the saddle

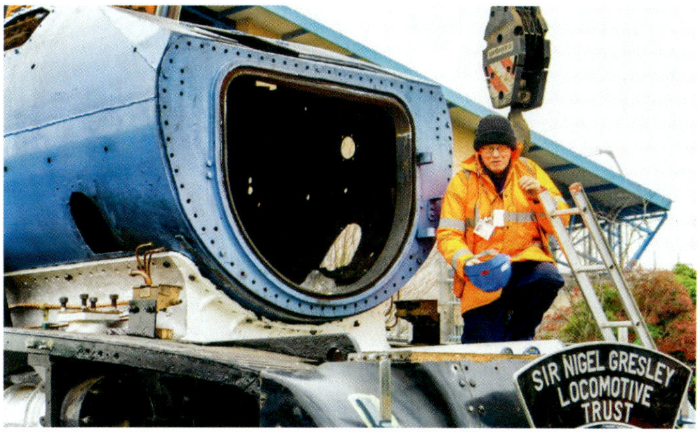

Darrin Crone poses for a picture as he climbs up to inspect the secured boiler on the saddle

moved and repositioned on the same workshop track as No.7 when all was completed. The shunts were carried out carefully and slowly with the team walking alongside both sides of No.7 and keeping a close eye on proceedings.

The rain was holding off and the transfer was going well. The initial move showed that No.7 was slightly out of square in the crane strops so some minor adjustments were made. Further minor adjustments were made by Andy Barwick grinding part of the ashpan so it did not foul the boiler. Eventually all were happy, and after Phil Slater, Richard Hawkhead and Roger Turnbull had applied a special heat resistant sealant to the smokebox saddle, which had been donated to the Trust by a friend of Nigel Franklin's, the boiler was successfully reunited with the frames and wheels. The boiler was secured into position with bolts between the smokebox and saddle, the pictures for which reminded me of those 60s pranks of seeing how many people you can fit in a mini or a phone box.

The sky was getting darker and a few spots had started to fall but that couldn't dampen the spirits of the assembled team as the multi axle rear steering lorry performed an amazing reverse out of the yard honking his horn as he left. I turned to a smiling Chief Mechanical Engineer Richard Swales and jokingly said, "Santa Specials then?" who equally jokingly replied "Yes, of course."

The heaviest rain had thankfully passed to the south of York and as night-time descended, the team were refreshed with another excellent brew from Peter Brackstone whilst waiting for the museum to close so the shunting could all be repeated and No.7 could be returned to the workshop. Now with all the wheels, a 4-6-2 pacific, and a boiler looking much more like the No.7 we all love.

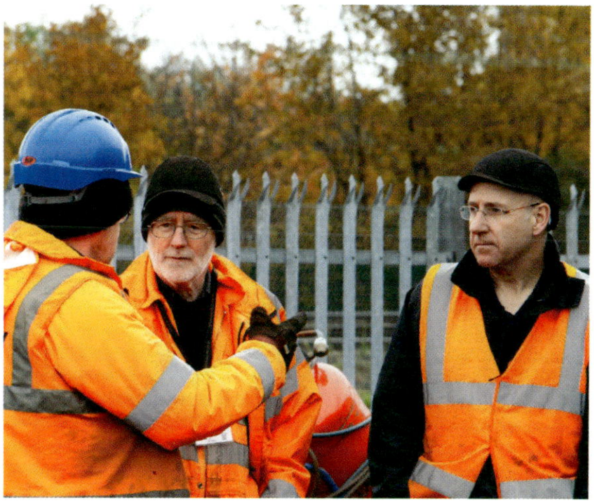

Darrin Crone with Peter James and Barry Horton

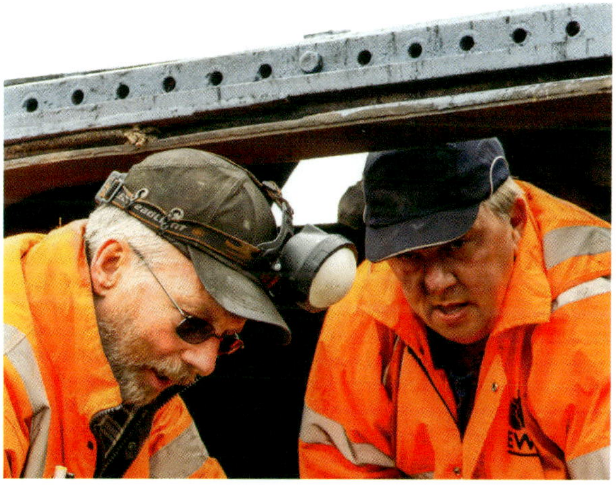

Roger Turnbull with Richard Hawkhead

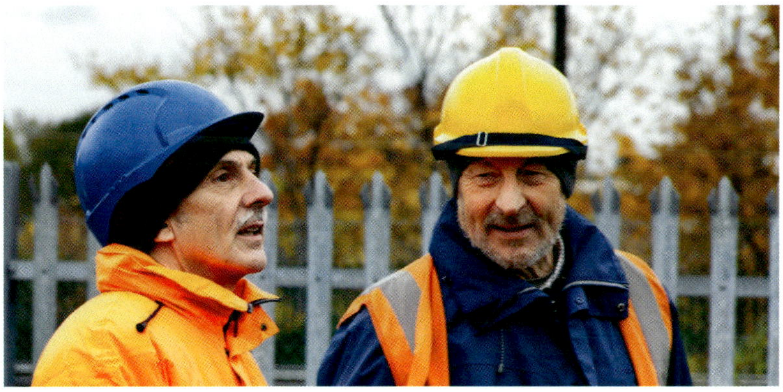

Tony Bickerstaffe with Mike Page

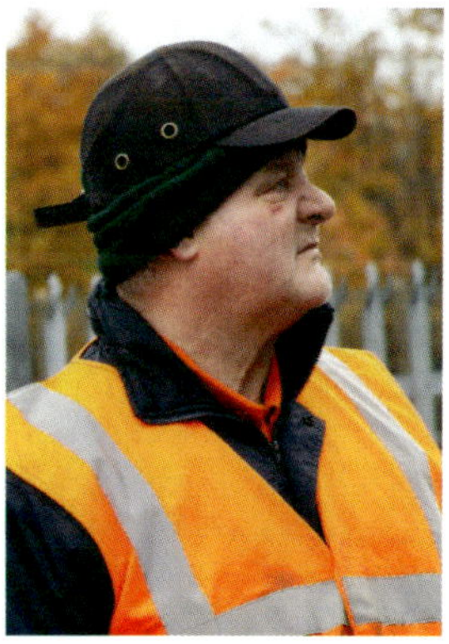

Trevor Wilford

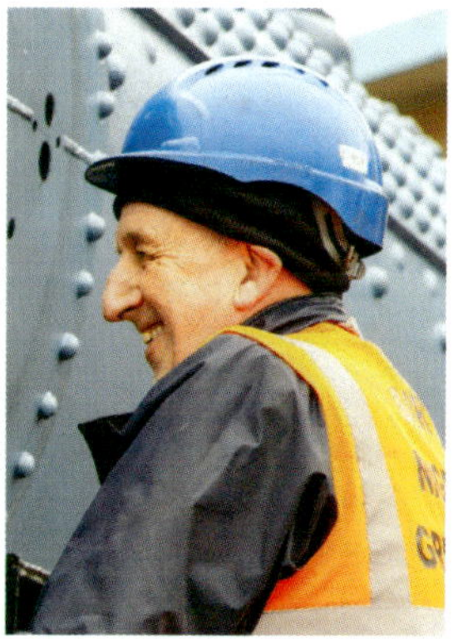

Nigel Franklin

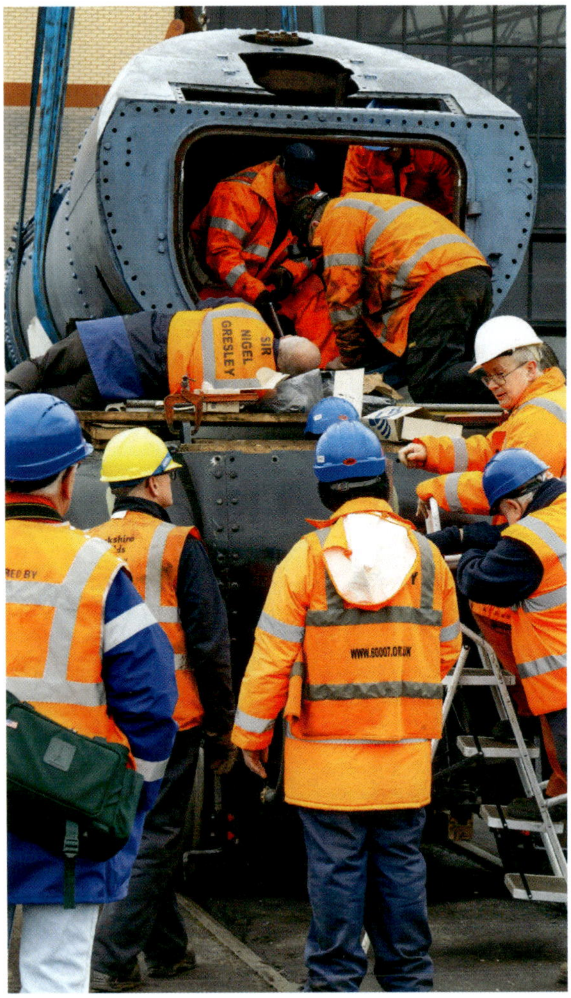

Looking a bit like a 'How many people can you get in a telephone box' competition the team start securing the boiler onto its saddle

Peter James

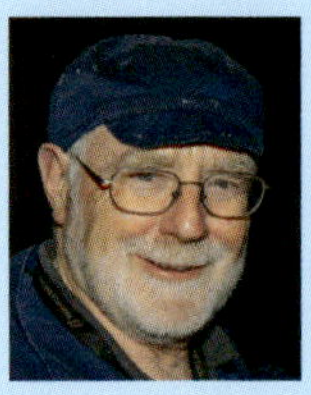

Peter's love of the railways was stoked by his move to York in 1973 to commence his training for his chosen career in nursing. When the NRM opened its doors for the first time in 1975 Peter became a volunteer, where he regularly came across No.7. In the 1980s he was involved in the overhaul of Duchess of Hamilton and in 1990 he joined the volunteer team at NYMR where he learnt to be a fireman, something he has been doing ever since.

Mainline opportunities came with spells on the Jacobite working the line beteween Fort William and Mallaig looking after NELPG's K1 (62005) and with West Coast Railways' summer Scarborough – York steam service. The latter resulted in an invitation to work on Flying Scotsman, an offer he couldn't turn down. "As it happened, I was coming up to 55, an age when I could retire from nursing, so it all worked out really well for me."

Peter adds, "For the last 15 years I have been employed by West Coast as both a fireman and a guard. On 5th July 2015 I was privileged to be the rostered fireman on 60007 from Newcastle to York and later that day from York to Grosmont, this being her last in steam movement on the network to date. Later that year, on 17th November, I was the guard on her move with 47798 from Grosmont to the NRM for overhaul. These days I still work for West Coast as required, I still fire regularly on the NYMR and I still volunteer at the NRM. I was able to be involved with the overhaul at York until the loco left for Crewe in September. Hopefully I will be able to help out on her again sometime in the future."

Nigel & Sheila Franklin

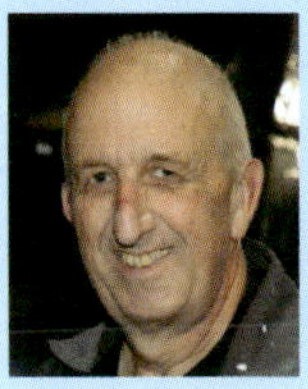

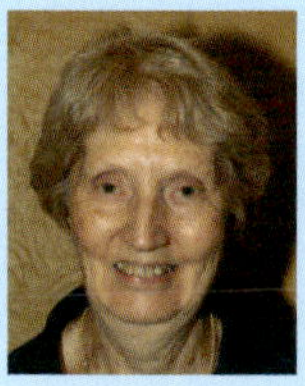

Nigel has vast experience in aviation engineering with Hawker Siddeley, Laker Airways, BAE, the Sultan of Oman's Airforce and British Airways before setting up his own aviation engineering company with his wife, Sheila, and carrying out contracts with companies such as Easyjet. They both started volunteering for the Trust after they moved to Yorkshire from the south of England in 2018 where Nigel was a volunteer on the Bluebell Railway.

Nigel adds "It's nice to work for Darrin and his team of skilled Engineers as they are a great bunch of guys with plenty of sharp wit and fast Northern banter..... I have enjoyed my time learning different things about rebuilding this great project."

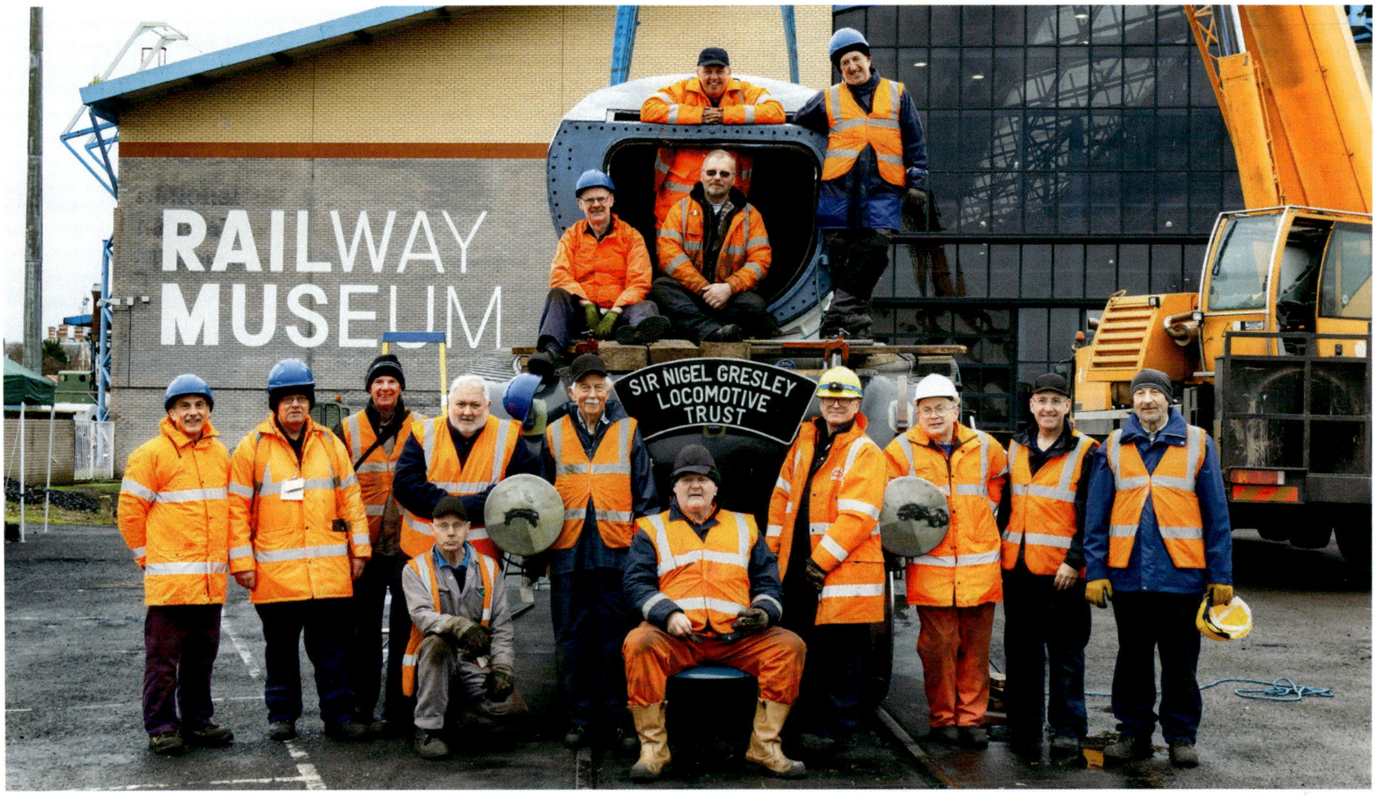

The Boiler Return Team LEFT to RIGHT - On boiler Richard Hawkhead and Nigel Franklin (standing), Phil Slater and Roger Turnbull (sitting) Front row Tony Bickerstaffe, Darrin Crone, Malc Bateman, Dave Lee (kneeling), Peter Ellis, Bob Shearman, Trevor Wilford (seated) Andy Barwick, Richard Swales, Barry Horton and Mike Page

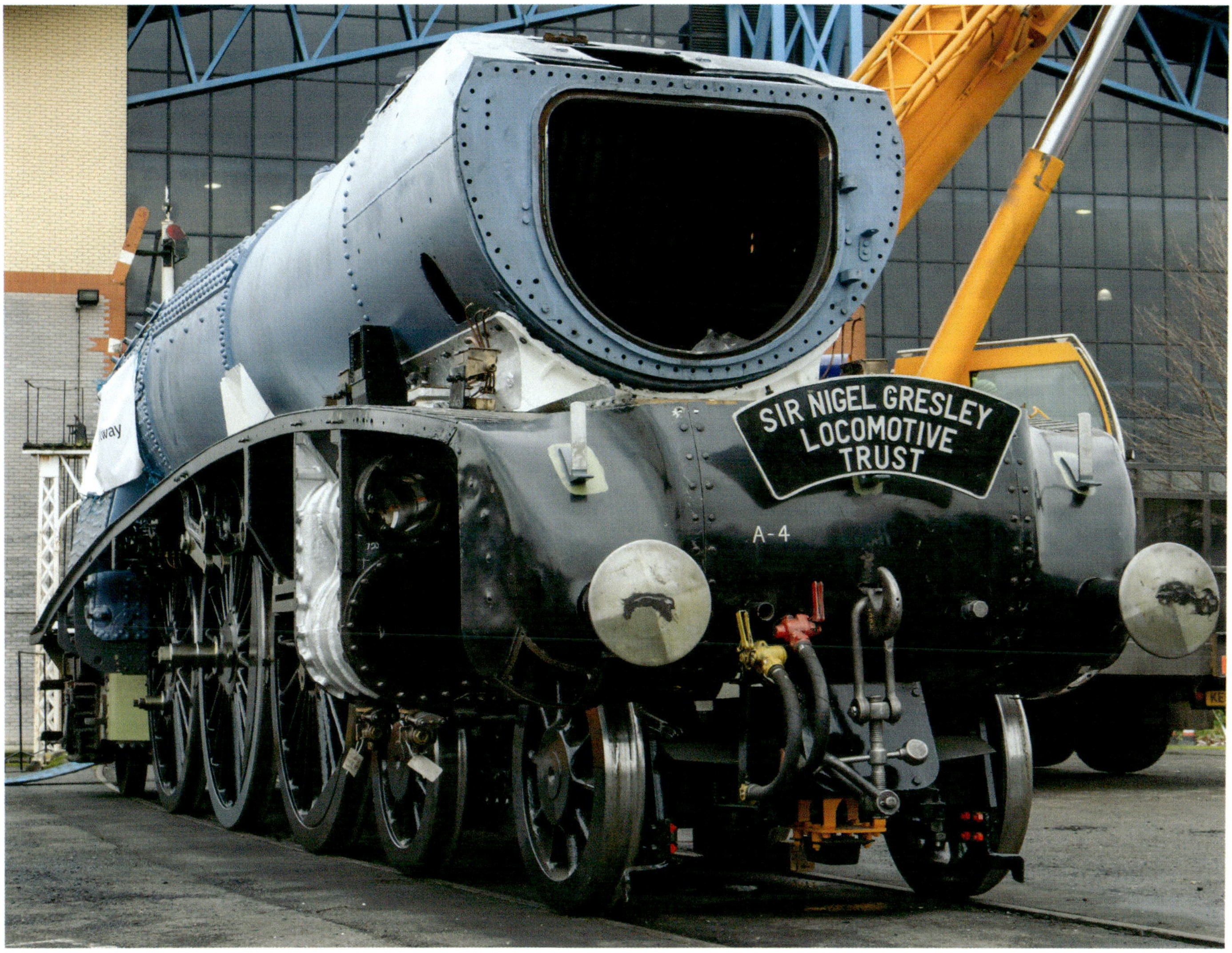

The boiler sits proudly back on top of the wheels and frames

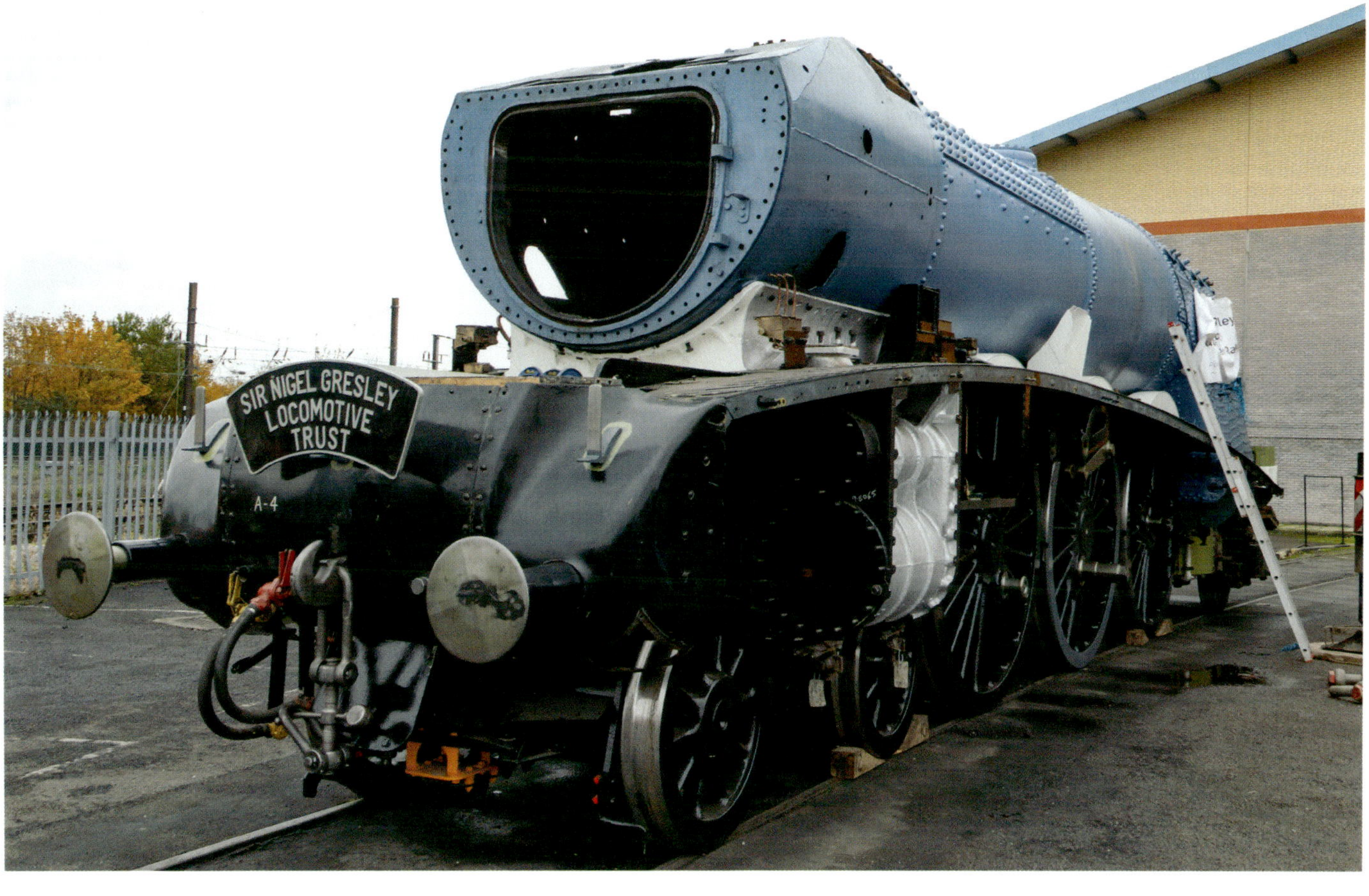

CARTAZZI

November 2019

My arm was getting stronger by the day and my physiotherapist had given me the all clear to drive again but I decided this side of Christmas I would limit that to short journeys and would stick with the trains for anything over 10-15 miles. Sounded a reasonable plan as armed (sorry, couldn't resist) with my E-ticket I drove the two miles to Hartlepool station, paid the all-day parking charge and set off to the platforms for my 08:53 train to take me on my next visit to the workshops in York. "Cancelled due to lack of train crew" was the message I found, the alternative being a replacement bus to Darlington and onwards with any train service to York. I decided this would turn into a marathon particularly as there was no sign of a bus, so I would trust in the physio's advice and drive to York. I must add, my sadiotherapist was right, my arm was fine. I managed the drive both ways and took all the camera shots relatively easily.

On arrival at the workshop carpark I found the "Duchess of Sutherland" in the prep bay steaming away very nicely, so much so that I could not see the ground as I approached the blue cabin and I nearly had another fall to contend with. The temperature in York was still below freezing and the steam from the Duchess was condensing onto the tarmac and instantly freezing, turning the ground into an ice rink. Now, I really was pleased I had bought that trolley bag for my camera as when I slipped the pull on the bag jammed the

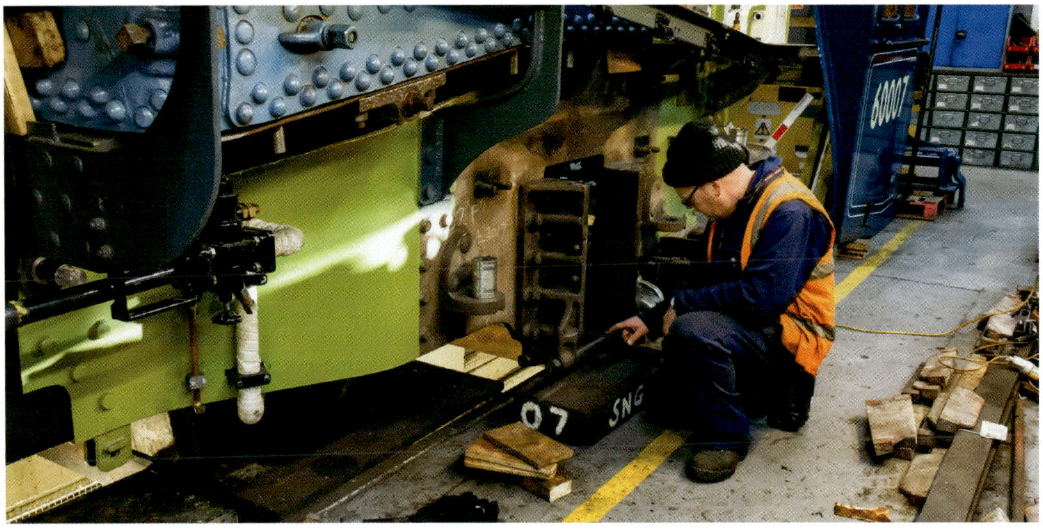

Darrin Crone inspects the bars across the cartazzi horns

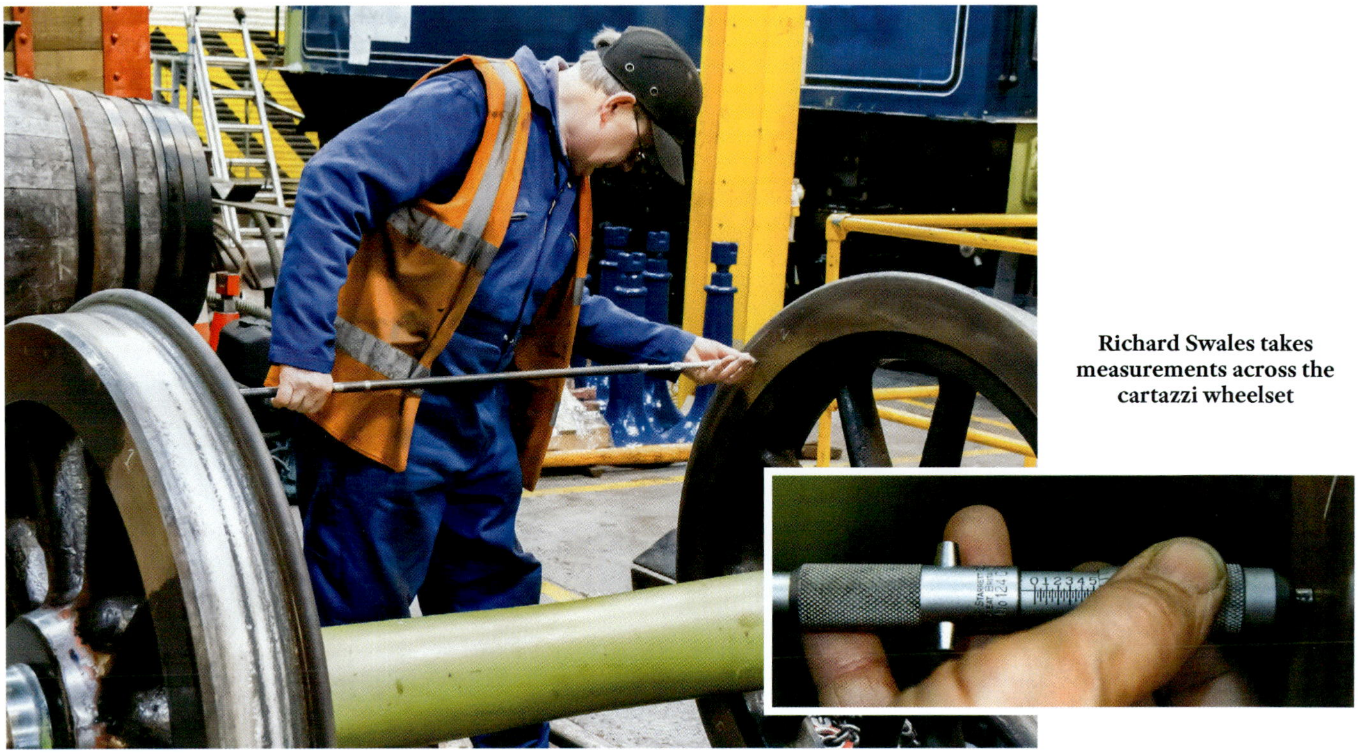

Richard Swales takes measurements across the cartazzi wheelset

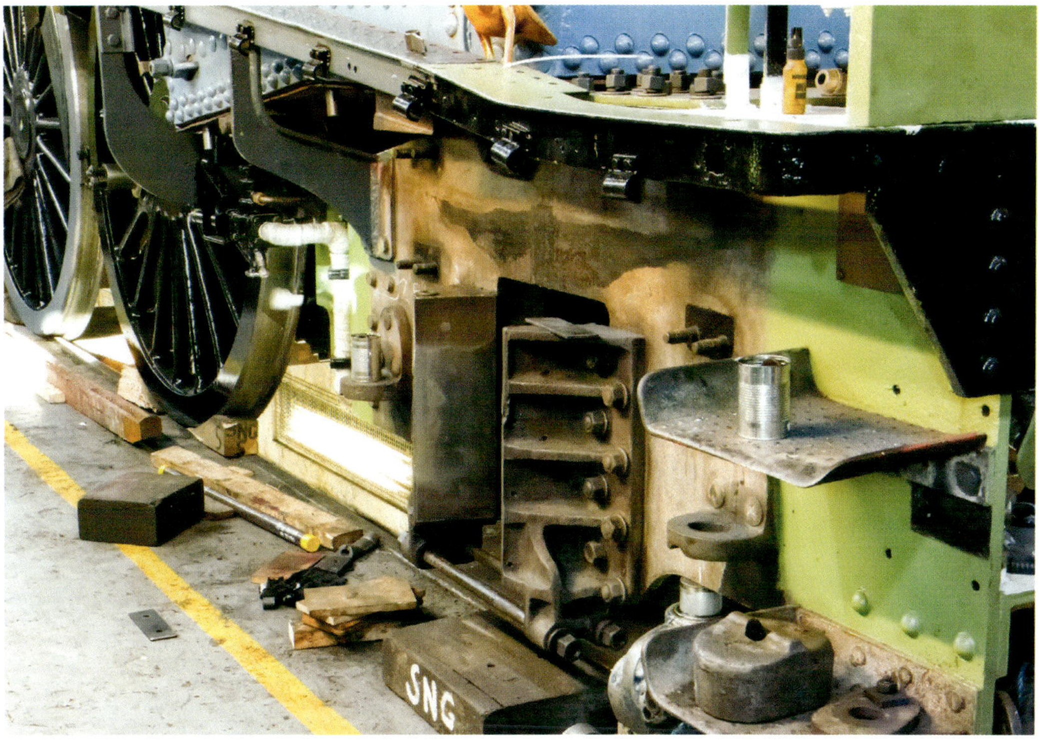

The cartazzi frame below the cab, unlike a pony truck on most steam locomotives which move around a pivot point, is a fixed frame with angled axleboxes guiding the wheels around curves

wheels and offered me a crutch to steady myself on. By this point of the morning I was beginning to think I should have stayed in bed.

When I finally arrived in the workshop, I saw Richard Swales working on the cartazzi wheels, No.7 being back as a 4-6-2 'pacific' turned out to be only temporary! The wheelset under the cab, on a Gresley pacific, is called a cartazzi and is named after the engineer who designed it and it was this wheelset's turn to be overhauled. So, out they came and No.7 was now, a 4-6-0.

The cartazzi is a very cleverly designed trailing wheelset which could be confused with a pony truck which it isn't, as unlike the pony truck it does not pivot. It rides the curves in the track by using angled horns in the frames and the axleboxes themselves have multiple angled brass plates which keep the wheelset self-centred and which control

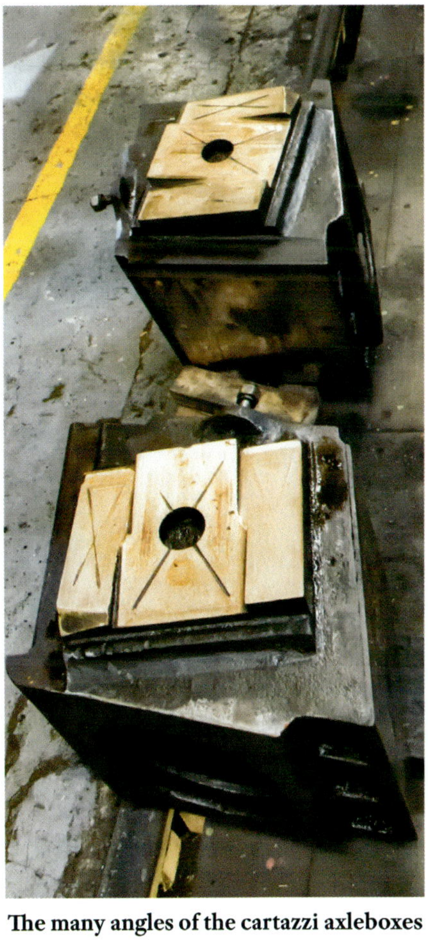

The many angles of the cartazzi axleboxes

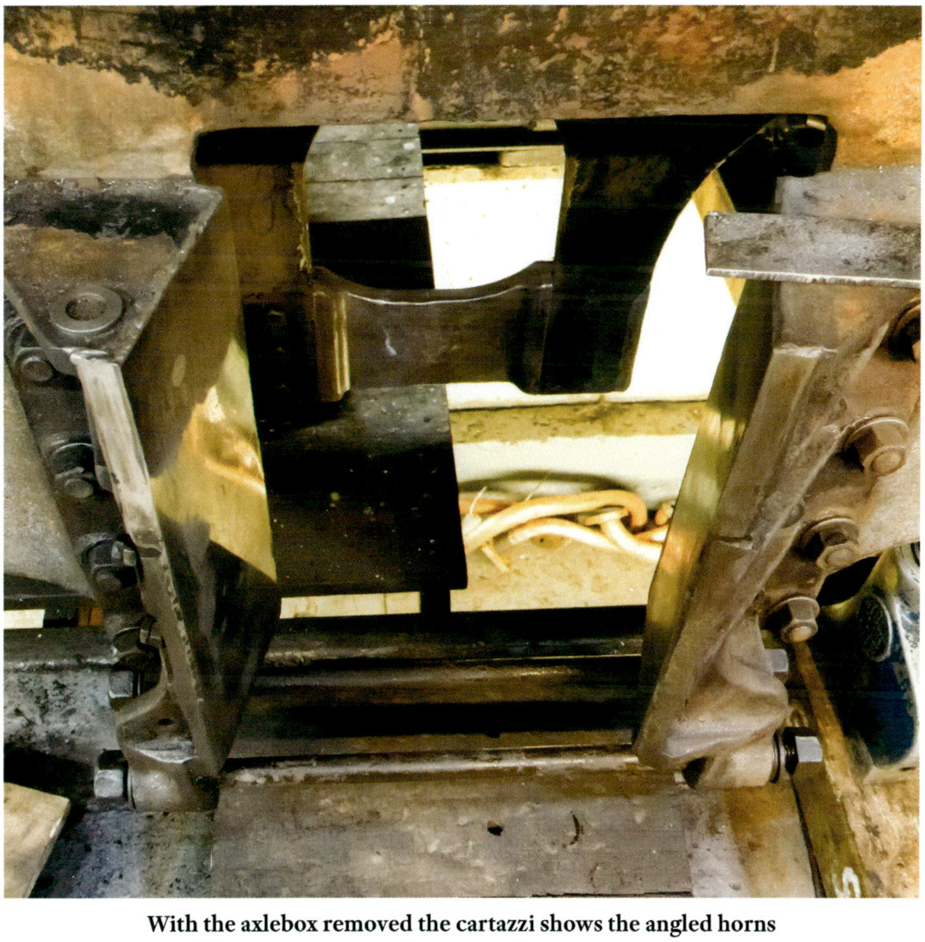

With the axlebox removed the cartazzi shows the angled horns

vertical and horizontal movement on curves. Richard was now busy taking lots of measurements to ensure the overhaul of the wheelset resulted in the best efficient operational use.

It was good being in the workshop on a Saturday as there are different faces to meet which I am keen to do and today Andy Evans and Chris Whiteley were in attendance - all working men who give up their weekends to volunteer to work on No.7. Andy Evans was working on brake pins whilst Chris Whiteley was renovating some pipe elbows.

This was the first time I had been back in the workshop since the boiler had been refitted and it was important that I got some close up shots of that. There had been many pics on social media taken from the gallery and I wanted to supplement them by showing some up close and personal shots. I decided to position the camera on the running boards and go for some HDR shots, for the photographers that worked out at an iso of 200 and at f16 three exposures of 2.5 seconds, 8 seconds and 25 seconds; yes, the workshop lights are not good for photography. This is something I have had to deal with and over the current four years of the

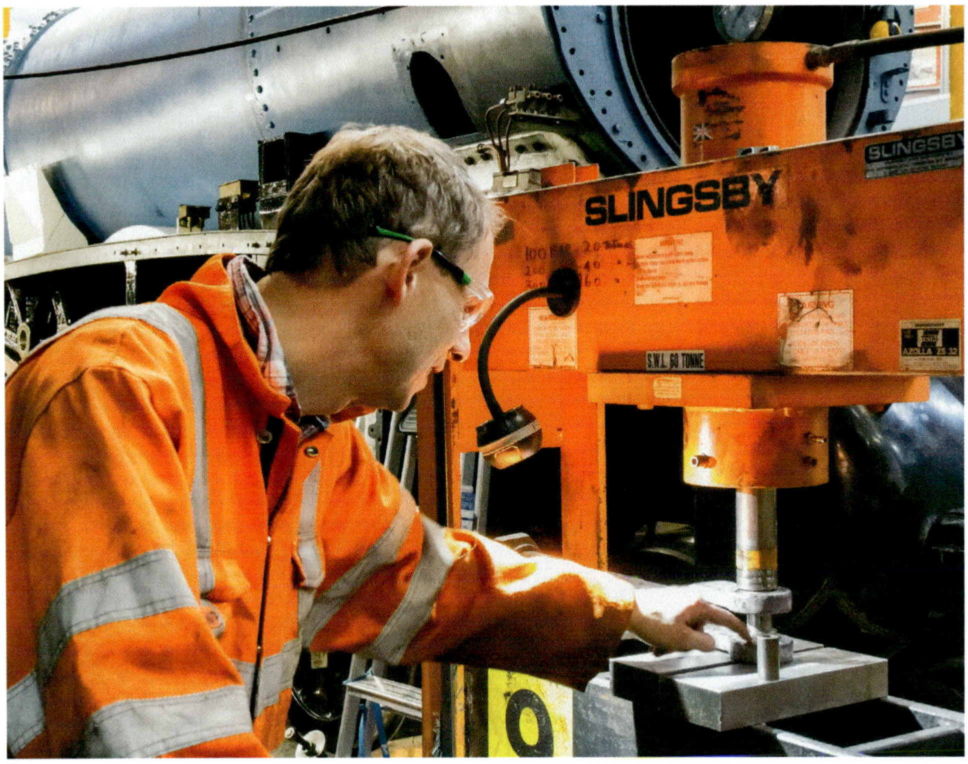

Andy Evans pushing bushes into place

Andy Evans

Software engineer Andy is one of Darrin's very valuable 'Saturday gang'. Still in full time employment he regularly gives up his weekends to volunteer on No.7 which is a great bonus as the majority of the team are mainly retired and thus able to fill the weekday roster spots.

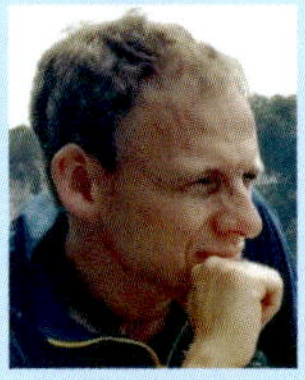

"It turns out that 20 years of software engineering experience wasn't a huge amount of help when it comes to renovating a steam loco, but it did give me a ready supply of excuses on why jobs didn't get done on time. Darrin Crone and Richard Swales taught me so much, which I am incredibly grateful for and working with the rest of the engineering team on such a great loco is a real privilege."

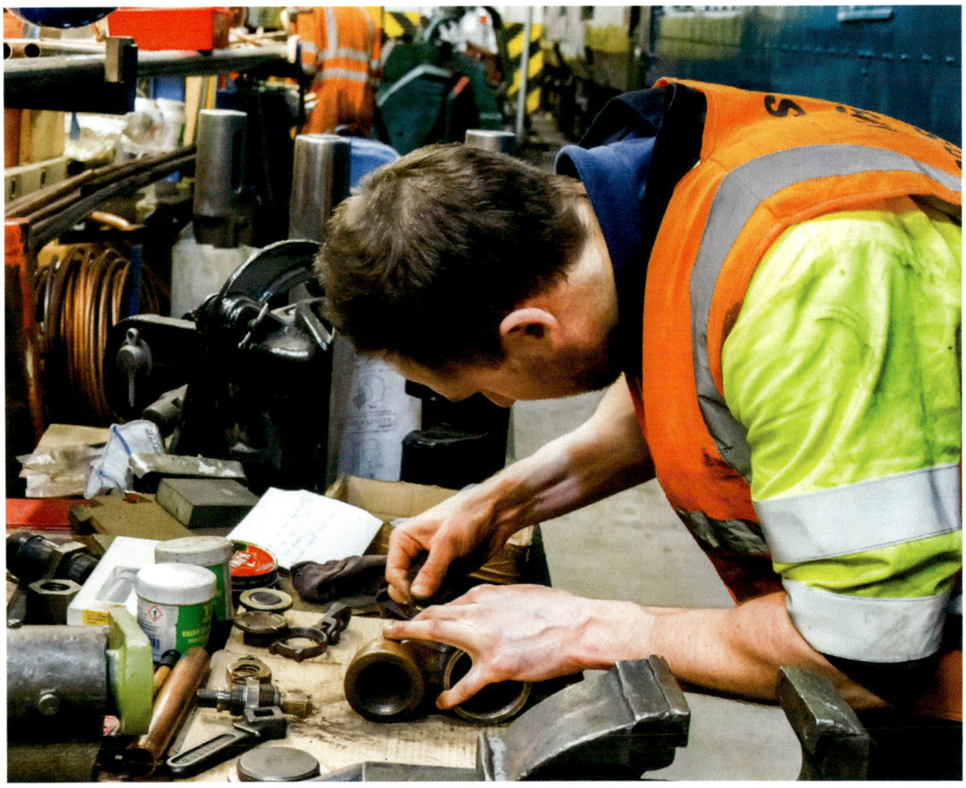
Chris Whiteley working on steam heat valves

Chris Whiteley

Working full time as a mechanical engineer within the power generation industry means that Chris is one of the 'Saturday gang'. Originally from a 'hands on' background, volunteering on Sir Nigel allows Chris the opportunity to get out from behind a desk and get his hands dirty again. Chris joined the Engineering team in February 2018 and has been involved in a variety of tasks, including the inspection and measurement of the boiler safety valves (which was in fact Chris' first job). Chris also played a significant part in the measuring of the tender wheelsets to facilitate the refurbishment of the white metal bearings.

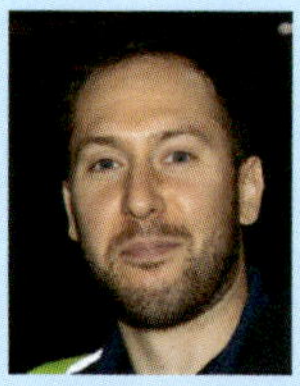

Volunteering gives Chris the opportunity to make a small contribution to a piece of England's engineering heritage, which he hopes will be appreciated for years to come. More importantly it has been an opportunity to work with and learn from a great team. Most of which he would unlikely have met, had he not decided to become a volunteer.

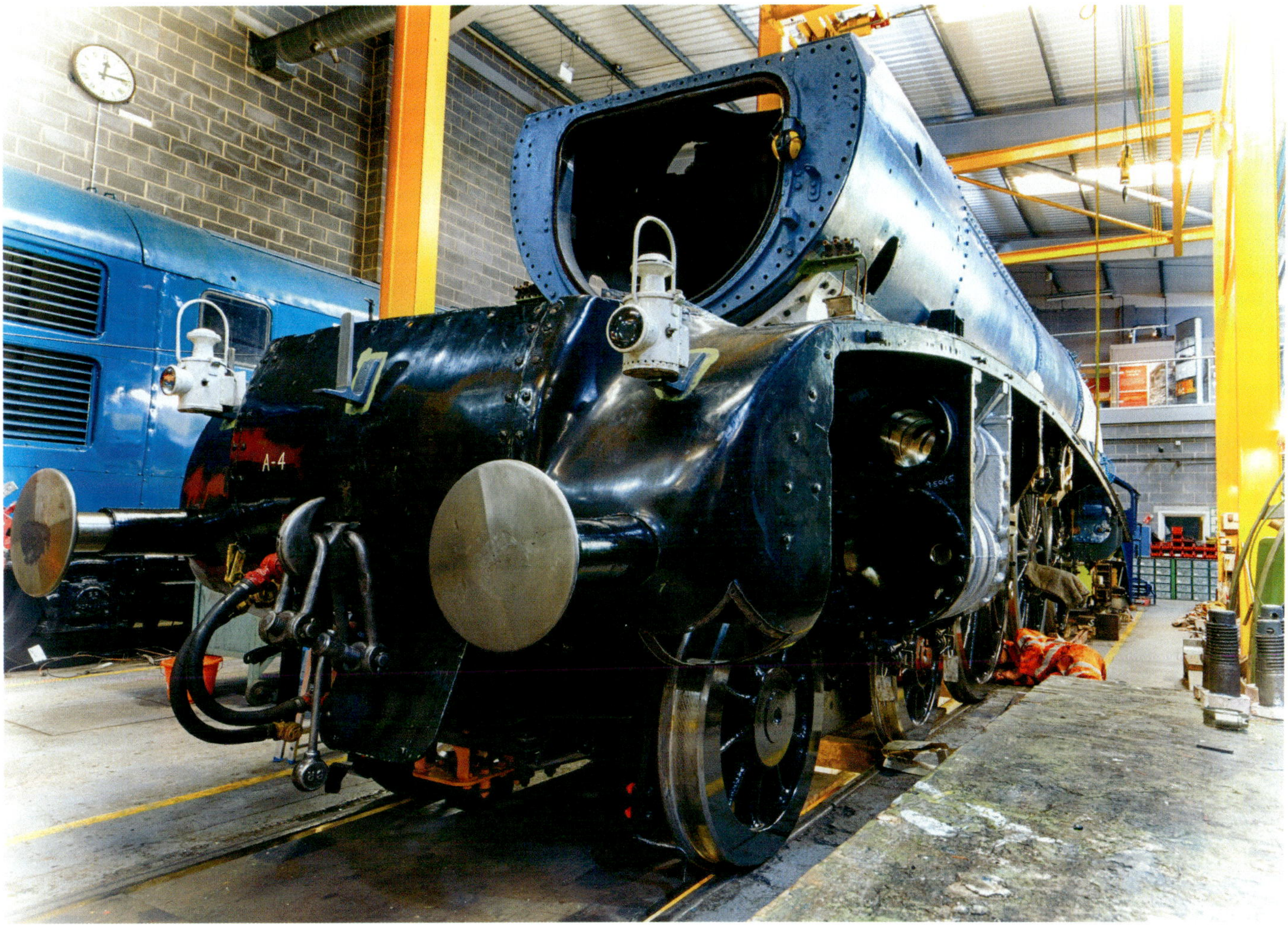

An HDR triple exposure image showing the newly fitted boiler on the frames

overhaul I have learned to use my tripod less often in search of getting more light into the lens through a longer exposure and going hand held with very high and noisy iso settings. It is quite remarkable what can be obtained with noisy (grainy in old money) images using modern noise reduction software.

"One of your five a day"

There was no 11am tea break on Saturday but an earlier lunch at 12 noon and an earlier finish at 4pm, so as noon approached I volunteered to replace the absent chief tea maker, Peter Brackstone, and headed off to the blue cabin ahead of the others to put the kettle on. Still taking the tea bags out of the cups when the team followed me, I was rebuked for not being quick enough and told I would not score highly on the Brackometer of tea making. However, nobody complained at the quality and when I passed around the cookies containing flecks of fruits, one of your five a day someone said, everybody was happy.

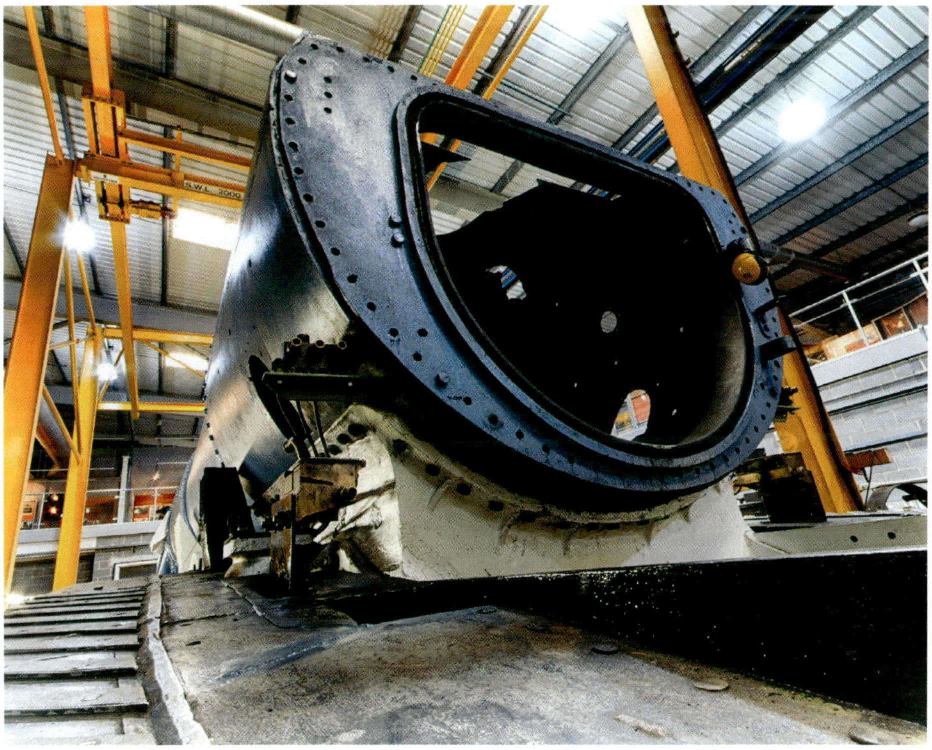

An HDR triple exposure image taken from the running plate

ROLLS & ROLLS

January 2020

"Do you not want that Tony?"

As is customary for my routine visits to the workshops, I contacted Darrin to ask him which day would be best for my first visit of 2020. This time it was by text, so to keep it brief I restricted my enquiry to less than a dozen words, and, also keeping it brief, Darrin's reply was his average number of words for a text message - one! That one word was "Thursday," and looking at the attendance log when I signed in on the day, I quickly realised why he had said Thursday. Darrin had given No.1 tea-maker Peter Brackstone the day off and my last exploits at standing in for Peter had clearly gone down well as I was asked to stand in again. (Although it may have been more to do with the planned work on the day which included starting to fit the lagging on to the outside of the boiler.)

Phil Slater and Roger Turnbull put on their personal protective equipment as this was classed a hazardous task due to the properties of the lagging material. After scaffolding had been erected along the sides of No.7, the overhead crane was used to lift the rolls of lagging over the top of the boiler before it was cut and fastened into position. The first roll was now successfully in position, and over the coming days the plan was that the rest would follow down the full length of the boiler.

There was a full squad in today which meant there were a variety of tasks underway. Bob Shearman was busy on the lathe, Tony Bickerstaffe on the cartazzi, Malc, as me with the tea duties, was also deputising for Peter but probably in a more productive way by taking up the painting duties, and Andy Barwick was working on a coupling end rod keyway. Andy was adjusting the width of the keyway in the rod end which, through natural wear and tear, had worn. This meant carefully enlarging it which, consequently, meant the key on the new bush would be larger in order to fit the new width. This was a delicate task and one which needed the attention of a top fitter such as Andy. Andy took time out to show me the new brass and white metal bushes, four for the two coupled drive wheelsets and two larger ones for the centre drive wheelset connecting rod big ends. I didn't know the costs of these, but I would imagine it would be a significant slice of cake from the budget.

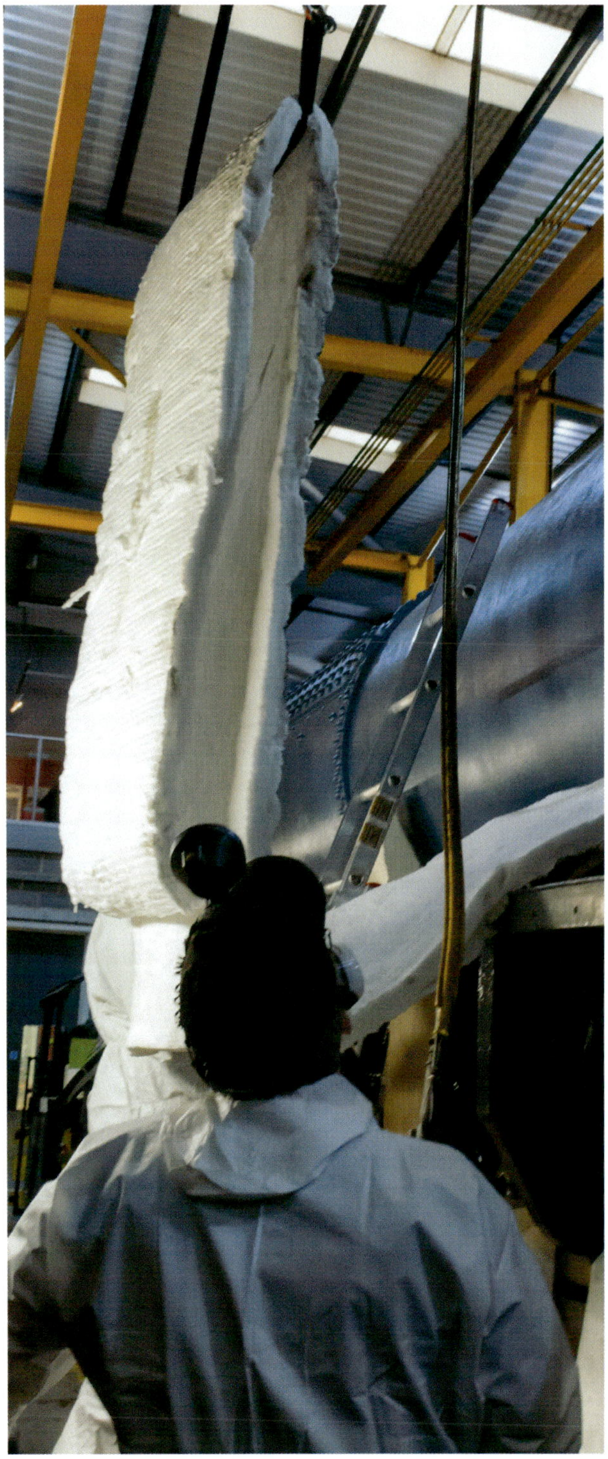

The first length of lagging is lifted over the boiler by the overhead crane operated by Roger Turnbull

Talking of cake - well Swiss Roll to be precise - when the team adjourned for lunch and a cup of my wonderful tea, as usual many delicacies emerged from the fridge and today there was the last two inches of what had been a significantly long swiss roll. This was quickly cut in half and the first half was soon gone, but what of the second half? A silence went around the table as the several sets of eyes focussed on the last remaining morsel. Who would be first to speak? Who would be first to admit they wanted it? Everybody was being far too polite to just dive in and take it and then finally Phil says, "Do you not want that Tony?" and added with a smile "You could do with putting a little weight on." Tony declined but didn't add the words "No, it's ok Phil, you have it." Pressure was mounting until, after another long silence, it was finally subdivided again by Roger who shared it with Phil. A debate then ensued where it was decided this should have been filmed and used as a student psychology test, with the camera going around all the faces and gestures and the students having to work out who would break the silence and speak first. The talents of the team go far beyond engineering!

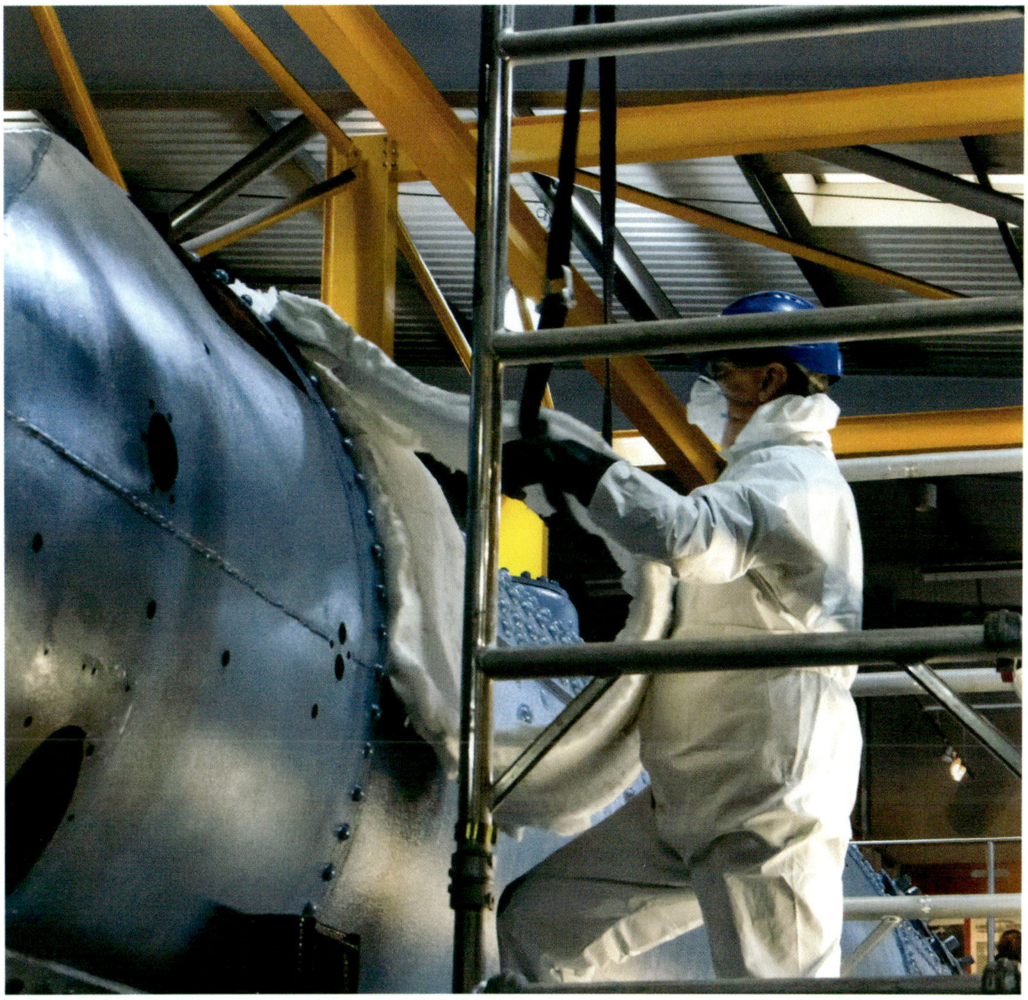

On the opposite side of the boiler Phil Slater receives the lagging and positions it around the boiler

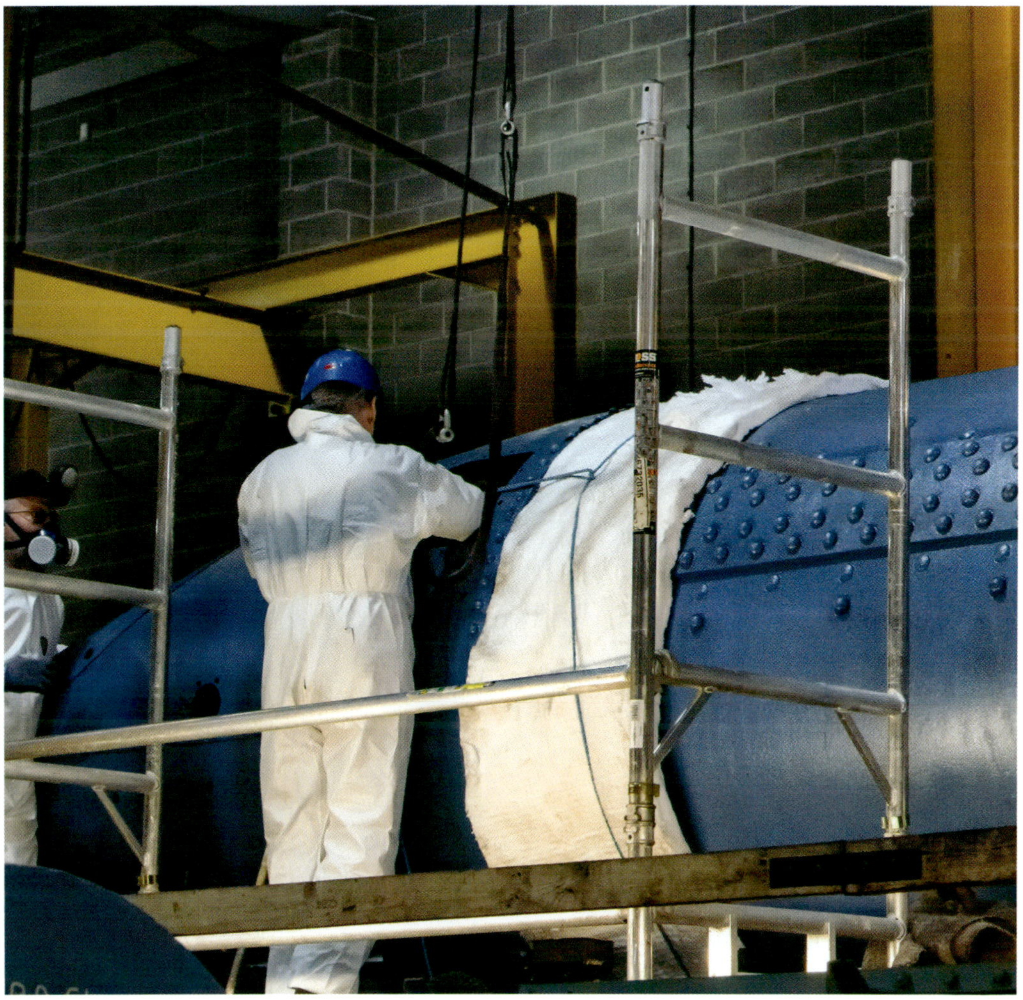

The first roll of lagging is secured in position around the boiler

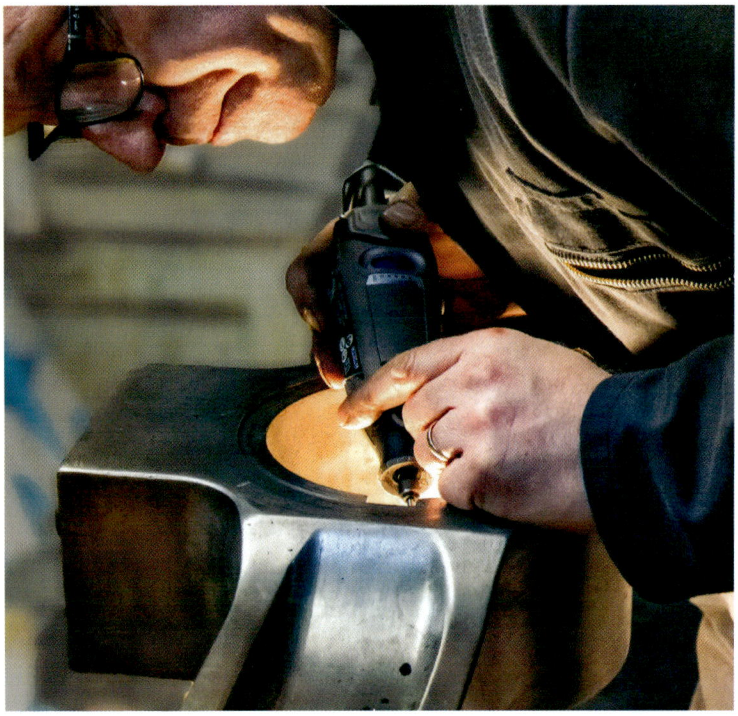

Andy Barwick carefully dresses the sides of a coupling rod end keyway

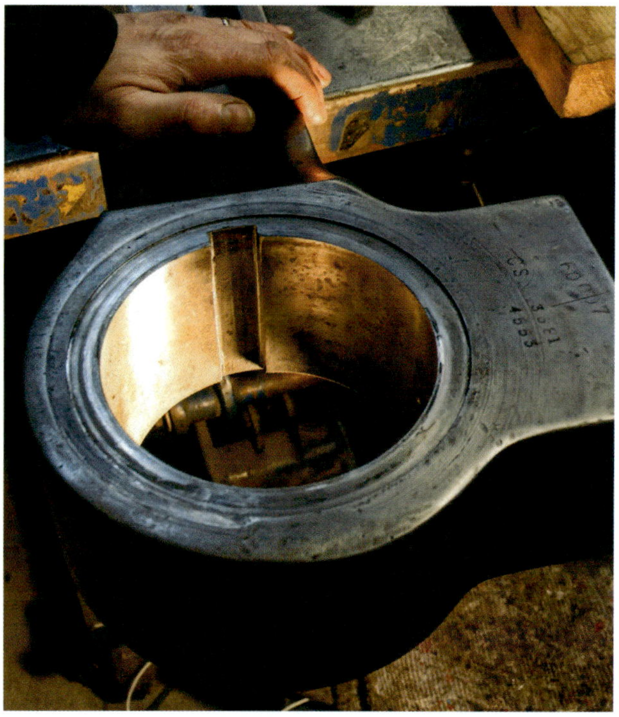

Coupling rod end showing the keyway into which the bush key locates and secures the bush in position

Brass and white metal coupling rod end bushes for the coupled driving wheels

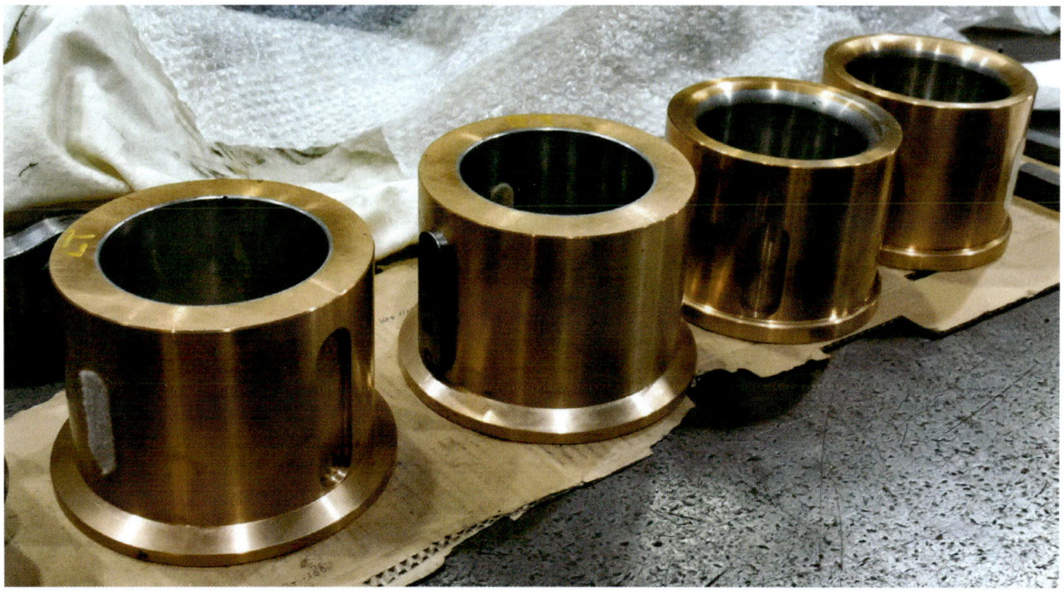

Big end bushes for the main driving wheelset

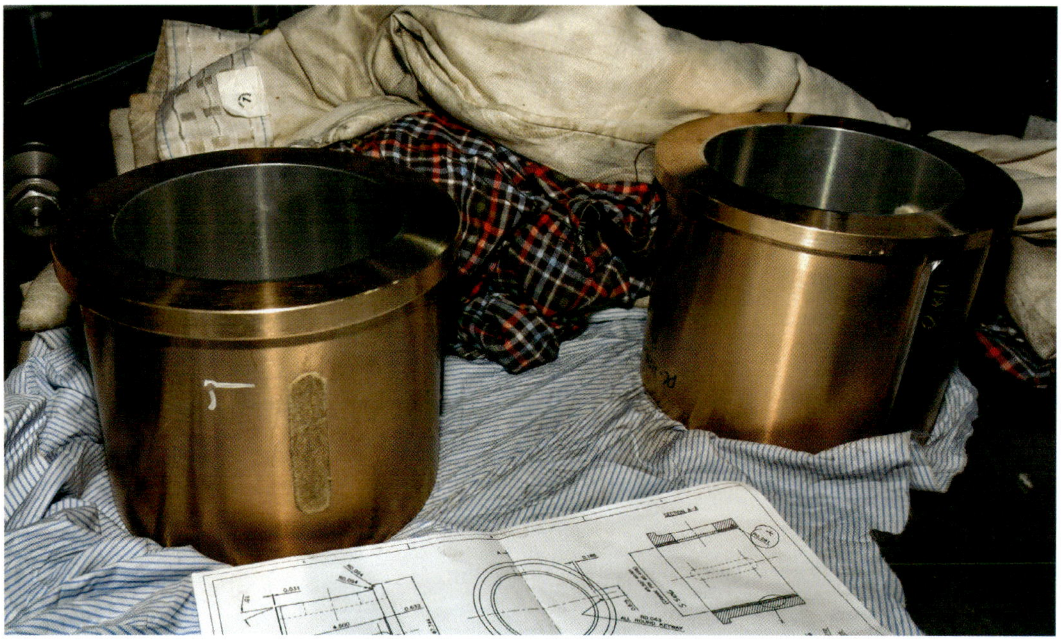

January 2020

I had captured images showing the start of the wrapping of the boiler in insulation and I was keen to get back and capture some which showed it finished. The pace at which things were now moving meant it wouldn't be long before that would not be possible, as the crinoline formers would be next to go on and then the streamlining which would cover the insulation from view. A quick visit would do the trick, so it was back on the 09:08 train and hopefully returning on the earlier 12:53 from York. Now, I'm all for being kept informed by the train manager of what the delays

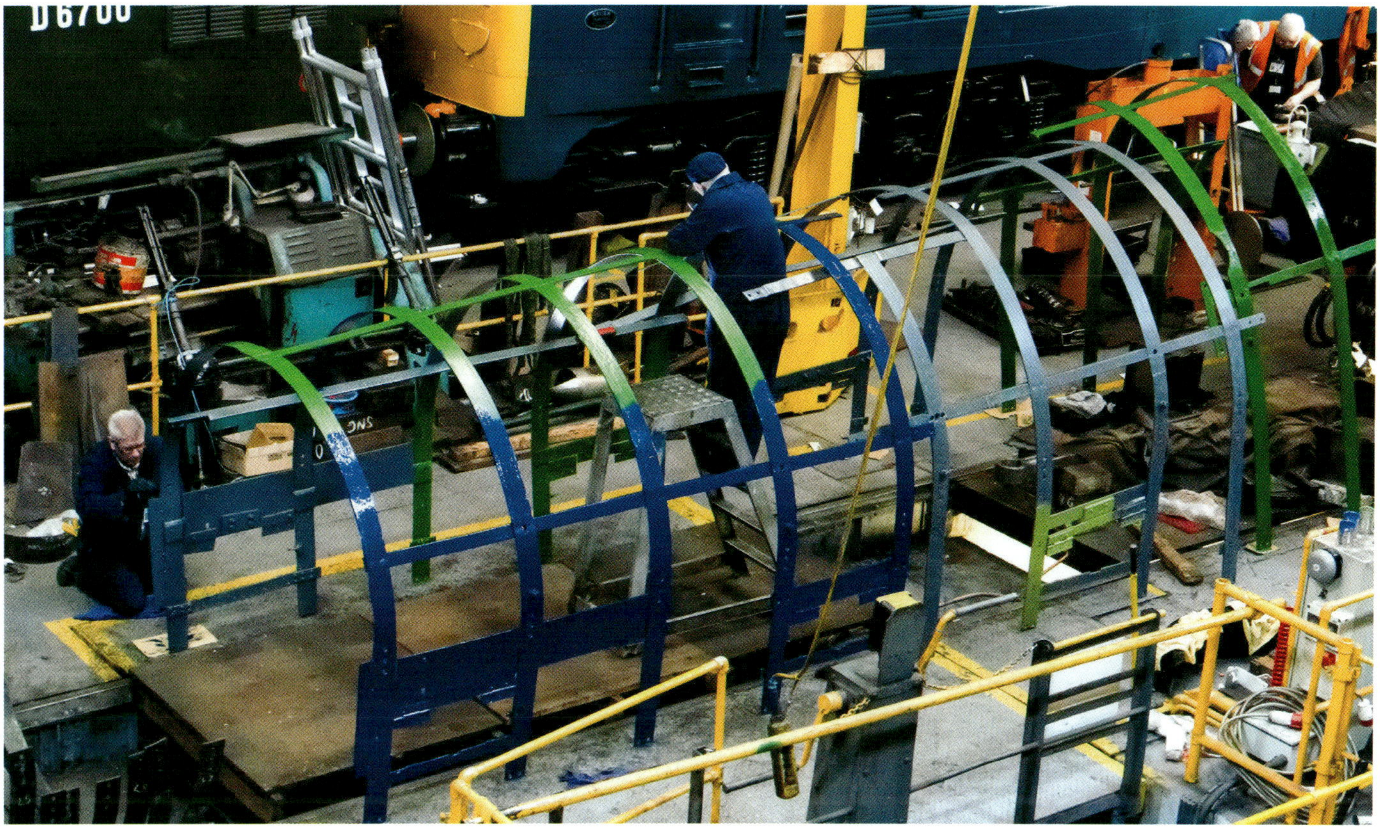

The streamlining's crinoline formers being painted

Alan Pitt

Having been involved with the team that put together the successful Heritage Lottery Fund application for the previous overhaul, during the period I was Honorary Treasurer of the A4LS/SNGLPT, it is certainly an eye opener being involved in the actual physical work this time round, even at my low skill level.

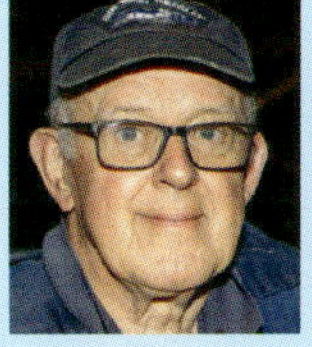

The banter in the Blue Shed at breaks helps keep spirits up.

I do enjoy needle gunning. Pardon? I SAID I DO ENJOY NEEDLE GUNNING!

As my memory becomes less reliable, I will not forget my time working on my favourite locomotive, helped by a print of the photograph that appeared in the Daily Telegraph of me in the firebox, on my living room wall.

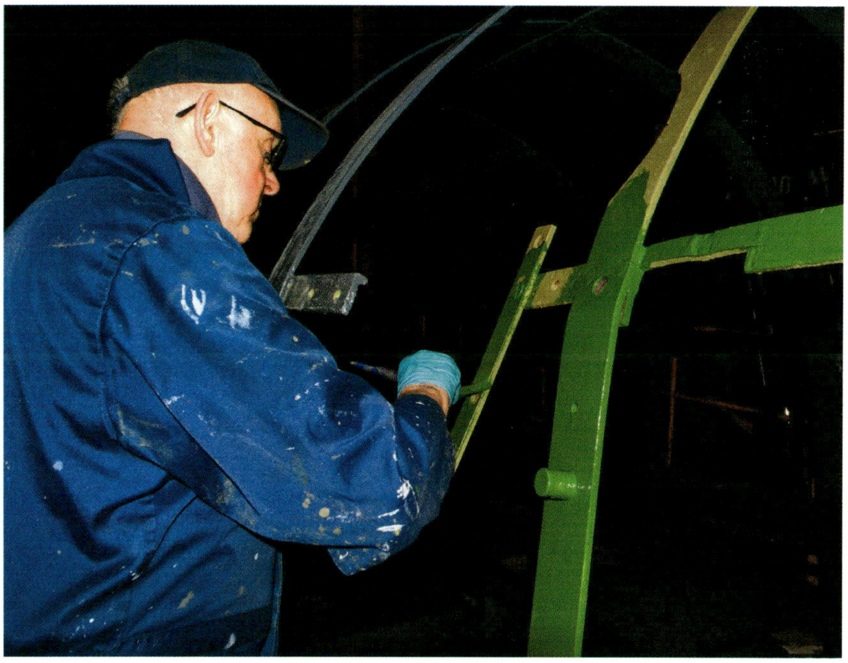

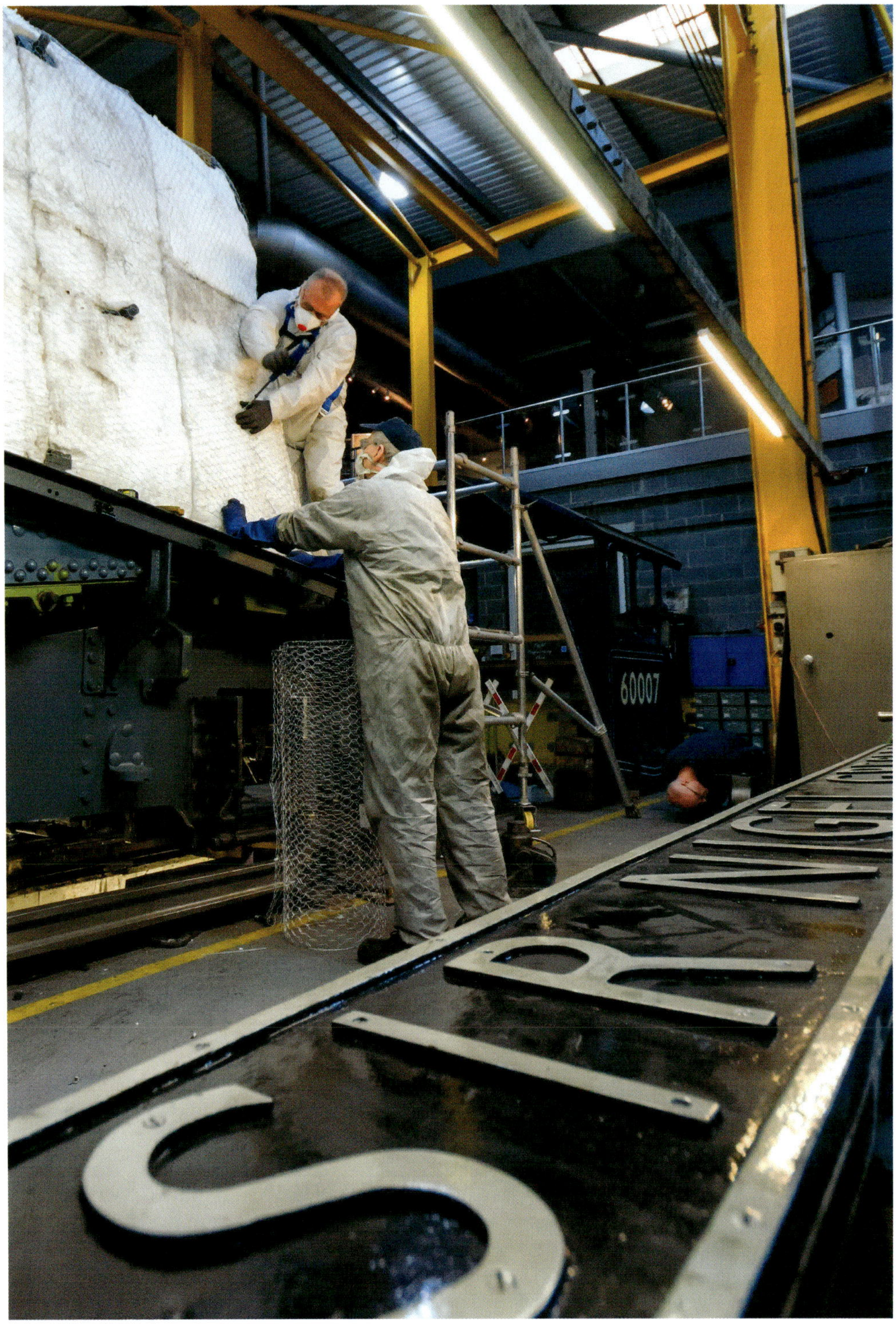

Roger Turnbull and Phil Slater attaching wire mesh to hold the lagging in position before the streamlining is refitted. In the background Andy Barwick seems to have lost something

and problems are these days but I could not help thinking of Reggie Perrin when I started telling people of my journey's problems as Reggie did in the famous tv comedy series from many years ago. This time it was a late running Saltburn to Darlington train in front of my Kings Cross service, an apology for the coffee machine breakdown again and no Wi-Fi.

The crinoline formers around the boiler were being painted today and the team were using up paint stock of various colours including blue and green, but as these formers would not be visible once the streamlining was refitted, it would have no effect on No.7's final colour. That didn't stop a few queries from the museum's visitors and social media asking about the colour in which No.7 would be returning to traffic.

Way back in early 2016 when the insulation around the boiler had been removed, it was a messy, dangerous and toxic job that required plenty of PPE (Personal Protective Equipment). Putting it back on with the modern product, supplied in rolls, was less of a problem but still required the PPE. Roger Turnbull and Phil Slater were heading up this operation and today I found they had completed the main barrel of the boiler and had even secured the insulation all in position around the boiler with wire mesh.

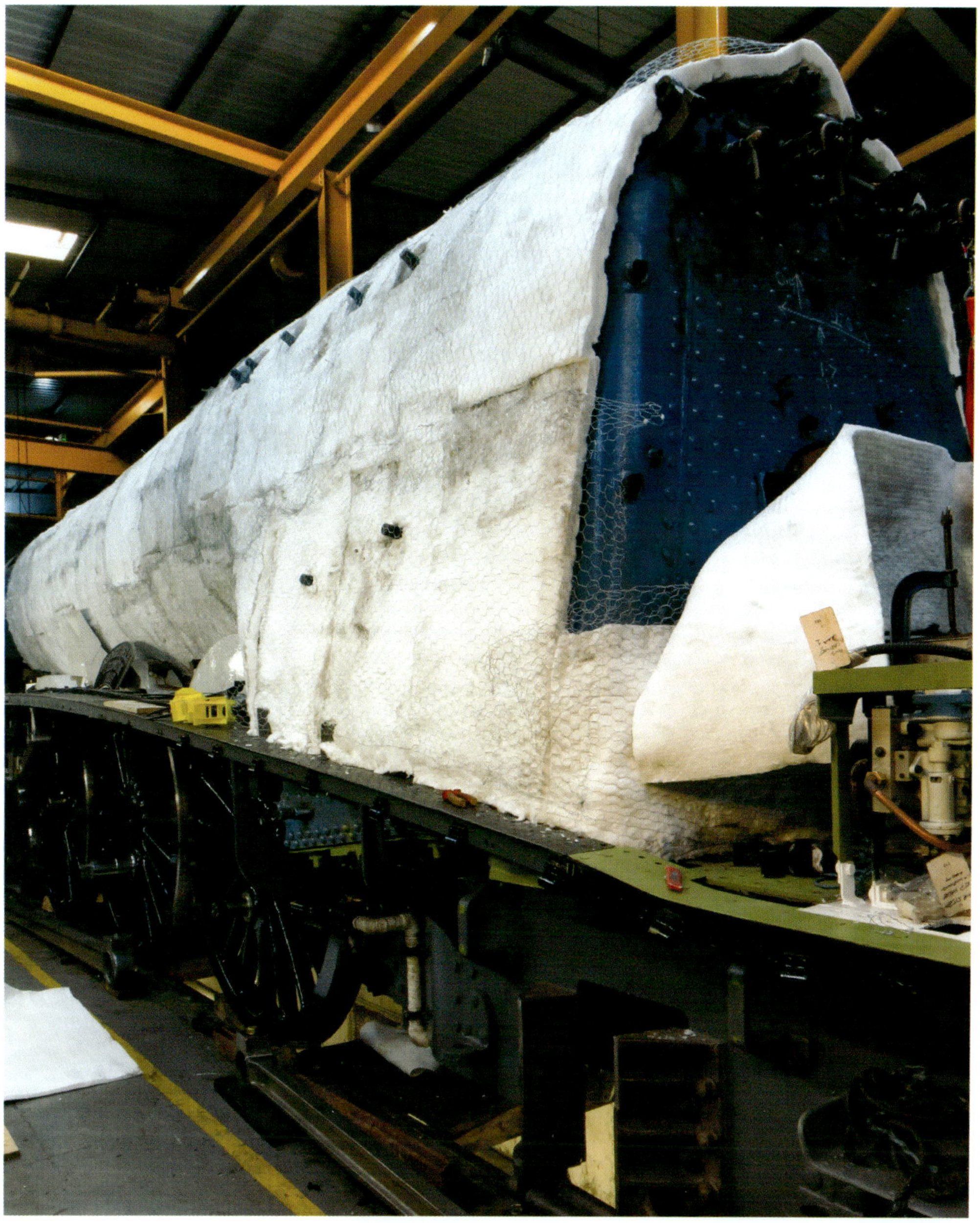

The lagging around the boiler barrel and firebox with wire mesh holding it in position before the crinoline formers are refitted

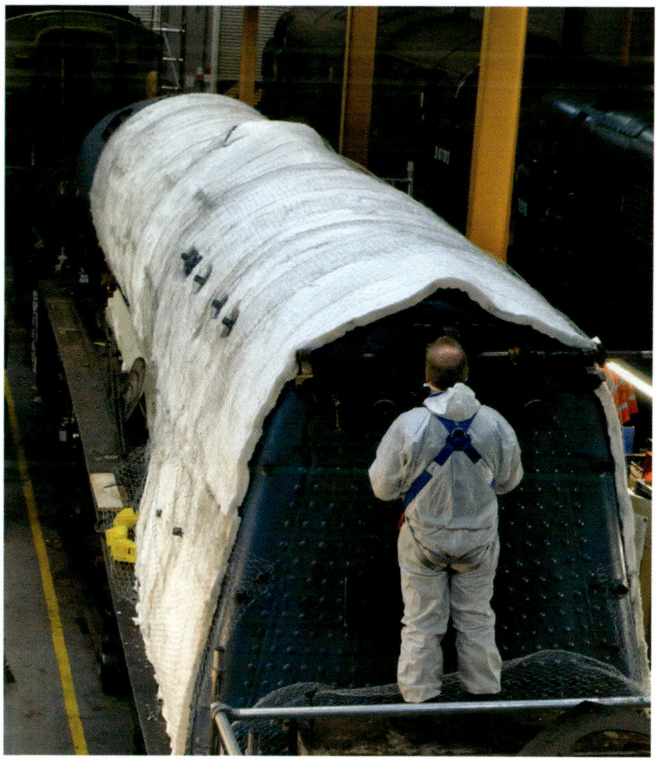

Roger Turnbull wears a safety harness as he works on the lagging around the firebox

Due to the number of volunteers working today and a shortage of space and chairs in the blue cabin, I had to forego the first and primary 11 o'clock tea break, as being a late morning arrival, I was allocated the 2nd sitting. This worked for me today as I had got the few images I wanted, so I departed straight back to the station from the break and was in good time to catch the early train home.

12 ON THE BUSHES

February 2020

There were a huge number of different tasks being undertaken in the overhaul and some had caught my attention more than others. Andy Barwick, working on the coupling rod bushes, was one of those - I had been fascinated by the precision work he had undertaken. Today these carefully restored bushes would be pressed back into the coupling rods ends using a press that exerted pressures between 8 and 15 tons depending on the size of the bush. The pictures here show a full sequence, starting with the careful alignment before pressing started, something you wouldn't want to get wrong as bronze and white metal bushes are one of the more expensive items on No.7.

CME Richard Swales was overseeing this critical moment carefully, checking alignment and taking final measurements before Andy, being assisted by Scott Middlemiss, commenced the press. Not wanting to miss the occasion Darrin shouted across the workshop "Have they started yet, Trevor?" to which I replied "Yes," and a small crowd then soon gathered. Being the youngest member of the team,

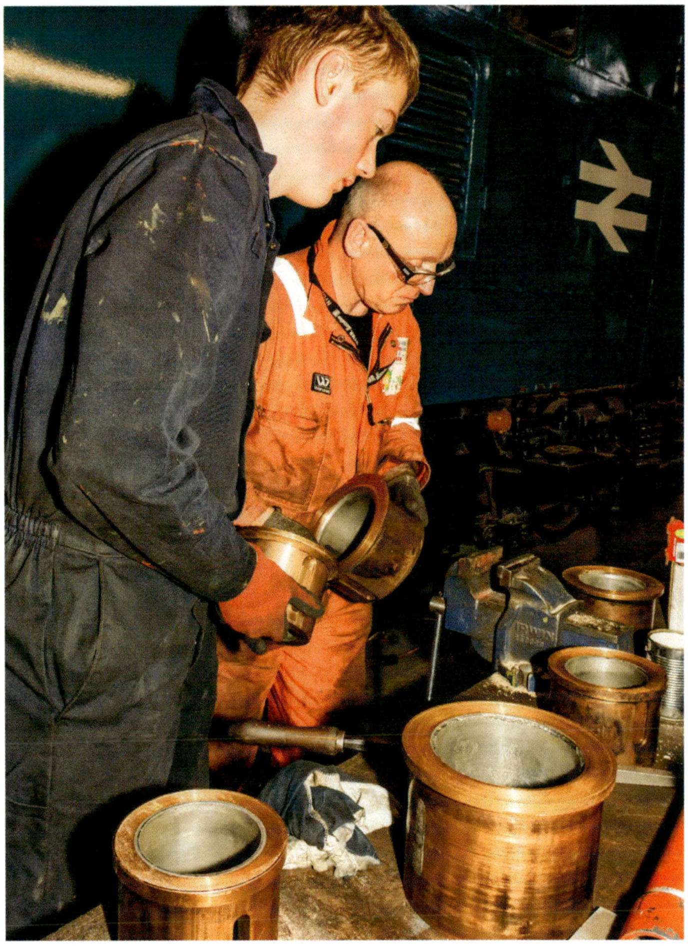

Scott Middlemiss and Andy Barwick selecting bushes

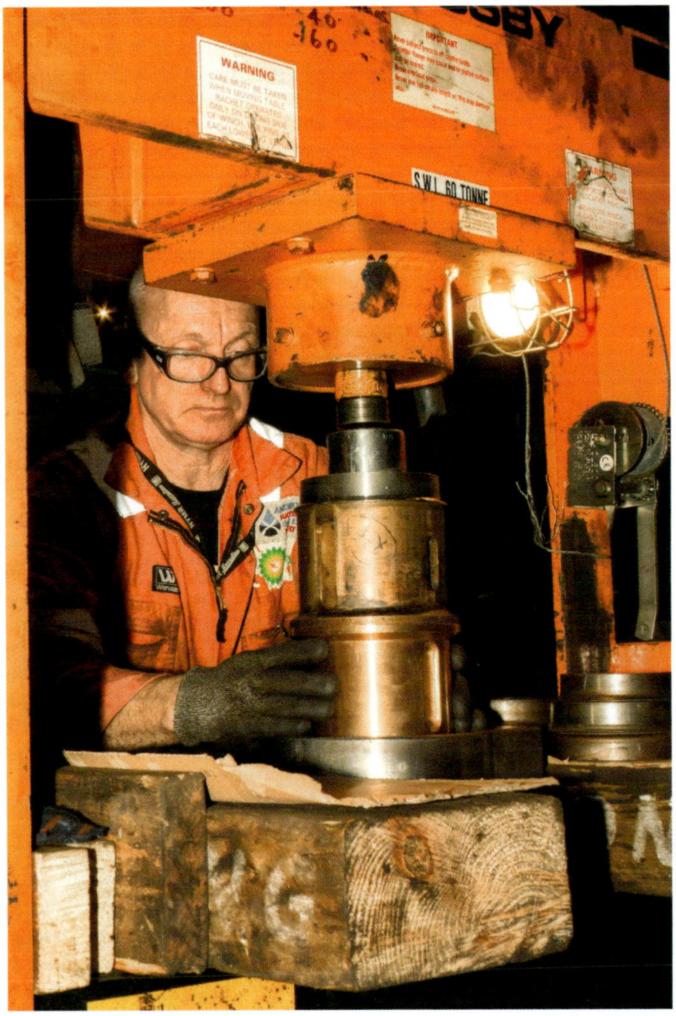

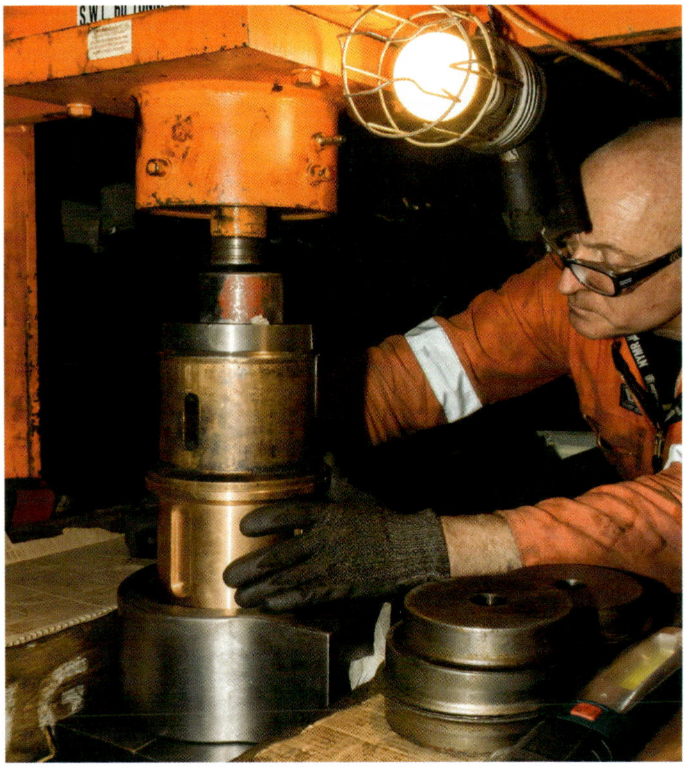

Andy Barwick checks alignment before the press starts

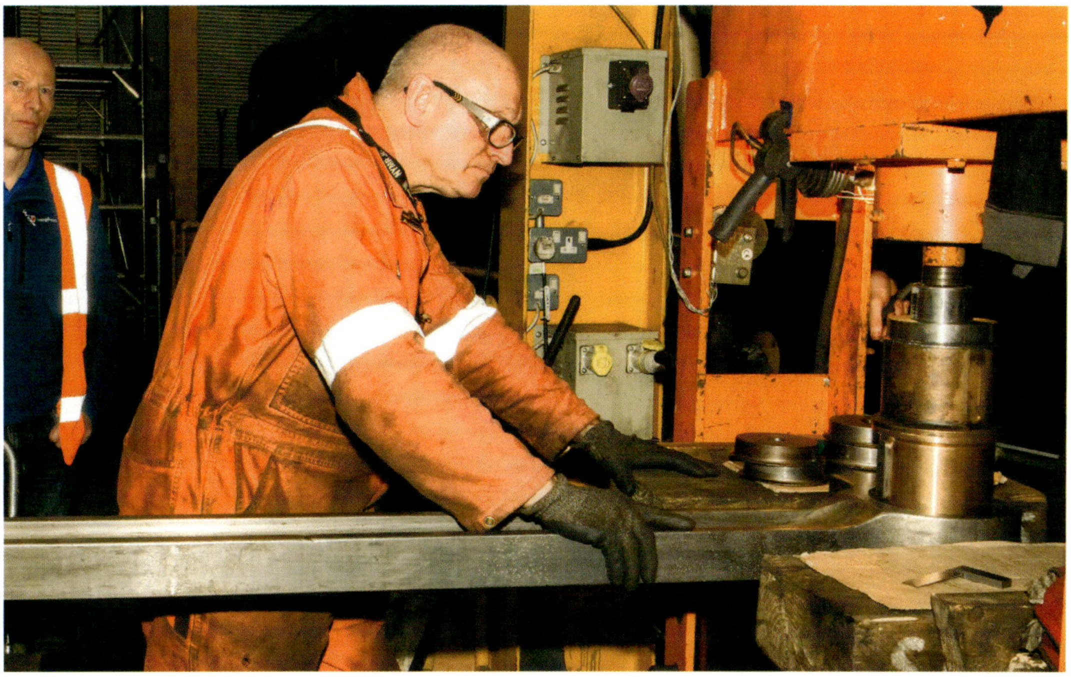

As the press begins Andy Barwick keeps a close eye on progress

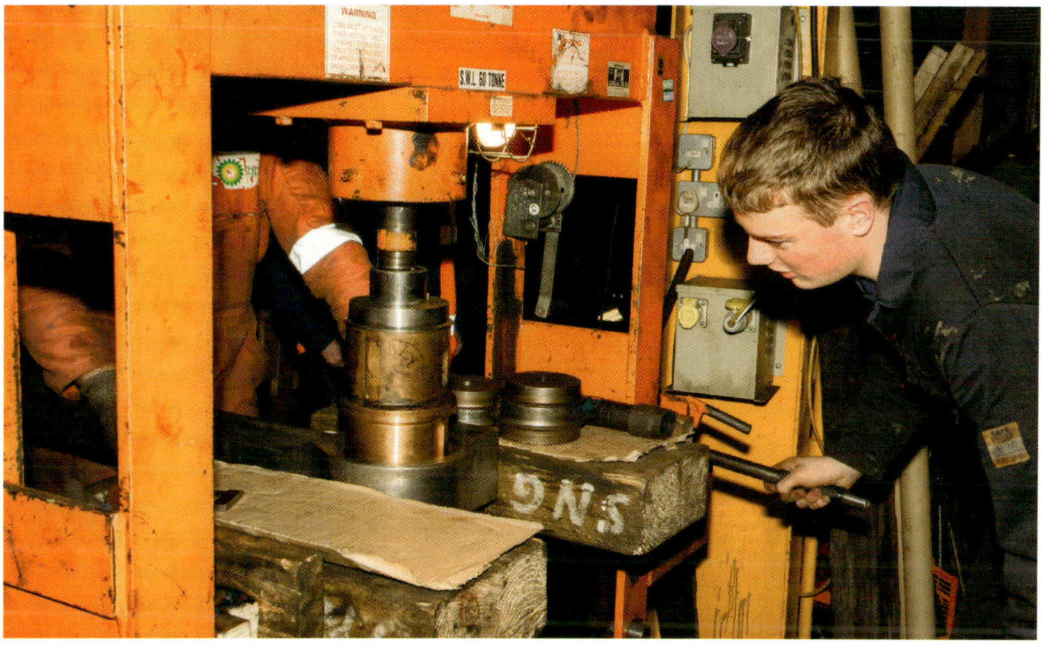

Scott Middlemiss pumps the hydraulics to press the bush into coupling rod end

Scott had the pleasure of pumping the press up to pressure and I have to admit I was rather envious of the effort he put in. I vaguely remember the days when I could have done the task, but had they asked me now it would have been a very long job.

The press went very well. The excellent fit of Andy's restored bush key in the keyway as the bush was pressed into the coupling rod brought plenty of smiles from the team which resulted in a very pleased Andy saying "We will be able to free wheel all the way from the top of Stoke Bank into Kings Cross!"

I had taken my good friend Nick with me today and decided to show him the view from below, somewhere I hadn't been myself for a while. So, armed with hard hats we descended into the pit and inched our away along, in crouching position, until we found a space just forward of the ash pan where we could stand up straight. I took the opportunity to shoot a new HDR image of the underneath, which when compared with the images I had previously taken, showed the enormous amount of work that had been undertaken which the general public from the viewing gallery do not get the opportunity to see. I was becoming aware that this was a problem which was manifesting itself on social media as recent posts had suggested that progress was slow, when it was actually the opposite. Great progress was being made, so I decided I would start releasing a few more images to social media highlighting work that is not clearly visible from the workshop's gallery.

This was another quick visit and my train times fitted perfectly with the work. Although we thought we had missed the earlier lunch time train and would have to get the mid-afternoon service, we returned to the station to find the earlier train sitting at the platform some 30 minutes after it should have departed. "North bound?" I said to the train manager. "Yes," he replied. "Running late," I said. "We weren't," he said, adding "Transport police are onboard, no idea how long we will be." However, we were soon on our way. Another little Reggie Perrin moment, I thought.

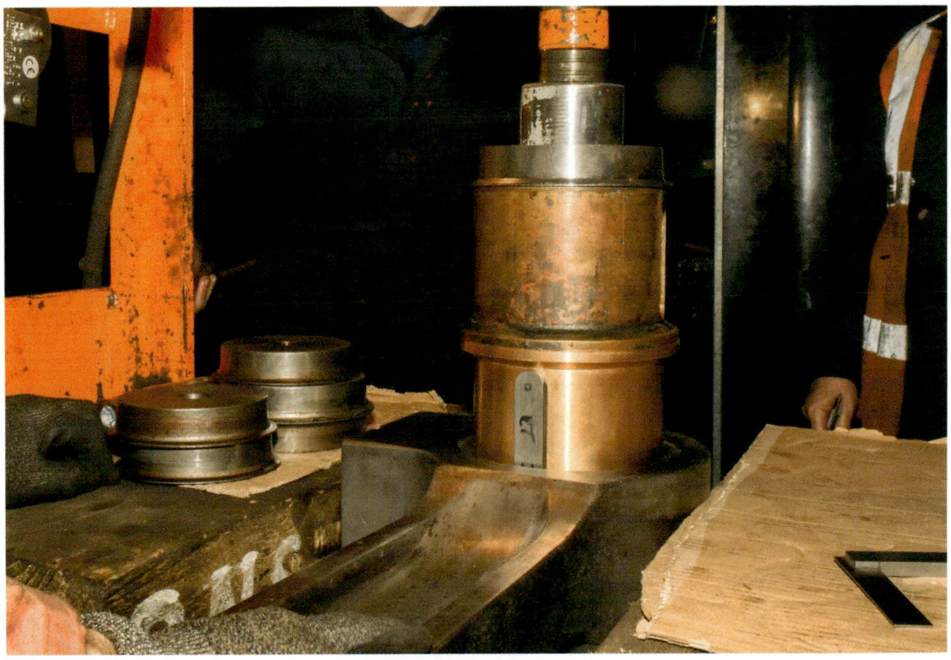

The bush starts to enter the coupling rod end

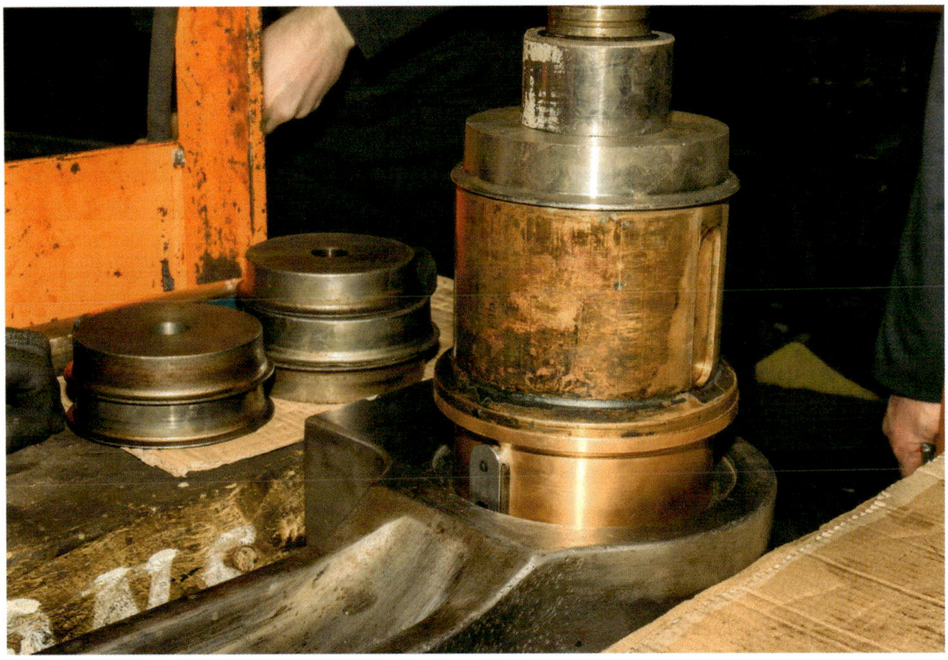

Just a few more centimetres and the first bush will be back in place in the coupling rod end

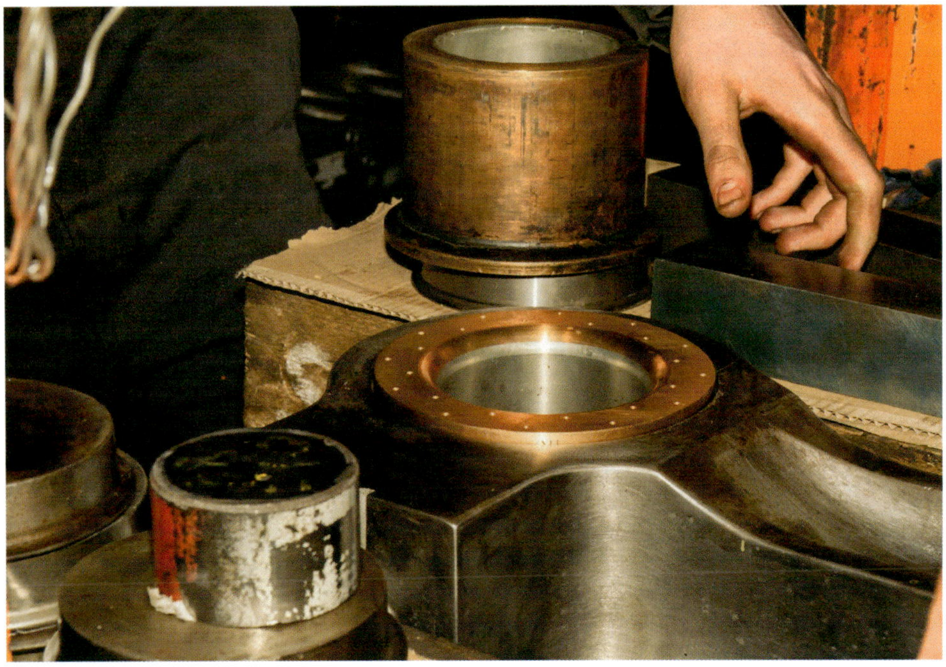

The first bush is back in position in the coupling rod end

"We will be able to free wheel all the way from the top of Stoke Bank into Kings Cross!"

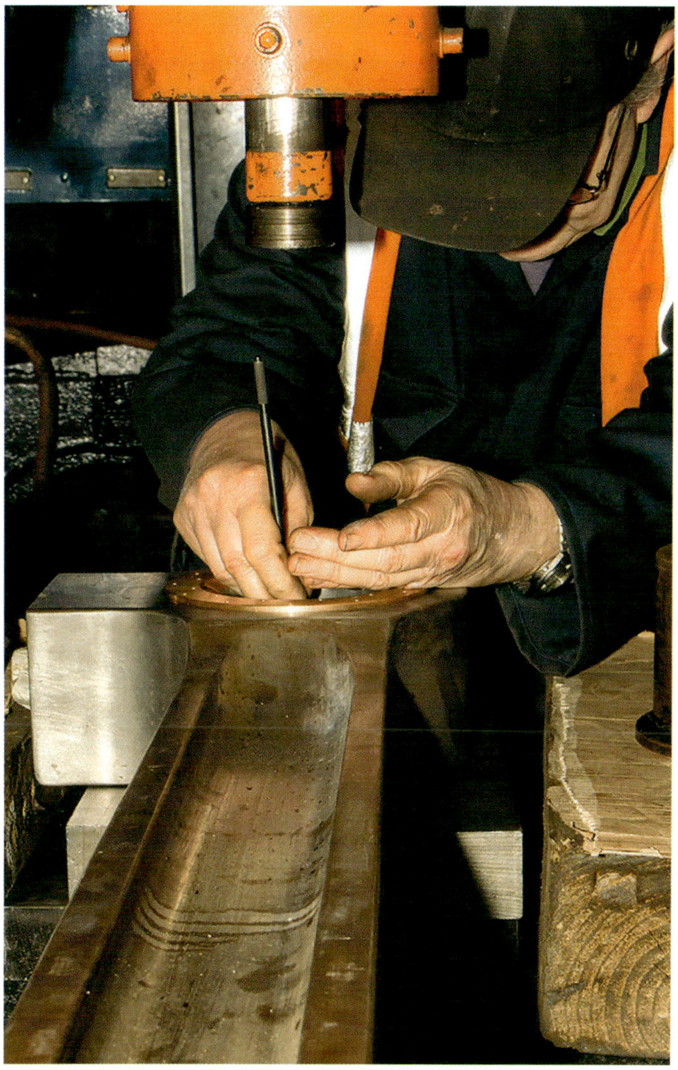

Cheif Mechanical Engineer Richard Swales takes some final measurements after the first bush is back in position in the coupling rod end

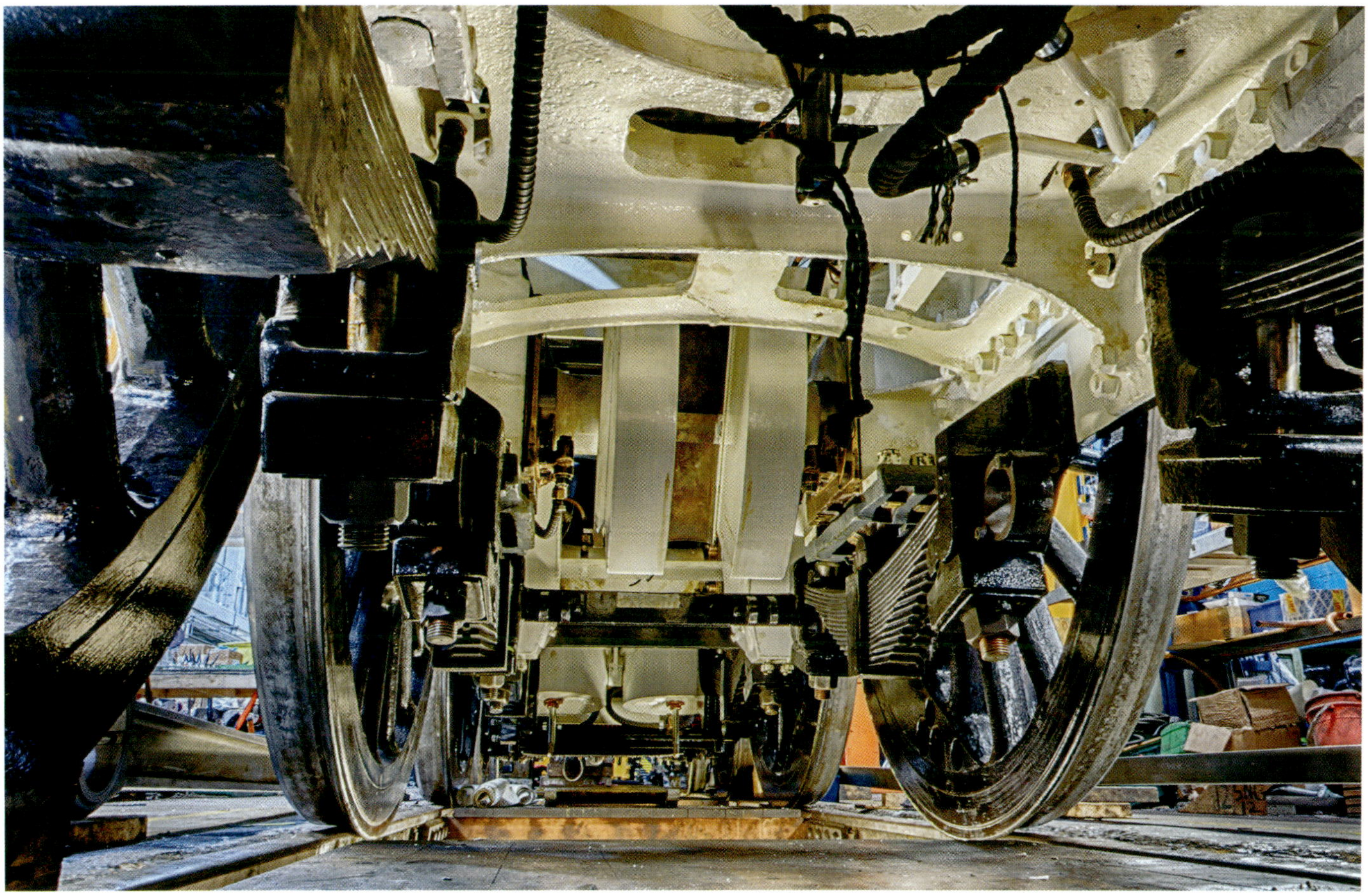

An updated image from the pit contrasts sharply with my first image from the pit on page 2

13 CORONAVIRUS HALT

March 2020

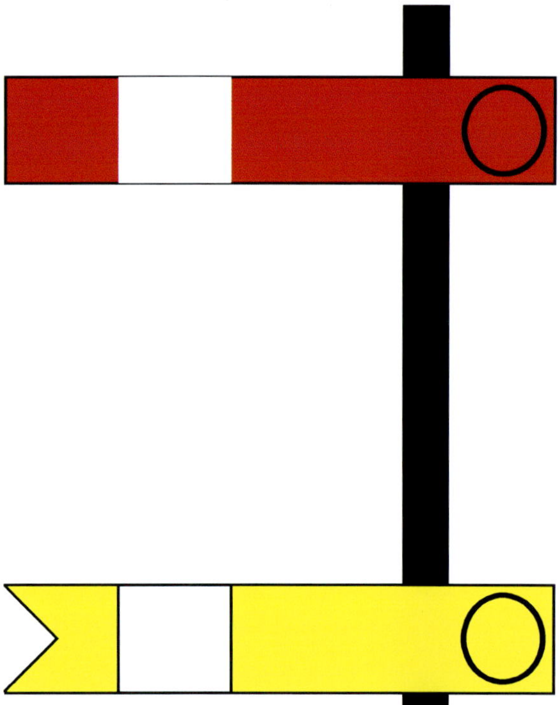

You know the feeling; you are sitting back in your coach seat watching the world race past the window and you are nowhere near your next stop, then, suddenly you start slowing down. This was happening on my recent visits to York as the effects of Coronavirus around the world was inching ever closer to the UK and it now seemed inevitable it would reach our shores when Italy and then Spain went into lockdown. The great progress that was being made was leading us to believe that a late summer return to traffic may be possible, but then we had a string of caution signals and we eventually reached a stop signal for an unscheduled stop at Coronavirus Halt.

The Railway Museum closed its doors in late March, but Darrin and team had a few more days to take some of the work off site. A controlled shutdown of work was necessary which was summarised by the following statement released by the Trust.

"COVID-19 Statement: As of 20th March, all work on the overhaul of 60007 at the National Railway Museum was stopped after a controlled cessation over the previous couple of days. Work at York will only re-start when it is judged safe to do so based on UK Government guidance. SNGLT encourages all our members, friends and supporters to please follow Government and NHS health advice during the present emergency.
Although work has stopped at York there will still be updates of work that continues to be done at various places off site (Note: off-site work only undertaken where fully compliant with all Government guidance)."

A new normal was to take place but it was important we all kept in touch, particularly as some of the work was still being carried out, compliant with government regulations, at various locations away from the workshops. Enter Zoom video meetings and the champion of it, for us anyway, of Richard Hawkhead, newly appointed volunteer Operations Manager to the Trust. With the patience of Job, Richard calmly and knowledgeably guided the team, aged mainly 65 years and above, through the process of video conferencing using the Zoom platform. Junior volunteers in the group, Tom Readman and Luke Perry, must have found this very amusing but to their credit refrained from commenting on it.

Joking apart, these were serious times and the meetings became especially important on checking the wellbeing of team members, a lot of whom were in the so-called vulnerable age group of being over 70. As I write this in June 2020, despite a few self-isolation scares, we have come through unscathed.

Regular updates on work progress and future plans were reported and a good deal of banter and craic ensued - a strong reflection of the 'blue cabin' lunch breaks and the remarkable camaraderie which the team had established since coming together in late 2015. Helped along by feature updates by various members including one from Richard Swales, who had been working on No.7 since it was housed at Philadelphia (1968 – 1977), and even a couple from me on my photography career, not to mention quizzes from various members, although the self-marking procedure needed a degree of honesty!

Regardless of the outcome from Coronavirus's halt on work we had to turn our attention to the actual return to traffic, so in May 2020 the Media Team was assembled to hold yet more video conferences where plans could be made. The opportunity to put forward ideas from the engineering team was met with a huge response and the Media Team had to hold weekly video meetings in order to work out a policy for the return to traffic which was to be proposed at the Trust video board meeting in June.

Most concerning of all to the Media Team was the longer lasting effects on heritage railways and indeed, the ability of the public to be able to attend events. A crystal ball would have been needed to figure out what to do, so it was decided to plan as if there was to be normality, as that is all we could do. We would simply have to do it Gradum ad Tempus (One step at a time).

August 2020

Government relaxation of the Covid rules had allowed the NRM to open its doors to the public in a controlled way but we still were not allowed to start work again in the workshop. Some progress had been made on smaller items that had been removed from the workshop before the lockdown and they had been distributed to all four corners of northeast England or as Darrin, writing in his updates notes on the Trust website says, "cold northern wastes of County Durham", " dark satanic mills of West Yorkshire", "tranquillity of North Yorkshire" and "down south in sunny Lincolnshire". Drive linkages, stepped washers, leather seals, knuckle pins and bronze bushes all received attention off site and I'm sure the technically minded amongst you will understand those terms. Most of you will probably relate more with the work that also took place on Train Protection & Warning System (TPWS) which is what rings the bell in the cab when the train is approaching a green signal and automatically applies the brakes if the train passes a red signal. Whilst volunteers continued working from home on the above several sub-contractors had been allowed to start work again and make progress on work such as the Cartazzi springs.

The team were still exchanging banter and jokes on social media and continuing with the weekly Zoom video conferencing calls hosted by Richard Hawkhead, the Zoom meetings providing an excellent platform for keeping everybody up to date with what work had been done.

But, with regards to working in the NRM workshop the home signal was still at stop.

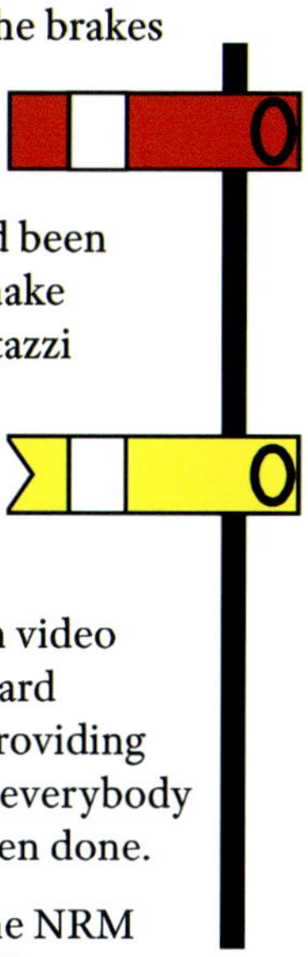

September 2020

Work continues off-site by sub-contractors who had acceptable Covid working arrangements and by individual volunteers in their own homes, but news was filtering through that a return to the workshops may be possible. This would require strict controls to be in place on many things, such as numbers of workers, working hours, use of tools, social distancing, use of the infamous 'blue cabin', the numbers allowed in at any one time and many more things. This all involved new risk assessments to be created by Bob Shearman and Andy Barwick, both of whom are experienced with the creation of these and who diligently set about the tasks to prepare for NRM approval. "There will be no communal tea making. All food, drink, cups, utensils etc must be personally brought in and taken home for washing", was a line that jumped out of one of Darrin's emails. This is a necessary precaution in these times, but a sad one to be invoked as the communal tea making, (even I had deputised for number 1 tea maker Peter Brackstone), was a highlight of many a day in the blue cabin with its camaraderie and banter. Now everyone was waiting for the home signal to be raised and later in the month it did when all the new procedures and risk assessments had been rubber stamped. The distant signal had not followed suit so it was with caution that we move forward once again.

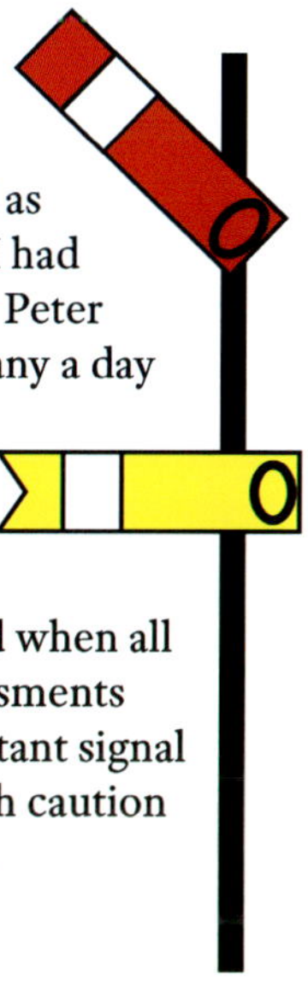

October 2020 - April 2021

October 2020 saw our tentative return to work in the NRM workshops. It was challenging, as the few members who were able to attend had to adapt quickly to the newly written social distancing risk assessments, and with so many key members of the team missing, some work could not proceed at all. Planning what could and could not be done required a fresh look at the critical path. Work was now progressing slowly but undocumented photographically by me as although Covid restrictions allowed volunteers to work I could not reconcile myself with the fact that my work at York could be considered as either necessary or essential, so I stayed away. I spoke frequently with team members and followed Darrin's reports which showed a plethora of jobs that could be done - some even in the home workshops of many members of the team. I won't hide my disapointment at not being there with my camera as I knew this would leave an inevitable gap in this book which no amount of summarising from information given by others could fill. But, on reflection, I tempered my disappointment by realising that this book is my story of the overhaul, and not being there during this period of time due to Covid was part of my story. It is about my interactions with others during the overhaul - human interest as opposed to technical detail. Darrin is far better qualified than I am on the technical side and his excellent website reports testify to this. Part of Darrin's report mentioned that due to Covid "morale had taken a

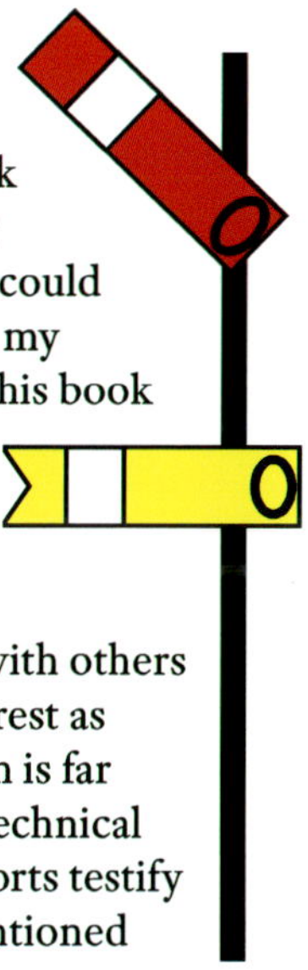

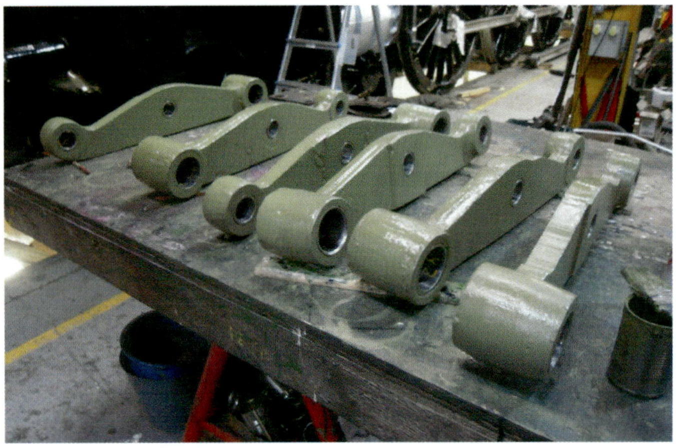

The loco brake hangers machined, bushed and painted to undercoat ready for refitting

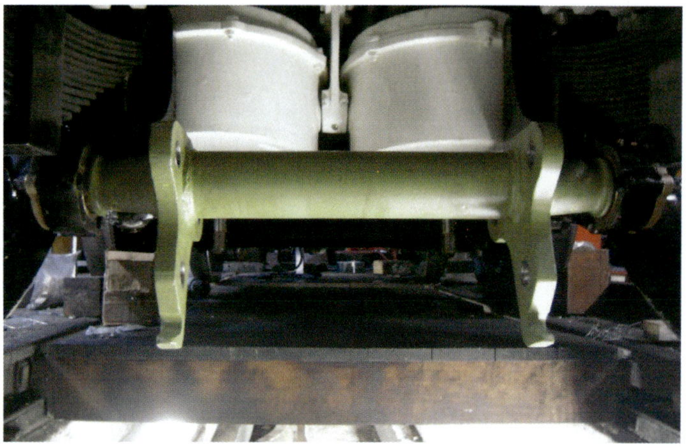

The leading brakeshaft in position

knock". I can vouch for that personally, but he also added "There's still plenty of laughs and determination to finish the job," something I was equally determined to do.

2021 arrived. 12 months earlier we had thought we would be back in traffic but now, Covid delayed, we had no idea when we would be back in traffic. Yes, it was true the relaxation of Covid restrictions had allowed some work to proceed but the number of volunteers allowed in the museum was limited and Darrin was also having difficulties working with suppliers, many of whom had staff on furlough. As winter edged into spring, not that anybody told the weather that as the snow continued to fall quite heavily, there was hope when the government announced the vaccination rollout and coupled with a steady decline in the infection rates a timetabled further relaxation of restrictions was announced. By mid-March volunteer numbers were slowly increasing as more of the team received their vaccinations and the museum, after signing off on our new risk assessments prepared by Bob and Andy, was allowing greater numbers to attend. A special mention here and another reference to the great camaraderie in the team is that not one single member had dropped out since the start of the pandemic.

Covid restrictions did allow house moves and my wife, Janice, and I had managed to complete ours in September 2020. We were now relocated in Yorkshire's beautiful Airedale close by what becomes the famous Settle & Carlisle line, cutting the distance to York in half but not the travelling time due to the slower road network and the necessity of having to change in Leeds if journeying by rail.

I was now desperate to make my first journey from our new home to York and get some images of the progress, but I could still not justify my return to the workshop, particularly as numbers were limited. I am grateful to Darrin and Richard for contributing images here to illustrate some of the many tasks that were being undertaken during my absence, such as work on the steam pipes, replacing some of the streamlined panels which were too badly corroded to refit, the great deal of work needed on the handbrake assembly due to historical repairs leaving it in poor condition, brake shafts, bearings, bushes, coupling rods and the cartazzi. Many of these tasks were not without problems and I again refer you to Darrin's excellent reports on the Trust website or for SNGLT members 'Chime' magazine numbers 199 and 200 for greater technical detail.

Although we were now making progress and the pace had increased, we were still passing distant signals set at caution and still peering ahead for a tantalising sight of both a distant and a home signal to be raised together. Safety was a number one priority and as the days passed by more and more of the team were receiving their Covid vaccinations giving new hope that some kind of normality would return to life and even possibly allow riding on a steam hauled train again!

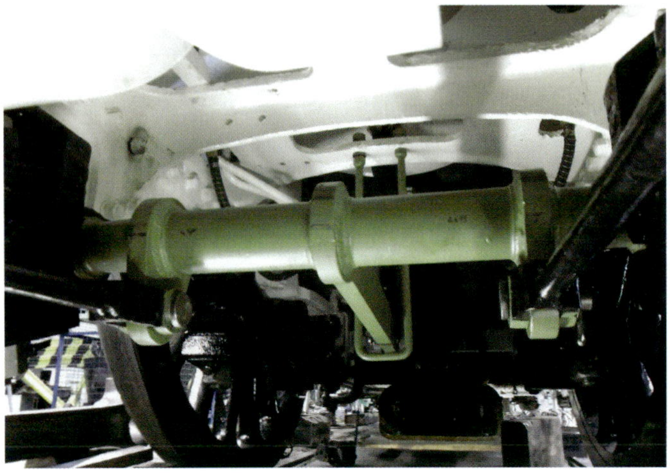

The refitted trailing brakeshaft

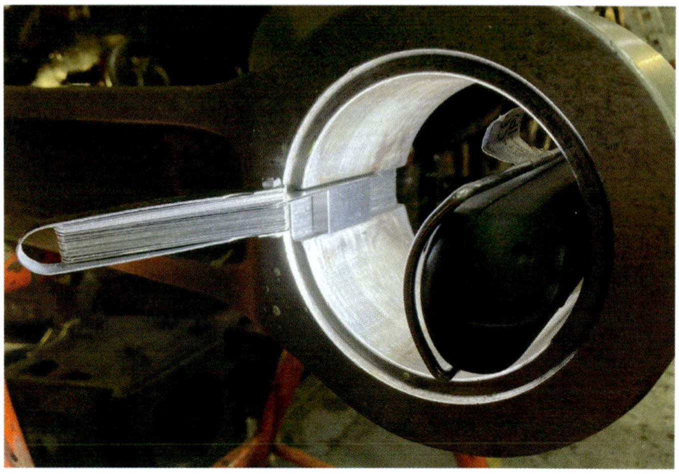

The keyway in the small ends of the connecting rods have been dressed by Andy Barwick and new keys made

BACK ON BOARD

14

April 2021

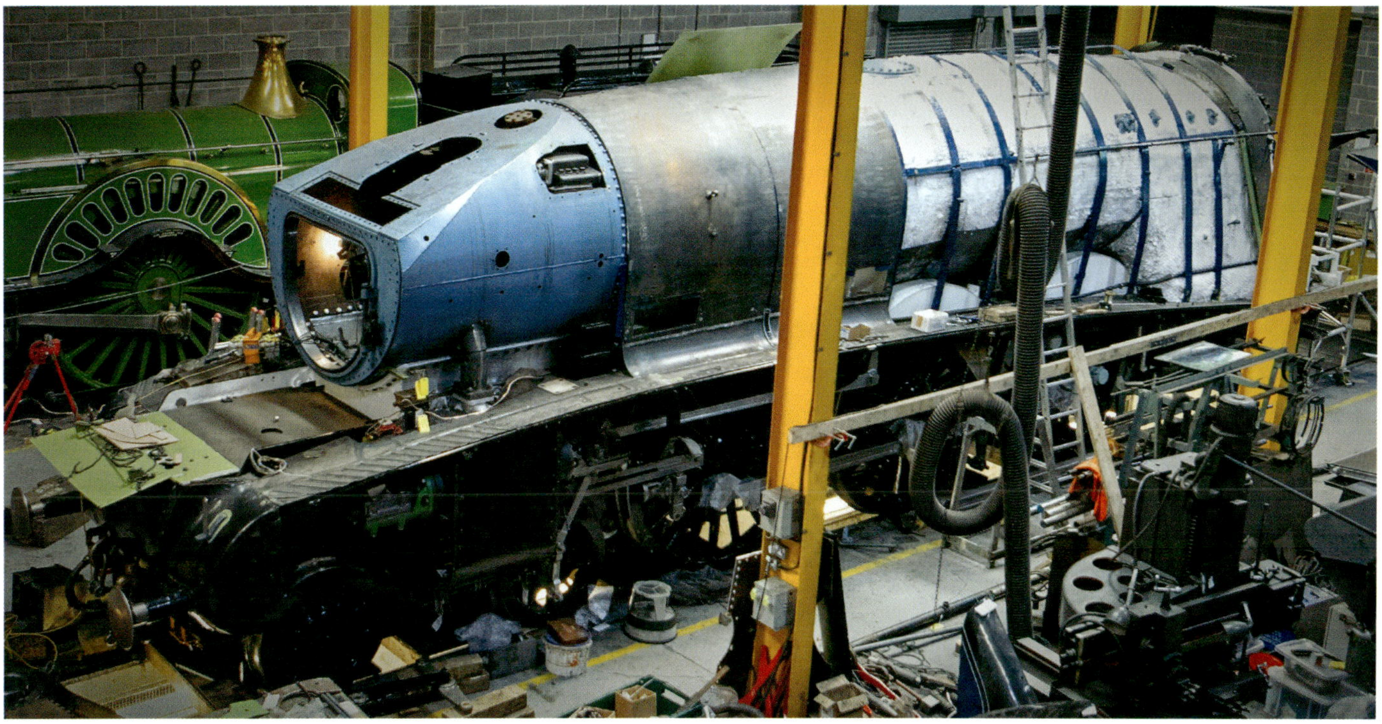

The 30th April 2021 had a strange feeling to it. Two days earlier I had had my second Covid vaccination and coupled with the improving statistics and further government relaxation of restrictions I decided it was time for me to return to the NRM and capture some updated images. I say strange as it was over a year since my last visit due to the Covid pandemic. It was also strange driving further than the few miles to the local supermarket or to meet up with family members outdoors - again local. However strange it felt it was good to be back on board.

Due to the restrictions limiting the number of volunteers and available attendance days the progress during my absence had been slow but there were visible changes, the most significant of these being the refitting of some of the streamline panels, valve gear and brake hangers. The less visible brake components necessitated a trip down into the pit with my hard hat on and my aching back complaining.

I guess there is not a single locomotive of any reasonable age that has not had some major components changed and is therefore not entirely in its original state and No.7 is not immune from that. Indeed, when purchased in 1966 by the A4 Preservation Society (that became the SNGLTrust in 1998) it was overhauled at Crewe during which it inherited driving wheels from No.26, and the same was the case this time with some of the main steam pipes to the cylinders and a few of the streamline panels. In both cases it was

Covid restrictions have resulted in socially distanced tea and lunch breaks on the workshop gallery with the blue cabin deemed out of bounds for more than one person at a time

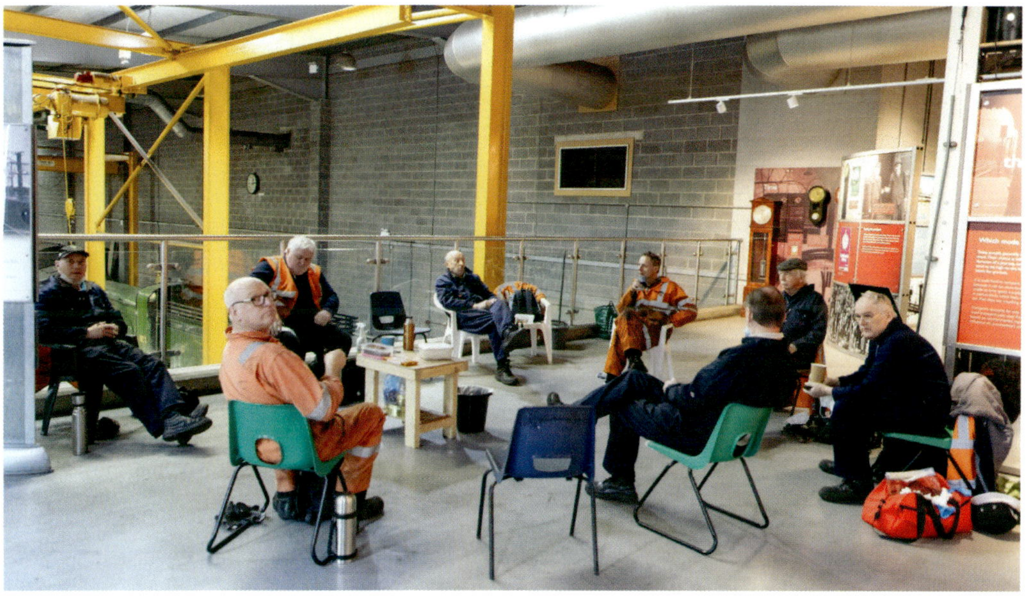

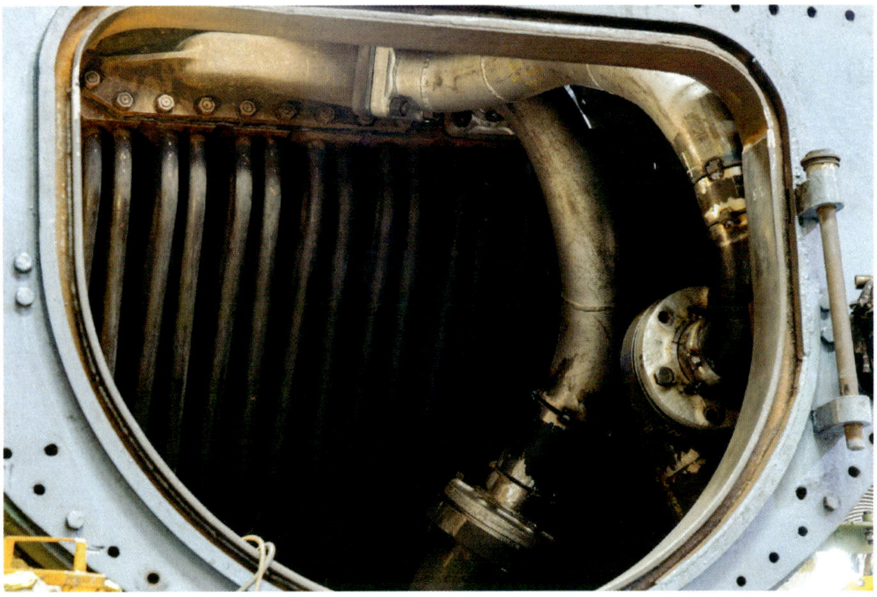

The middle and left hand cylinders' new steam pipes in position inside the smokebox

considered that the condition of these components was too poor for them to be refitted and an exchange was necessary

Dave Lee was working on the steam pipes to the cylinders, and after taking some images from in front of No.7 looking up into the smokebox at the pipes, he invited me to join him up the ladder to get a closer look. I thanked him but declined on this occasion as the aforementioned backache formed part of my risk assessment, together with the risk of damaging my camera equipment by climbing around and setting up, as opposed to using a longish lens! I am certainly not as young or agile as I once was which is also true of a great deal of the volunteers and I am extremely impressed with how well they all manage. Darrin frequently jokes he had not realised being a care home manager was in his job description! Dave was working on the left and middle main steel steam pipes which are now ready for the final welding. It was decided to renew the right hand pipe, so the existing one was tacked in place to provide a pattern for new sections of pipe to be made by the pipe bending contractors.

My next invitation was still challenging for me but this time I accepted when Andy Evans invited me into the pit to see the progress on the brake shafts. With hard hat on I felt like

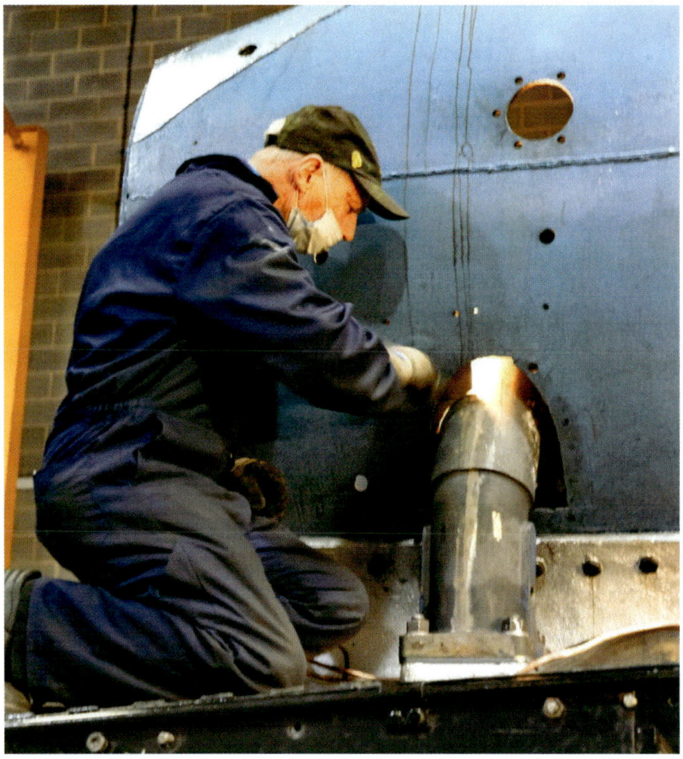

Dave Lee working on the left hand cylinder's steam pipe

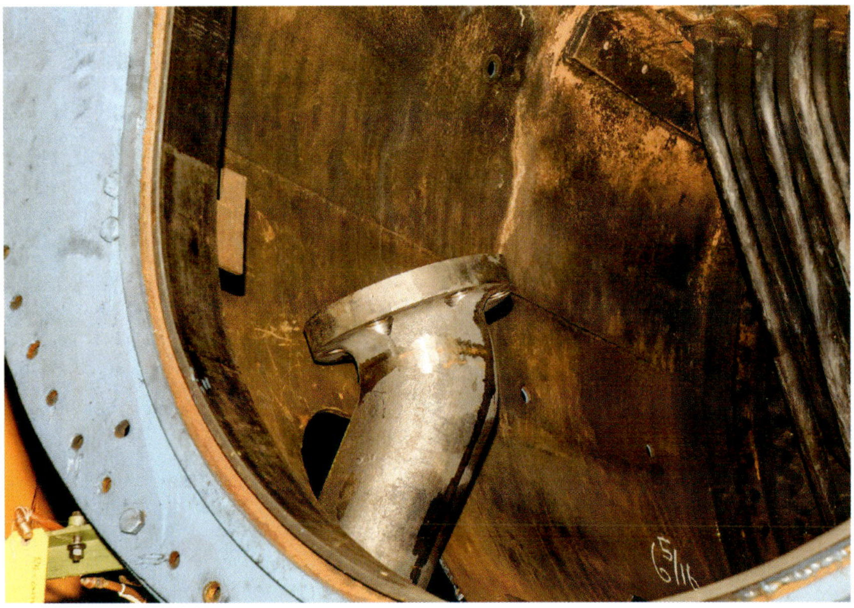

The right hand steam pipe awaits a new section of pipe from the pipe bending contractor

The camera rests on the pit floor and points directly upwards through the open ash pan door into the fire box. I managed to take a 'selfie' of a kind as there is not much room to get out of the way

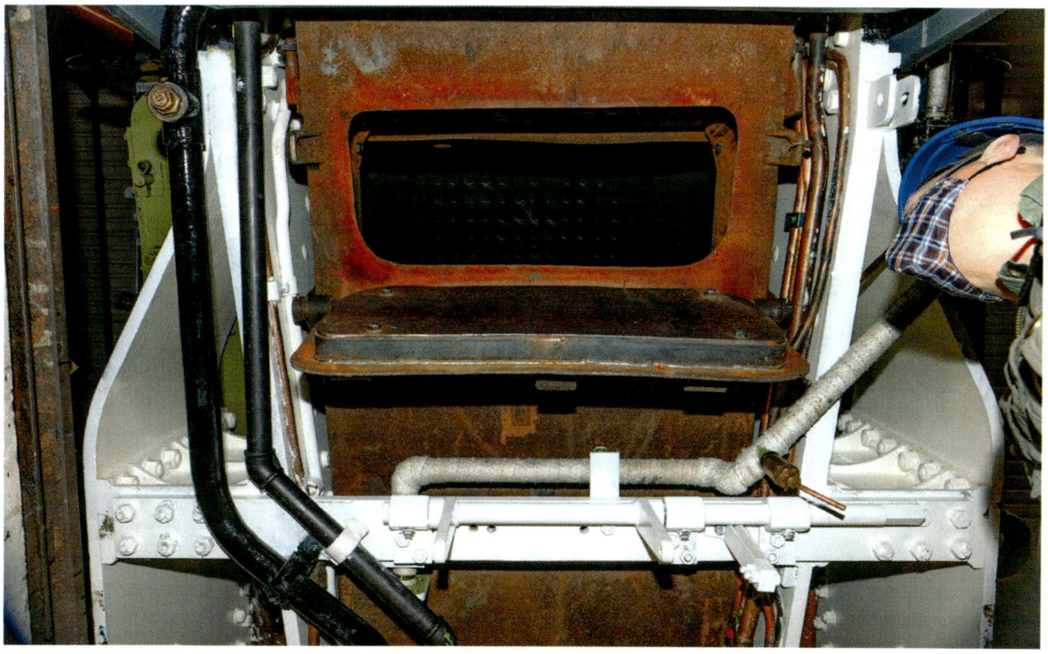

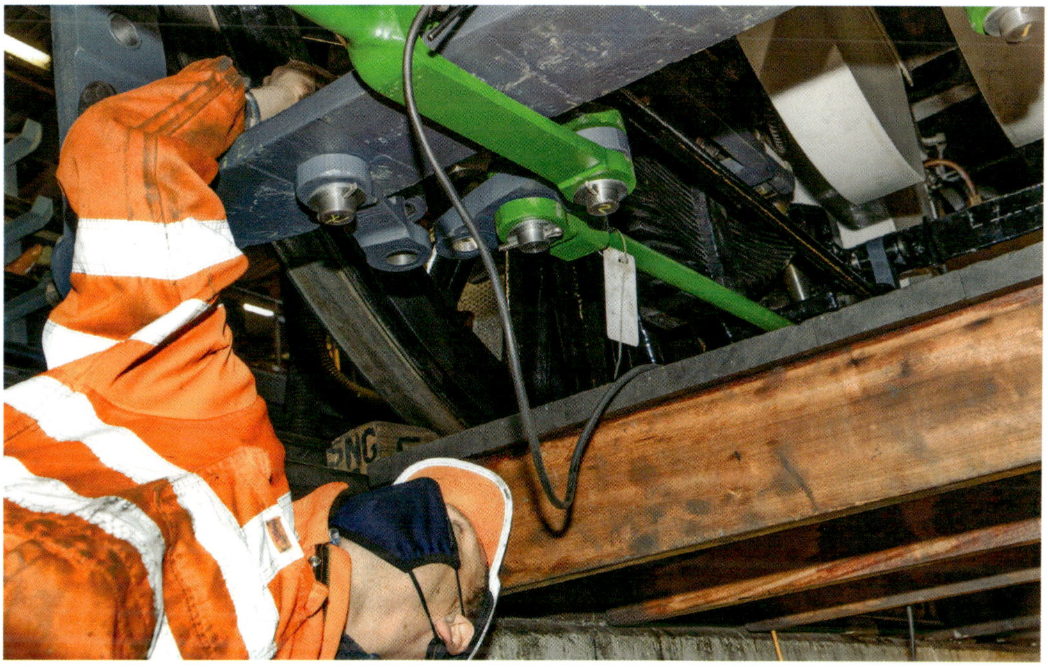

Andy Evans assembling brake shaft linkages in the confined space of the pit below the main drive wheelset

a caver as I descended the steps into the pit and crouched to get even lower as the available headroom is further reduced by several wooden pit boards between the rails. So, I ended up crawling to get the shots. On my crawl forward I noticed the ash pan trapdoor was open so, after placing my camera on the pit floor and shooting blind with the camera pointing directly up, I took the opportunity to capture that shot as it also shows the complex pipework underneath which has required much work from the Peter Ellis led pipework team. Now Andy is considerably younger than most of the team and I was extremely impressed when he joined me by sliding in from the side between two of the main drive wheels. This is a claustrophobic workspace, so I am always pleased to see the daylight when I emerge from the pit, but I am not deterred; the risk photographers take to get the shot!

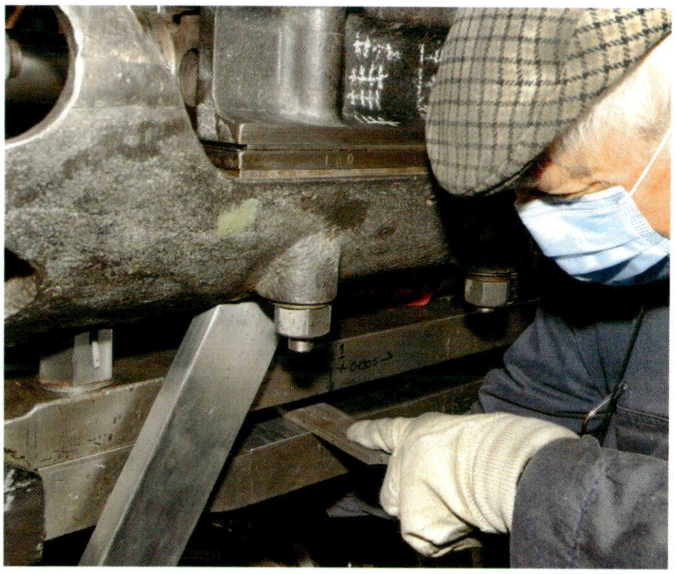

Bob Shearman makes micro adjustments with a fine grade file on the slide bar spacing. Note the chalk marks helping him keep count

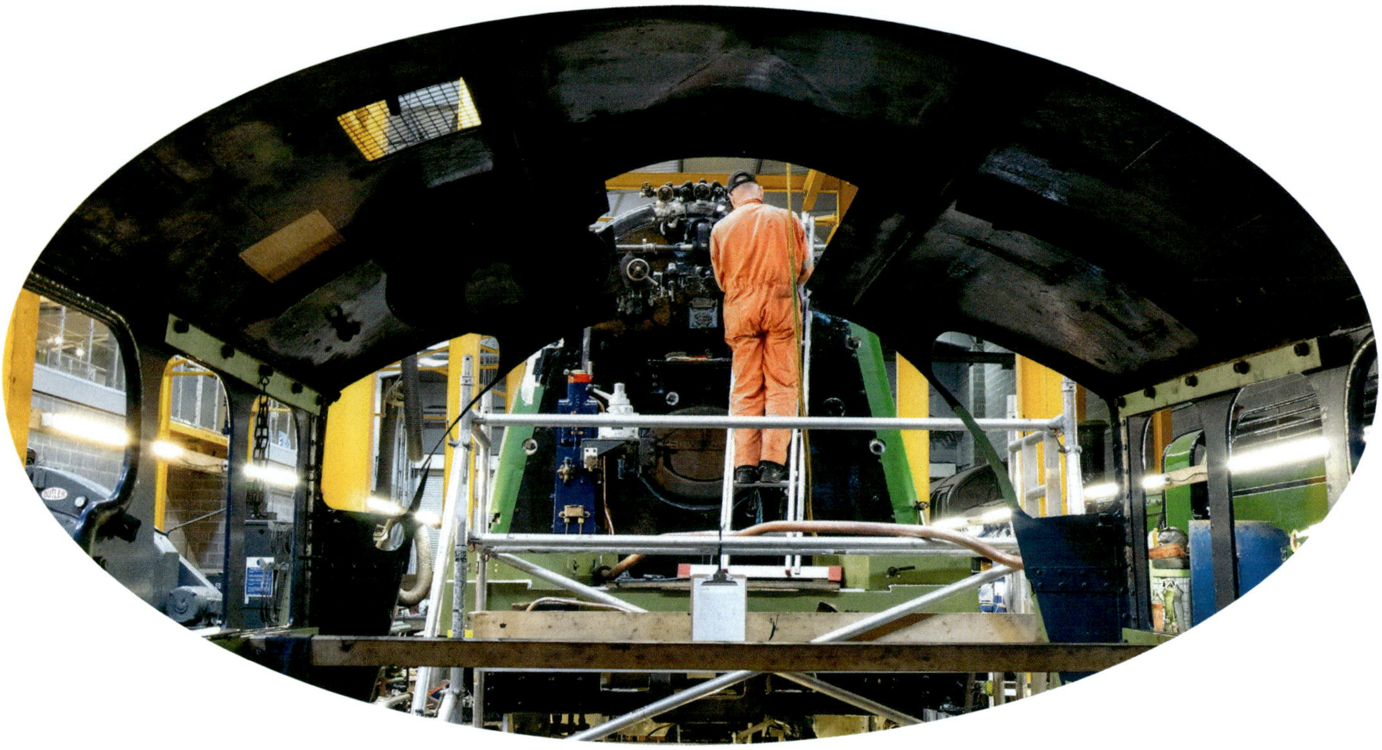

The view through the removed cab sides and roof of Andy Barwick working on the re-assembly of cab components. The centre section of the cab roof had been removed for painting which enabled a better view of the work

Bob Shearman had caught my eye as he was relentlessly filing away on the left-hand valve gear slide bar - well it was not so much Bob but the chalk marks counting off days on the casing. This, Bob explained, was not days in a solitary confinement prison cell but his way of counting the number of strokes his file was making, such is the precision required.

With the cab roof and sides still removed there was good visibility to get shots of Andy Barwick assembling components in the cab, as he was working today on the regulator. The cab would of course have several new components when finished, as now to be able to run on mainline railways, electricity would be introduced into the cab to run things such as TPWS (Train Protection and Warning System) and OTMR (On Train Monitoring and Recording).

For various and unavoidable reasons some of the team had been missing from the workshop for longer than just the Covid pandemic period, so it was good to catch up with Rod Thomas and Mike Page today. Rob was working with Peter Ellis and Malc Bateman on various pipework components that were proving problematic to refit and Mike was working on the new electrical work mentioned above.

Breaks, due to Covid social distancing restrictions, could no longer be taken in the blue cabin, so dedicated space was allocated on the gallery for SNG use only. However my work for today was complete so I set off for home leaving by the Covid assigned exit in the workshop. This exit meant a long walk around the outside of the NRM to reach the engineering car park instead of a direct exit through the NRM and just my luck, despite it being the driest April in many years, it was pouring down, so I was thoroughly soaked when I finally reached the car.

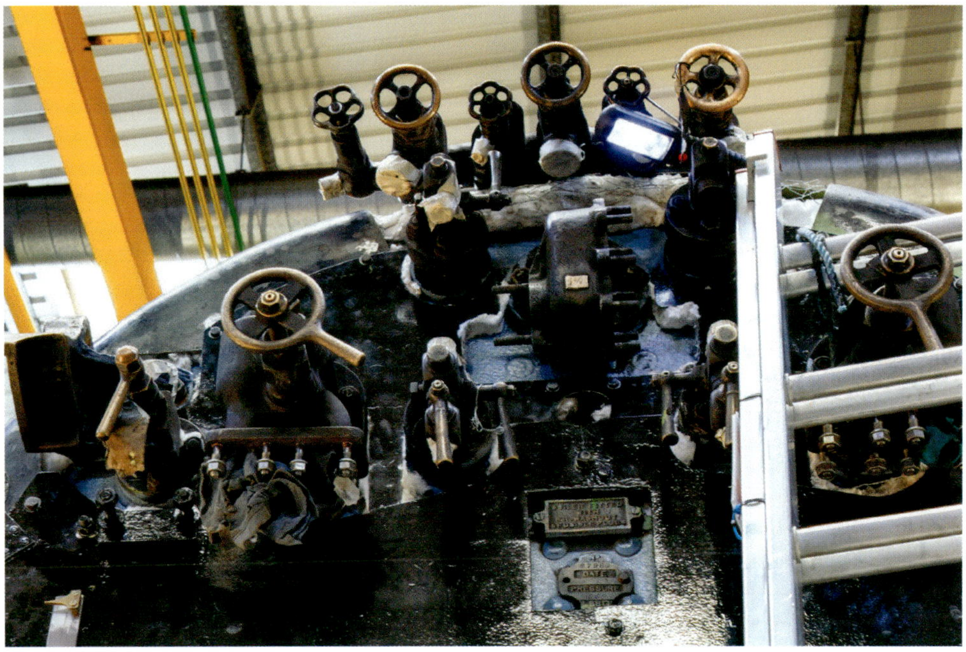

Many of the cab components are now back in place

SPRINGY THINGS

May 2021

"springy things that don't want to spring the way you want them to"

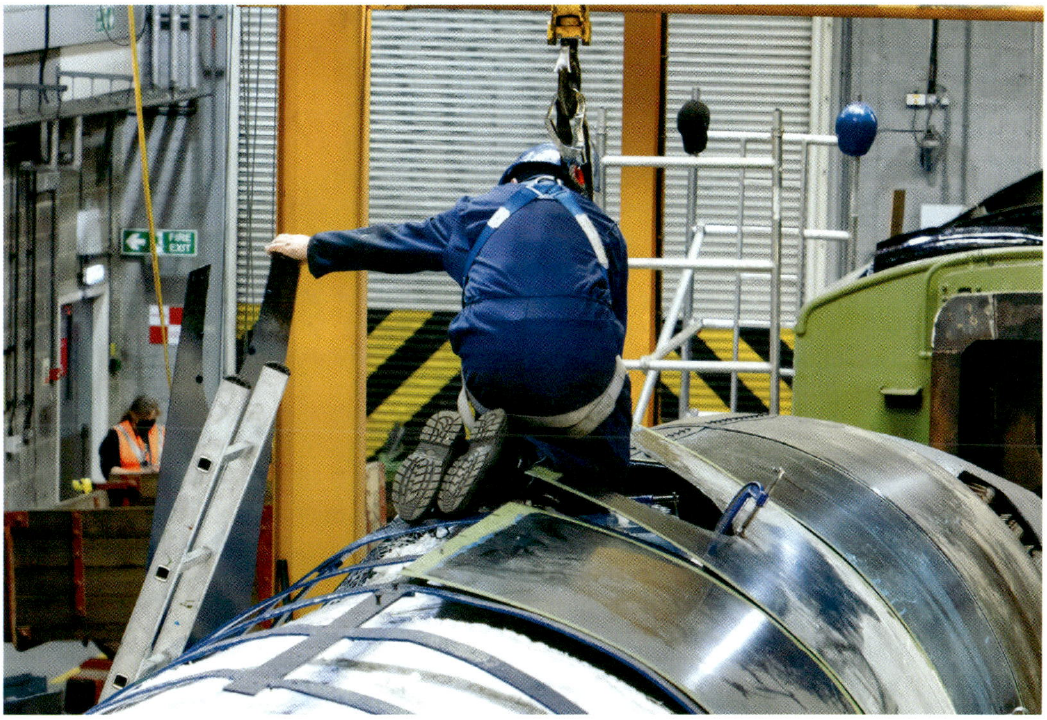

Darrin Crone fixing the springy streamline cladding panels with steam dome cut out sections around the boiler

A quick scan down the week 21 roster told me I should go to the workshop on Tuesday as one of the volunteers with whom I had not been able to catch up much was working that day. I informed Darrin of what I was proposing to do, and promptly got the response that for one reason or another that particular volunteer was now the only one available on Tuesday, so he had had to cancel work that day. However, a quick switch to Wednesday worked out quite well as Darrin planned to fit some of the cladding, providing a good picture opportunity as seeing the outer skin back in place was a very visual sign of progress towards the eagerly awaited return to traffic.

Removal of the cladding was the first task I had photographed way back in January 2016. The panels had all been stored ready for refitting but on inspection some of them were badly corroded, which required the manufacture of replacements and it was one of these that the team

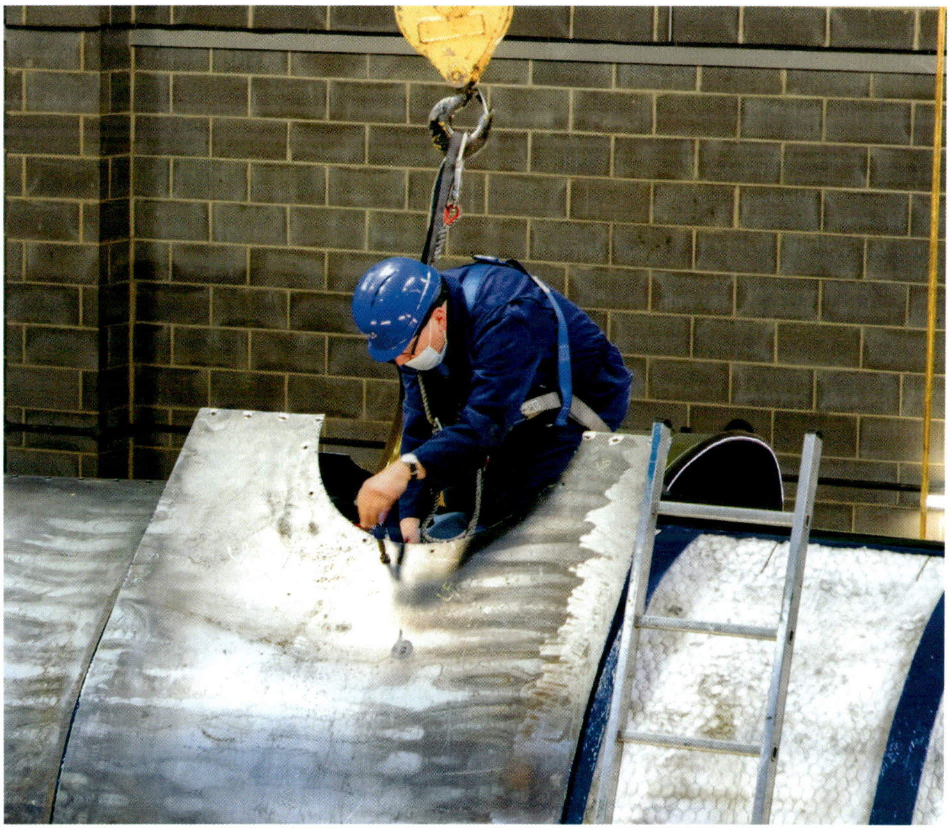

The new left hand side panel around the steam dome is fixed into position by Darrin

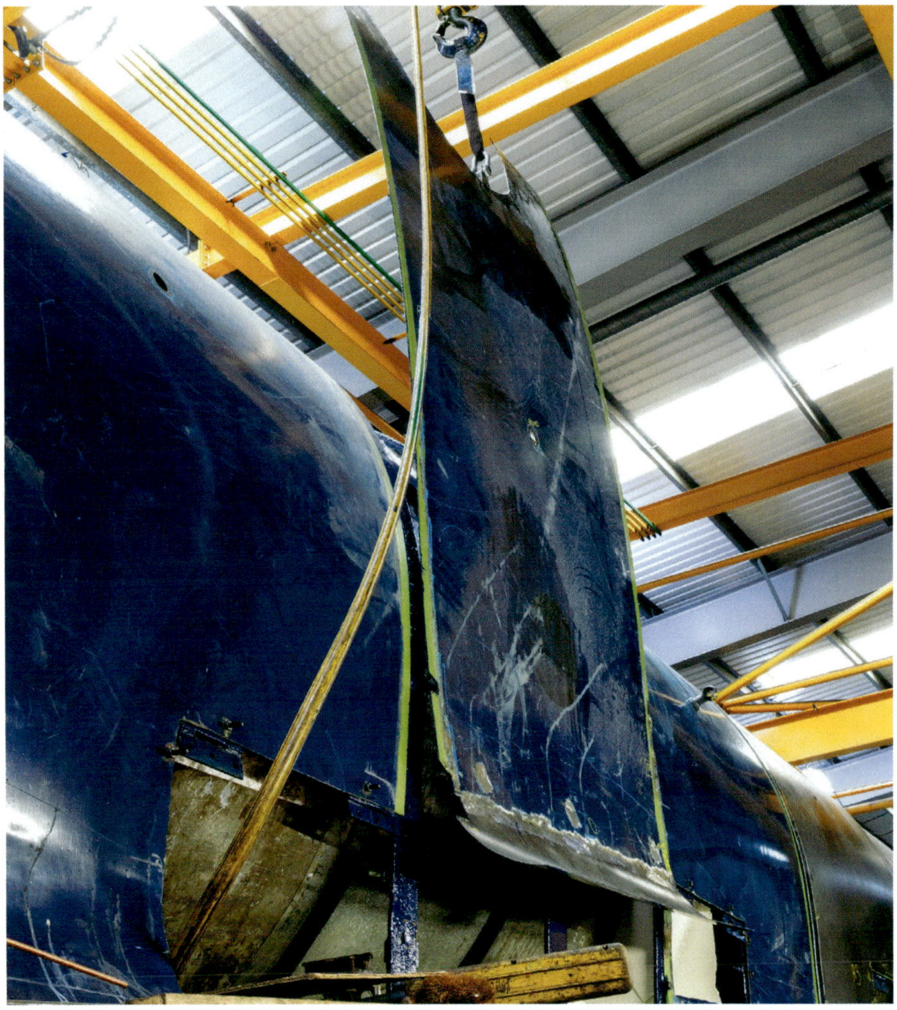

With the overhead crane taking the weight the right hand side panel with the cut out section for the steam dome is lowered into position

were fitting today. As it was heavy and awkward to handle, the overhead crane was used to take the weight and help position it for fixing by Darrin who, with a good head for heights and a safety harness, clambered up his ladder onto the top of the boiler aided by Trevor Wilford and Richard Hawkhead. Darrin's description of the cladding panels being "springy things that don't want to spring the way you want them to" encapsulated the awkward manoeuvres, but he was ecstatic when everything lined up and the fixings went home first time. This new panel was to the left hand side of the steam dome; the right hand side was an original panel with some sections repaired and this was also successfully refitted.

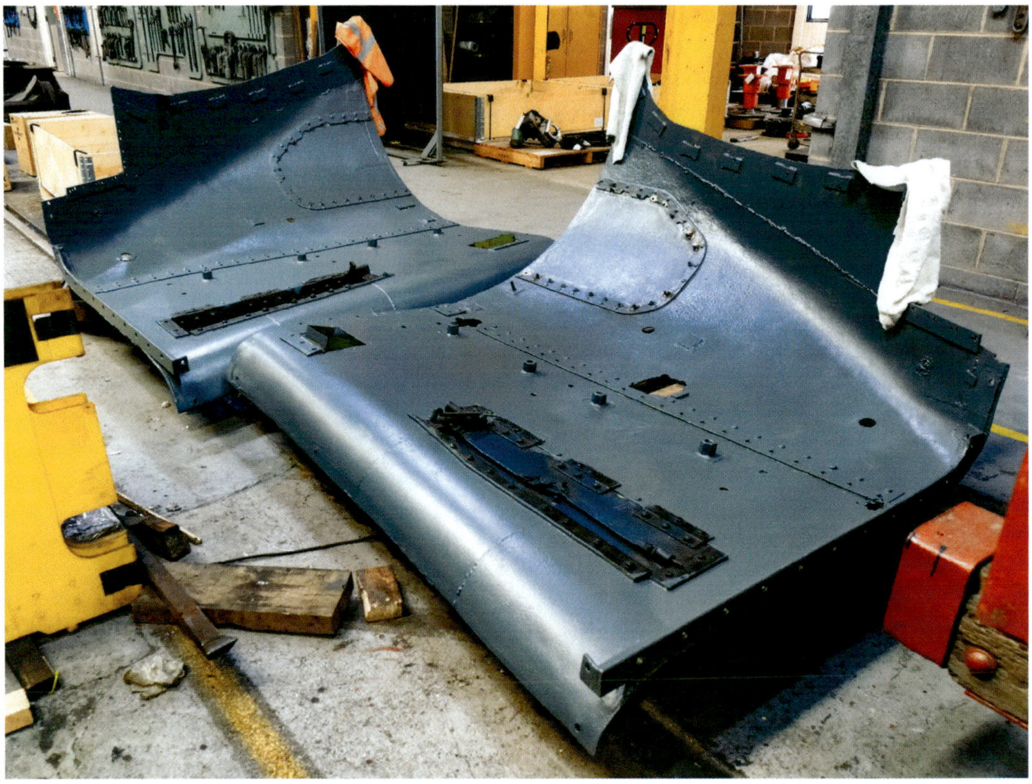

Streamlined panels for the smokebox lie awaiting their turn to be refitted

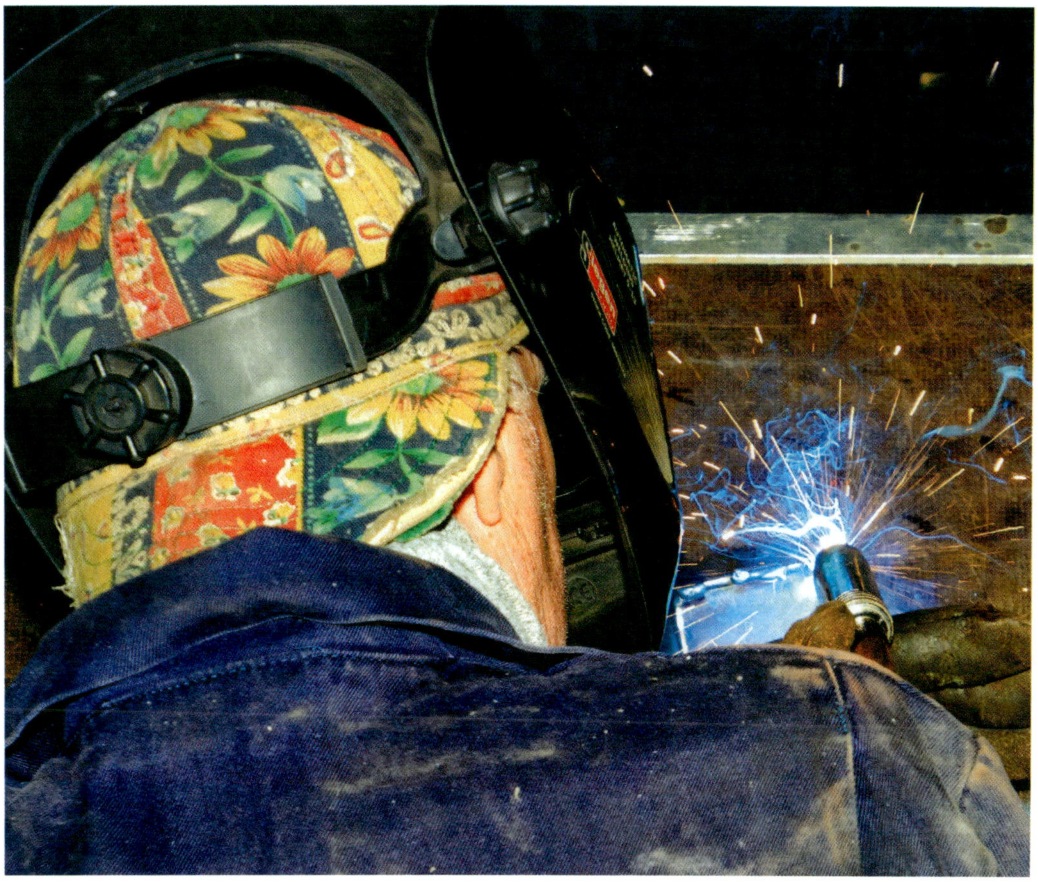

Dougie Blyth welding tender components

Richard and Trevor were being as flexible as the springy cladding panels today as they were also working with fashion conscious Dougie Blyth, who was welding some components together on the tender which is where I now turned my lens. Dougie's very fashionable cap was what caught my eye as I carefully averted them away from his welding arc light and set up a remote shot from above after climbing up the ladder onto the platform at the front of the tender. There was plenty of banter going on between the three tender team members as Dougie had Richard assisting and Trevor was wedged into the tender's corridor with his grinder, insisting that back in the day people were smaller than they are now. Hmmm, I think it best not to comment further on that as I would probably have to agree with Trevor's view.

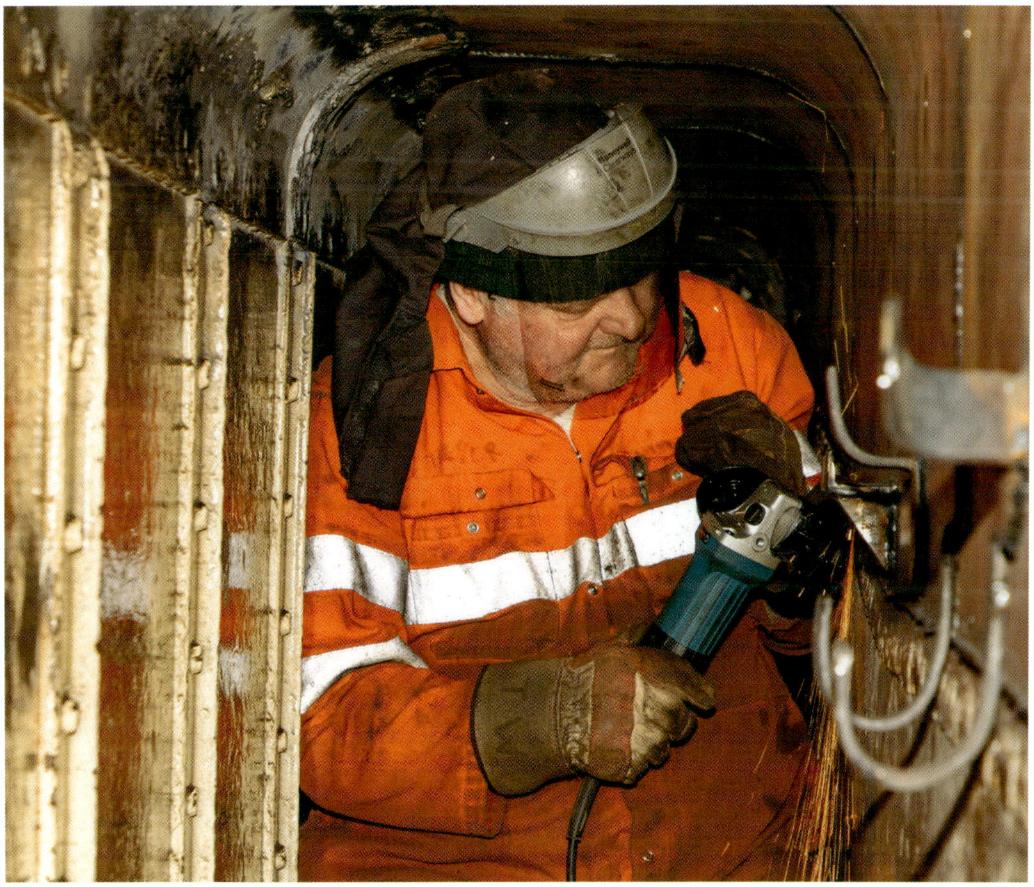

Trevor Wilford working with a grinder in the cramped conditions of the tender's corridor

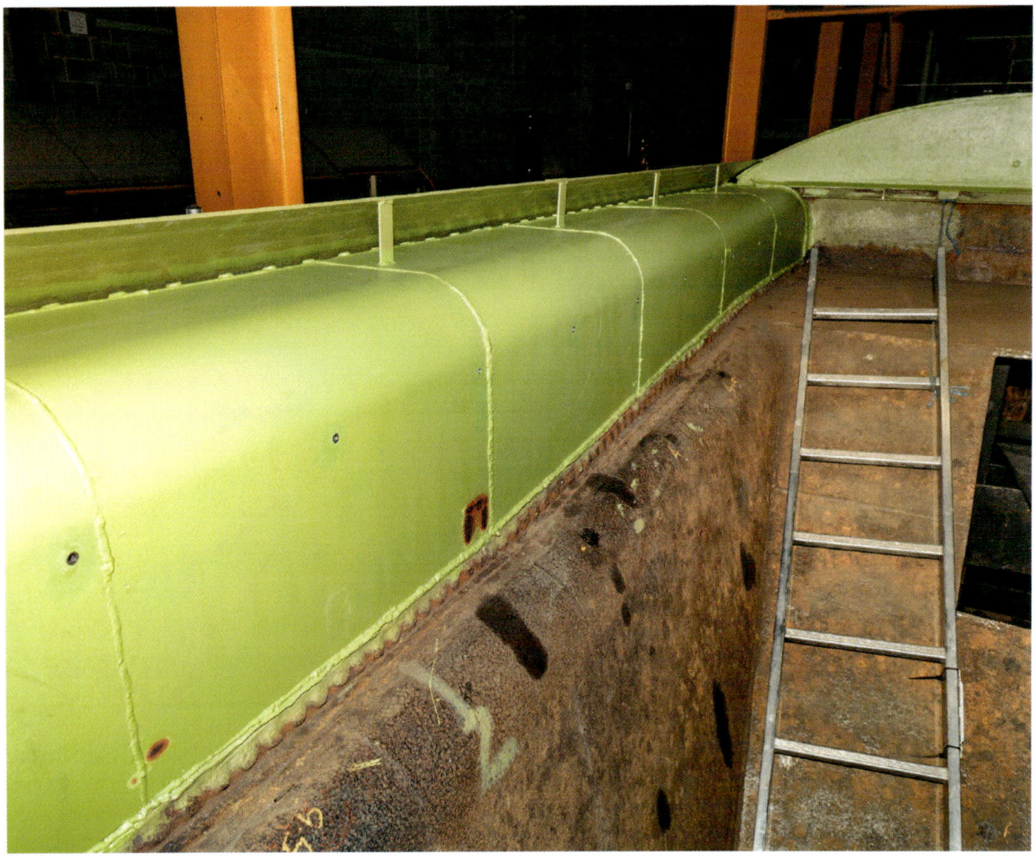

The newly fabricated roof of the corridor tender from inside the coal bunker

Whilst on the platform I stepped inside the tender's coal bunker to get a shot of the corridor roof which, due to corrosion, had been mainly cut away and replaced. The new sections were fabricated by Trevor and welded in place by Dougie Blyth, and now painted with primer and undercoat their work looked very smart to my albeit untrained eyes.

My final picture shows the right hand side cylinder with valve gear assembled to check everything lined up and fitted correctly. On my next visit it would be disassembled awaiting the fitting of the piston rings and then refitted for a final time.

I left mid-afternoon but not before I had a chance to look over some old photographs of the workshop dating back to the 1950s, with much clearly still recognisable today, and there was a definite tinge of sadness to think that with plans to repurpose the space soon to go ahead, No.7 would be the last locomotive to be worked on in the workshop. At a different moment in time, I fondly remember sneaking around the space that was once York Motive Power Depot (50A) whilst trainspotting and I can't help thinking what a loss to history the new plans will cause.

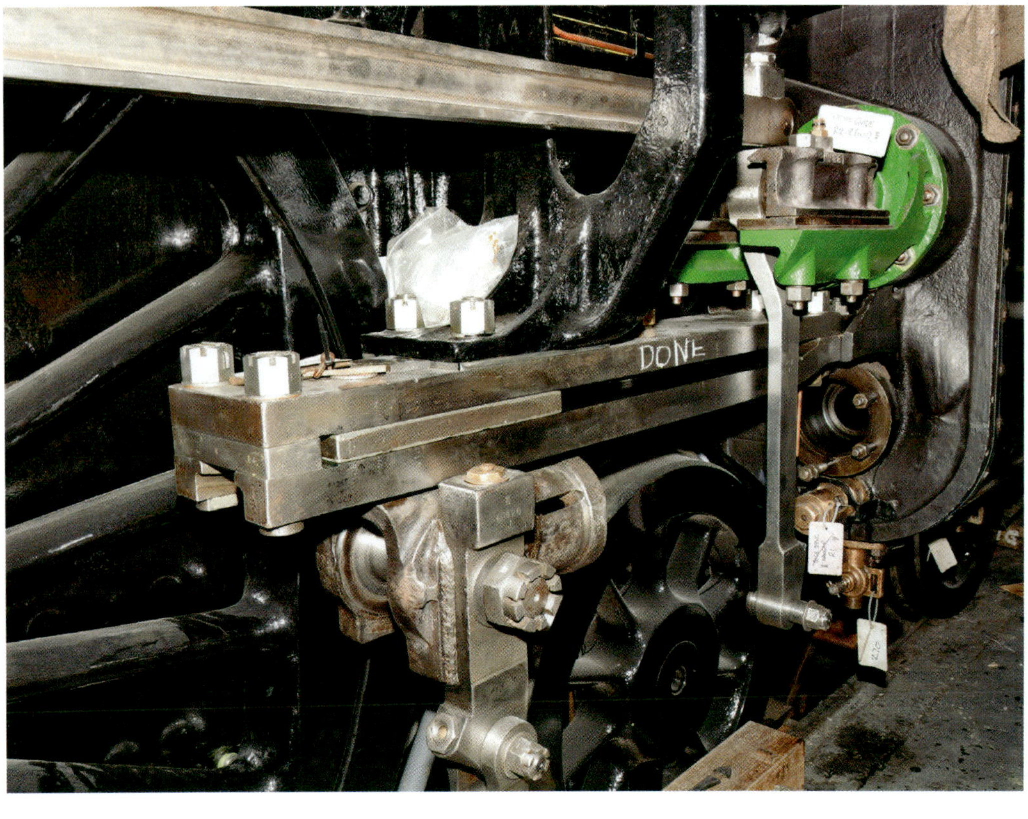

The right hand side valve gear assembled to check for fit and which would soon be disassembled to facilitate the fitting of the piston rings

MOVING ON

15

June 2021

Work on and off No.7 was gathering momentum with several meetings taking place to map out the final dash to the finishing line. The decision by the Science Museum Group to close the excellent and famous York workshop by the end of 2021 and repurpose the space would probably not coincide with our yet undetermined completion date so, after discussions with LSL (Locomotive Services Limited), it was decided that a move to their equally famous and excellent workshops in Crewe would be made. As access to a boilershop and boilersmiths was necessary for hydraulic testing and commissioning and No.7 would be based at Crewe when hauling LSL charter services the decision to move there made a lot of sense. When the build was complete, we would be hauled to the York carriage sidings by rail and then transferred onto road transport to complete the trip to Crewe. Anybody connected with railways would see the workshop closure as a sad occasion, so it was probably fitting that the locomotive named after the LNER's most famous engineer would be the last locomotive overhauled there.

As the work was moving at a fast pace, I knew I would have to keep up by increasing my attendance days, so June 8th was the first visit of the month. Darrin met me at the side entrance to the workshop and quickly directed my attention to the left hand cylinder valve which had been dismantled - a step backwards surely so what was wrong? The answer was nothing, in fact it was very much the opposite as Darrin went on to explain that this small step backwards was to lead to a giant step forward for the return to traffic. Everything had been assembled and checked to ensure all the various components fitted and worked correctly together, which they did. So now the cylinder valves would be taken apart again to allow the valve piston rings to be fitted before they were reassembled for the last time prior to the steam tests taking place. My picture shows the valve components removed and the piston valve now installed with the rings in place. CME Richard Swales was to carry out the full reassembly later in the week and, wanting to capture this, I arranged with Darrin to return on Saturday to meet up with Richard.

Darrin was continuing his work on installing the streamlined boiler cladding panels, and today with the help of Peter Brackstone and junior volunteer, Tom Readman, he was

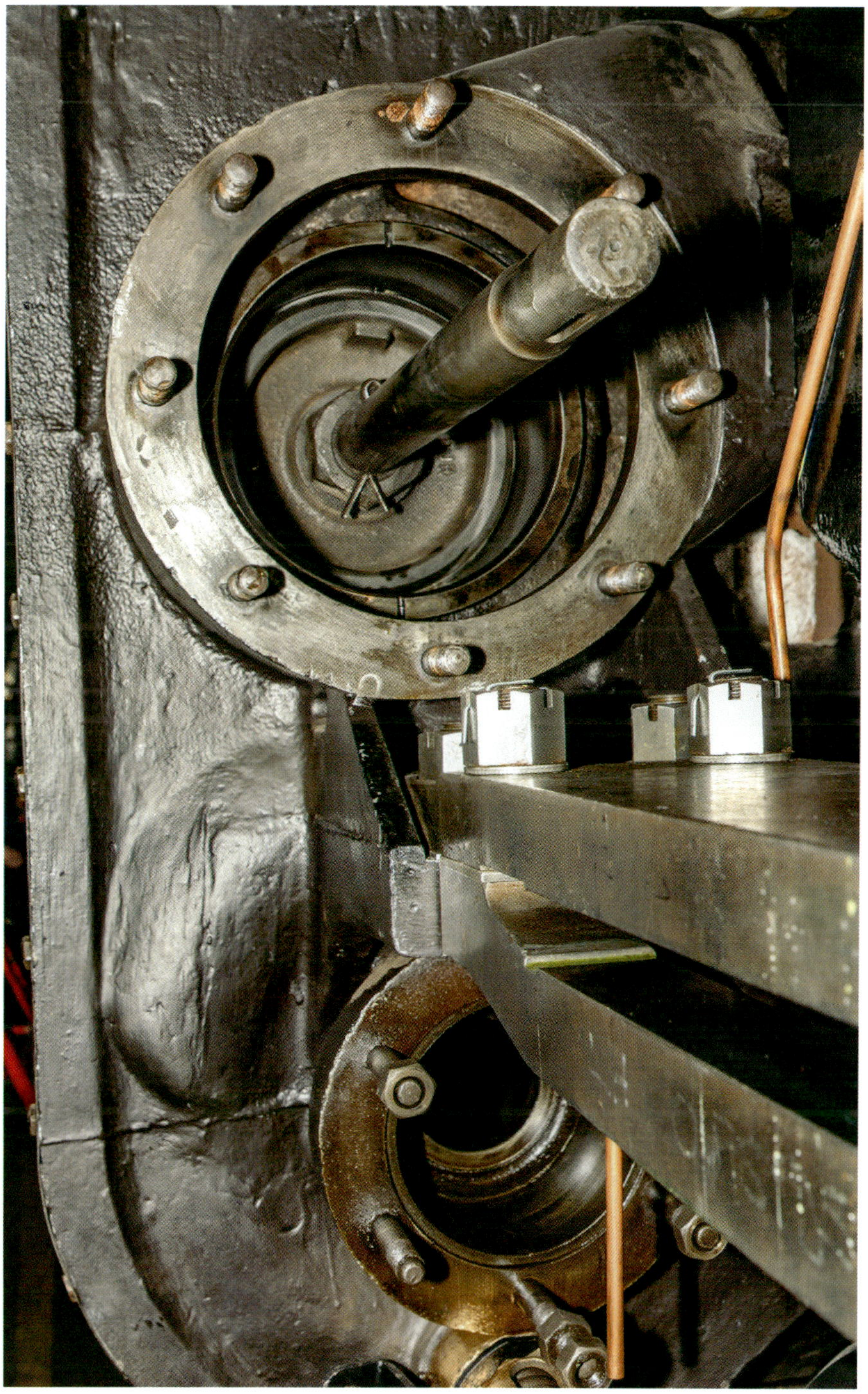

The left hand side cylinder showing the rings now fitted to the valve head at the top of the picture and ready to be reassembled for the last time

juggling with a very long and very difficult to control boiler strap which fits around the boiler between the edges of any two panels. The image I have used here shows the strap somewhat tamed and being fixed at its left hand side position.

All the work seemed to be happening on the left hand side of No.7, as Bob Shearman was patiently working on setting the gap between the cartazzi horns. This was meticulous work and needed a lot of care and patience, something Bob had in spades. This Cartazzi horn liner was being ground to give the correct horn gap for the axlebox, and as the leading horn liner had become worn, the gap had to be customised which made it a very tricky job. Bob, not for the first time, gave me another long exposure tripod shot image capturing the sparks and also deliberately blurring his movement to

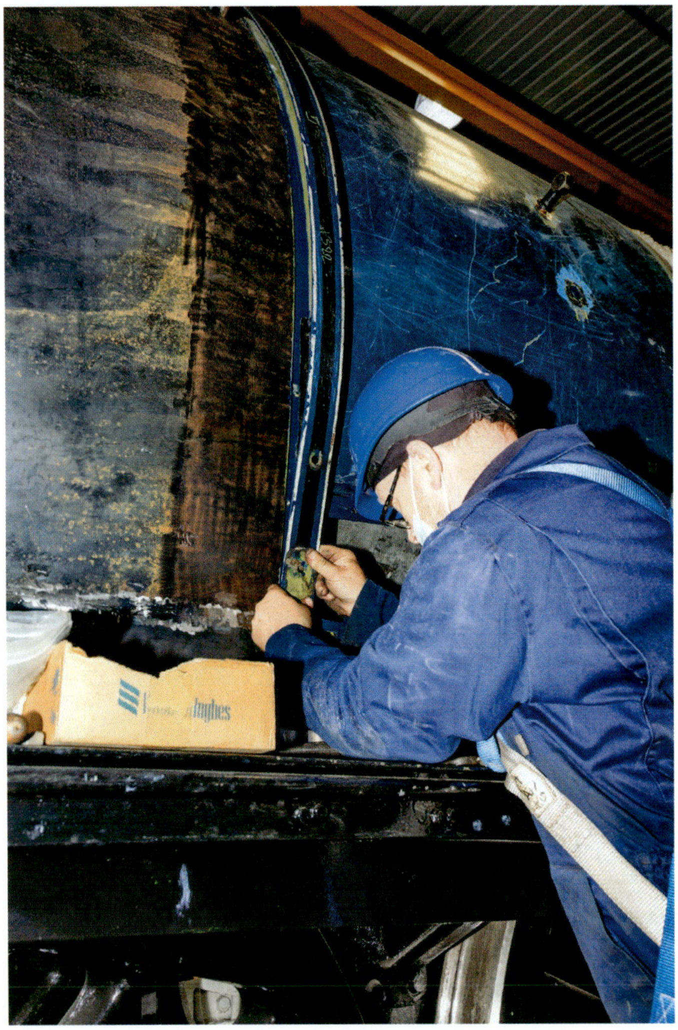

Darrin Crone fitting a band over the join between two sections of the streamline cladding

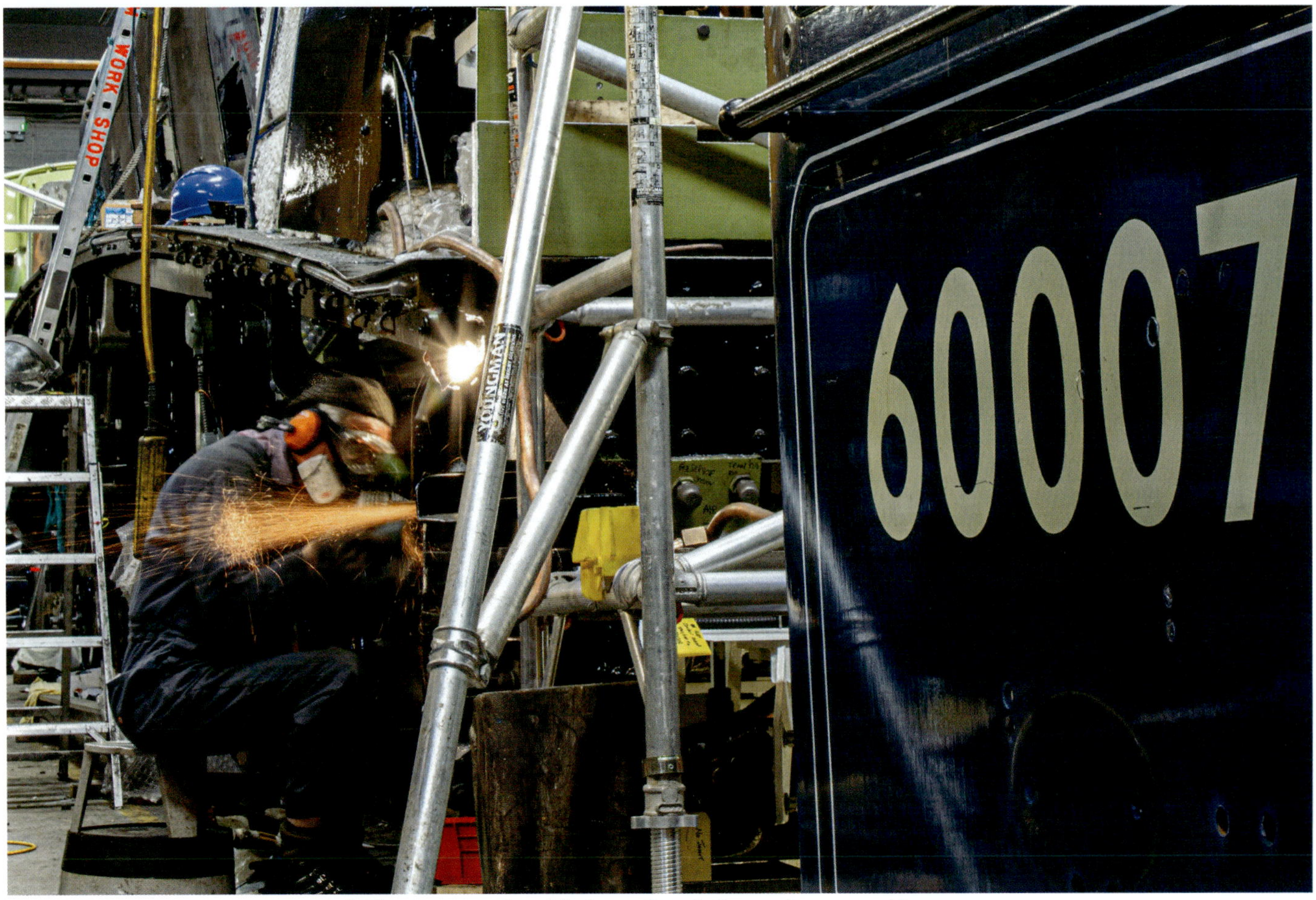

Bob Shearman works with the angle grinder on the cartazzi horn stays

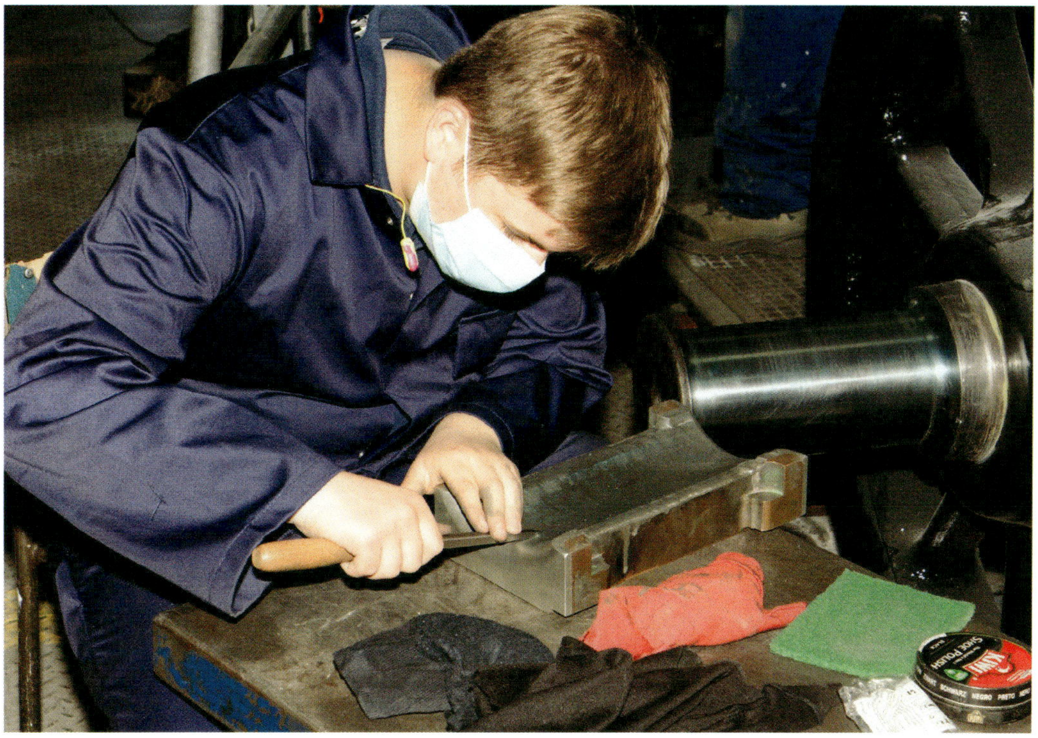

Tom Readman smooths out high spots on the cartazzi axle bearings

enhance the action whilst he used the angle grinder.

Having helped Darrin with his boiler strap, Tom had returned to his task for the day of smoothing out the high spots on one of the Cartazzi axle bearings and, having just left school, I thought how well he was following in Bob's footstep at being meticulous and patient in his work.

Still maintaining the left side theme, after passing Peter working at his paint station on the multitude of components requiring his paint brush, I found Nigel Franklin finishing bodywork filler repairs and inside the tender corridor Alan Pitt also carrying out finishing work in preparation for the tender to be ready for painting.

Due to Covid our mess area was no longer in the blue cabin but on the workshop gallery which was now closed to the public for our dedicated use only. This was a terrible shame as members of the public could no longer see the progress being made on the overhaul so, as on my last visit, I decided to fill the void of the many missing social media images that were usually being taken from the gallery and take some myself for publication. After taking these images I was joined by the team as lunch time beckoned. They were unusually

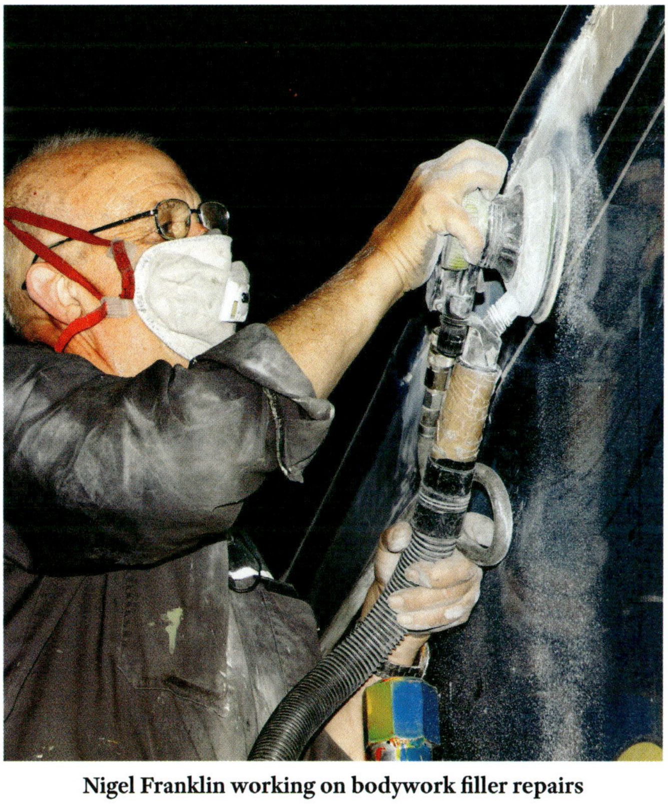

Nigel Franklin working on bodywork filler repairs

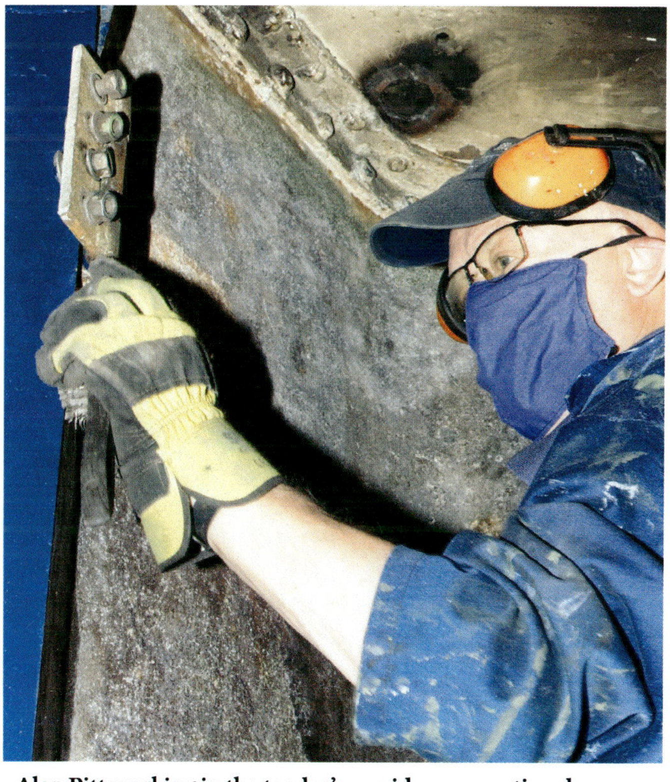

Alan Pitt working in the tender's corridor connection doorway

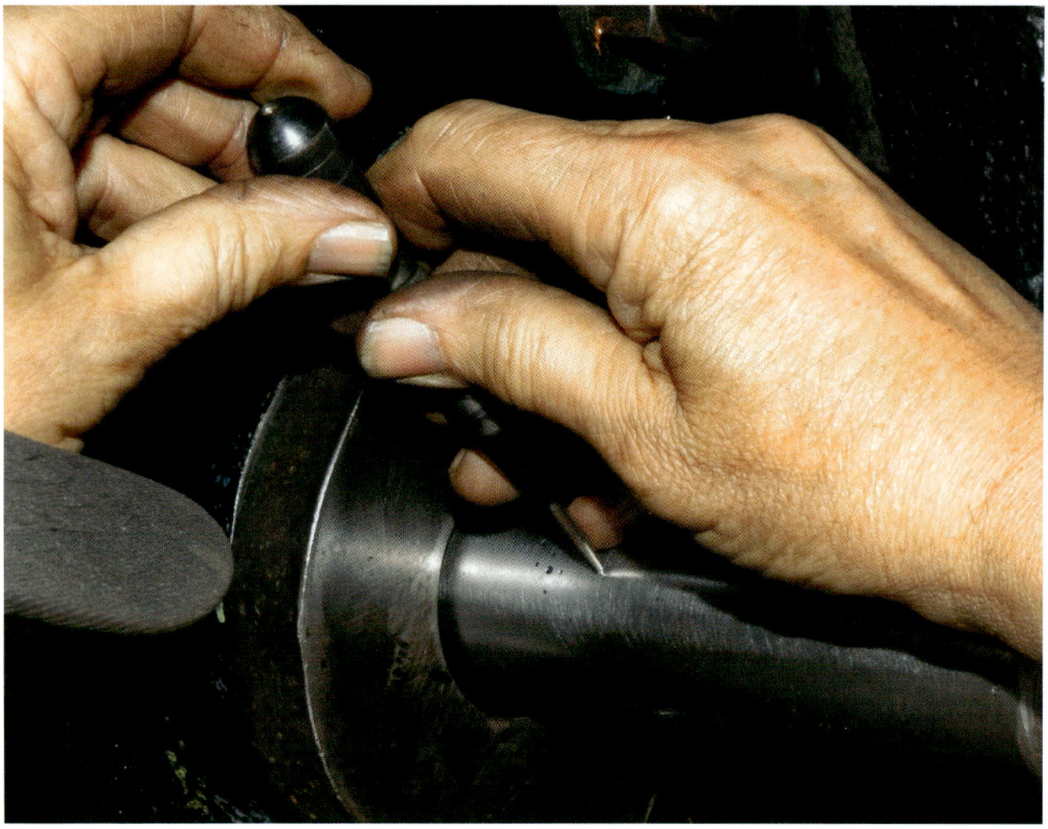

Having positioned the left hand cylinder dead centre Richard Swales put a pop mark on the spindle which will tell him the valve position when fully reassembled.

quiet during this break prompting Darrin to comment to me that, "This is exciting," which broke the ice and prompted a full group chat which ranged from RAF Typhoon fighter planes to the weather in the Orkneys and Shetlands and the medieval rituals still held in those lands, which further led to a bizarre discussion on the different methods of cremation practised around the world. I set off for home after lunch reminding myself of what a unique bunch of individuals had been brought together by the overhaul.

As arranged I returned to meet up with Richard Swales on Saturday to catch up with the reassembly of the valves. Richard had fitted the piston valve rings on the right and left hand side valves and reassembled the right hand side, leaving the trickier middle until last as this required working from the pit below No.7. He was now working on the left hand side assembly and getting its position dead centre was the crictical task he was undertaking. Sir Nigel Gresley and his assistant, Harold Holcroft, had designed what is known

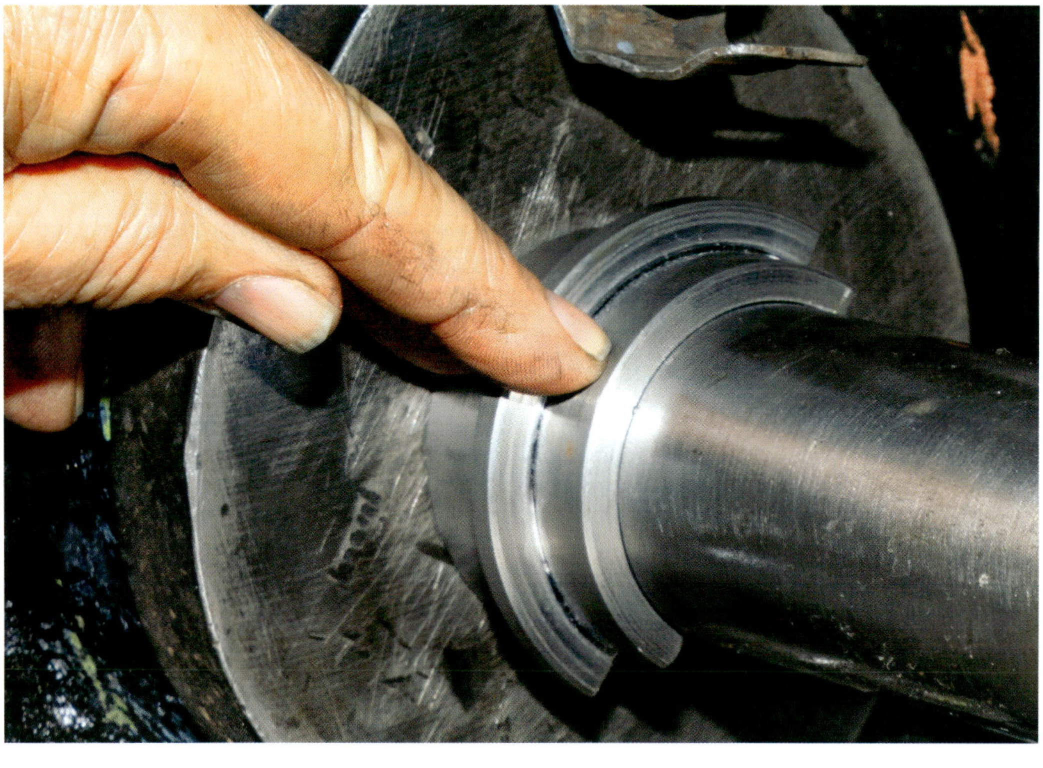

Richard points out the pop mark on the spindle after other spindle components are temporarilly positioned

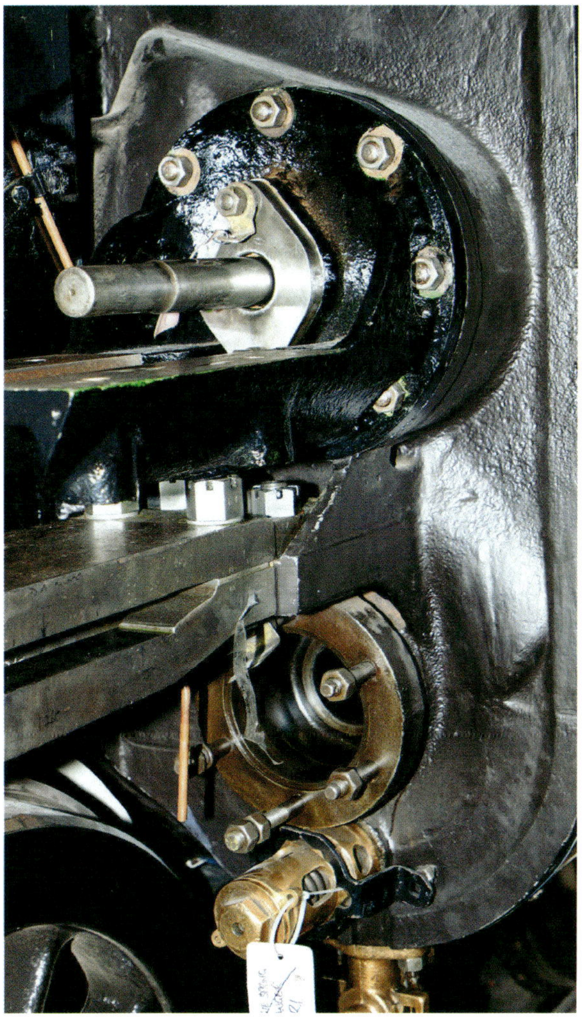

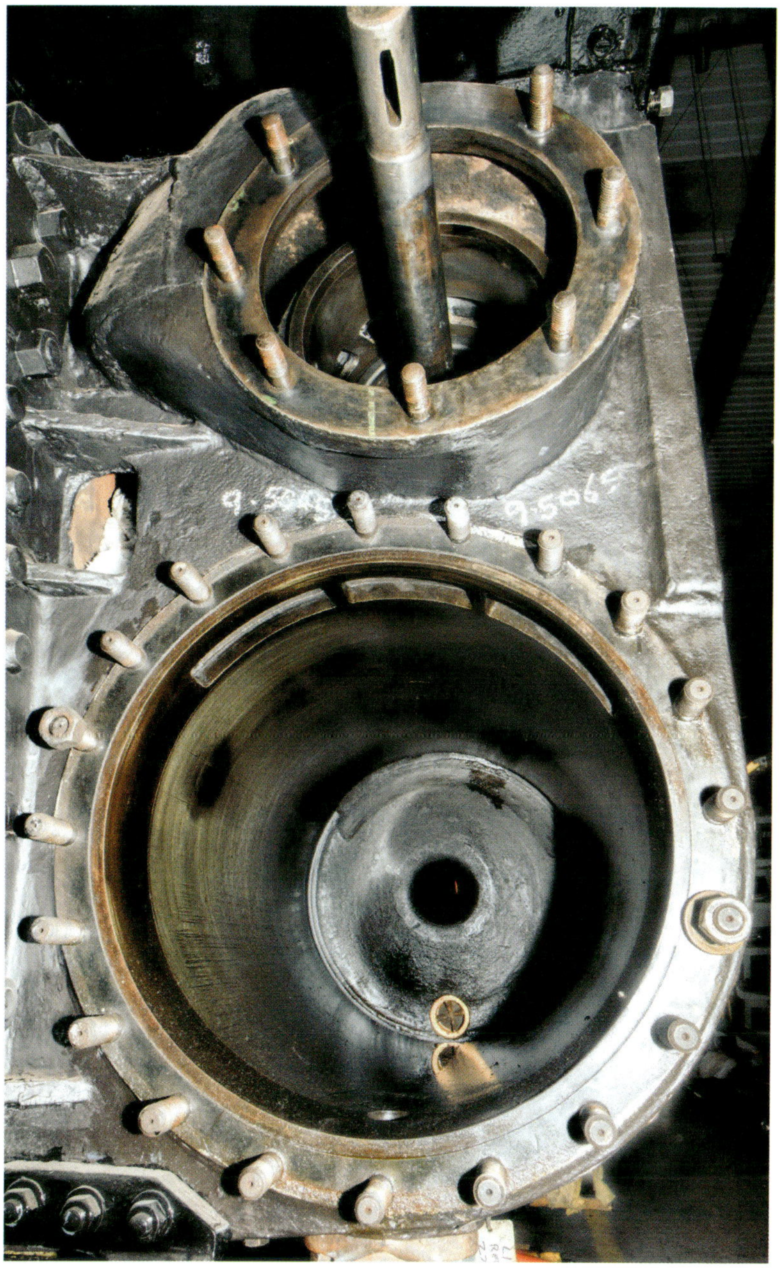

ABOVE - The back of the right hand cylinder and valve
RIGHT & BELOW The left hand cylinder and valve with valve spindle in place, all in pristine condition and awaiting reassembly

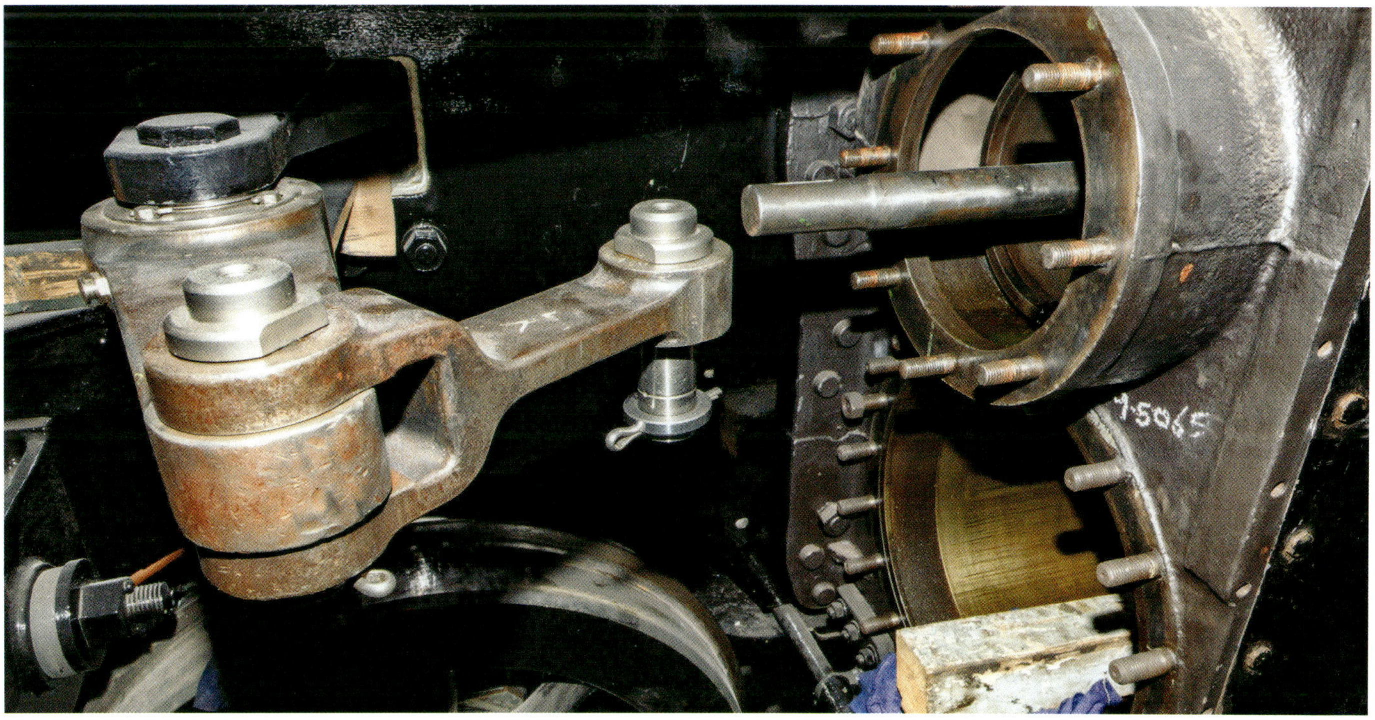

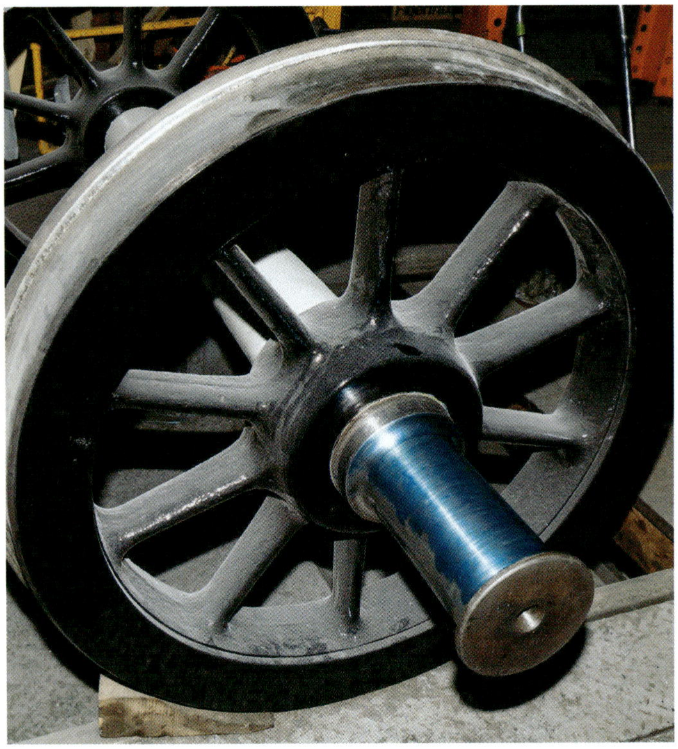

Engineers Blue on the Cartazzi axle stub

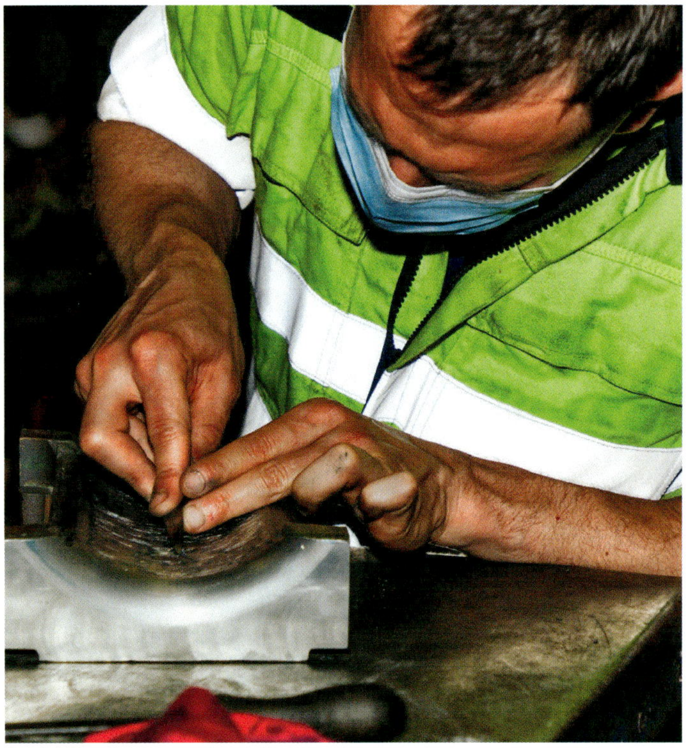

Chris Whiteley removes high spots on a Cartazzi axle bearing

as the "conjugated valve gear", which allows the middle cylinder of a three cylinder locomotive to dispense with valve gear by connecting with the outer cylinders through a 2 to 1 lever and an equal lever. Being 360 degrees in a full cylinder cycle each cylinder stroke should be set 120 degrees away from the other two. With the front of the left side valve casing open, Richard was able to position the valve dead centre and at this point make a pop mark on the spindle at the back of the valve which would tell him the valve position when reassembled. Setting this valve dead centre then allows the setting up of the other two valves, one fully open and one fully closed, 120 degrees between each position.

Also working today was Chris Whiteley who was continuing the work on the Cartazzi axle bearing which junior volunteer Tom Readman had been working on earlier in the week. Andy Evans, along with Nigel Franklin, was applying finishing touches to the tender in readiness for painting and Nigel's wife, Sheila, who had joined the painting team, was finalising the painting of components that Richard needed to complete the reassembly of the cylinder valves. Richard would continue the reassembly work in the week ahead when the paint had dried and hardened so I planned to call in again the following Saturday for my third visit of the month.

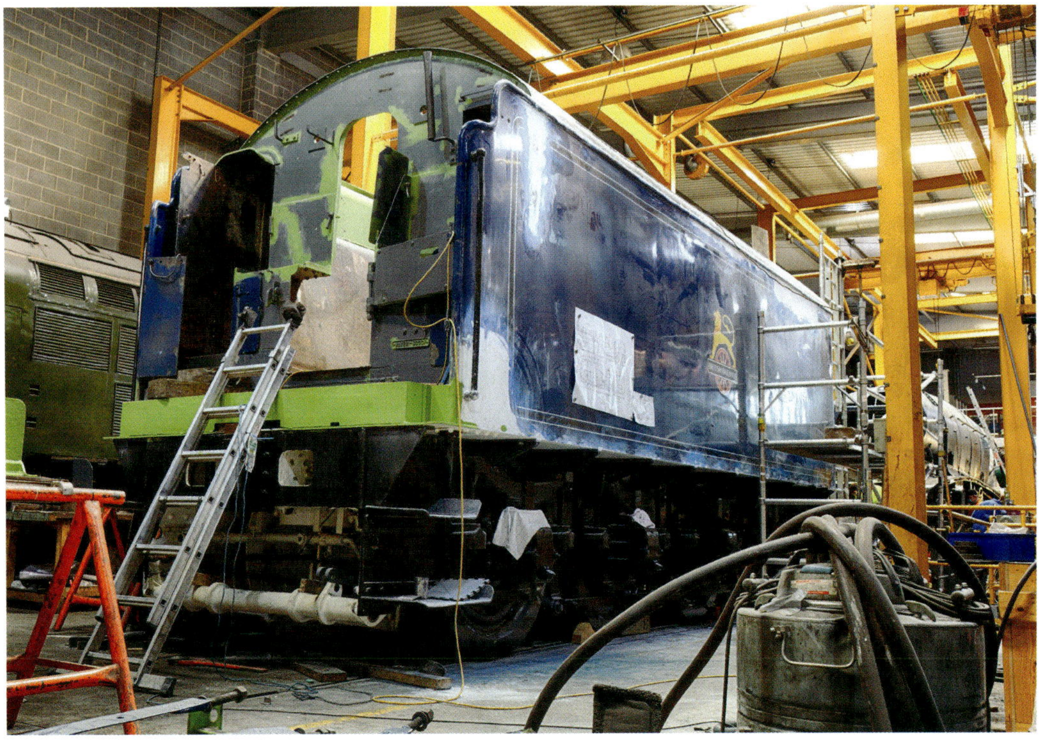

Progress on the tender bodywork is evident as it is nearly ready for painting

LORD OF THE RINGS

16

June 2021

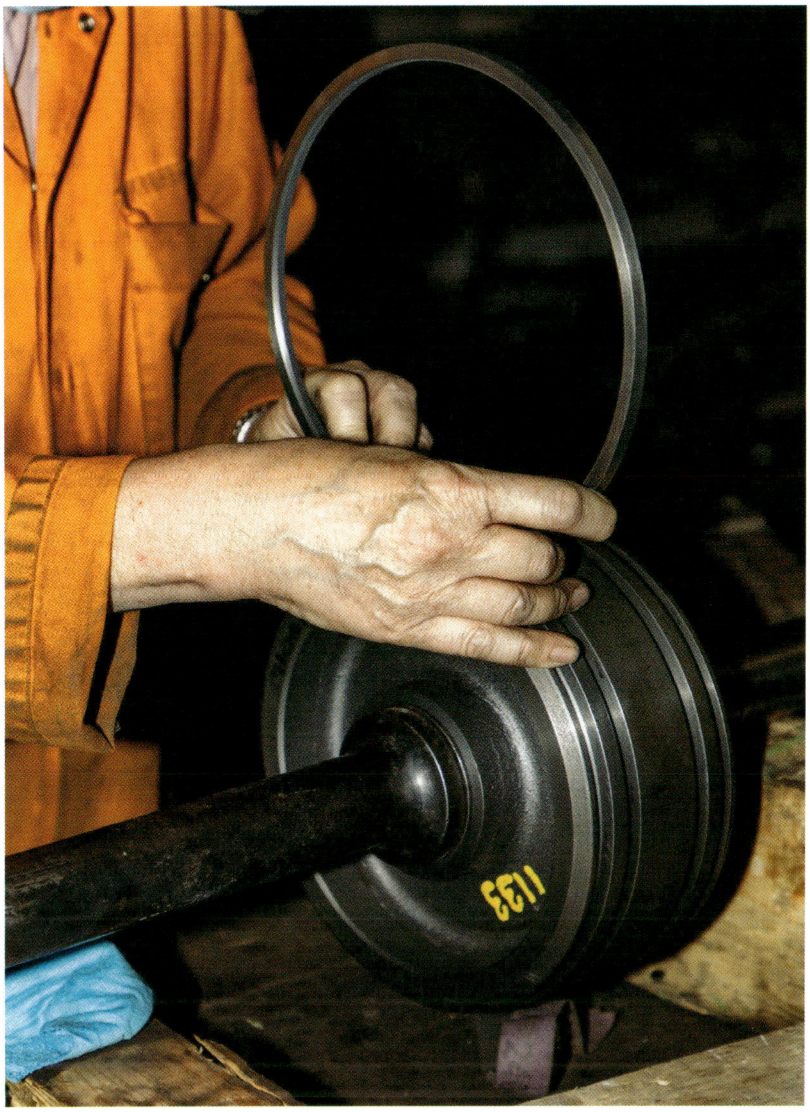

A valve ring is checked for fit

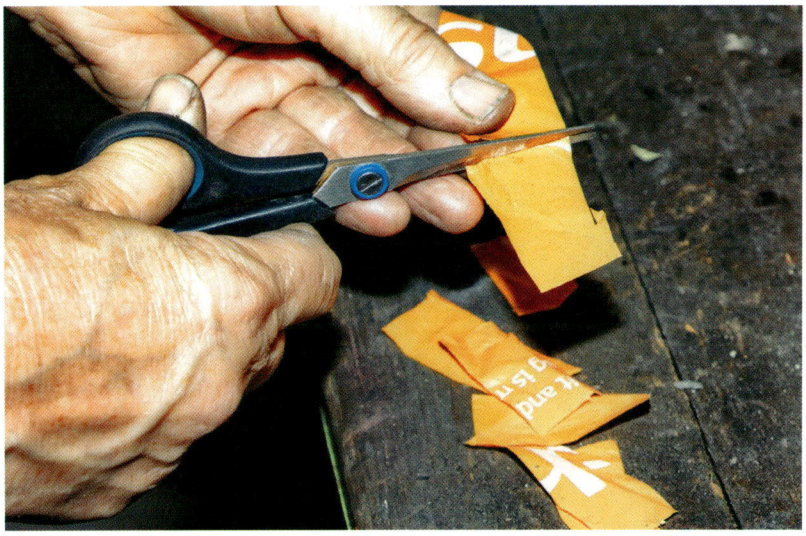

A plastic bag cut up into small pieces which will aid the grip of the ring in the valve head

"Ah, this is where the plastic bag comes in"

I had an appointment with the Lord of the Rings, aka Chief Mechanical Engineer (CME) Richard Swales, on my third visit to York of the month. Whilst my earlier visits in June had covered the refitting of the valve spindles with the rings fitted, I hadn't actually captured Richard fitting any rings to the valves but, today I would. What I didn't expect to see were a pair of scissors and a Sainsbury's plastic shopping bag in Richard's collection of tools and parts! I should add other supermarkets bags are equally available and suitable.

A quick tour of the workshops to see what else was going on revealed Darrin getting to grips with a particularly difficult section of the cladding at the base of boiler's left hand side firebox end, and Peter Ellis creating a bracket which would allow an oil drip feed pipe to supply oil to a piston shaft without fouling as had previously been a problem. Nigel and Sheila Franklin, with Malc Bateman, were applying undercoat paint to the tender. But it was Richard who was to get the focus of my 24-105mm lens attention.

Each end of the valve spindle is fitted with four cast iron rings, eight per valve, and being a three cylinder engine, that's twenty four valves rings in total. The much larger cylinder pistons have two rings each so a further six rings are required. The rings are not complete rings as they have a short space at the ends but they were larger in diameter than the valve heads, so how are they held in the grooves without wanting to spring

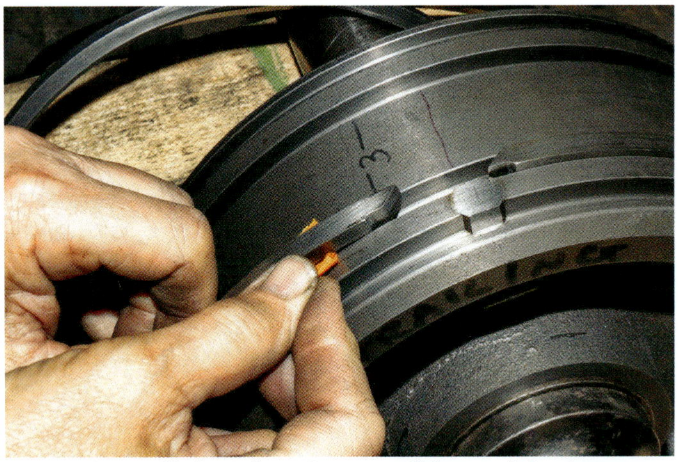
The small plastic bag pieces are slid around the ring

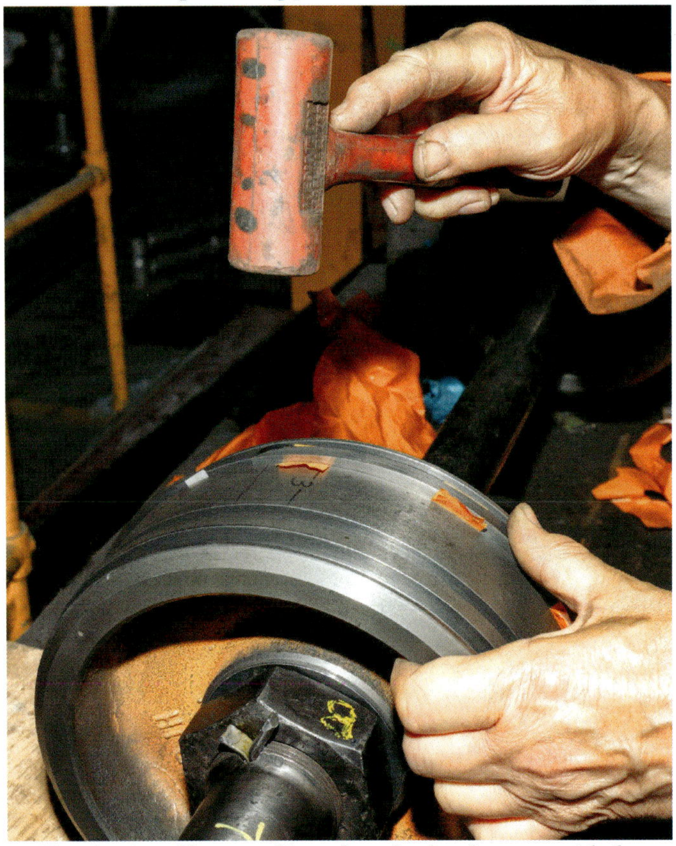
The ring is softly tapped into the valve head groove with the plastic bag pieces gripping and holding it in position

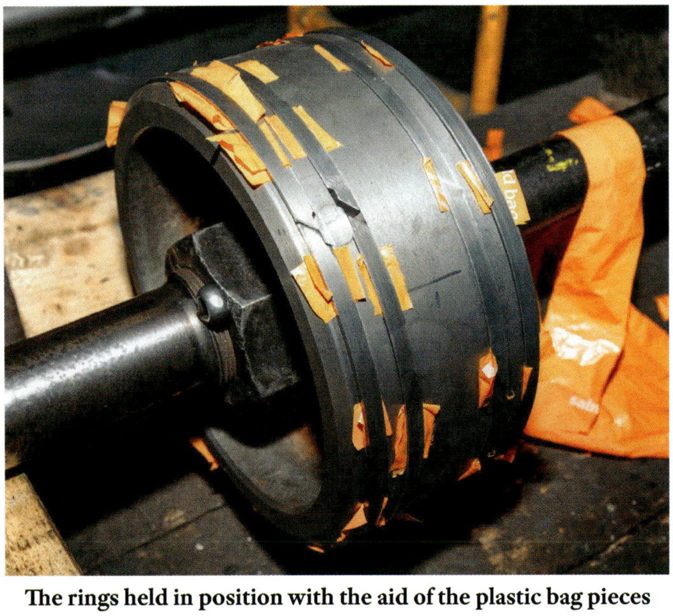
The rings held in position with the aid of the plastic bag pieces

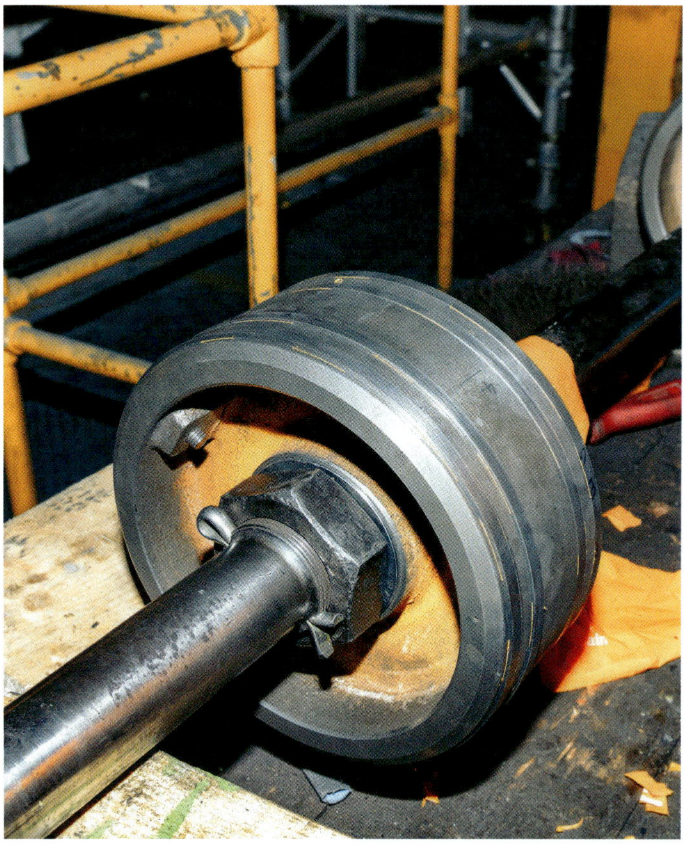
The valve head with the rings fitted and excess plastic removed

out to their natural diameter making it almost impossible to fit the spindle back in the valve? "Ah," Richard said, "this is where the plastic bag comes in." Richard cut the bag up into small square pieces so they would slide between the larger outer ring diameter and the valve head surface which, when the ring was rubber hammered into the groove, would create a tighter fit, thus restricting the ring's natural tendency to spring out of the groove. Upon first steaming of No.7 the plastic would be vapourised.

The work on the right hand valve and cylinder was now complete, and the left hand cylinder's valve was complete and was just awaiting the fitting of the piston. The valve Richard was working on today was for the middle cylinder, and having experience of how challenging the enviroment when working under the locomotive in the pit was, I asked how hard getting the very heavy valve and cylinder piston back in position would be, particularly as you would be working above your head in an area where there is no access for the overhead crane. "With difficulty," was the answer, "and with some of the strongest members of the team." Clearly a job for Roger Turnbull and Peter Brackstone then!

"With difficulty....... and with some of the strongest members of the team"

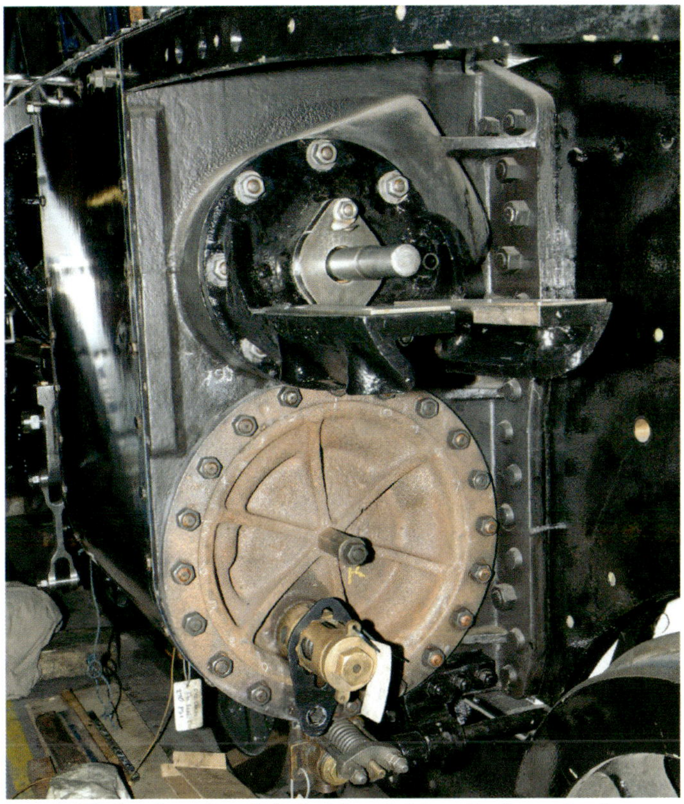

With the rings fitted the left hand side cylinder and valve

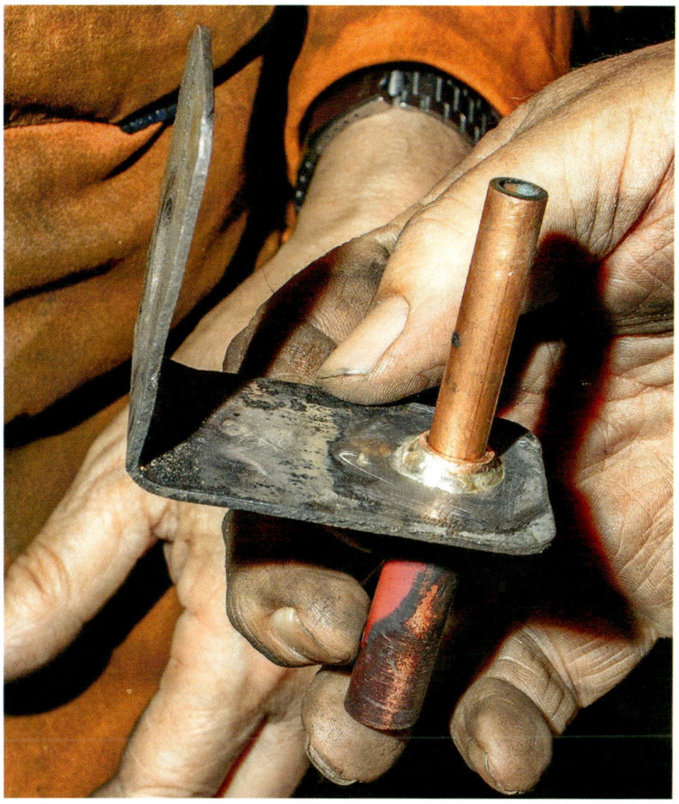

Peter Ellis shows Richard his newly created oil pipe bracket

Peter Ellis was close to finishing his oil pipe bracket and joined up with us to seek Richard's approval. "Is this what you're looking for?" he asked Richard, and after a close look "No", was the answer. There was a long pause as a disappointed Peter pondered a response and then, after Peter started explaining his thinking, Richard added, "It's actually much better than what I was after!" Smiles all round.

"Is this what you're looking for?"

"No..........Its actually much better than what I was after"

It was close to lunchtime but I had captured what I wanted for the day so I was going to leave early and forego the highly popular and often humorous lunch debate, although the morning tea break had resulted in some toilet humour as the discussion turned onto the SNGLT support coach's toilet tank modification, or should I say installation, as you can no longer flush directly onto mainline tracks.

At an ever increasing pace No.7 was starting to come back together and with plans to fit the iconic streamlining around the smokebox soon, it certainly would be looking closer to the much anticipated return to traffic.

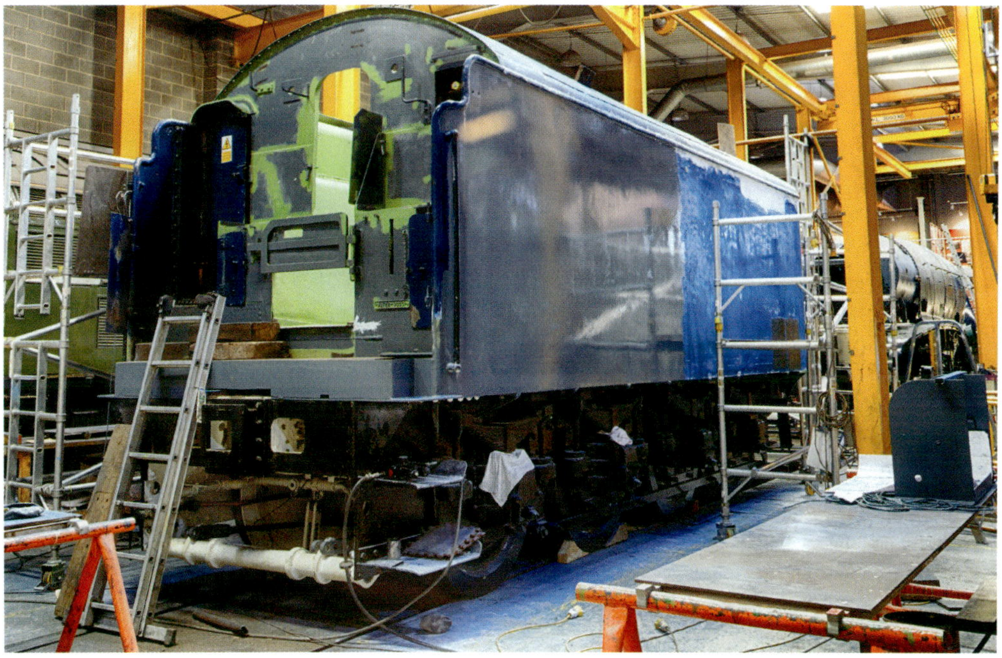

Undercoat paint has now been applied to the front half of the tender

17

ROUTED TO CREWE

July 2021

As previously mentioned the NRM had plans to redevelop the workshop which meant No.7 would need to be completed elsewhere. The logical place for that was at LSL in Crewe as they would be taking on the day to day maintenance when back in traffic, so it was simply bringing forward the co-operation already planned with them. With the move planned for late August the overhaul would need to be essentially completed ready for a short rail transfer across York and then onwards by road to Crewe where hydraulic and steam tests, de-snagging and painting would be undertaken. The clock was ticking and to meet the challenge the engineer team of volunteers were putting in extra days.

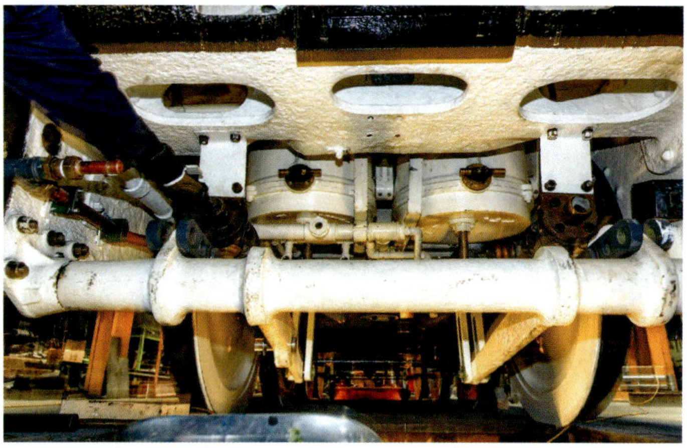

Richard Hawkhead points out the water valves below the tender

Darrin had requested some images of the work being undertaken on the steam pipes in the smokebox but first I wanted to catch up with Richard Hawkhead who was finishing off a few last jobs on the essentially completed tender and he took time out to show me some of the immaculately restored components under the front of it. Richard's arm is pointing out the water valves which supply water from the tender to the fireman controlled injectors on the locomotive. The white bar is linked to the handbrake mechanism for use when the locomotive is stationary and the cylinders behind are vacuum brake cylinders for braking when moving. Every aspect of the tender was now looking in exceptional condition, a credit to all who had worked on it.

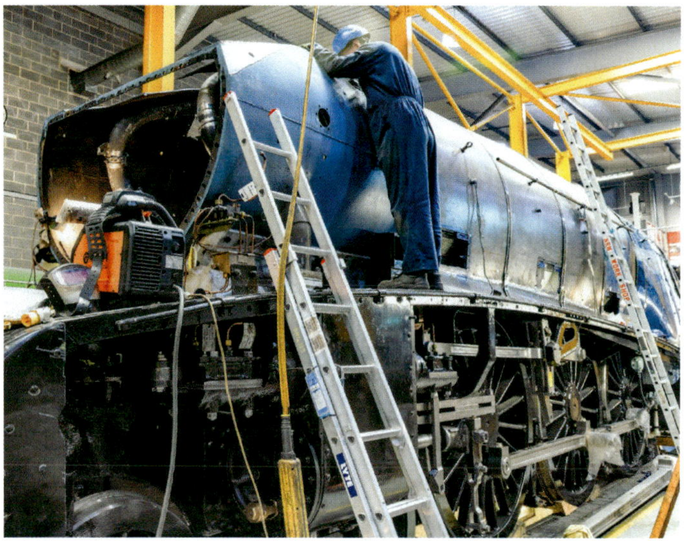

Dougie Blyth (above) and Dave Lee (below)working on the steam pipes

Returning to the locomotive I found Dougie Blyth and Dave Lee working on the main steam pipes in the smokebox and with more cladding back in place along with the final reassembly of the cylinders and valves there was very visible progress on view. It would not be long before the blast pipe and chimney could be fitted and soon after that the famous streamlining at the front and the cab at the back. With the assistance of a specialist welder from LSL some welding work was being completed on the steam pipes today and some he would take back to Crewe to weld and fit when No.7 arrived there.

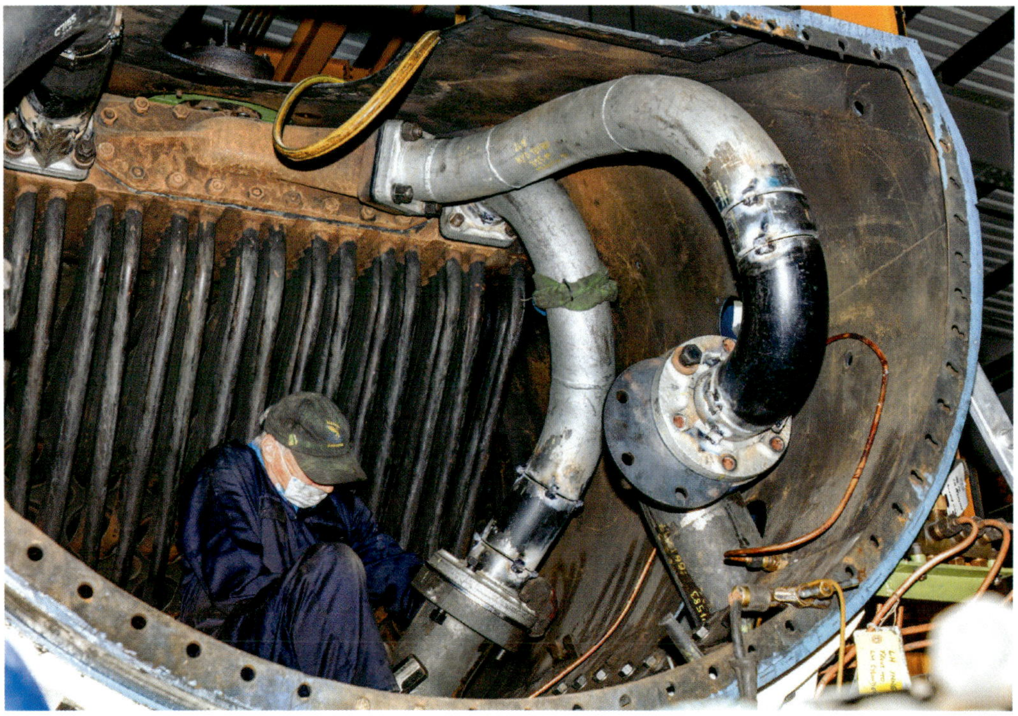

THE LAST DROP

18

July 2021

"Have you left it in gear?"

July 29th 2021 was not only to be another big step towards the return to traffic but also another final chapter in the history of 50A (York Motive Power Depot now the NRM workshop), with the refitting of the Cartazzi wheelset using the wheeldrop for what would almost certainly be the last time with the impending redevelopment of the workshop into a 'Wonderlab'. I am not against progress, but it is a difficult commodity to measure when balanced against a loss to history as the past is swept away. I accept the wheeldrop is not a Euston Arch but as I write this it seems that there is no Woodham Brothers Ltd. scrapyard of Barry rushing to save this historically iconic piece of plant.

Darrin had recruited an extra member to the team today, with the addition of Iain Hodgman who was equipped with his video cameras to record the event and I was promptly told to look after him as I was his boss - so much for retirement! Although I strongly believed that history is best recorded with moments frozen in time by photographs, I did not dismiss the use of video, but it was always a secondary medium to me, until I watched Iain's edited videos a few days later and just maybe then I dropped the adjective 'strongly'. Please check his work on 'You Tube.'

It was deja vue again as the team had to move No.7 into position over the wheeldrop using muscle power, except this time with the wheelset being the Cartazzi under the cab an extra effort was needed as that was the furthest away from the wheeldrop. This was a tough ask for the team which was made even tougher as the tender was parked in front of No.7 and had to be pushed out of the way first! Having done this before on several occasions for the bogie and the drive wheels, you would think the team would have been quite proficient at the task, but with everything being Covid delayed, it was quite a while since the last time. My comments about "Have you left it in gear?" drew a few sneers. After a few false starts it was decided a coxswain calling stroke with "Down, down, down" would be employed, which resulted in everybody pushing or working

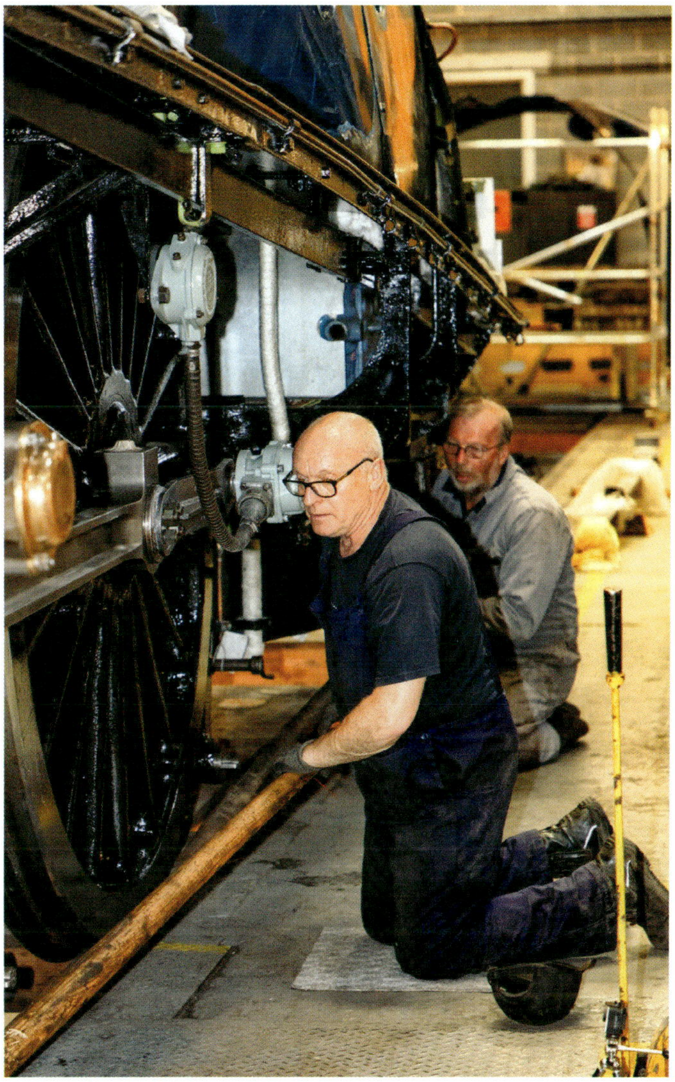

Andy Barwick (left) and Roger Turnbull use pinch bars to move No.7 forward over the wheeldrop

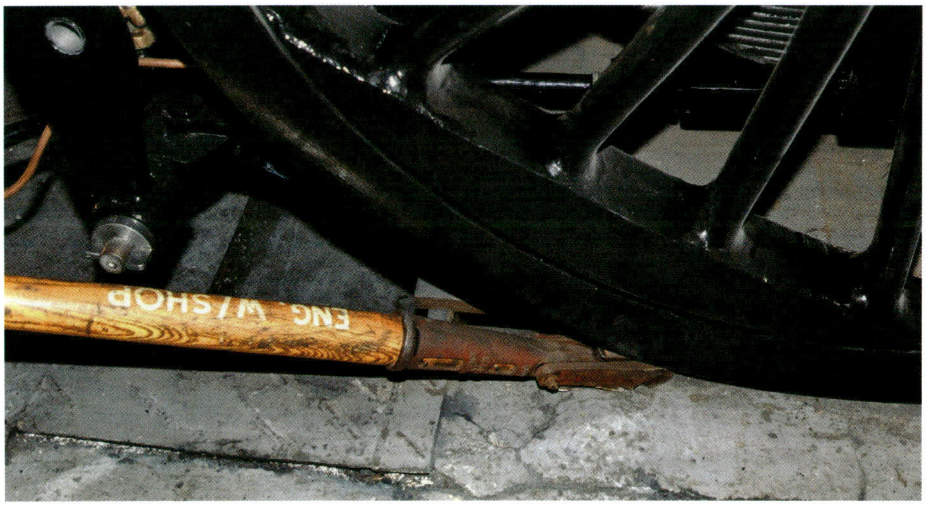

A pinch bar inches No.7 slowly forward over the wheeldrop

a pinch bar at the same time. No.7 was soon on its way towards Edinburgh. Getting closer, the call came that there was 4'8" to go, and then a little later 2'10" to go until finally all was in alignment over the wheeldrop.

With No.7 now in position it was possible to lower the wheeldrop's section of track to the bottom of its pit and move it horizontally into a parallel position so the wheelset could be lifted by the overhead crane and lowered onto it. Succesfully completed, "Lunch," was the call from Darrin and everyone trooped off to the SNG mess area on the gallery.

"You would think everybody came here just to chat," Darrin proclaimed

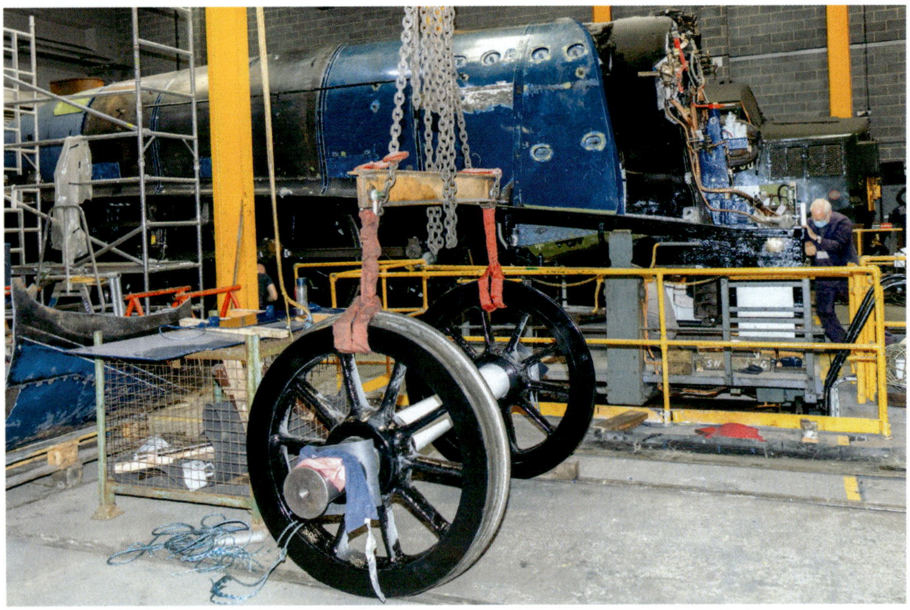

The Cartazzi wheelset is ready to be lifted into the wheeldrop

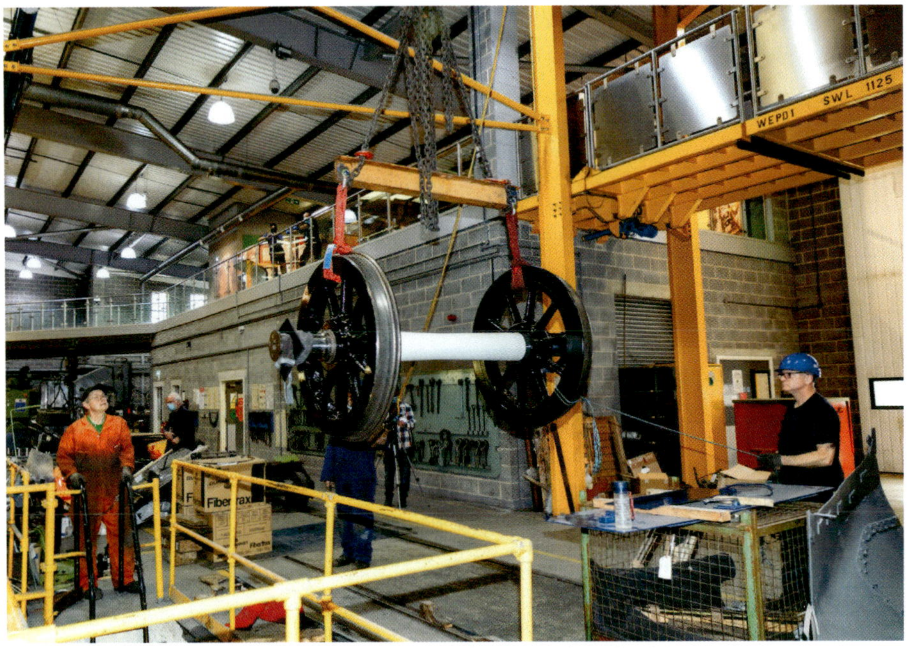

The Cartazzi wheelset is lifted over the wheeldrop

"You would think everybody came here just to chat"

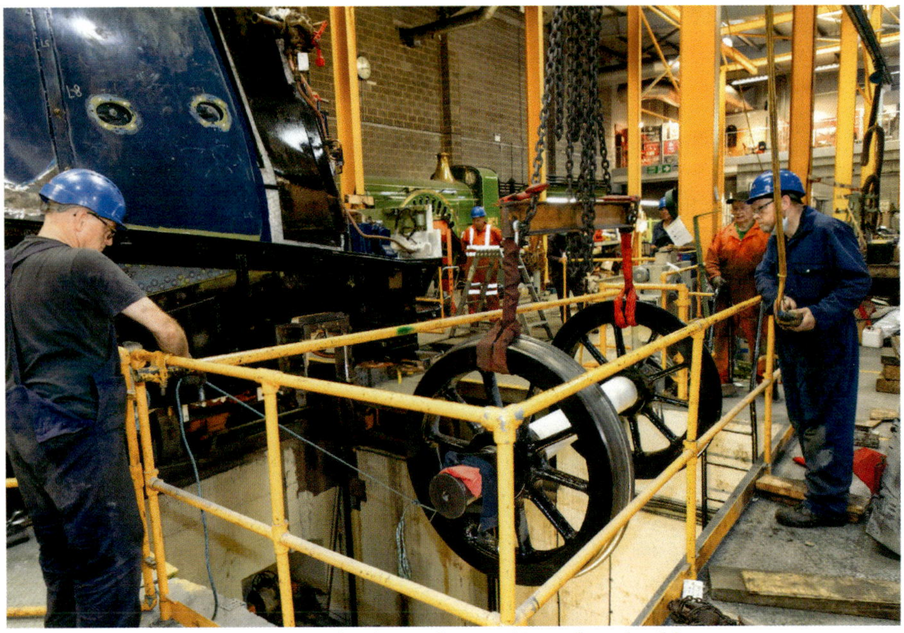

The Cartazzi wheelset is lowered into the wheeldrop

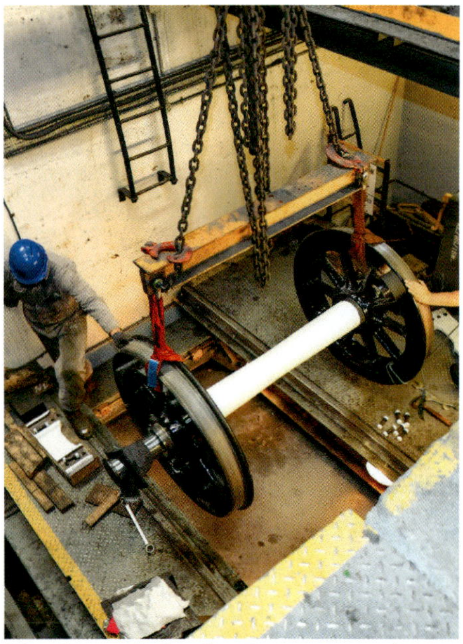

The Cartazzi wheelset is manoeuvred onto the rails in the wheeldrop

in his motivational announcement that lunch was over. The wheelset was now in the pit, but before it could be raised and reunited with No.7, the two exceptionally heavy axleboxes had to be fitted using the overhead crane again to position them on the axle of the wheelset. All went well and we were soon ready for the lift which, being a critical alignment to get accurately into the Cartazzi horns, was monitored by CME Richard Swales on one side and Darrin on the other, moving quickly at first and finally an inch at a time. The rest of the team then assembled below the wheelset and refitted the horn stays to complete an excellent day's work. Well not quite complete, as Darrin said "Well you know what is next, don't you all?"

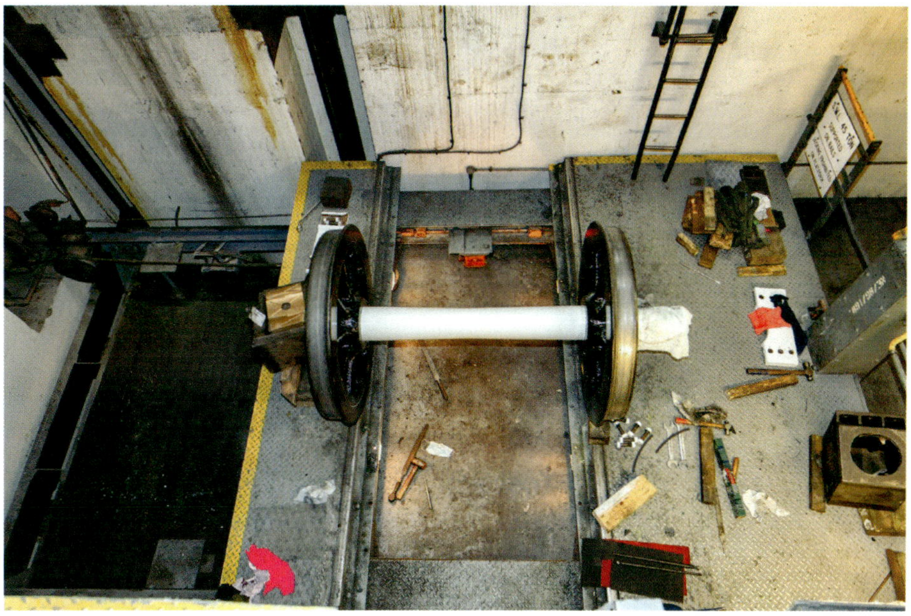

The wheelset sits in the bottom of the wheeldrop

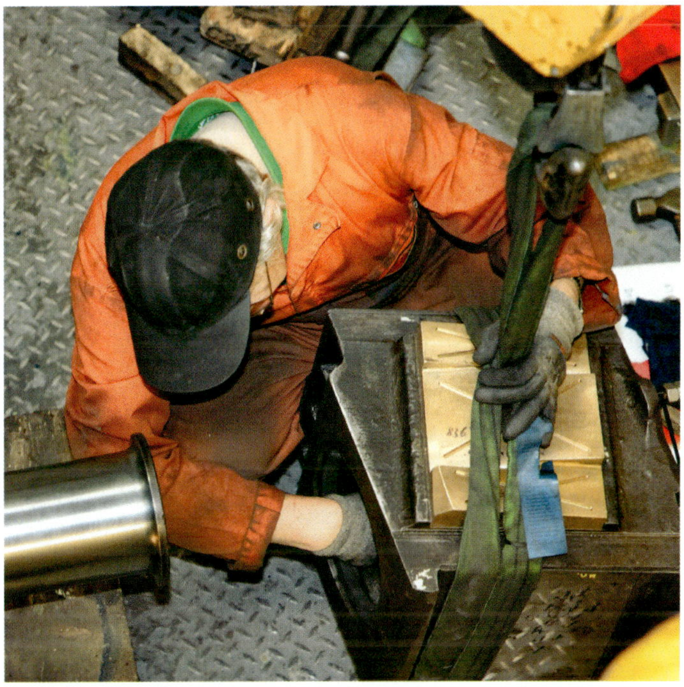

Richard Swales prepares a Cartazzi axlebox for fitting on the wheelset axle

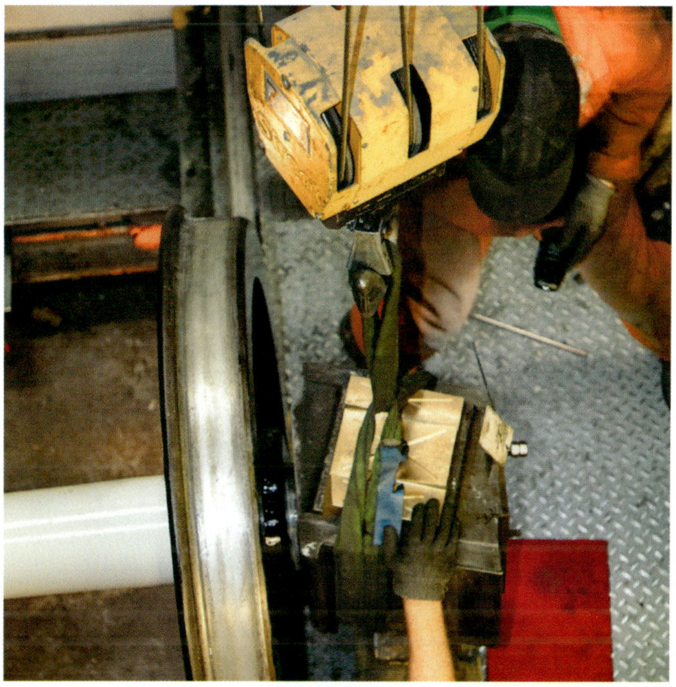

A Cartazzi axlebox is fitted to the wheelset axle

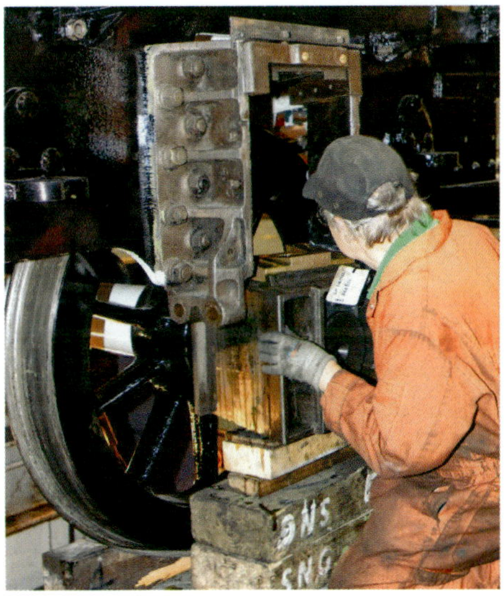

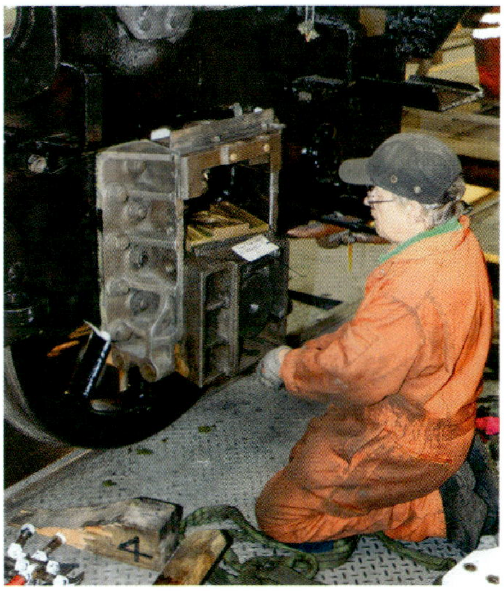

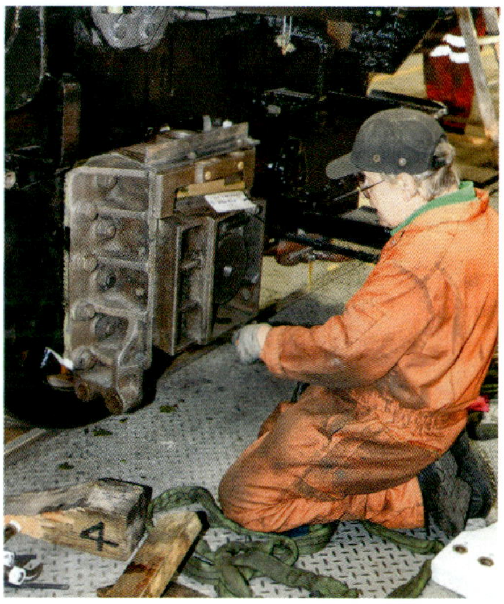

From left to right, the wheeldrop is slowly raised into the Cartazzi horn

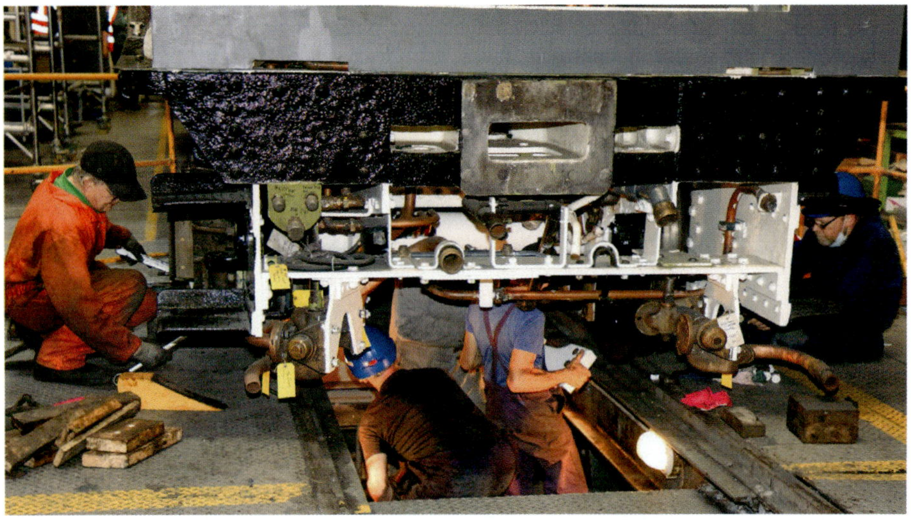

In the pit Andy Barwick is ready to secure a horn stay, being lifted into position by Tony Bickerstaffe, Richard Swales (left) and Darrin Crone (right) oversee the operation

"Well, you know what is next, don't you all?"

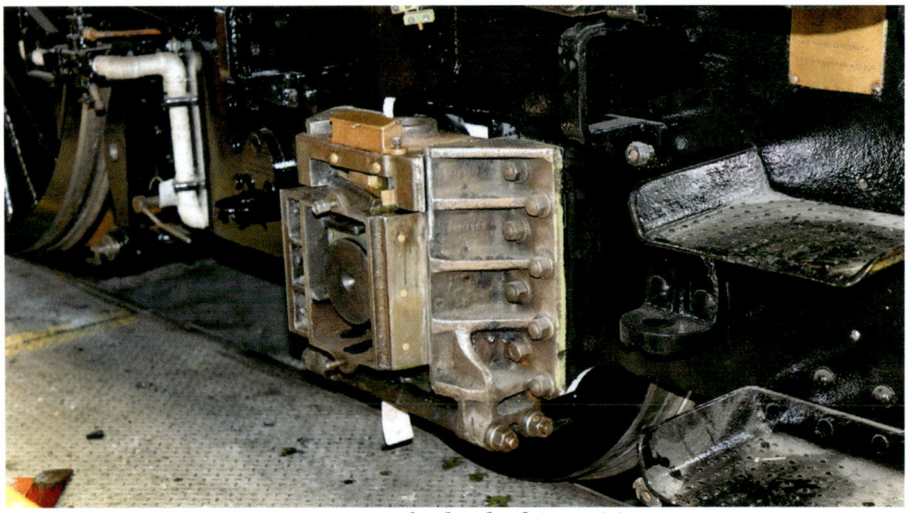

The Cartazzi wheelset back in position

Yes everybody did. No.7 needed pushing back along the workshop track, this time in the direction of Kings Cross together with the tender of course. Audible sighs echoed around the workshop.

Pictures all in bag so to speak, I said my goodbyes making my usual excuses of having a lot of editing to do and besides my dodgy back would not have helped me contribute much to the push. Iain stayed on and captured the shove back in a speeded up video which, I have to admit, a still photograph could not have replicated - another speed record!

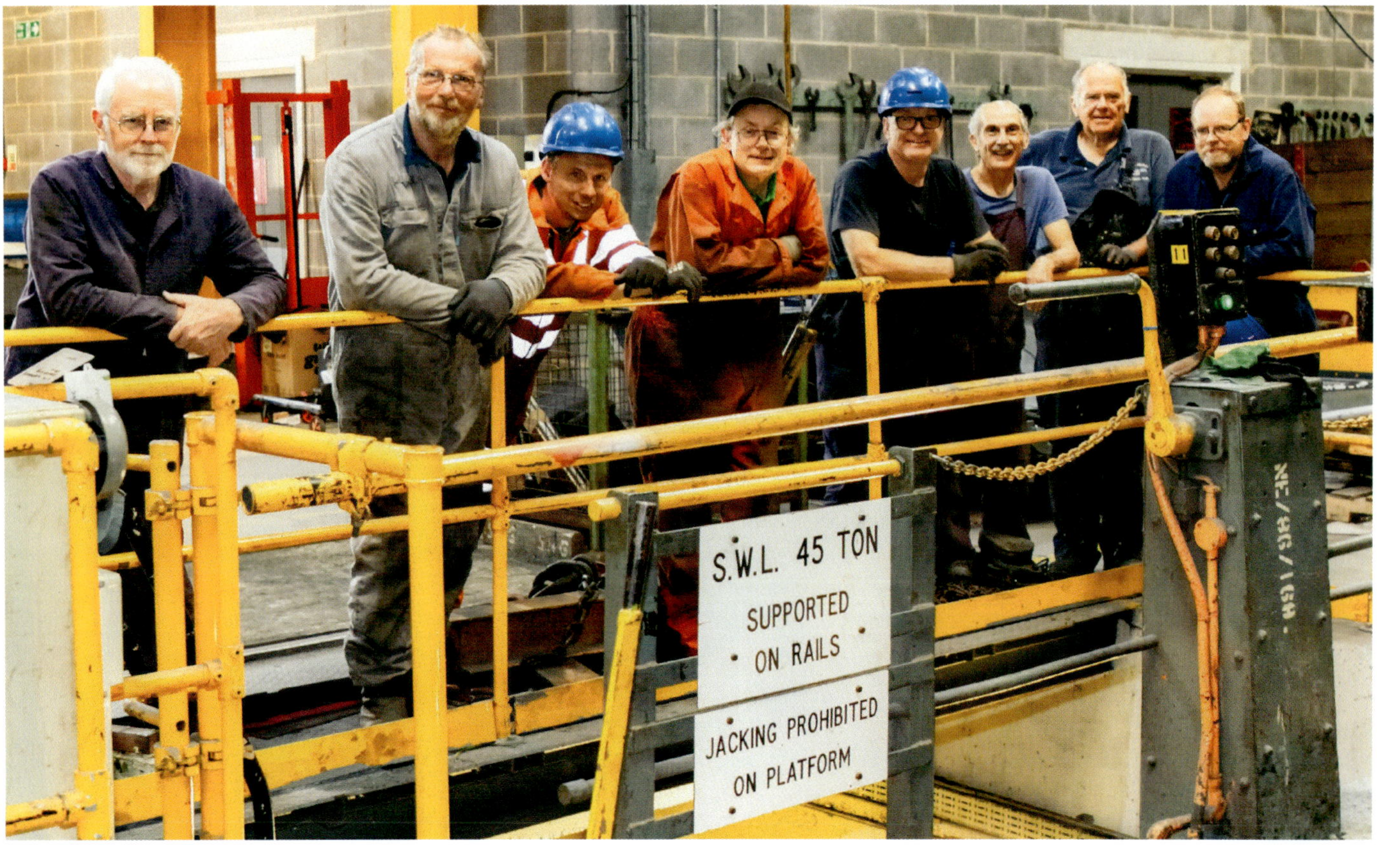

The wheeldrop team for the day, left to right - Peter James, Roger Turnbull, Andy Evans, Richard Swales, Andy Barwick, Tony Bickerstaffe, Peter Brackstone and Darrin Crone

CABBED

19

August 2021

"I wa day dreamin other day"

I broke my usual routine of arriving in time for the morning tea break today, as the cab was to be reunited with the back plate and I knew the lads would be all over this from the off, so I made sure I was on site by 9:00am. I arrived just in time. Phil Slater and Roger Turnbull already had the overhead crane hooked up to the cab and were ready for the lift. With no time for morning pleasantries I told Roger not to wait for me as I would catch up, which I did quite quickly, eventually ending up with a sequence of shots from various angles showing the lift in position. Relative to today's events, it took very little time to move into position, but what did take up some time was aligning the bolt holes on the cab sides with the holes on the cab floor. New insulation was covering up the holes and once the excess had been cut away to reveal them, lining them up needed quite a lot of persuasion. The youthful dexterity of junior volunteers, Tom Readman and Scott Middlemiss, who had now joined Phil and Roger, proved quite an asset in reaching into tight spaces around the cab floor. My back was aching watching them and I was only pressing the shutter button! To hold the cab sides in position once the holes were all aligned needed a bit of improvisation with G clamps and blocks of wood, but with CME Richard Swales overseeing the operation the team once more excelled with their problem solving skills.

At lunch conversation honours were awarded to Tony Bickerstaffe, who amused all when he announced in his broad Yorkshire accent, "I wa day dreamin other day" adding that he wondered why carbon dioxide, which has two oxygen atoms, was dioxide and not bioxide. Nobody had the answer but Tony was able to tell us as he had researched the answer and discovered it is the difference between Latin and Greek. Tongue in cheek, Darrin quickly thanked Tony for raising the level of the conversation and we moved on to the three wheeler cars our dads used to own as well as very expensive vintage motor cycles probably none of us could ever own. It was time to get back to work.

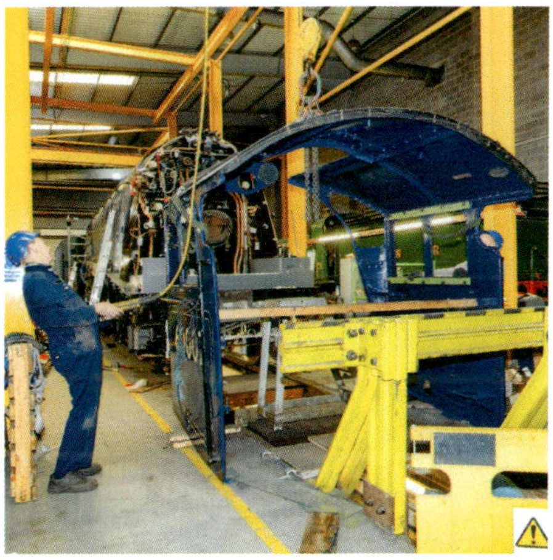

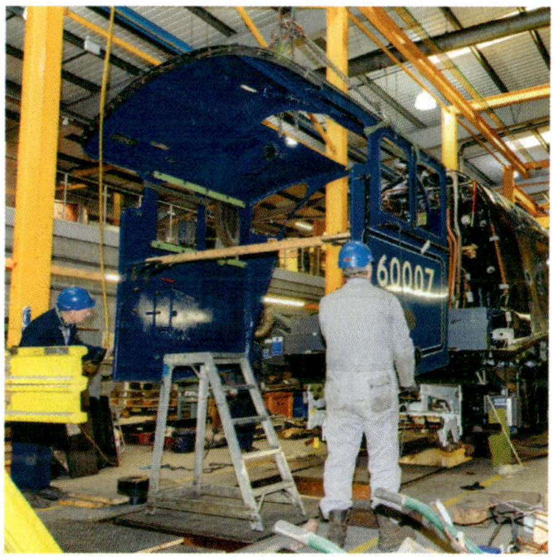

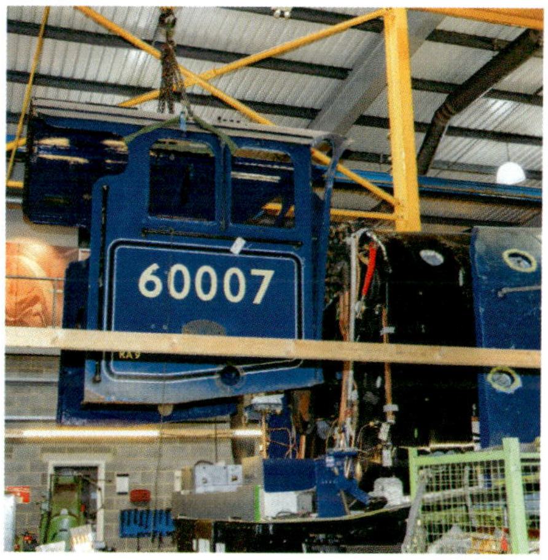

The sequence of pictures shows the cab being lifted into place by Phil Slater operating the overhead crane controls and Roger Turnbull guiding the cab into position

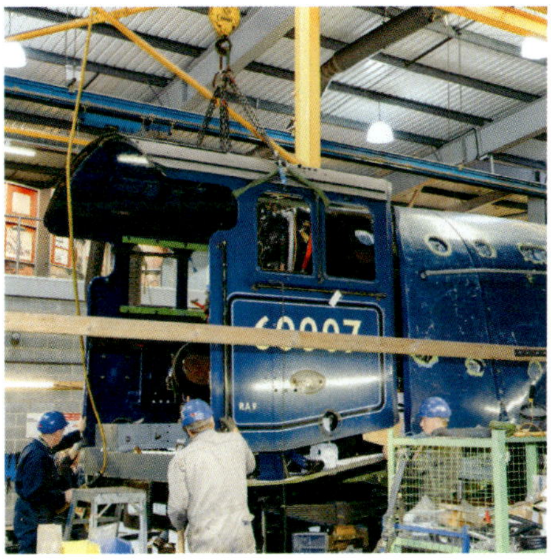

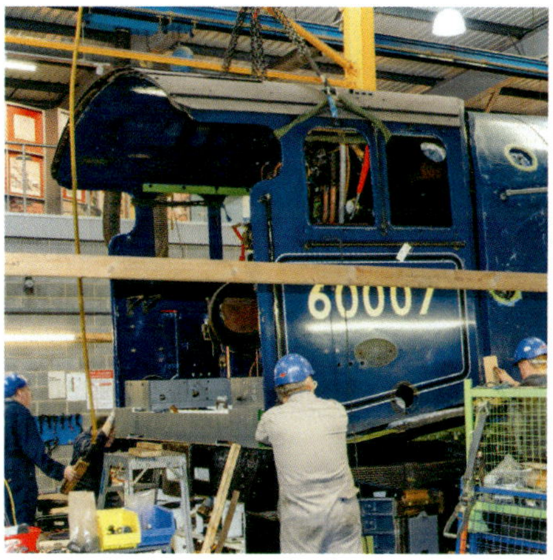

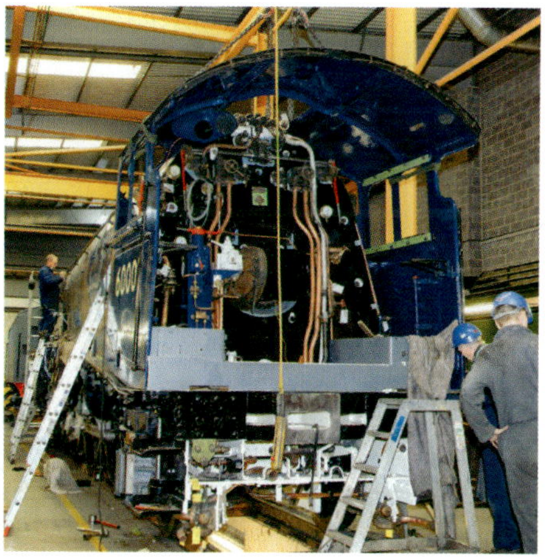

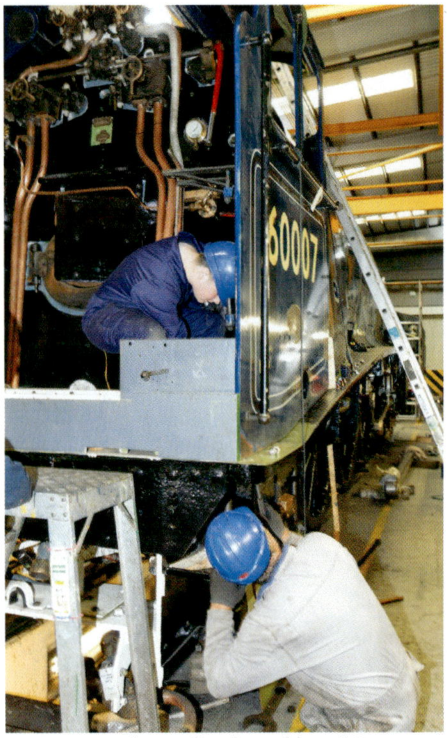

Tom Readman (in the cab) and Roger Turnbull fixing the cab into position

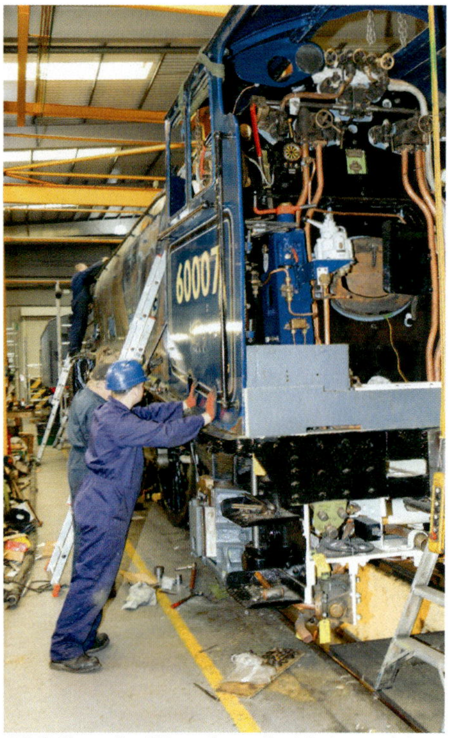

Phil Slater holding the cab side in position

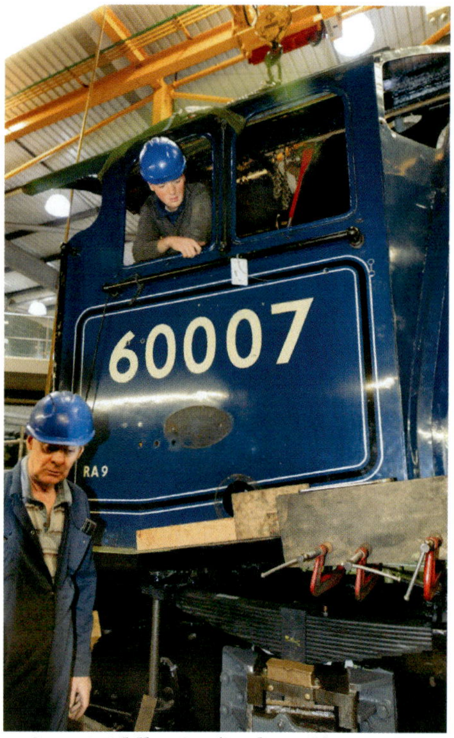

Scott Middlemiss (in the cab) and Phil Slater fixing the cab in position with the help of some wood blocks and G clamps

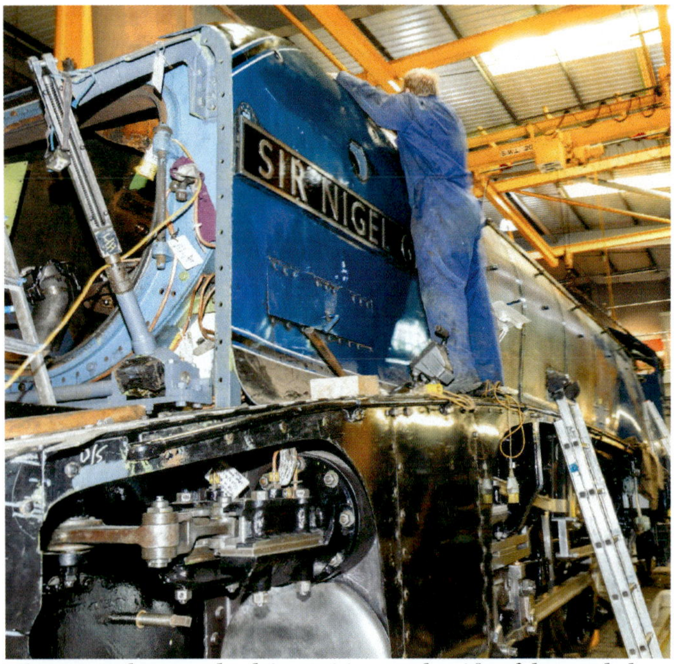

The name plates are back in position on the side of the smokebox

With the cab team still fixing the cab into position I wandered to the front and found Darrin and Tony working on small diameter oil feed pipework and I noticed the name plates had been taken out of safe keeping and fitted back onto the sides of the streamlining - a sure sign we were getting very close to completion of the overhaul and nearer to the steam and hydraulic tests which would take place at LSL in Crewe. I asked Darrin when the funnel would be fitted and got a swift rebuke. "Funnel! Funnels are on ships." I really should know better. Darrin said the very front streamlining panels and the cod mouth would be fitted in the next couple of days and maybe er, the chimney, so I got some final shots from the gallery of a nearly complete No.7 and arranged a swift return in a few days' time to get even more nearly complete No.7 images.

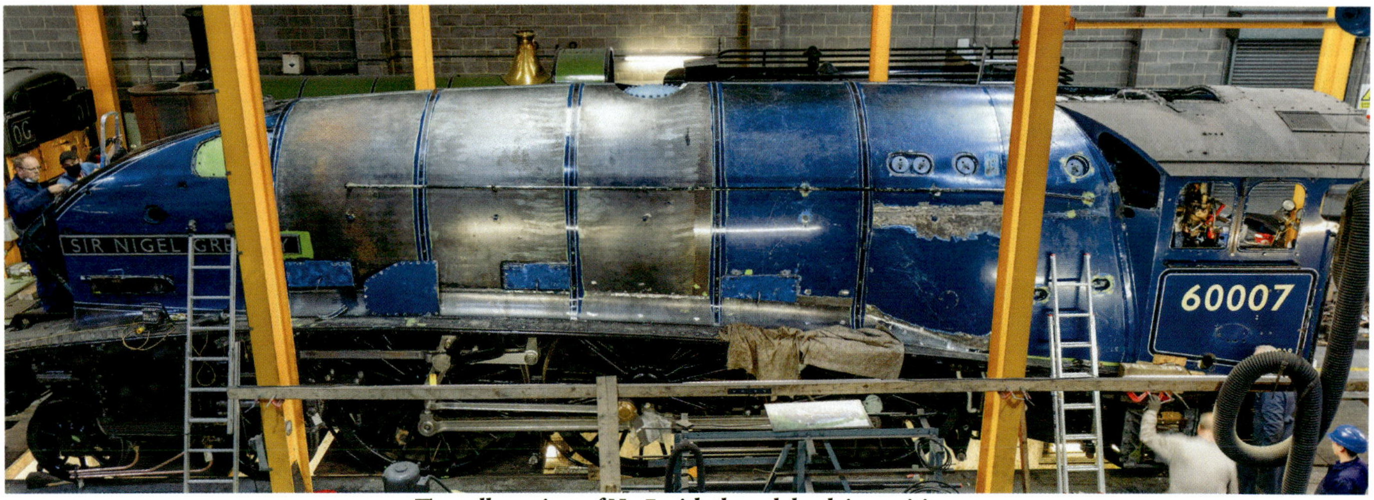

The gallery view of No.7 with the cab back in position

TAKING SHAPE

20

August 2021

My second visit in August to the NRM was down to the rate of visible progress; the nose had now been fitted and I needed images for the press and social media and this book, of course. However, a read through Darrin's weekly report really emphasised just how rapidly work was moving forward, but a lot of it was less visible.

Although major work on the tender was essentially finished there were some tasks still taking place. 21st century technology was being introduced with Global System for Mobile Communications-Railway (GSMR) being installed by Mike Page, effectively replacing the need to communicate with the signalman by a telephone on a signal post with a radio. White gloss paint had been applied in the tender corridor and black in the tender coal space by Nigel & Sheila Franklin, after Alan Pitt had applied the final coat of undercoat. Work was also being finalised on the various tender brake components. The number of tasks being undertaken, even now over five years into the overhaul, still amazed me.

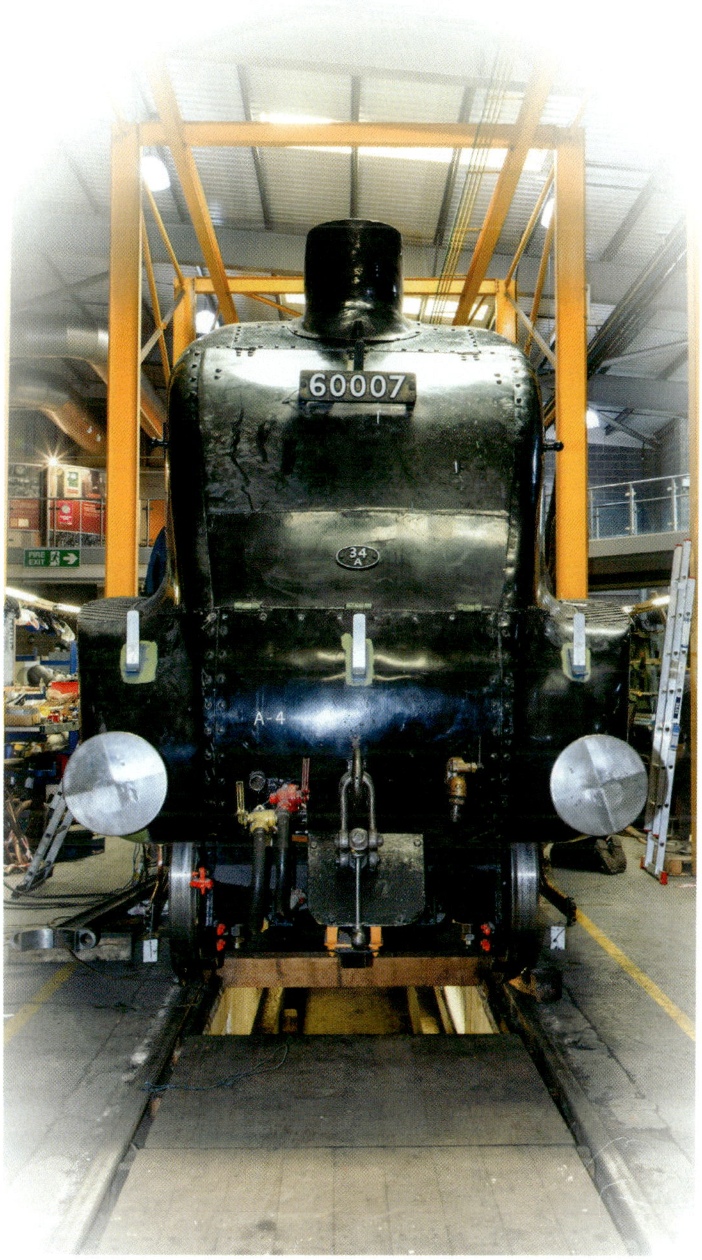

It is a great shame that access to the magnificent facilities in the NRM workshop will soon no longer be available as the museum goes ahead with its own redevelopment plans. It was becoming very evident to everybody in the team that this would be the last overhaul in this famous workshop, which had previously serviced so many iconic locomotives. It seems very fitting that the locomotive which bears the name of the LNER's most famous engineer was to be the last locomotive to be overhauled at the former York MPD 50A, consigning a significant chapter to railway history. The new plans are designed to inspire future engineers and are backed by the success of similar projects, so, with the heavy hearts of an older generation, I think everybody in the team hopes the future goes well for the museum, whose support to the Trust during this overhaul has been invaluable.

A few days ago I had captured a nearly visibly complete locomotive and I had returned today to capture an even more nearly complete locomotive. Today my attention was at the front where the cod mouth doors had been fitted. Bob Shearman and Dave Lee had the doors wide open to enable them access to their areas of work, but I needed them shut for the camera, and the guys were more than happy to oblige by cranking them shut for me. But wait, we need the cowl on the chimney so Peter Brackstone was despatched to fetch it. The copper cowl arrived on the back of a wheel barrow and it is quite amazing how large it is when you see it close up. It is not particularly heavy but it is cumbersome and awkward to handle so the overhead crane was used to lift it into position, and with Dave on top of the boiler and Bob operating the crane they soon had the task completed. But wait, "We can put the shed plate on the front," said Darrin and Peter was despatched again to retreive the 34A plate of Kings Cross shed from the store and that was soon fitted. I have not disrupted the team's work over the years of the overhaul as I was trying to get on with my job quietly without stopping theirs. However, today I had the opportunity to call the shots (sorry I couldn't resist) as we all wanted No.7 to look as complete as possible for the photo shoot. So, I was now in full swing setting up the image, "Can we close that panel?", "Can we put the cover plate on there?", "Can we move those ladders?". My polite requests went on, but the team drew the line when I requested some coal and water! I finally got the camera into action and was pleased with the results. No.7 is far from complete but it now looks well on the way.

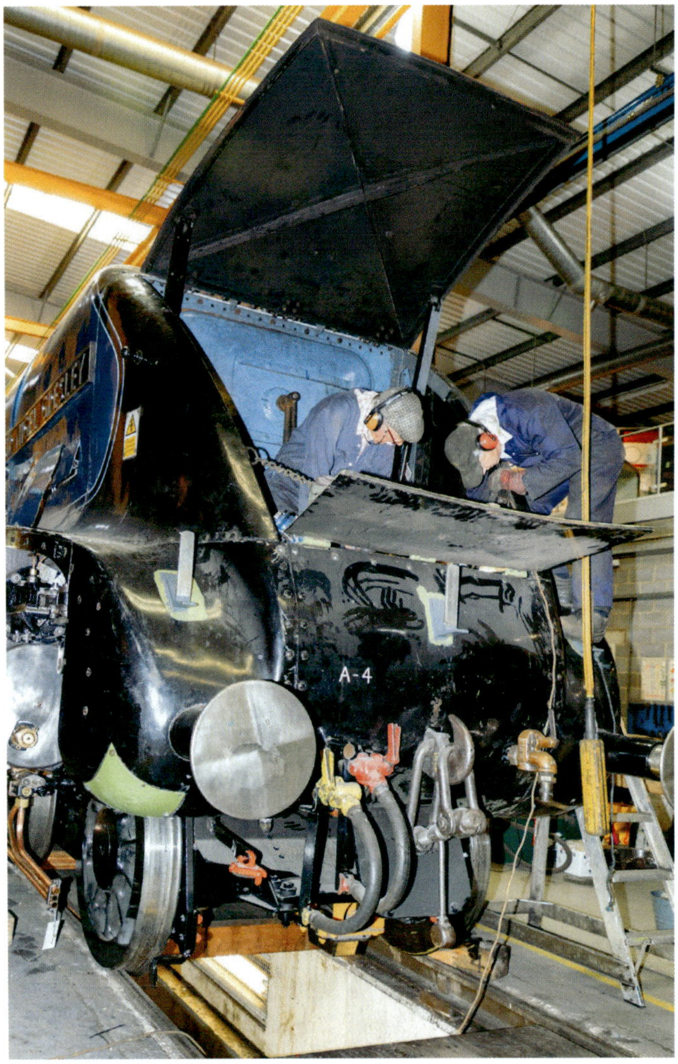

Bob Shearman (left) and Dave Lee working on the cod mouth doors

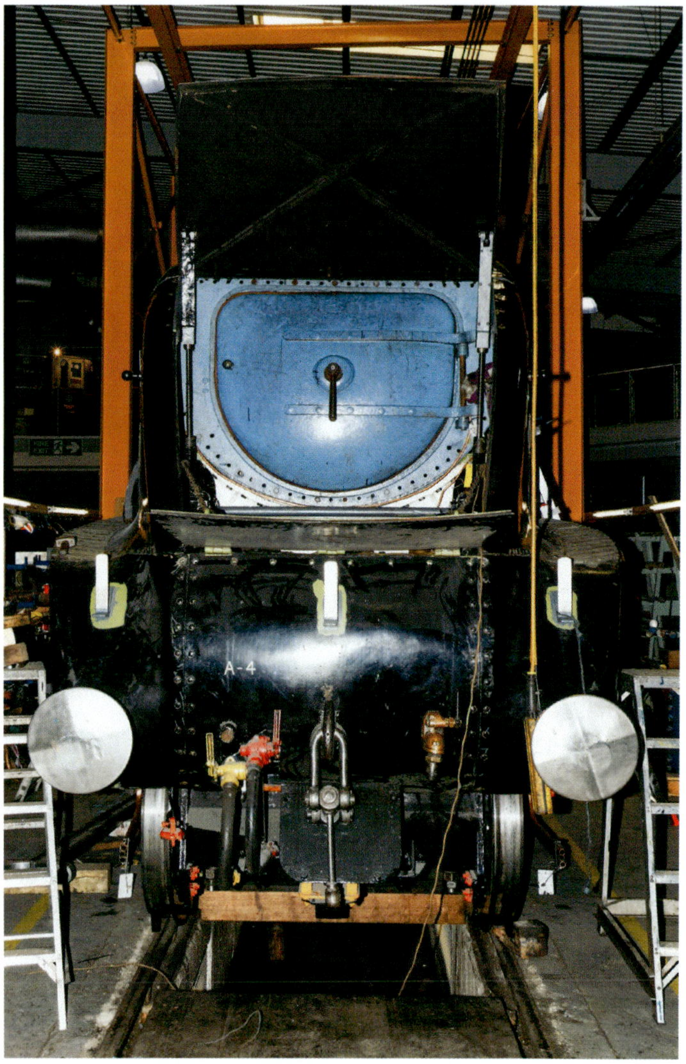

The cod mouth doors fully opened revealing the smokebox door

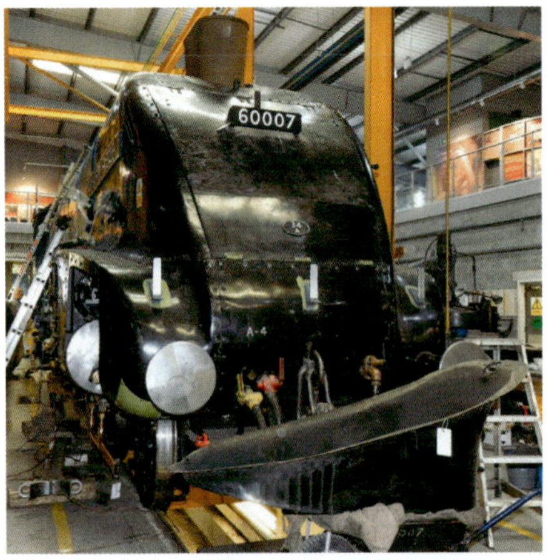

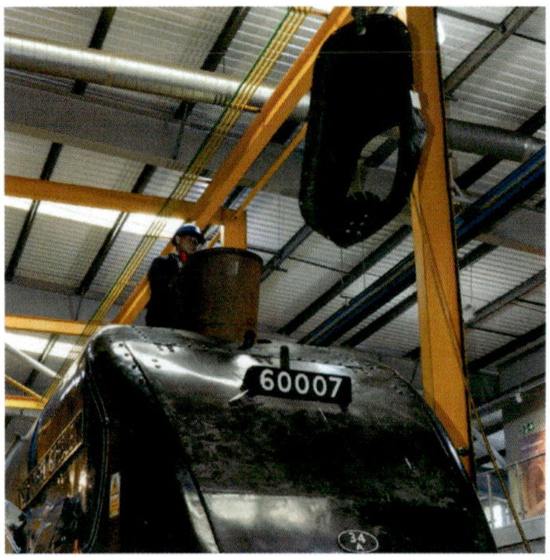

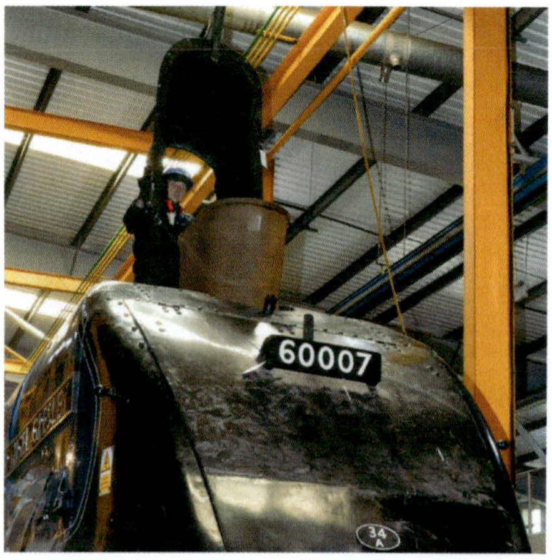

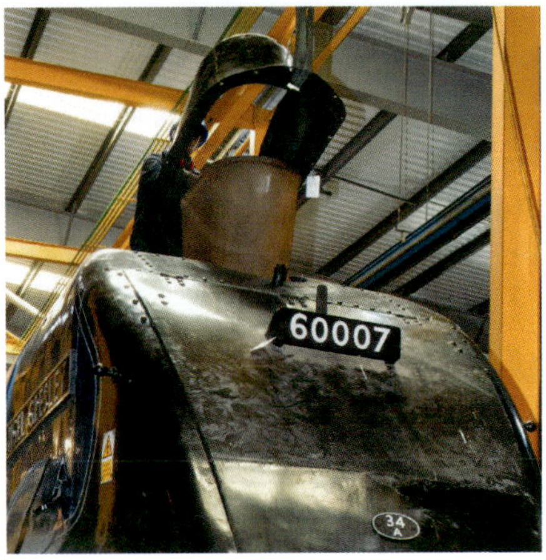

Sequence of pictures showing the fitting of the chimney cowl by Dave Lee (on top) and Bob Shearman (out of shot) operating the crane

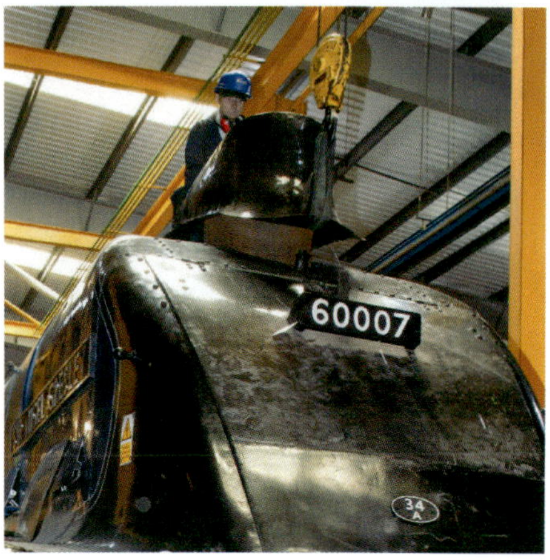

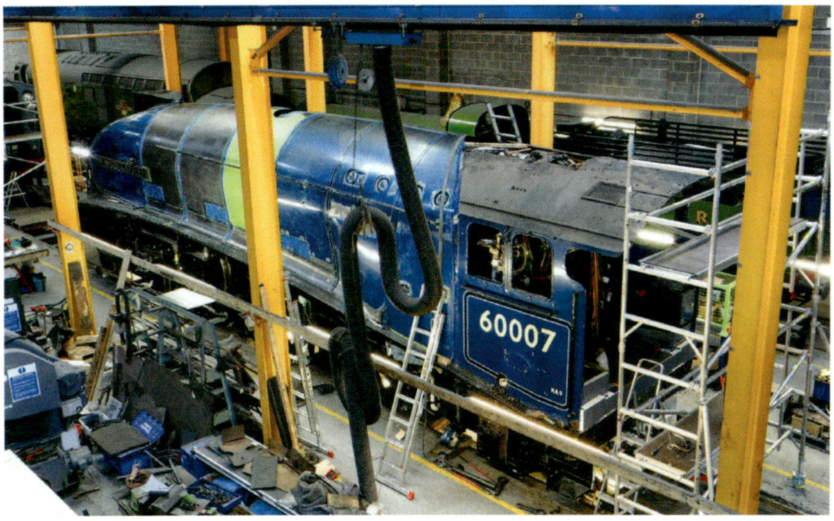

NRM Gallery view showing No.7 now with all of the streamlined cladding fitted

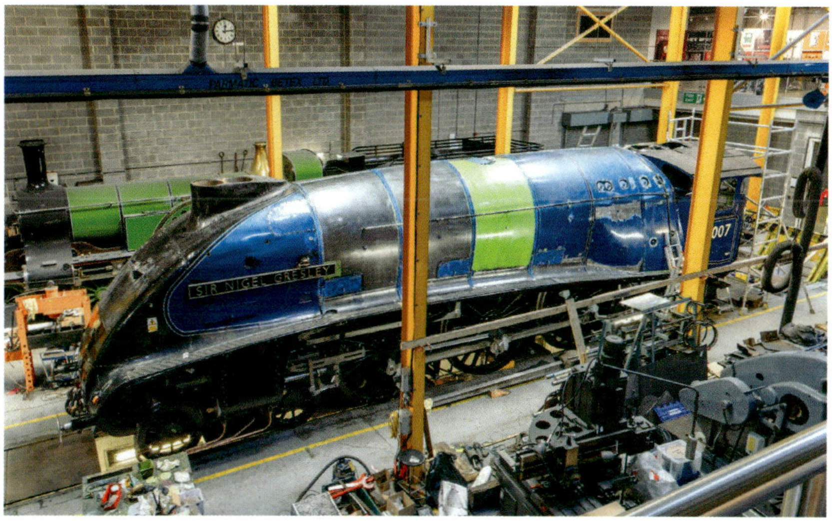

No.7 complete with cladding panels, front streamlining and chimney cowl all in place

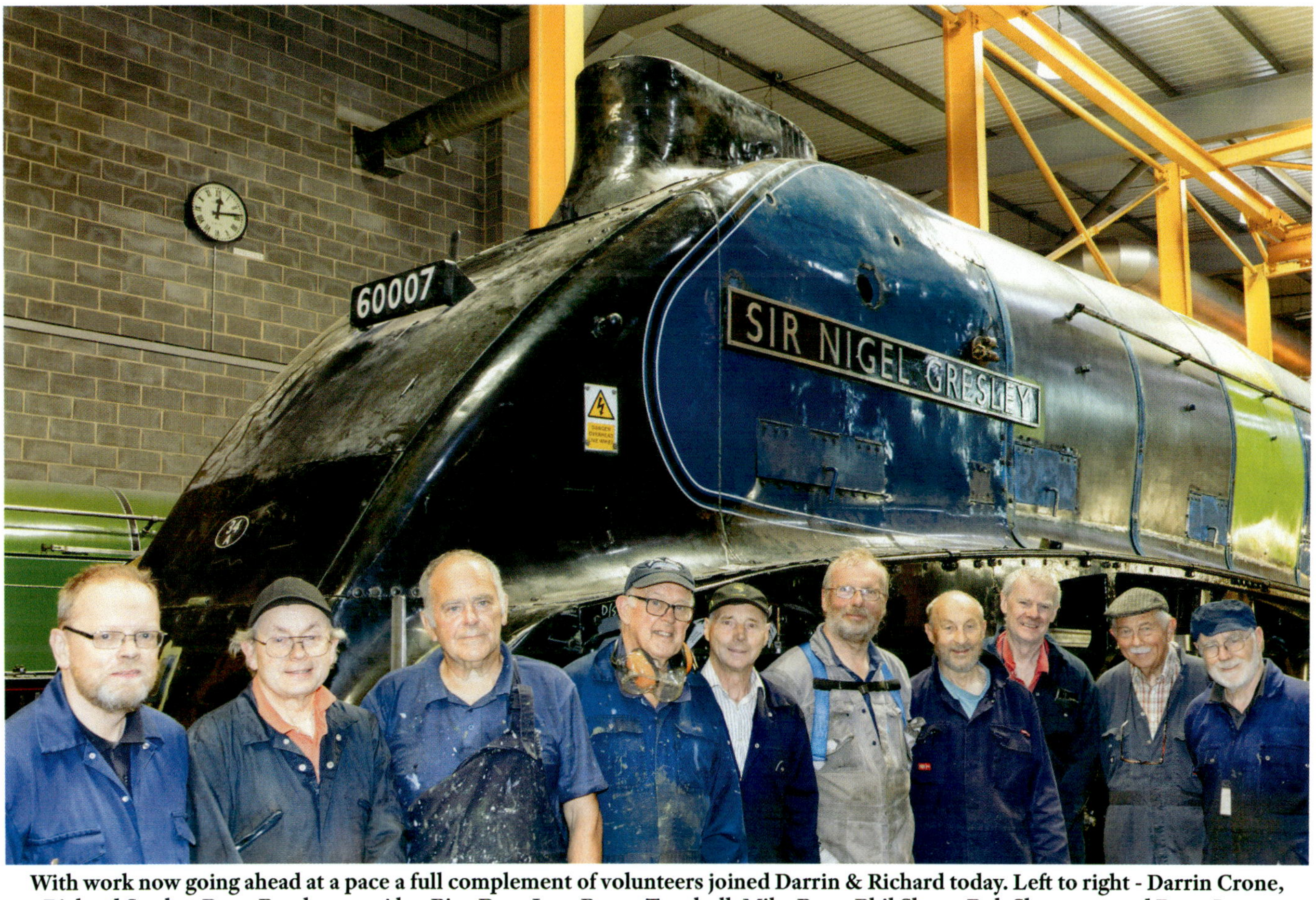

With work now going ahead at a pace a full complement of volunteers joined Darrin & Richard today. Left to right - Darrin Crone, Richard Swales, Peter Brackstone, Alan Pitt, Dave Lee, Roger Turnbull, Mike Page, Phil Slater, Bob Shearman and Peter James

OPEN DOORS

August 2021

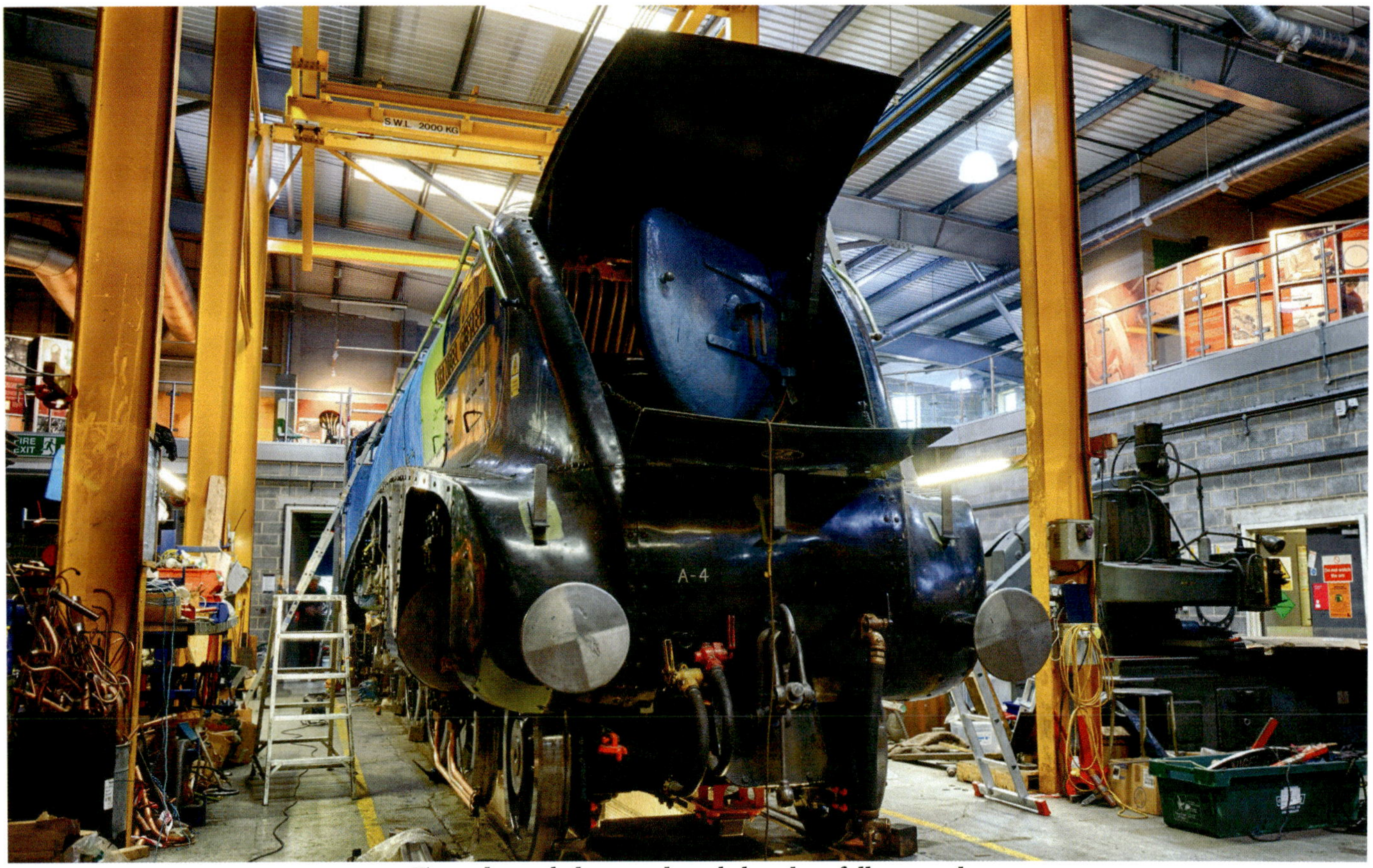

The cod mouth doors and smokebox door fully opened

The cod mouth door is the fishy term used to describe the doors which open upwards and downwards at the very front of an A4 locomotive, and are only found on streamlined locomotives such as No.7 as they allow access to the smokebox door which on other locomotives is not hidden behind streamlined cladding. When open I think they resemble more a shark than a cod or, for those that can remember the James Bond film "You Only Live Twice", the satellite capturing spacecraft. Today they were open as was the boiler door so I took some HDR triple exposure

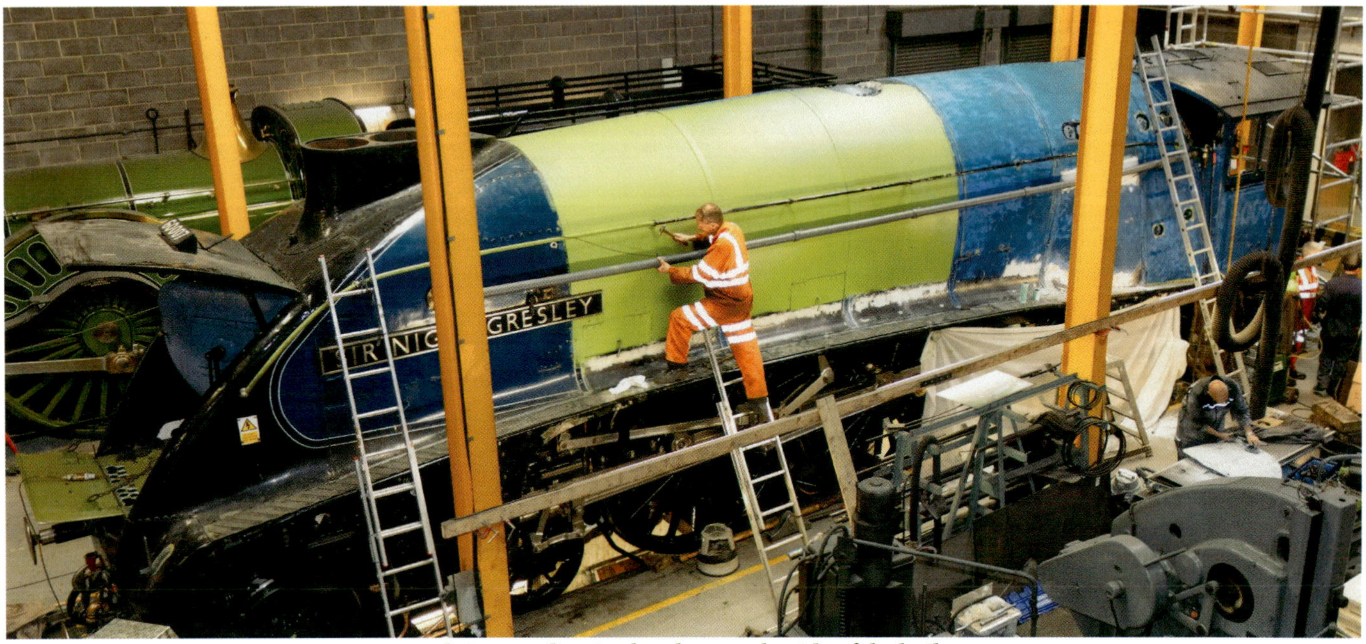

Andy Evans adjusting brackets on the side of the boiler

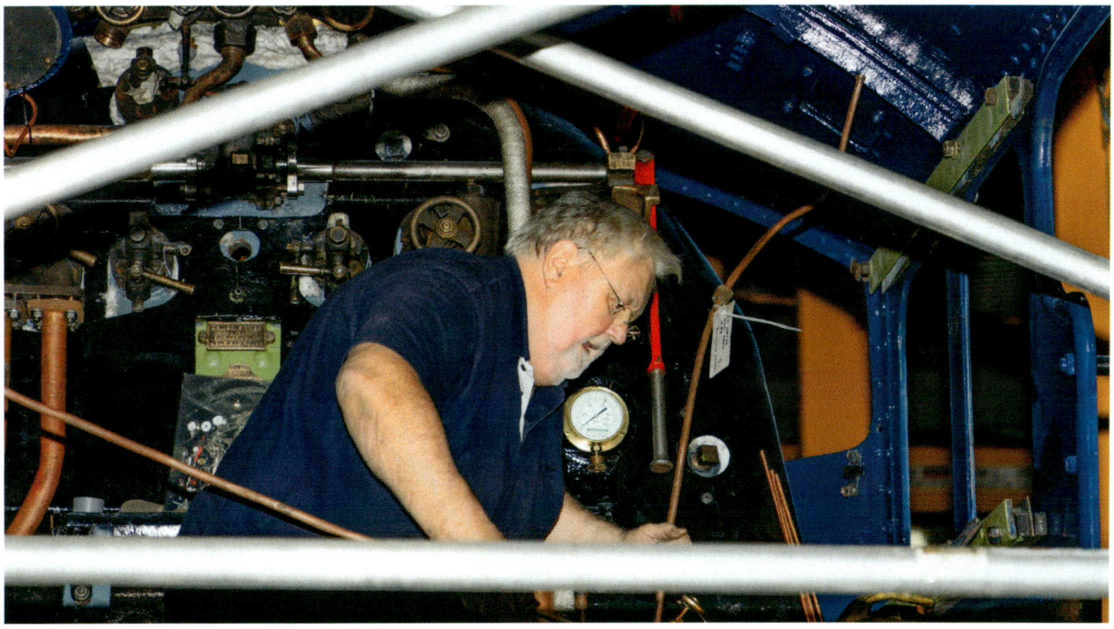

Peter Ellis performing pipe gymnastics in the restricted area of the cab

shots which, along with the help of a single light bulb inside the smokebox, enabled me to get detail in the dark interior whilst not over exposing the surrounding workshop. Another shot taken from the gallery shows the open doors seemingly capturing the main drive wheel of NRM's GNR No. 1 Class 4-2-2 Stirling Single on the adjacent track, whilst Andy Evans completes some work on the recently fitted boiler hand rail. Green primer paint has been applied to the new cladding panels, thus identifying them as replacements for 'beyond repair panels'.

Peter Ellis, to whom Darrin had awarded a silver medal for pipe bending gymnastics, was working on the footplate in the cab putting the final pieces of his pipework jigsaw puzzle together. Not being a great deal of room to move around when Peter and I are on the footplate together I took a couple of shots from behind the gantry scaffolding bars which resulted in the inevitable jokes of being found behind bars and, of course, in bars! I asked Darrin why Peter only won the silver medal and he replied the gold wasn't available as the standard needed to be higher. I couldn't help but remember my old school teacher from many years ago who never awarded an 'A' to anybody so as to encourage you to keep trying harder, so Peter had my sympathy.

I moved down the track to the tender, which was looking in superb condition, where I found Chris Whiteley in the rear corridor connection fitting some rubber seals around the cabinet which houses the GSMR radio. The tender now housed a lot of electronics and on my other image you can see the control box for the GSMR on the left next to the corridor. The red handle on the right hand side operated the famous A4 water scoop for collecting water from the troughs between the tracks back in the day of the non-stop Kings Cross to Edinburgh runs, and is now no longer operational.

Final preparation of the paintwork was taking place with new surfaces receiving primer and existing paintwork being rubbed down, erasing the famous 60007 numbers from the cab sides, and, as like an old vintage car which requires some body filler, some areas were filled and sanded down. Next week would see the undercoat applied and what a dramatic transformation that was going to bring.

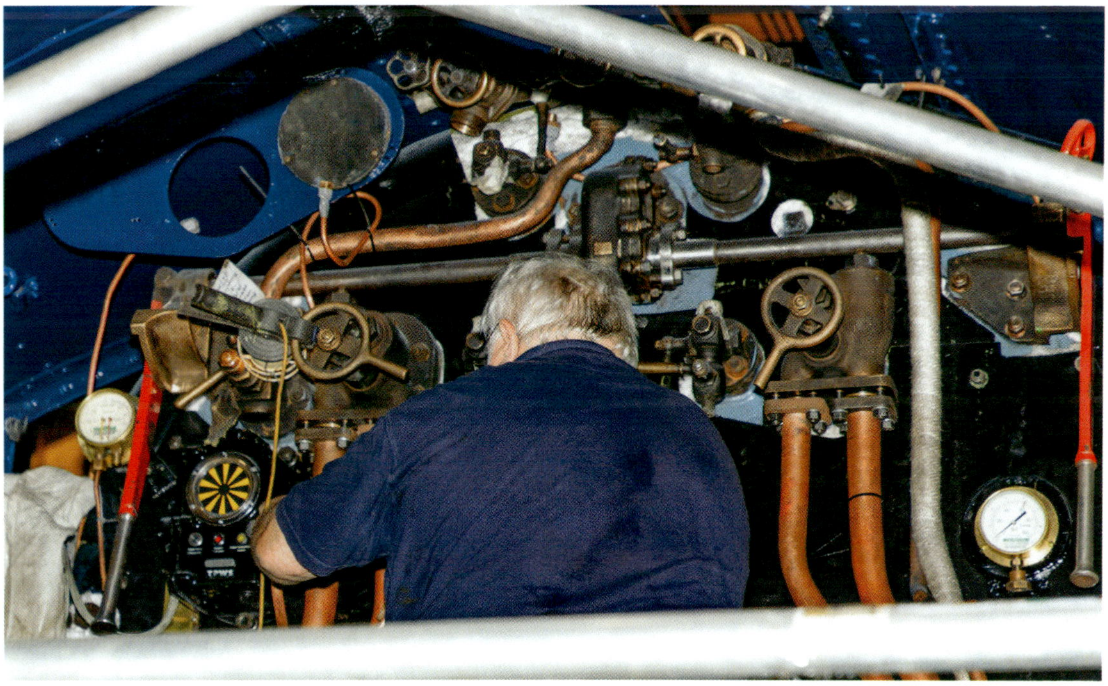

Peter Ellis working on pipework inside the cab

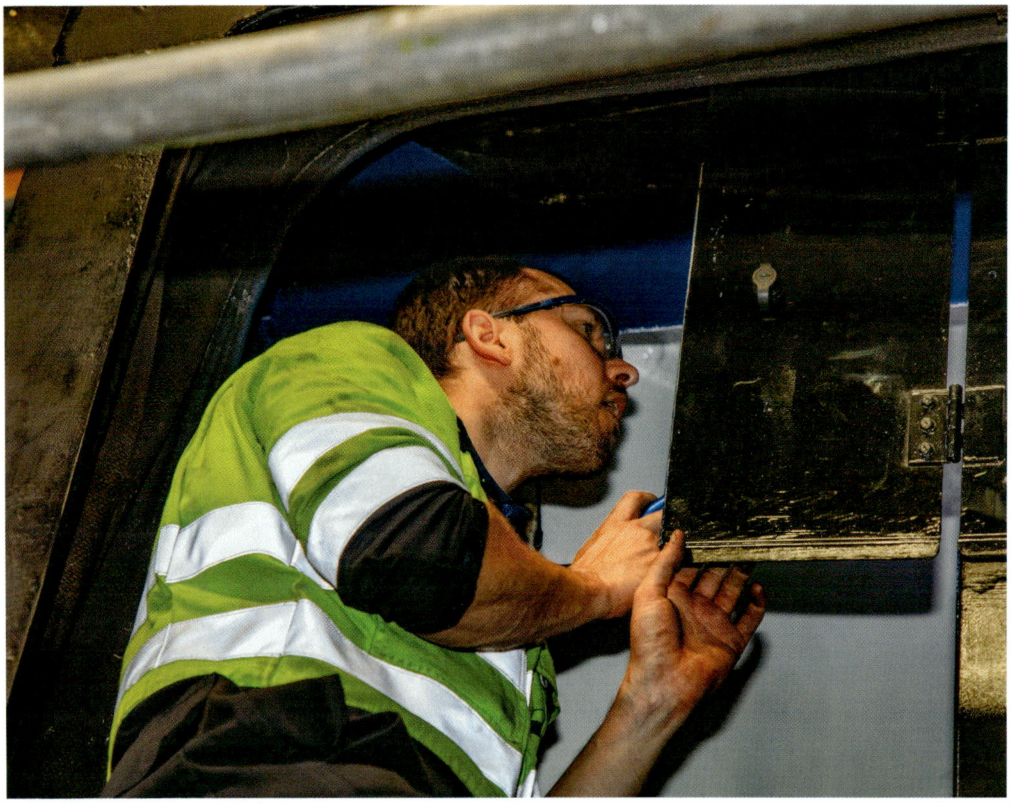

Chris Whiteley checking the recently fitted waterproof rubber seals on the GSMR radio transmitter in the tender corridor

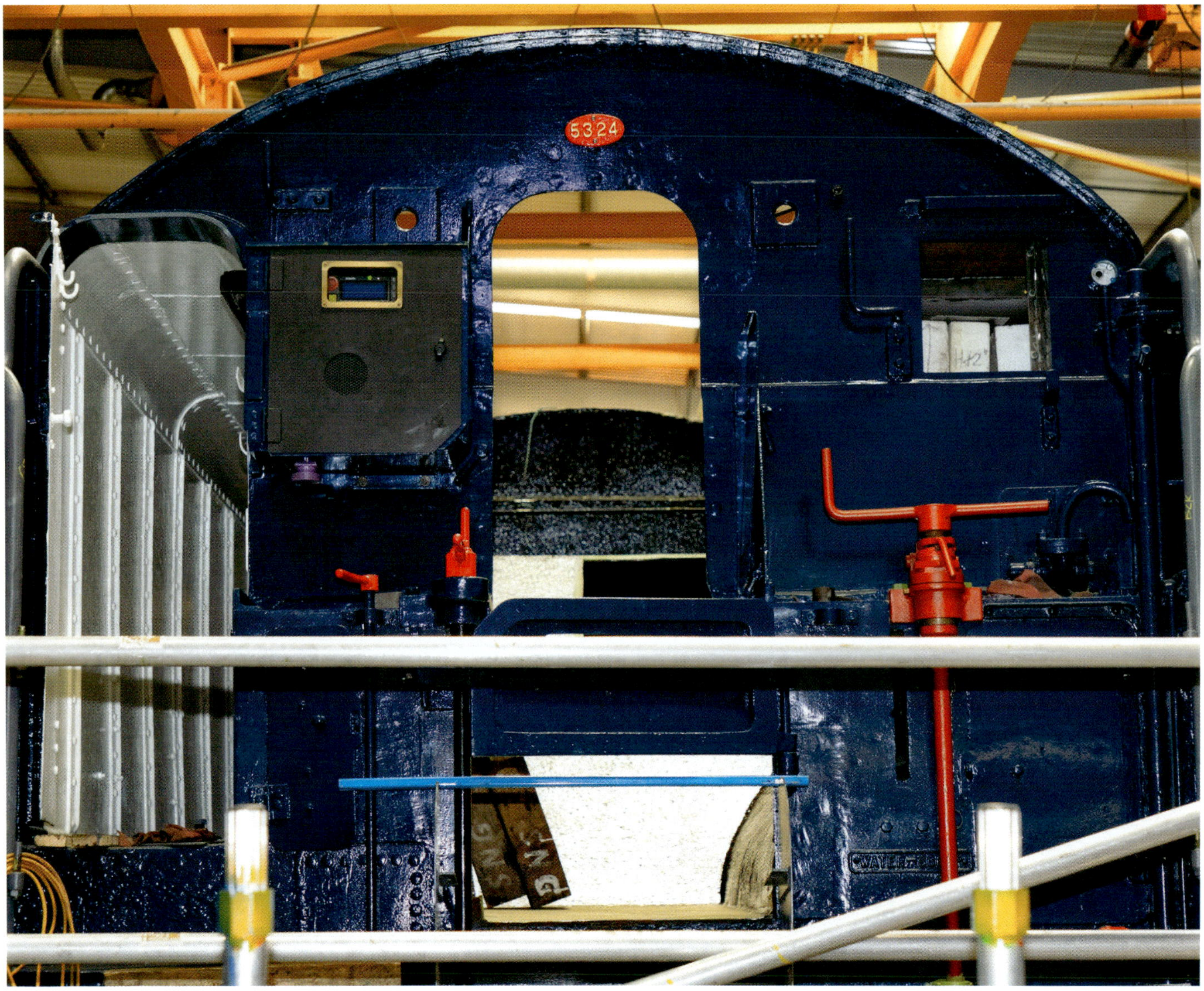

The freshly painted front of the tender with the corridor door open, the GSMR radio control panel and the refurbished water scoop control

August 2021

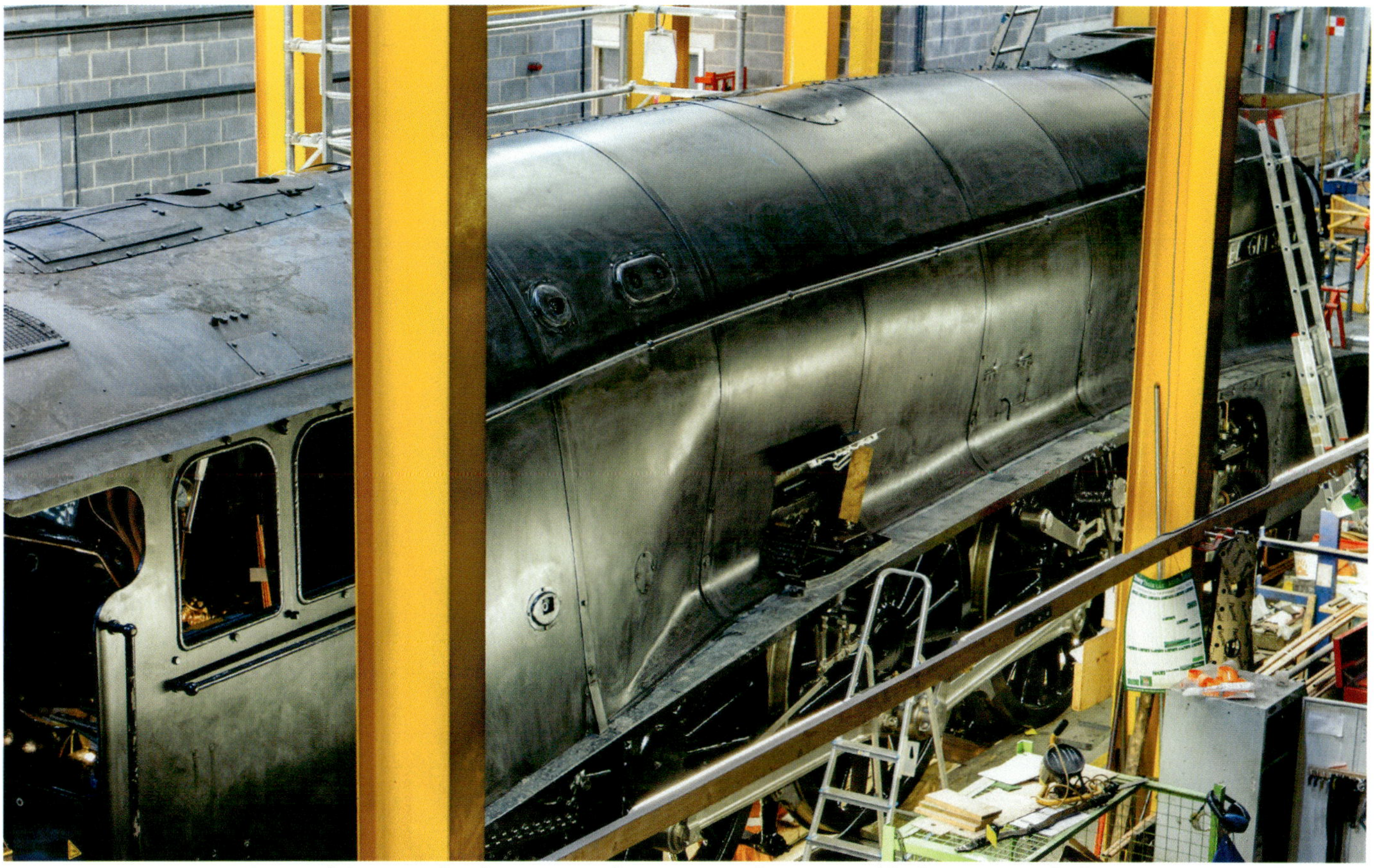

Readying No.7 for the move outdoors, a coat of protective black paint is applied

I needed to call Darrin to check how the planned locomotive painting was proceeding but I couldn't get hold of him, which is not unusual as he has been quite busy over the last few years. So I called Bracko (Peter Brackstone) and before we had had chance to exchange pleasantries I could hear Darrin shouting in the background, "Is that Trevor?". "Yes, yes," replied Peter, adding "He wants to know if he should come in tomorrow." "YES, YES, I'll call him back," was Darrin's exhuberant reply. And WOW what a transformation. Darrin video called to give me a quick preview of the excellent paint job that the team, with the help of specialist locomotive painter, Ian Matthews, had undertaken.

The Trust membership had voted quite emphatically to restore No.7 into her British Railways Express Blue which would be applied in LSL paint shop at Crewe, so this was a protective coat of black paint for the interim period whilst she is exposed to the elements outside during the wait and move to Crewe. The result was the first time any A4 in preservation had appeared in WWII black livery.

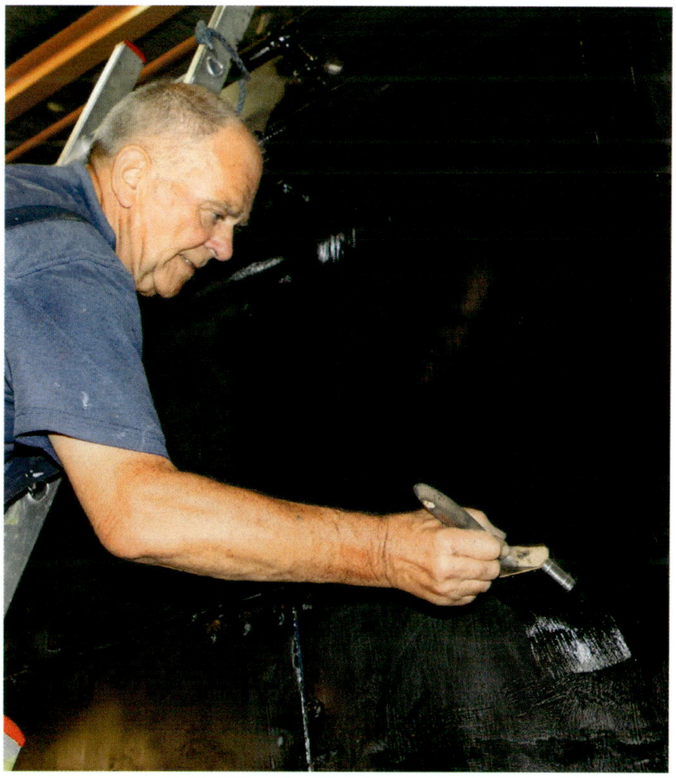

Peter Brackstone applies a protective coat of black paint to the boiler sides

Unusually for me I arrived after lunch the following day to give the paint team as much time as possible before I started taking pictures. Peter, who had led the paint team throughout the overhaul, was applying a second coat to the middle of the boiler's left hand side with Ian to his right and Nigel Franklin on his left appropriately painting around his namesake's nameplate. The tender had already been painted but there was a bit of touching up to do and here I found Malc Bateman applying red paint to the water scoop mechanism the renovation of which he had also been a major player in.

It wasn't all painting today as several engineering/ electrical tasks needed completing which Darrin, Richard Swales, Mike Page, Andy Evans and Bob Shearman were endeavouring to get finalised before the big move out of the workshop, which had now been scheduled for the following Wednesday. I arranged with Darrin which days would be best to attend in the next week and it soon became apparent that everyday would probably be best as it was going to be a busy week.

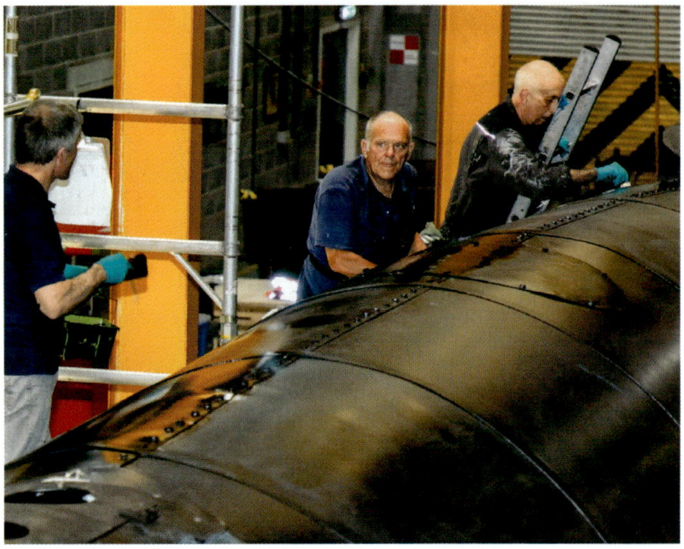
Specialist painter Ian Matthews (left) painting the boiler cladding with Peter Brackstone (centre) and Nigel Franklin

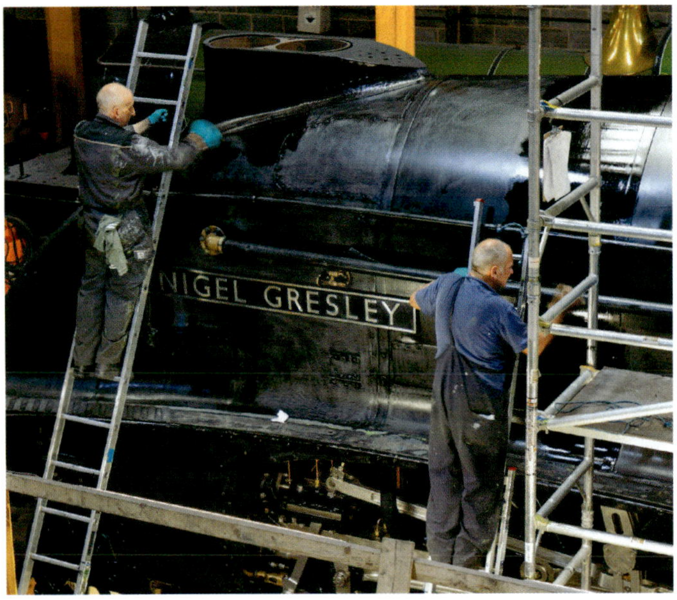

Nigel Franklin and Peter Brackstone applying the protective coat of paint

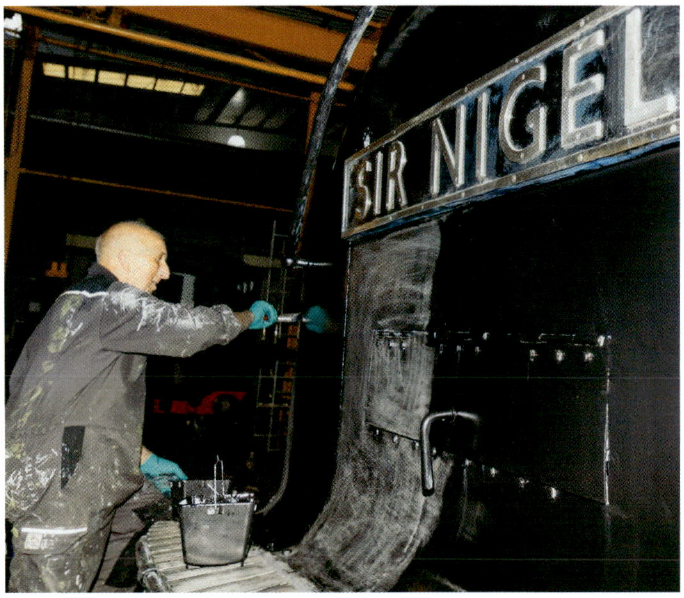

Nigel Franklin paints the front streamlining

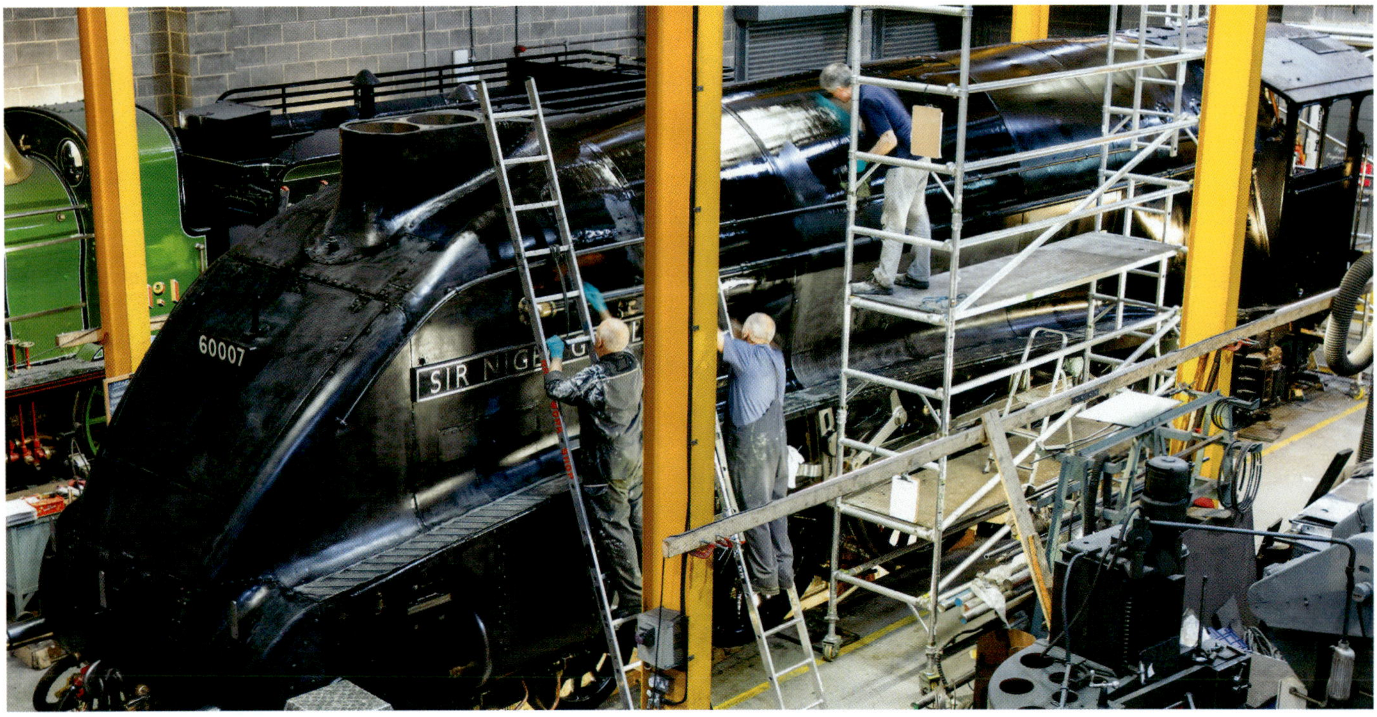

Nigel Franklin (left) Peter Brackstone (centre) and Ian Matthews applying the protective coat of paint

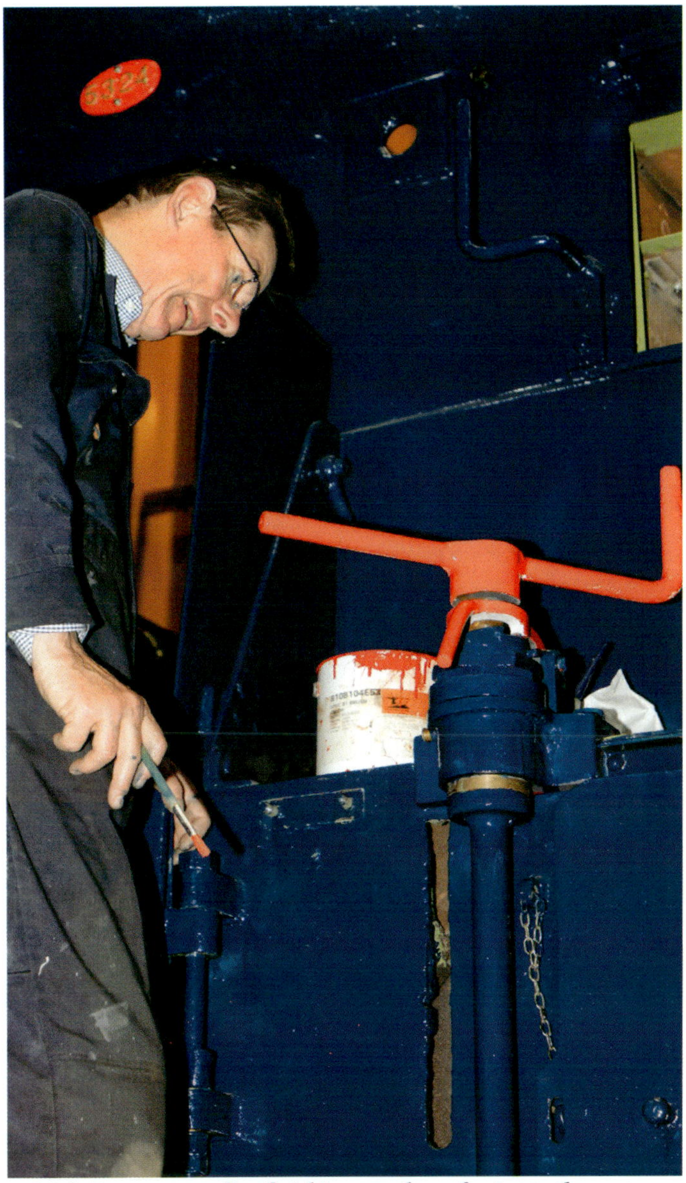

Malc Bateman applies finishing touches of paint to the water scoop handle

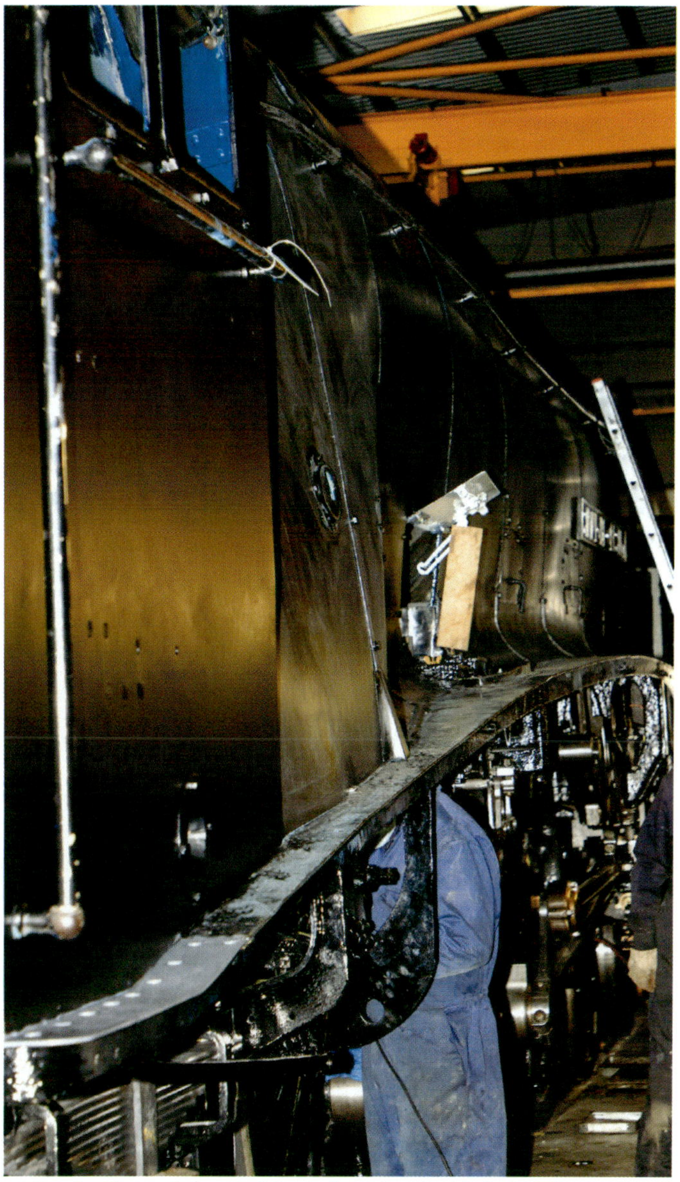

Darrin Crone is lost in his work under the right hand running board

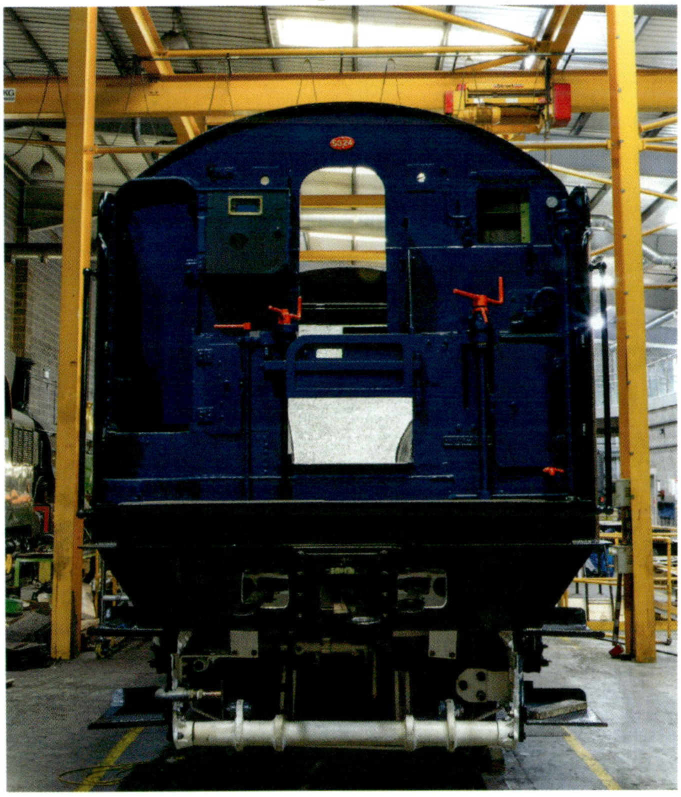

The paintwork has been completed at the front of the tender offering a clear shot with the scaffolding now removed

22

INTO THE LIGHT

August 2021

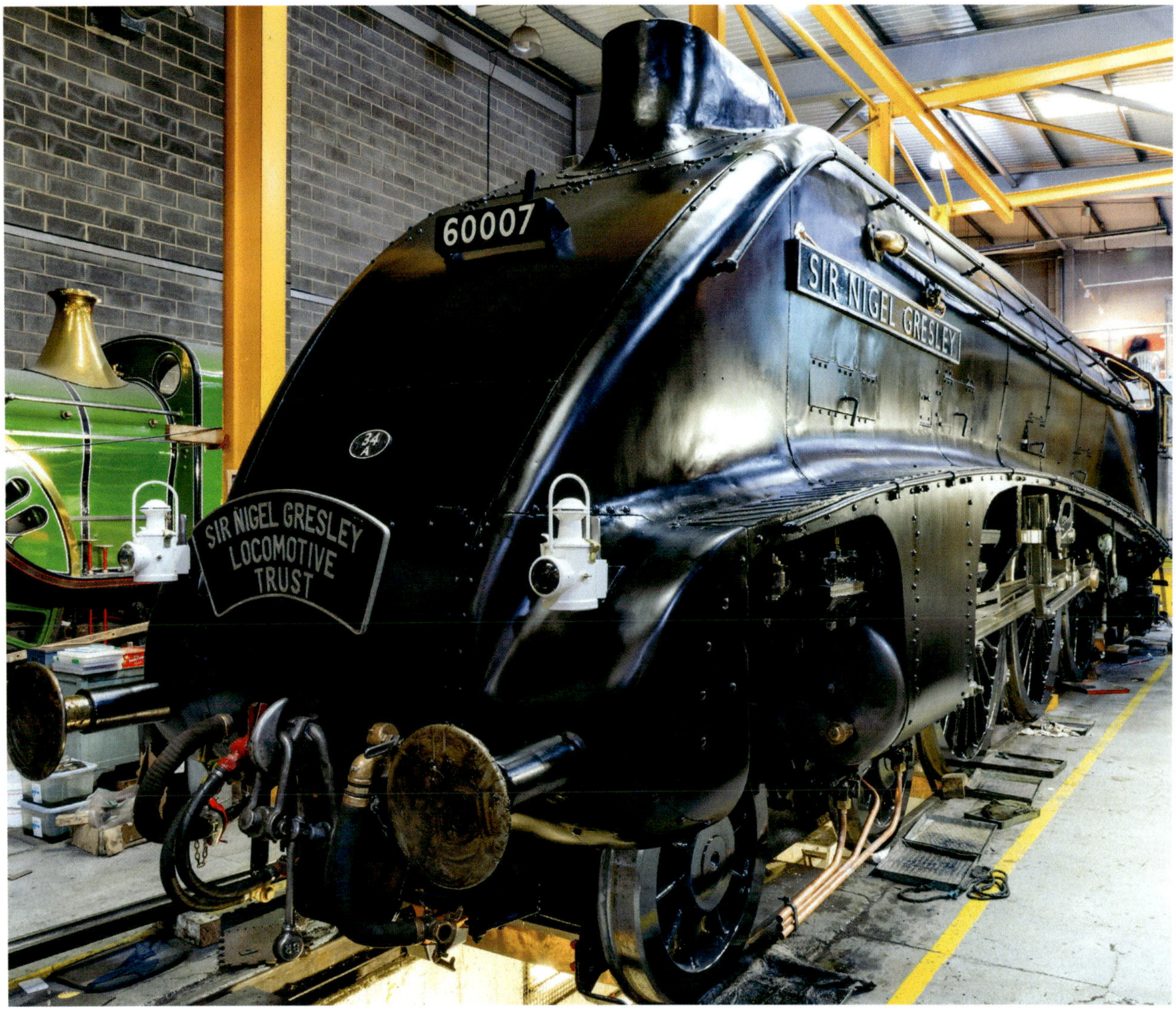

Ready for the move outside

"You coming in today?" was the text I received from Darrin around 1pm. "Yes, around three," I replied nearly adding "as agreed yesterday" but I thought I wouldn't. The poor guy didn't need reminding of that, he was busy enough as this was a huge milepost day in the overhaul. No.7, after a fantastic fast run by the engineering team over the last three months, was ready to venture out into the outside world again. I arrived around three as promised and set about getting some final shots in the workshops of what would be the last ever locomotive overhauled there before the NRM went ahead with their redevelopment plans. It was to be a day to remember for more than just us as it was the end of an era for the former York 50A workshop.

As the tracks leading out of the workshop pass through the museum's Great Hall, the move could not take place until it closed at 5pm so some last minute jobs were being undertaken and I found Barry Horton and Rod Thomas working on the reinstatement of the footplate in the cab, which raised a smile later in the day when Darrin noticed the completed work and shouted "Hey, there's a floor in here!" As Darrin is a great perfectionist, the inevitable humorous question shouted back was, "How are you spelling floor Darrin?"

Andy Barwick, ably assisted by Bob Shearman, was busy cranking the oil primer on the right hand side of No.7 as there was a lot of pipework that now needed air pumping out and filling with oil. Richard was keeping a close eye on the progress, monitoring it at various points. This was a physically demanding job that seemed to take forever to complete, which would not be the case in the future after the pipes had been properly primed from being empty and full of air.

The NRM tannoy announcements of the museum closing for the day soon arrived and as many of the team were now just waiting for the shunt to start, I spent some time taking

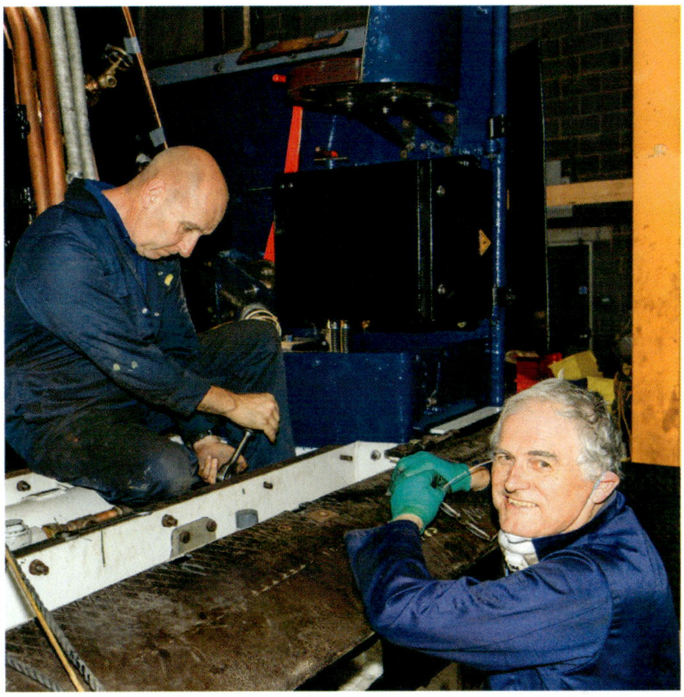

Barry Horton (left) and Rod Thomas working on the footplate floor

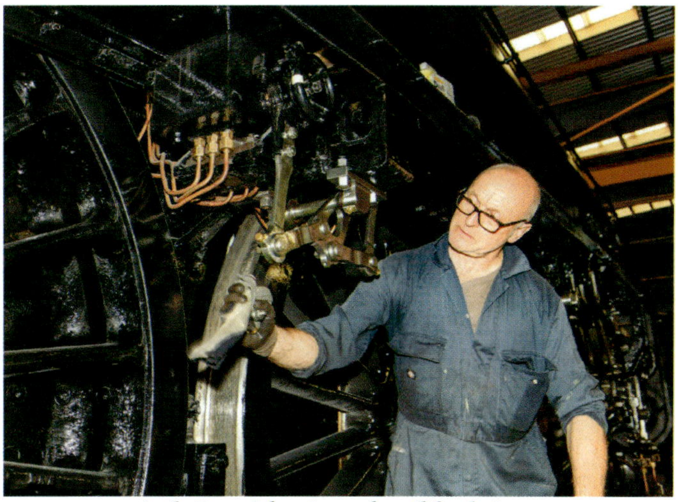

Andy Barwick primes the oil feed pipes

"How are you spelling floor Darrin?"

shots of individuals standing in front of No.7 until the task commenced. I could see the daylight flooding in through the great doors of the museum, which had now been opened and then the sound of the museum's 09 diesel shunter as it approached to shunt out the exhibits which were sharing the track. The last exhibit to be moved was very appropriately the Class 47, 'Prince William', which had hauled No.7 from Grosmont on the NYMR to the NRM in York at the start of the overhaul back in November 2015.

First to be shunted outside was the tender and then No.7 herself, but a brief pause had been agreed with the museum's Rail Ops staff to enable a group shot in front of No.7. After

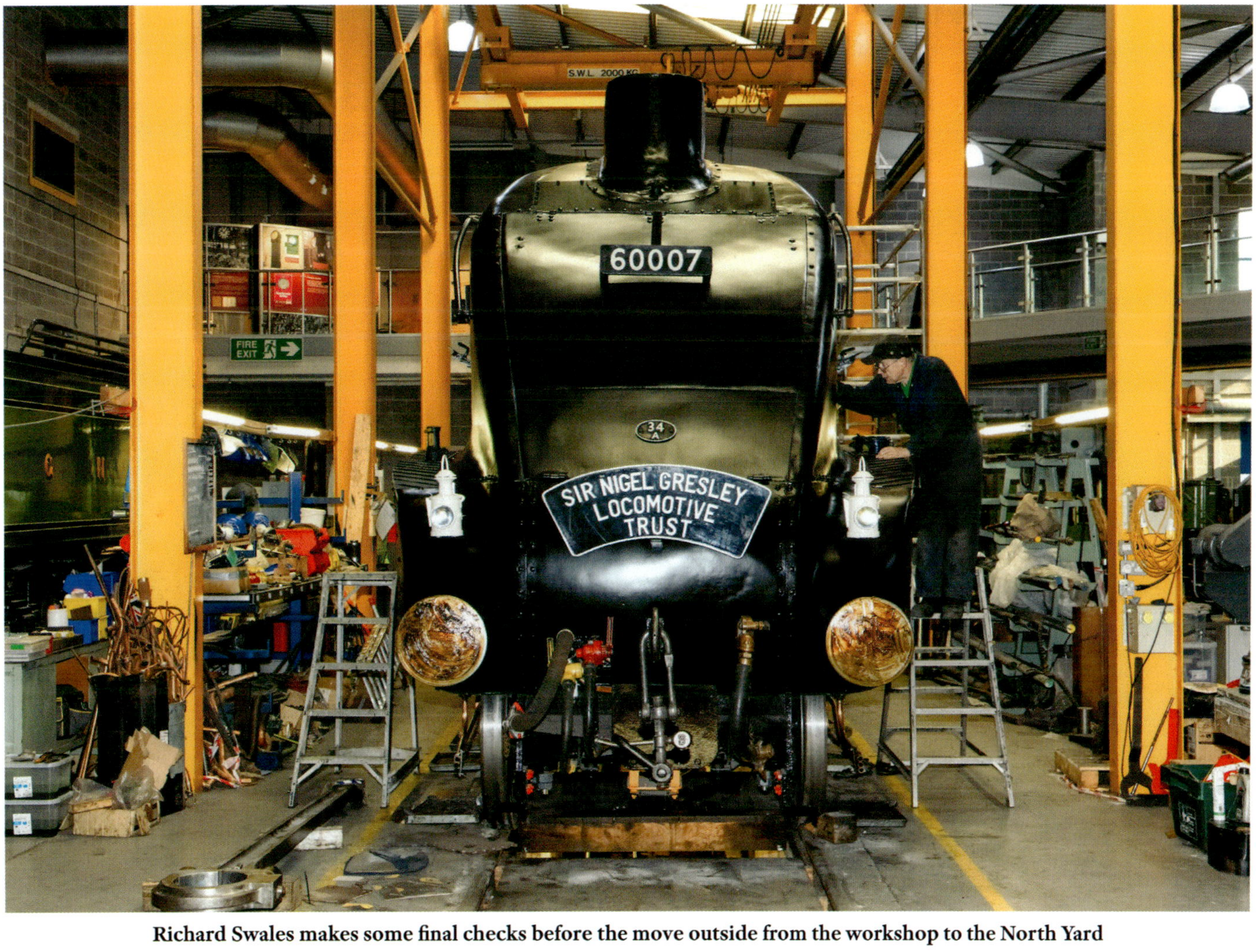

Richard Swales makes some final checks before the move outside from the workshop to the North Yard

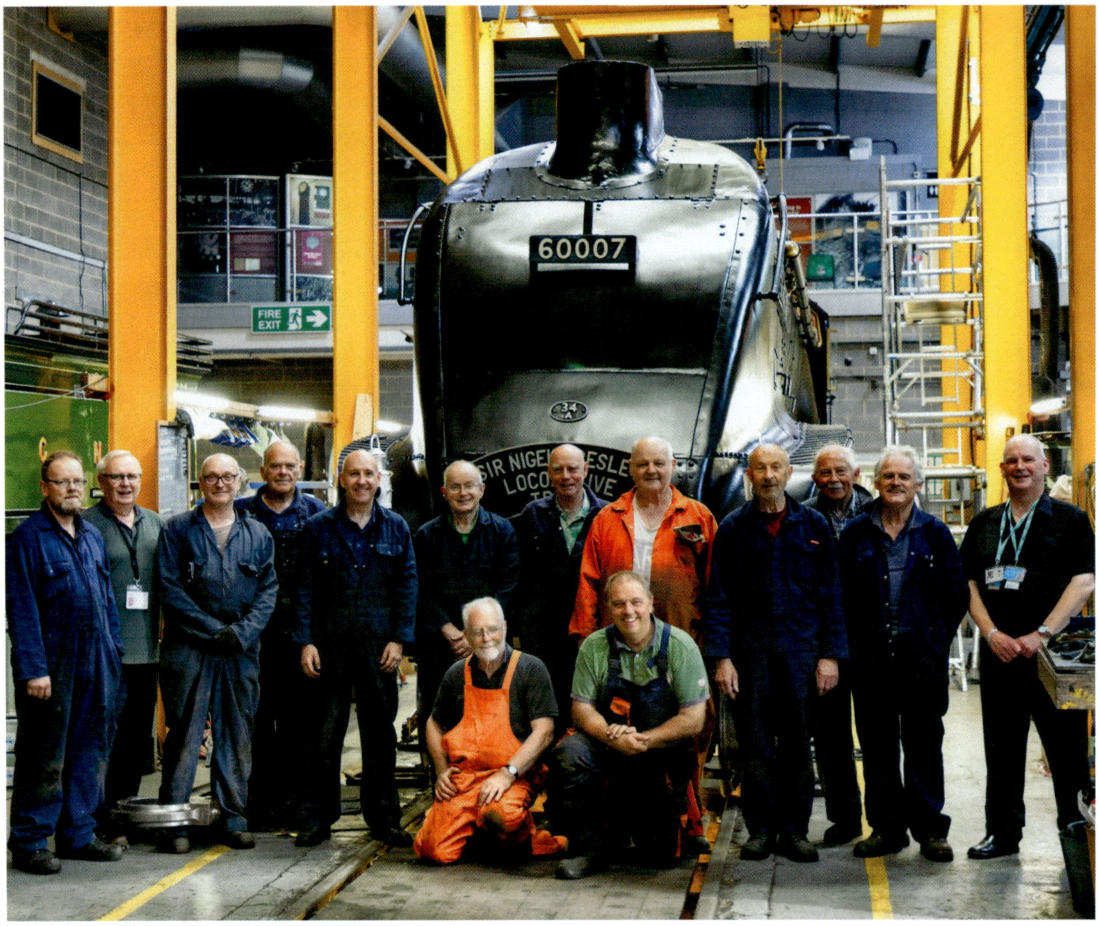

The team on the day of the move outside. Left to Right - Darrin Crone, Trevor Camp (Me, in the picture for a change) Andy Barwick, Peter Brackstone, Barry Horton, Richard Swales, Doug Blyth, Trevor Wilford, Mike Page, Bob Shearman, Rod Thomas, Gary Ward (NRM) Front - Peter James and Richard Hawkhead

all this was a very significant departure not only for SNGLT but also for the NRM as this was the departure of the last ever locomotive from the historic workshops following an overhaul. Once the pictures were taken everybody but essential staff stood back in a safe location as the shunter slowly approached and was coupled up to the front of No.7. As I have mentioned before I think photographs capture moments in history in ways video doesn't but again I found myself thinking this was a time for a video as when No.7 started to move forward my pictures did not convey the

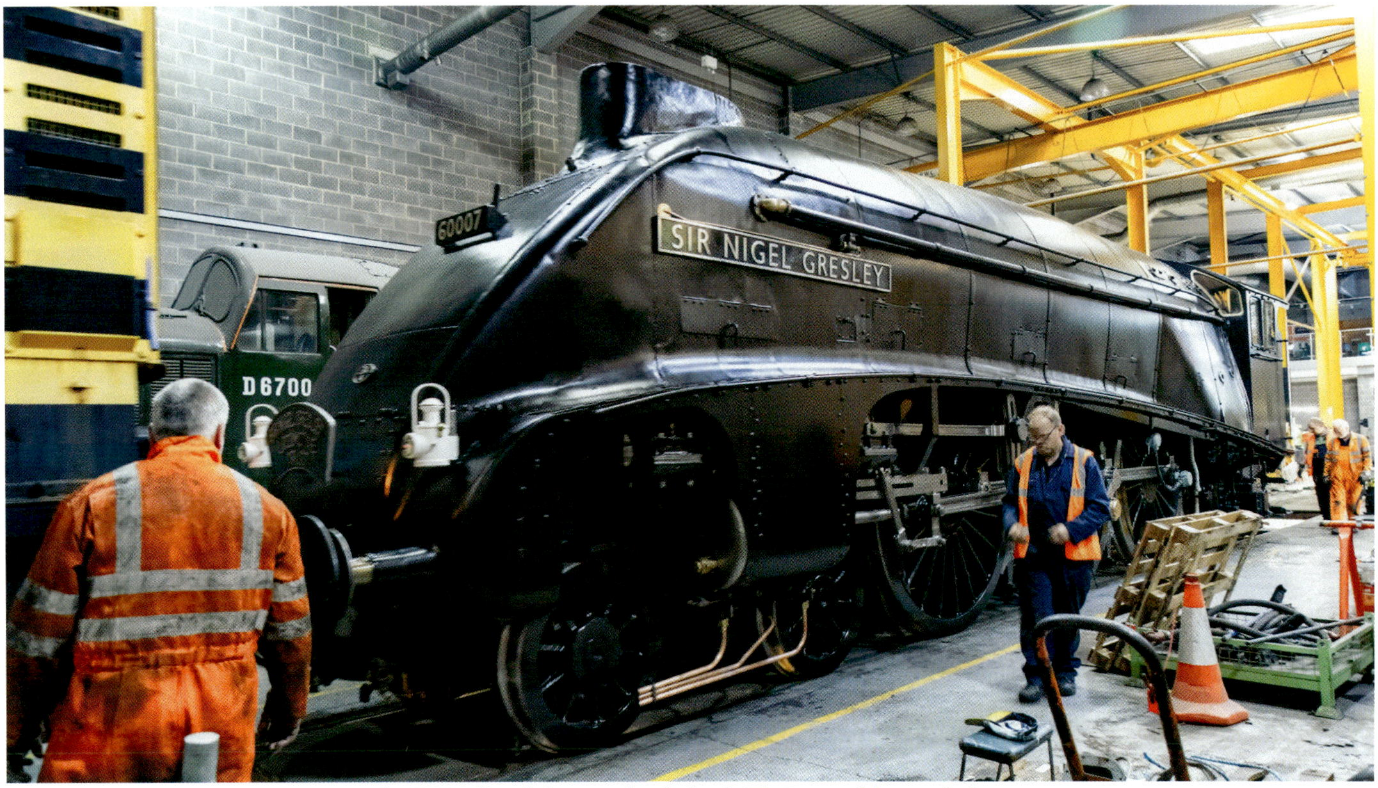

The NRM diesel shunter slowly hauls No.7 out of the workshop

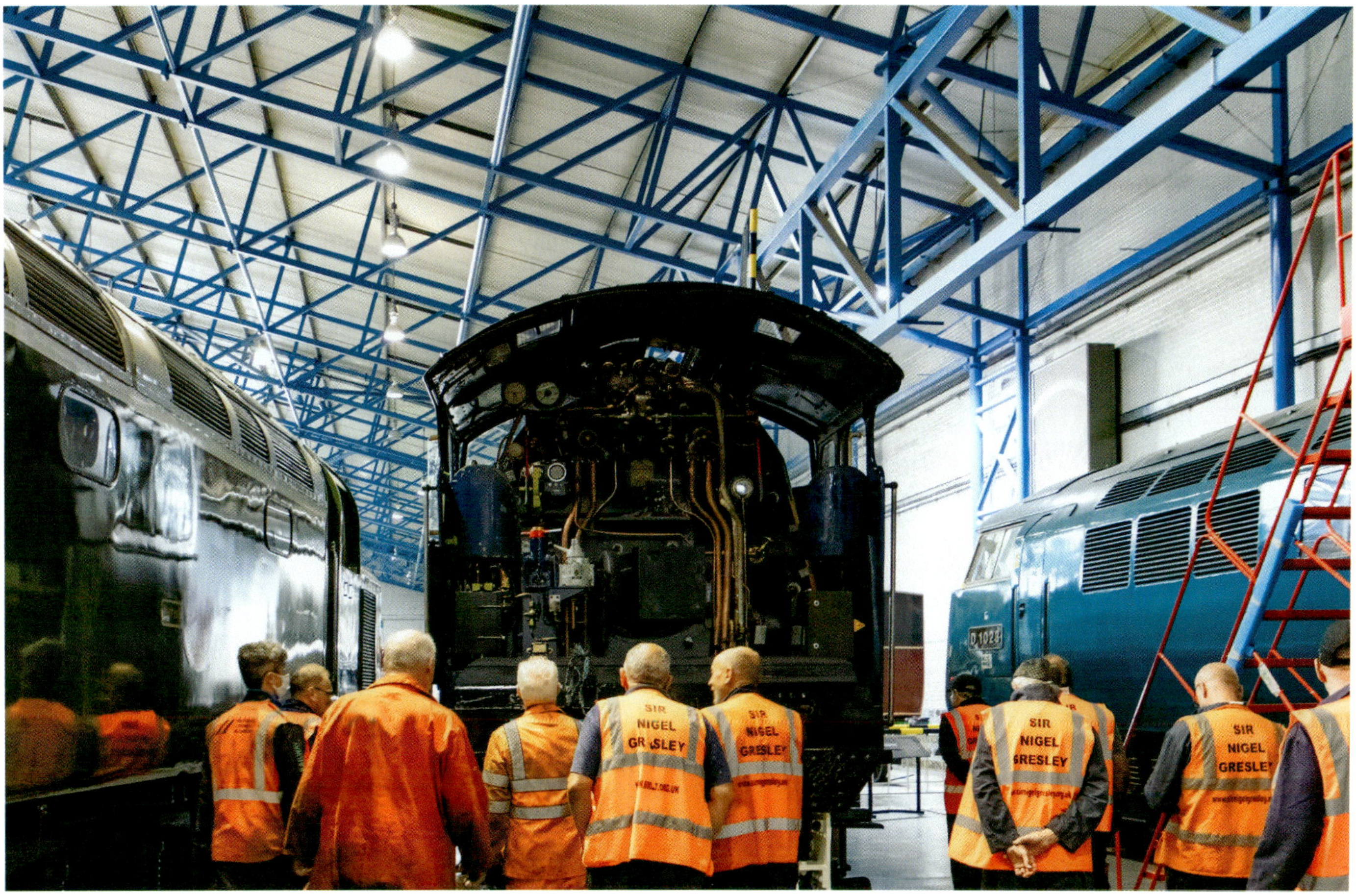

No.7 heads through the Great Hall towards the daylight

moment as well as video would have done.

With Darrin and Richard walking alongside checking all of the movements and listening intently for any moans or groans, of which the beaming smiles from them both conveyed that there were none, No.7 slowly passed through the Great Hall, followed by the team, and out into the bright sunlight of the North Yard. The tender had been shunted onto an adjacent track which put it into the correct position

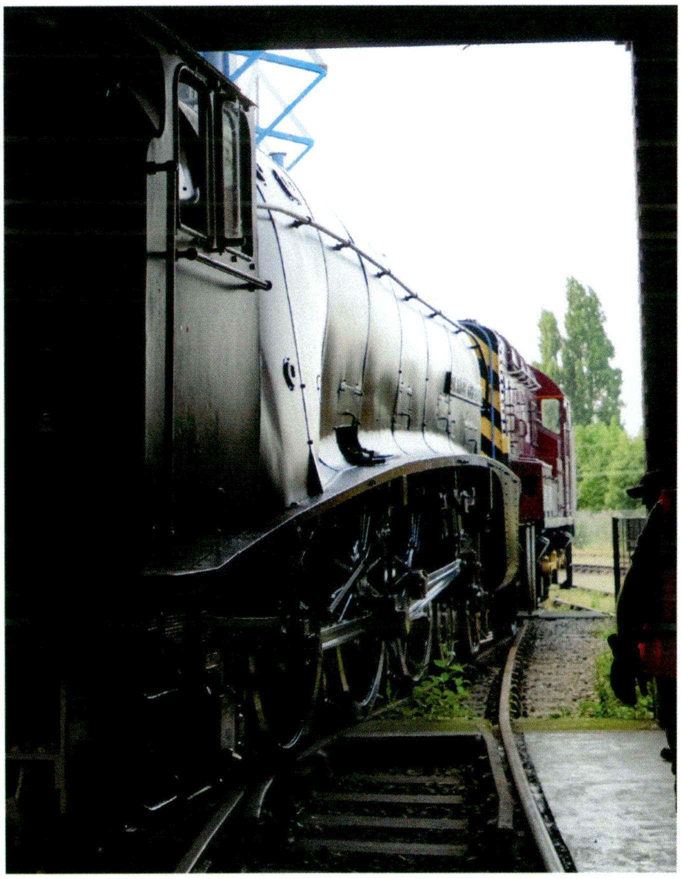

Into the light through the Great Hall's doors into the North Yard

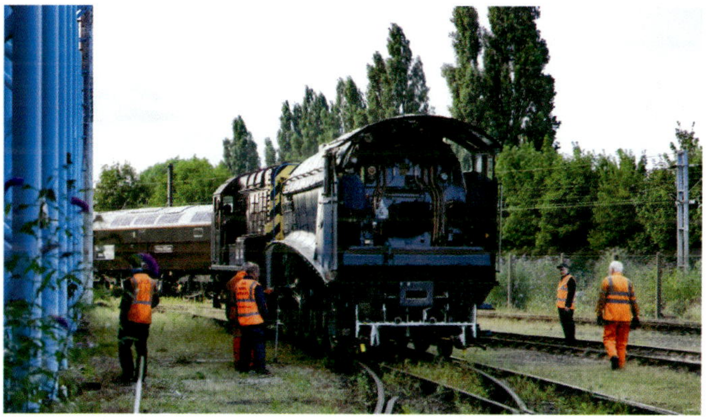

No.7 is hauled forward across the pointwork in the North yard

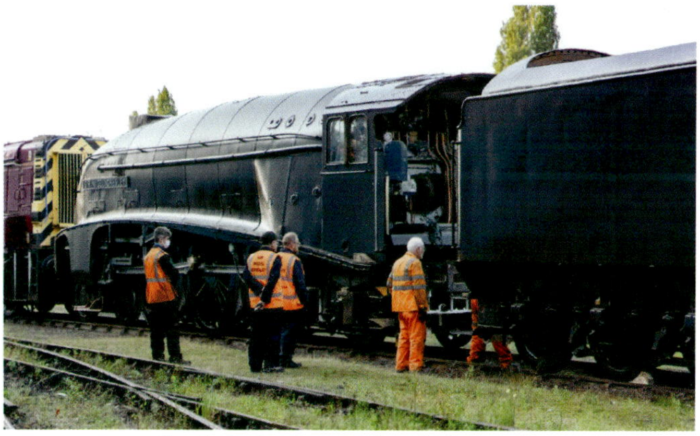

Slowly being reunited with the tender

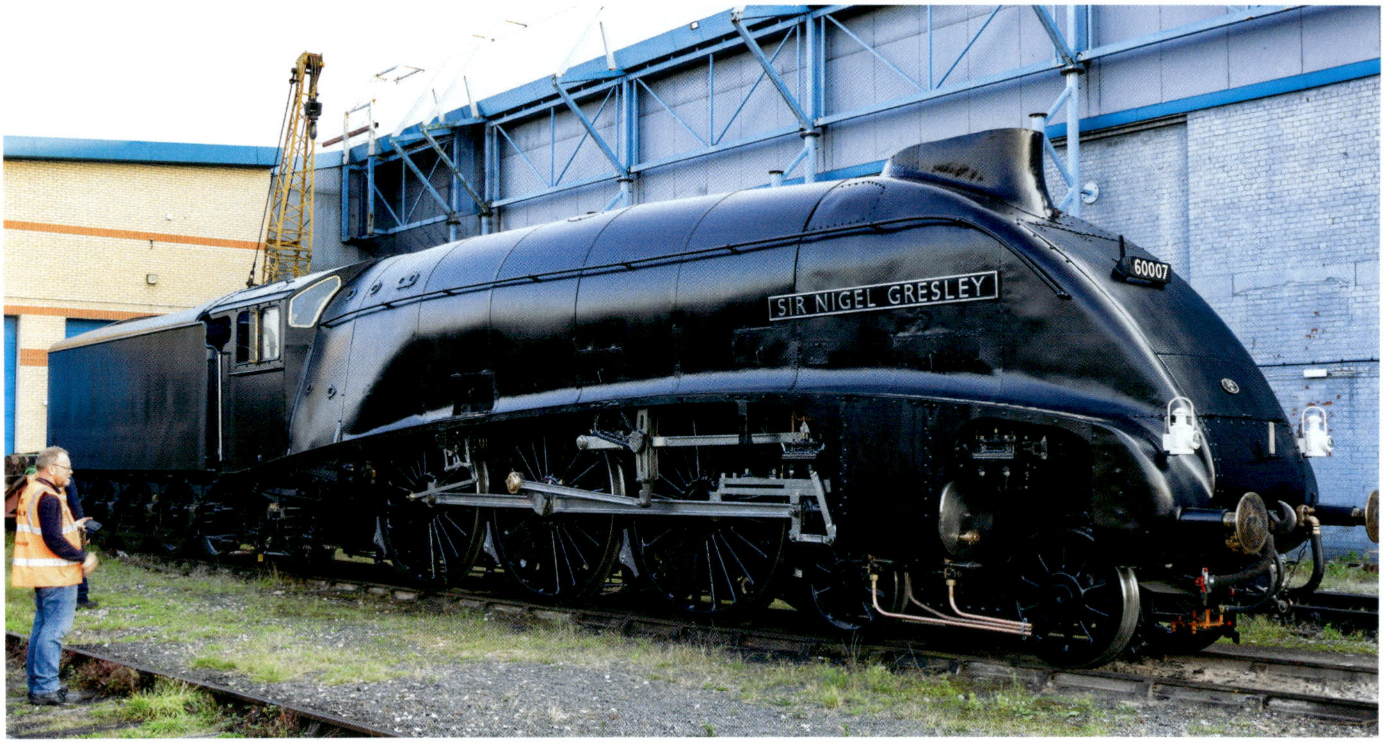

Darrin Crone is a proud man as he gets a proper first look at the reunited loco and tender back out in the daylight

for it to be reunited with the loco, and this was soon done. The museum's exhibits were then shunted back inside and the great doors closed, signalling the late end of the day's proceedings. Not quite the end for me and the team since this was the long awaited photo opportunity and a good hour or so passed as not only my camera shutter clicked frantically, but so did those of many others, including a couple of young enthusiasts who had somehow found out about the unpublicised move and were clicking from behind the trackside safety railings. This brought a smile to my face as it reminded me of my teenage days in the 1960s.

One of my favourite shots was caught here as I captured Darrin standing alone looking at No.7, immersed in his thoughts of the last six years and being clearly very proud of what he was surveying.

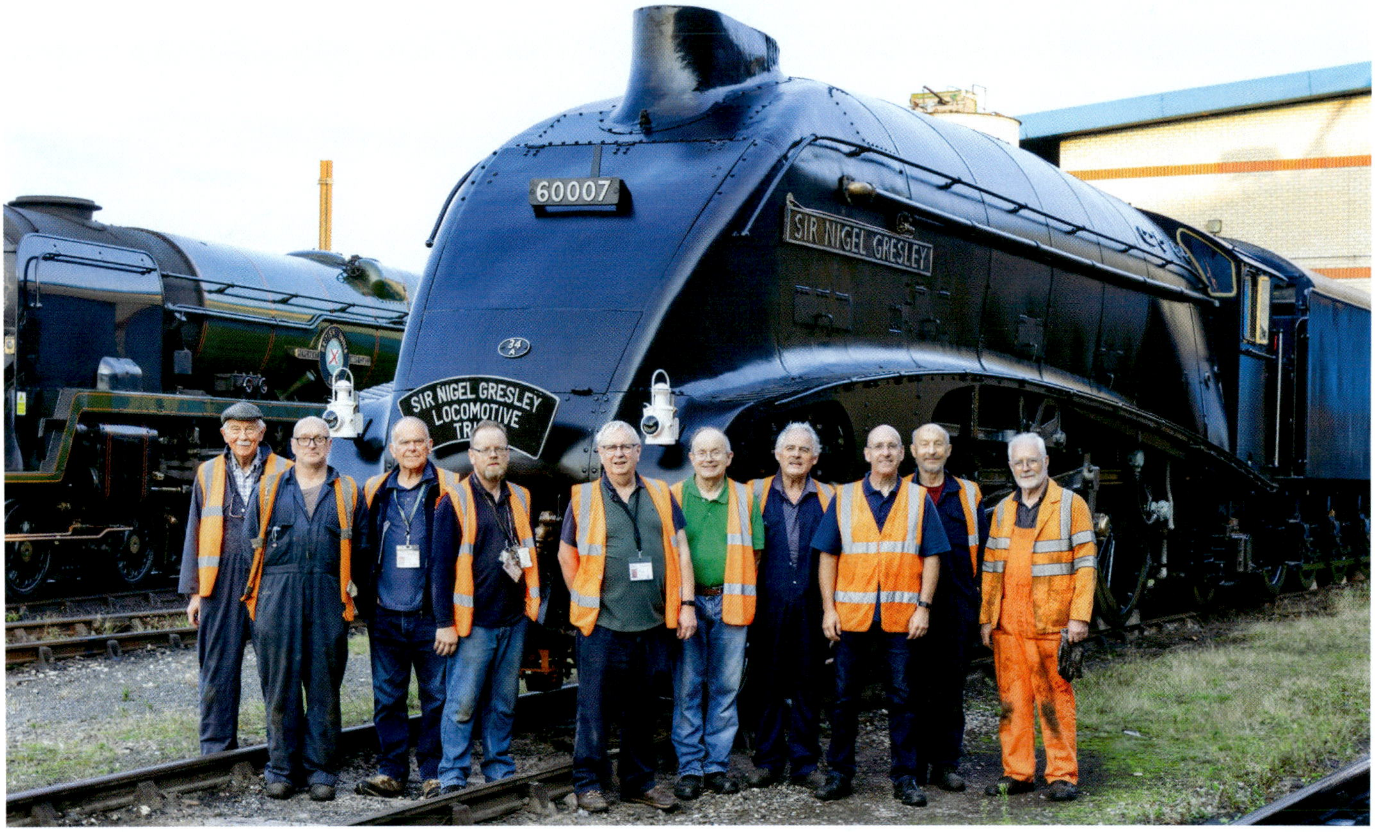

The team on the day of the move. Left to Right - Bob Shearman, Andy Barwick, Peter Brackstone, Darrin Crone, Trevor Camp (Me again!), Richard Swales, Rod Thomas, Barry Horton, Mike Page, Peter James. Merchant Navy 'British India Line' in the background

ROLL OUT

23

August 2021

Now outside with locomotive and tender coupled together, it was time for a mini celebration as, with No.7 looking complete, it was a good time for the team to take the opportunity to get a proper look. In the workshop it was difficult to get a clear view with the overhead cranes, gantries, ladders and scaffold platform all hindering the view. So, Saturday 28th August was set aside as a family day for all the engineering volunteers and their families to come and get a close look, arranged with permission from the NRM, for people to enter No.7's stabling point in the North Yard. Around 60 people attended including the Trust Chairman, Nigel Wilson, and several Trust board members.

The weather in the previous few days had been overcast but bright and dry, which is fairly good for what I knew would be difficult in bright sunlight due to the position No.7 was in. Guess what, today the bright overcast weather was replaced with wall to wall blue sky and blazing sunlight which was shining directly into my lens when I was at the front of the locomotive. Oh well, not much I could do but just get on with it. I took ¾ shots from the front and the back. The images from the back, with the sun now behind me, also posed problems when groups of people were in the shot, as with No.7 painted black the exposure was set for a dark subject, which meant the faces of the people facing me became over exposed. Sounds like I am making excuses. Well I am, as I would have much preferred the filtered even light of the previous few days. The ideal shot would have been square on from the side but unfortunately the perimeter fence was preventing me from standing far enough back for my 24mm lens to get the full length of loco and tender. Another day will surely present itself.

The big change that was causing a great deal of interest was of course the temporary painting in a protection against the elements coat of black paint, which had now had decals applied to the side of the acronym LNER and the number 4498 which were both used during the war years. The opportunity was taken to do this as no A4 had ever appeared looking like this since that period in history. The Trust's membership had voted for a return to the pre-overhaul livery of BR Express Blue and this would be applied in the paintshop of LSL in Crewe prior to the return to traffic.

A great day ensued with Peter (Bracko) Brackstone supplying copious amounts of tea and Bob's wife, Jackie Shearman's cakes.

No.7 was now the last loco to be overhauled in the NRM workshops and the doors had closed on another chapter of history at this famous location. There was still work to be undertaken whilst outside and next week there would also be an inspection from the body responsible for approving the transportation move. A date could then be set for the short move by rail to the York carriage sidings where loco and tender would be transferred to road haulage for the move to Crewe. Yet another adventure was on the horizon.

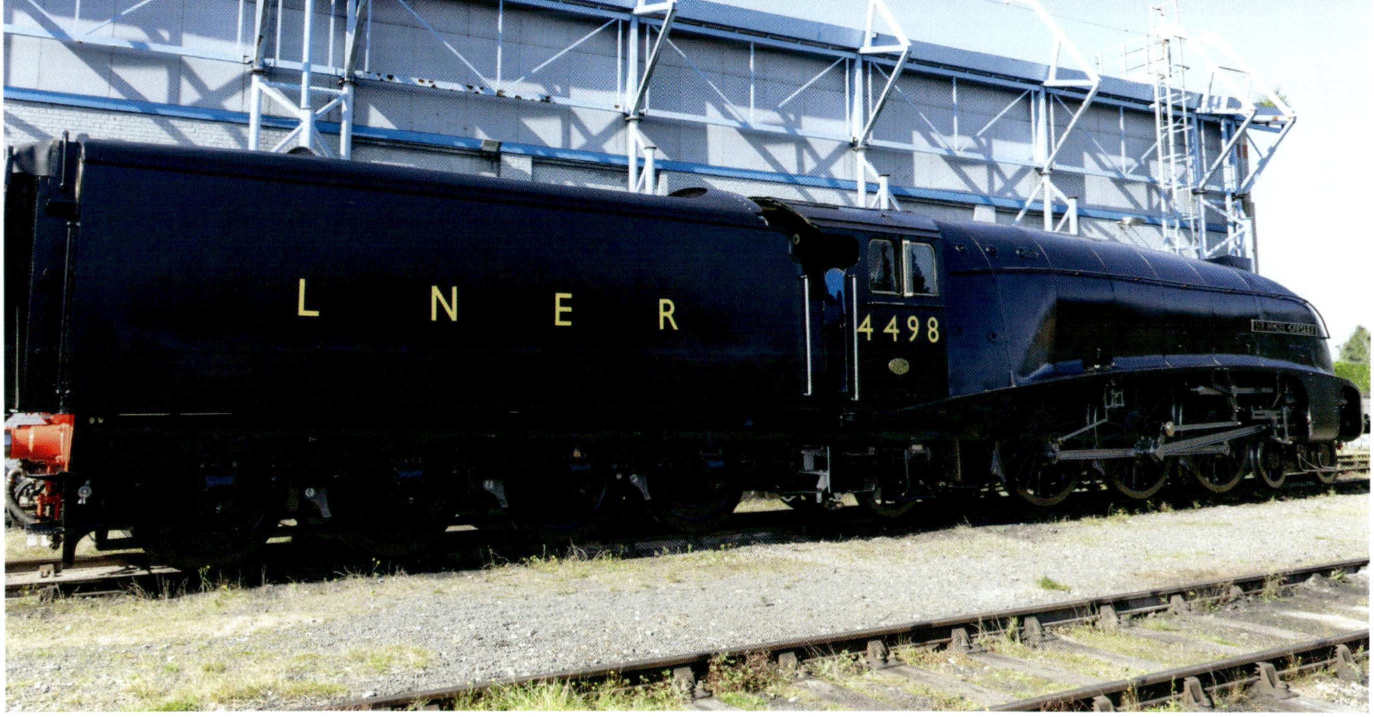

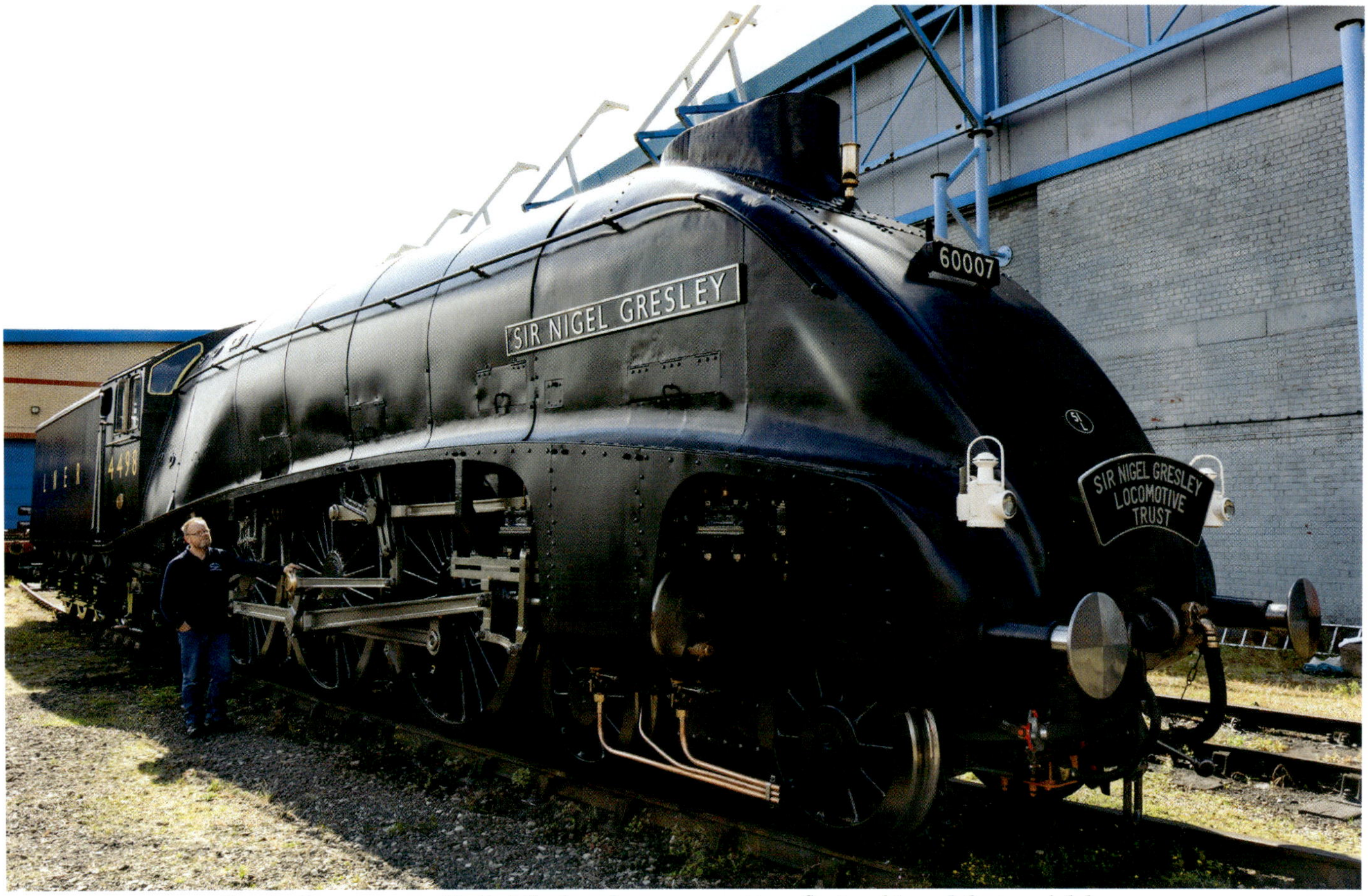

Darrin Crone surveys No.7 whilst he takes a moment to reflect on his and the team's labours

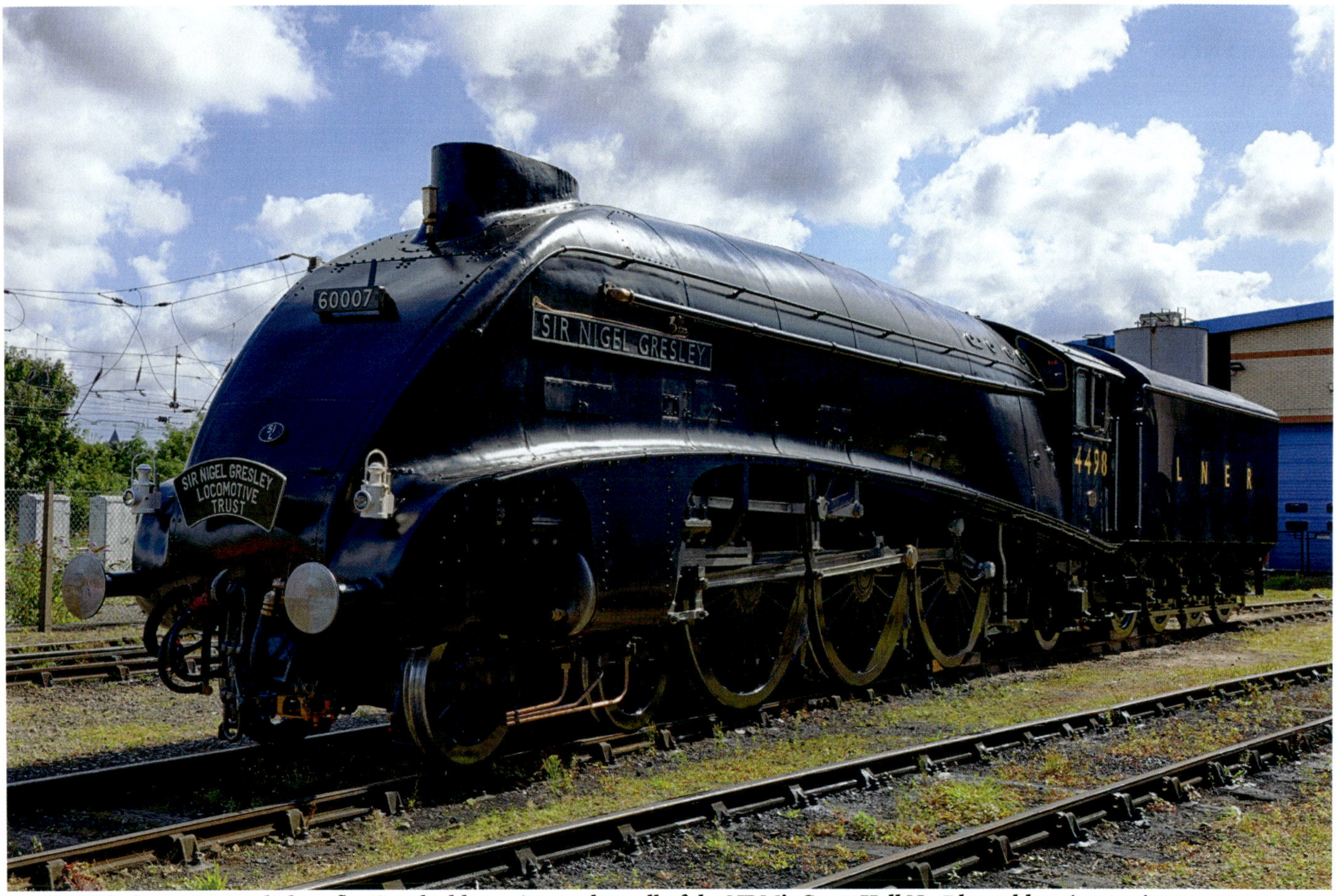

With the strong sunlight reflecting the blue paint on the wall of the NRM's Great Hall No.7 has a blue tinge to its temporary black paint work

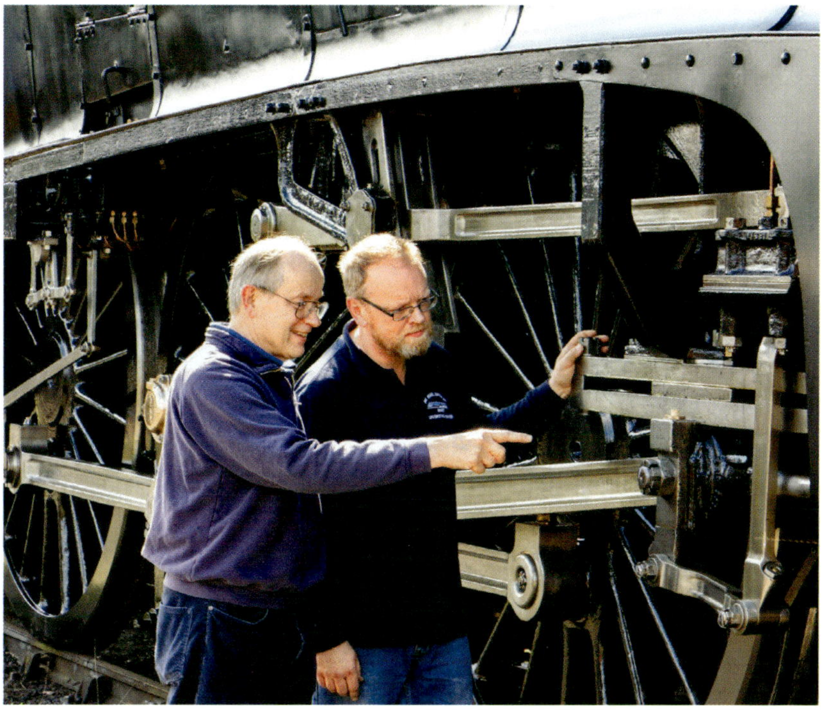

Richard Swales and Darrin Crone share a humorous moment

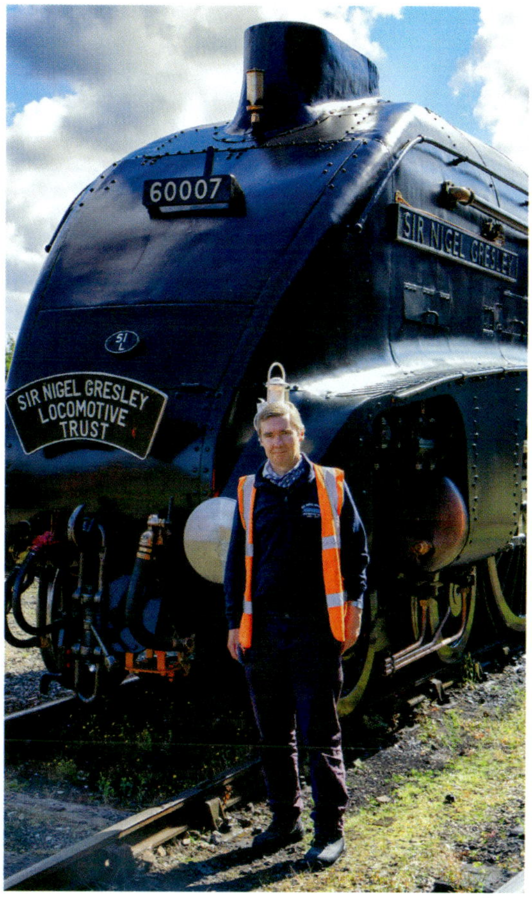

Trust Chairman Nigel Wilson

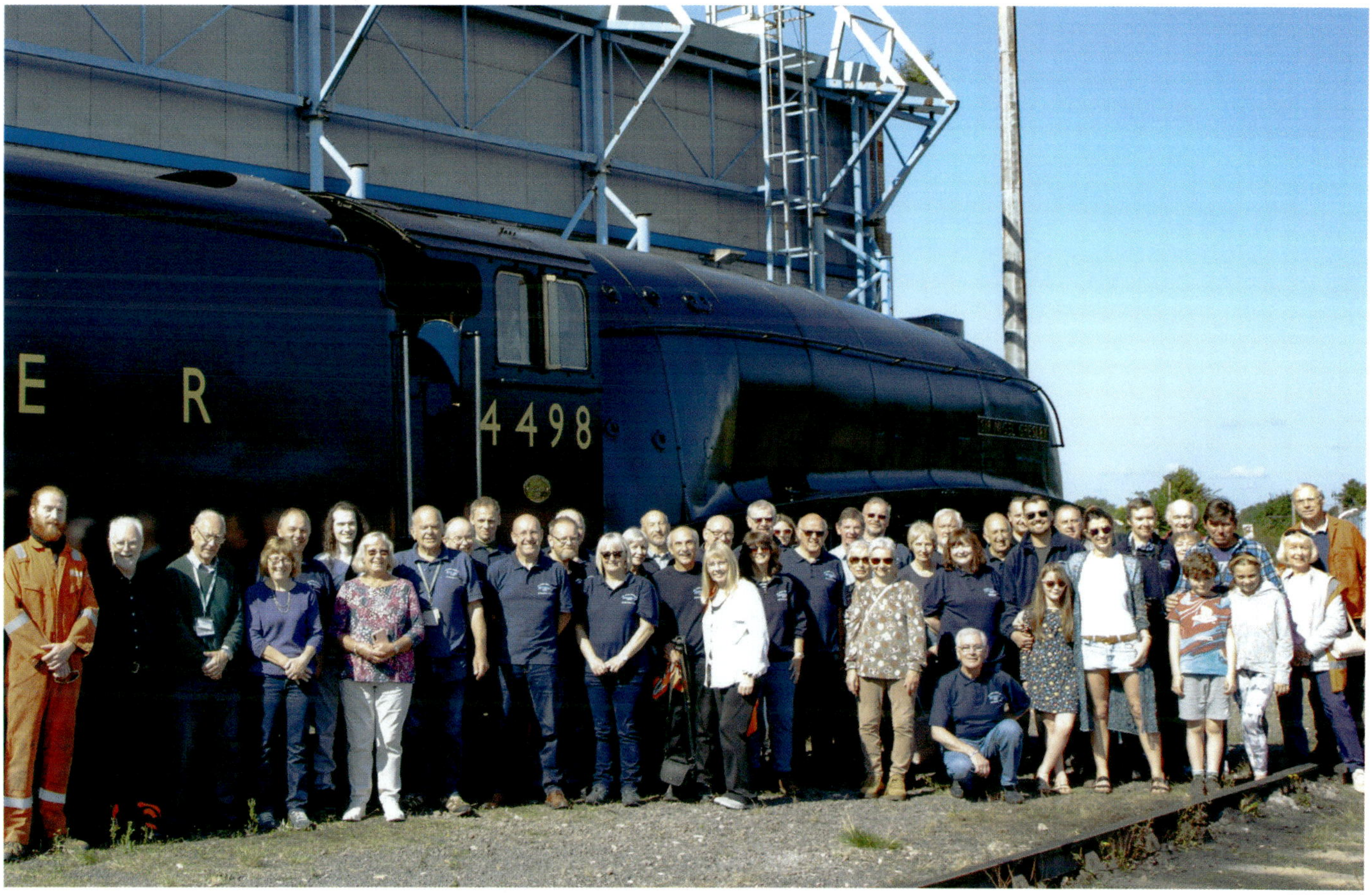

Engineering team with family and friends

SLACK TIME

September 2021

There had been a few high and lows over the last six years since coming out of traffic in September 2015. The biggest low, of course, was the result of the long delay and uncertainty created by the Covid pandemic and the highest high we had just reached with the almost complete locomotive being brought back out of the workshop. With the move to Crewe it was, for some, a farewell, as for various reasons not all of the volunteers would be travelling to Crewe to finish the overhaul.

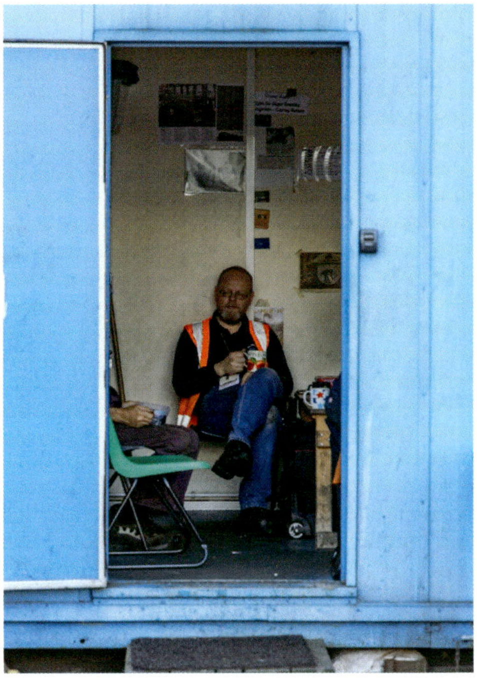

Extended tea breaks in the Blue Cabin whilst a date for the rail move was awaited

Bringing the team together for a photo day had resulted in lots of print orders from the team for me to sort out, keeping me quite busy at home after a few weeks of several trips to York. It also resulted in a big media interest from online and print organisations whom I supplied with images for publication. Amongst the team there were six who had been members since before the move from Grosmont in 2015 and were still in the team. They became known as 'The Stayers', and now with all this slack time there was a chance to bring them all together for some group pictures, something we weren't able to do on the photo day as Bob Shearman was missing due to prior commitments.

As we moved into the second week of September, we still had not got the go ahead for the move to Crewe. This was a low for a frustrated Darrin and the remaining team as they were keen to get on with the work and every day idle at York added a day to the return to traffic date.

Whilst this delay was a low point, I had a high with a phone call telling me that one of my images had made the front cover of a top selling railway magazine - I was now looking forward to a copy falling through my letterbox.

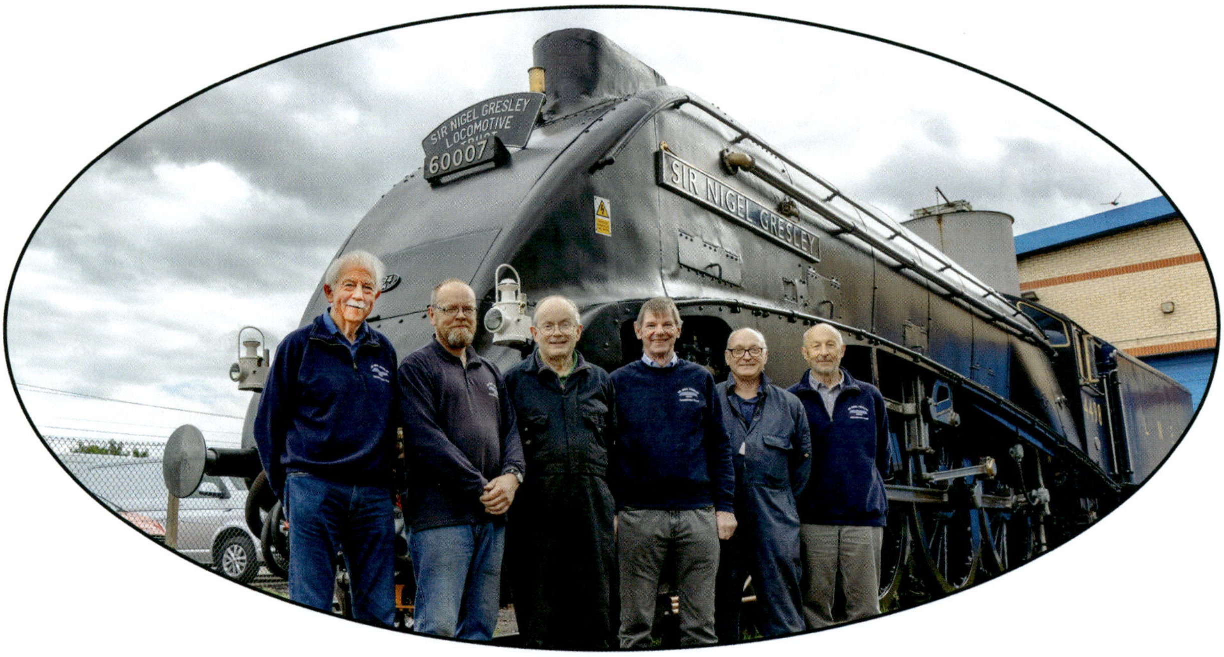

'The Stayers'

Sir Nigel Gresley Locomotive Trust

Left to Right - Bob Shearman, Darrin Crone, Richard Swales, Malc Bateman, Andy Barwick, Mike Page

The six Trust members who worked through all of the overhaul in York's NRM workshop 2015-2021

24 THE CONVOY

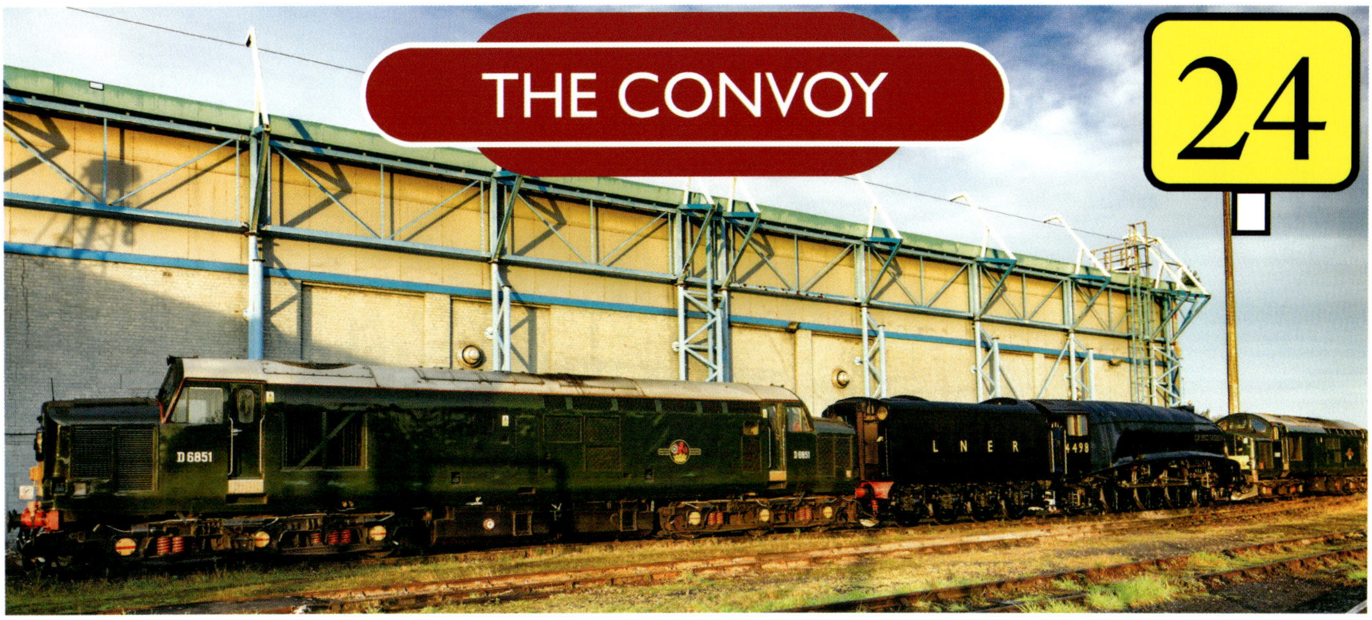

September 2021

After almost a month parked outside in the NRM North Yard, we finally got all the boxes ticked with the various other parties for the rail move out of the NRM to the Holgate Engineering carriage works, from where the road transport would take over.

Now, some of you will be wondering why we needed a rail move of a couple of miles if we were going to Crewe by road. Why can't we just go by road? The answer is, the low bridges that pass under the East Coast Mainline (ECML) at each end of Leeman Road in York, so we have to leave the yard by rail. Next question - if leaving by rail why can't we go all the way to Crewe by rail? There are several reasons, but in essence No.7 is still not complete so is not ready for a long rail journey. However the short move across York will have been very helpful to Darrin and Richard.

I got the call from Darrin that the diesel transport for the rail move had been booked for Monday 20th September and sure enough it soon popped up on Real Time Trains website. Darrin added that they would be in the yard early evening for the move to happen in the early hours. Oh dear, I totally understand the reason for an overnight move particularly along a very busy daytime ECML, but it would be challenging taking photographs in the dark. So I planned to get shots as the convoy passed through York station and get some help from the station lighting. That is of course if the convoy went via that route as there are two ways to get to the carriage works. Whatever happened and whichever picture I got, I had to be there, so I accepted the challenges. However, Darrin's next phone call raised my hopes of getting a decent image, as the move was going to happen at 6am and it's late September, so the sun will just about be coming over the horizon, meaning some light will be available. This is getting better. Finally came a third call - the LSL driver had told Darrin he would be onsite at 08:45 am, so this now this put a smile on my face, as I would have proper daylight to work with and, as an added bonus, a cloud free sunny morning was expected.

The morning duly arrived and I was up, yawning, at 05:30 am and arrived in York at 07:30 am. The forecast wasn't wrong, as it was a beautiful sunny morning and the low sunlight was creating wonderful light in what photographers call the 'golden hour'. Darrin and Richard arrived at the same time as me and Andy Barwick soon joined us, with the two LSL drivers arriving a little later.

As the move required a reversal, LSL had provided two Class 37 locomotives to top and tail No.7, in order to avoid time consuming run arounds which wouldn't be well received on a busy ECML. This convoy of three made a striking image in the morning sunlight and I was particularly grateful that there were no other rail vehicles in the yard obstructing the view.

Now the disappointment. As I mentioned above, there are two ways to reach the carriage works and I was hoping for the move through the station as I reckoned I could take a picture of the convoy leaving the yard, and then stretch my legs and be on the footbridge at the station in time to get a

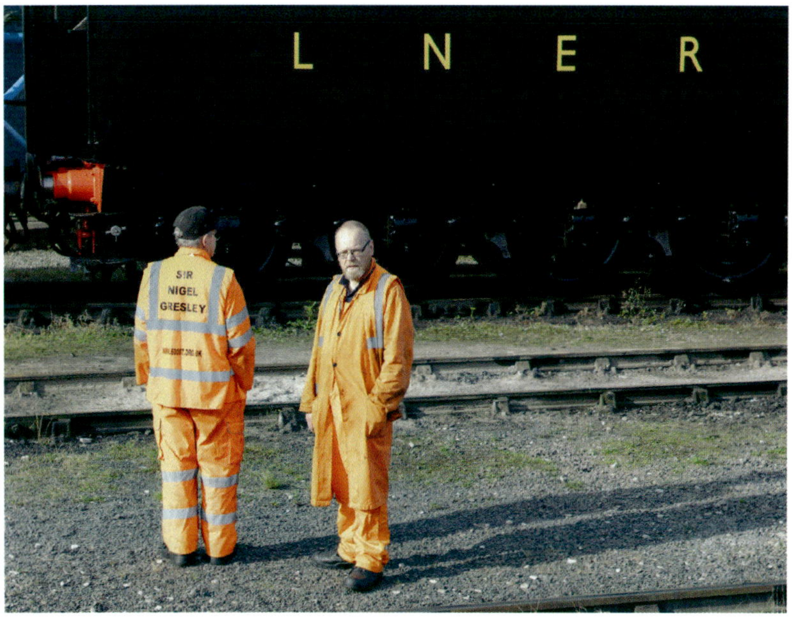

Richard Swales and Darrin Crone wait for the moment to climb aboard No.7 for the journey away from the NRM

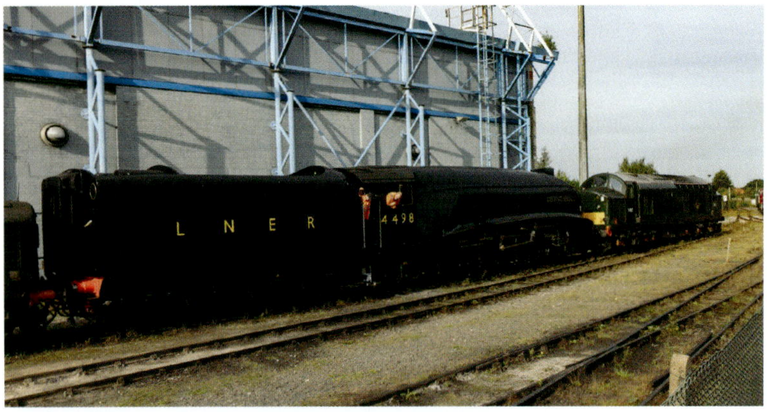

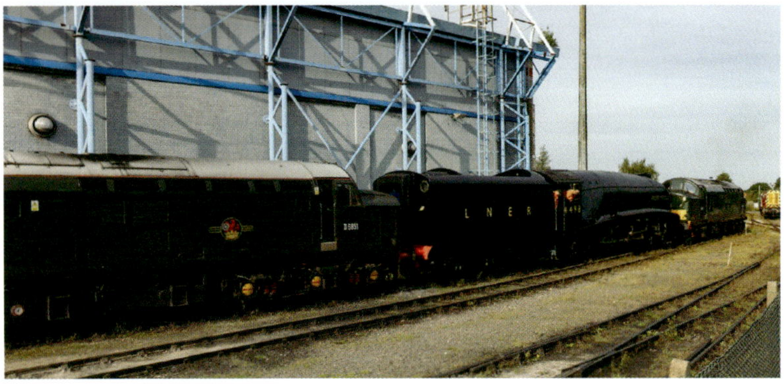

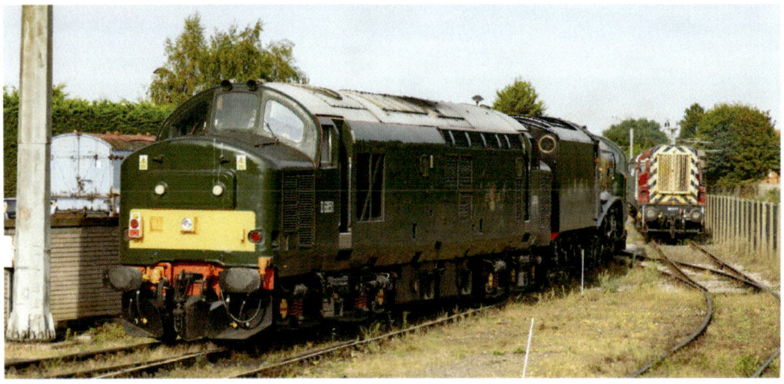

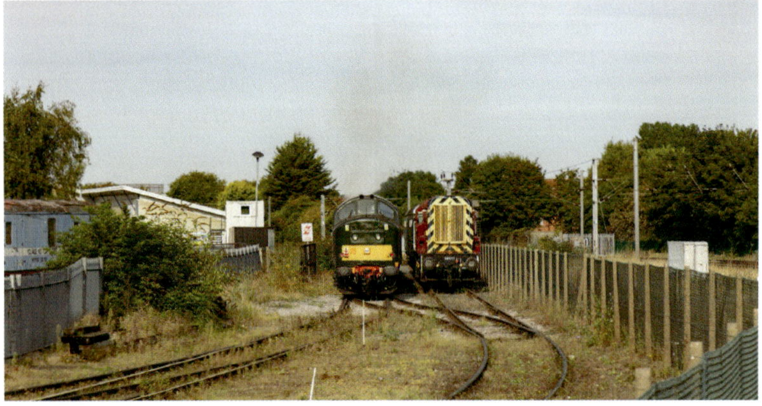

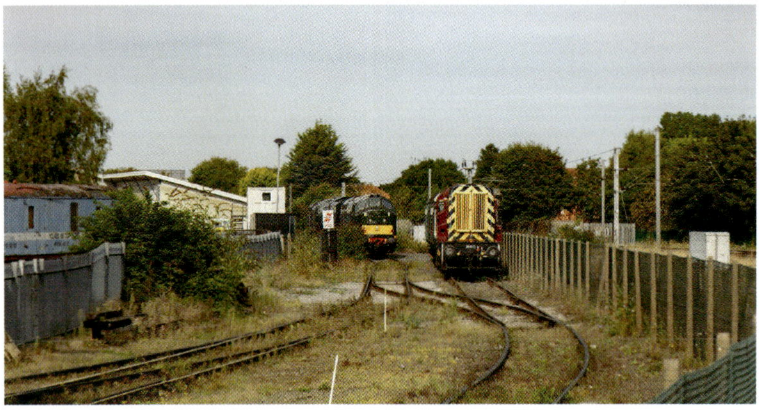

second shot of it passing through the iconic York train shed. But my hopes were dashed, and again, I totally understand the reasons. The other route via Skelton Junction was to be used as it only involved one reversal, not the three required on the station route, and also most of the journey would avoid running on the ECML.

It was what it was, so I had to settle for getting what I could, and with the help of Andy and Richard, who brought out some steps to give me a better viewpoint, I took up my position and readied my camera. I couldn't be too disapointed, as I had the best seat in the house for taking pictures and there were a handful of other snappers on the other side of the yard who had to settle for shots which offered them poor angles and the challenge of shooting through the railings.

The time had come; the diesels were running and the two drivers climbed onboard their respective locomotives. Darrin was joined onboard No.7 by Richard and Andy and after a very short wait, the convoy slipped away out of the yard with the noise of my camera shutter slowly rising above the fading noise of the two English Electric Type 3s (Class 37s). Sadly, all this was done without great fanfare which didn't seem fitting as this was the last locomotive ever to be overhauled in this famous and historically important workshop. And that was it. After the best part of six years No.7 had left the NRM. Besides the handful of enthusiasts behind the fencing on the far side of the yard, the only witnesses to the event were myself and a member of the NRM Rail Ops team. I walked back to the blue cabin as some geese, flying south for the winter, broke the now silent North Yard with their honking. After Darrin sent me a text to confirm that the signallers had routed the convoy via Skelton junction and I definitely wouldn't get my station shot, all that was left for me to do was to collect my gear together from the Blue Cabin and head on home. I wasn't going to the carriage works today. I would get the images there when the road move took place.

I know it is not remotely the same, but as I closed the Blue Cabin door behind me when I left, it reminded me of the days when I worked at the Sunderland Echo and I saw the image of the sign on the door when Monkwearmouth Colliery closed, (now the site of Sunderland AFC football ground, The Stadium of Light) - "Last Man Out Turn The Lights Off". So I guess I felt a little bit sad walking away from the NRM, as it had been a second home for a number of years now. Still, Crewe was on the horizon so it was onwards and upwards as another major step had been taken towards the return to traffic.

Sequence of images showing the convoy departing the NRM's North Yard for Holgate Engineering where No.7 will be transferred to road haulage and continue the journey to Crewe

"TRANSPORTE EXCEPCIONAL" 25

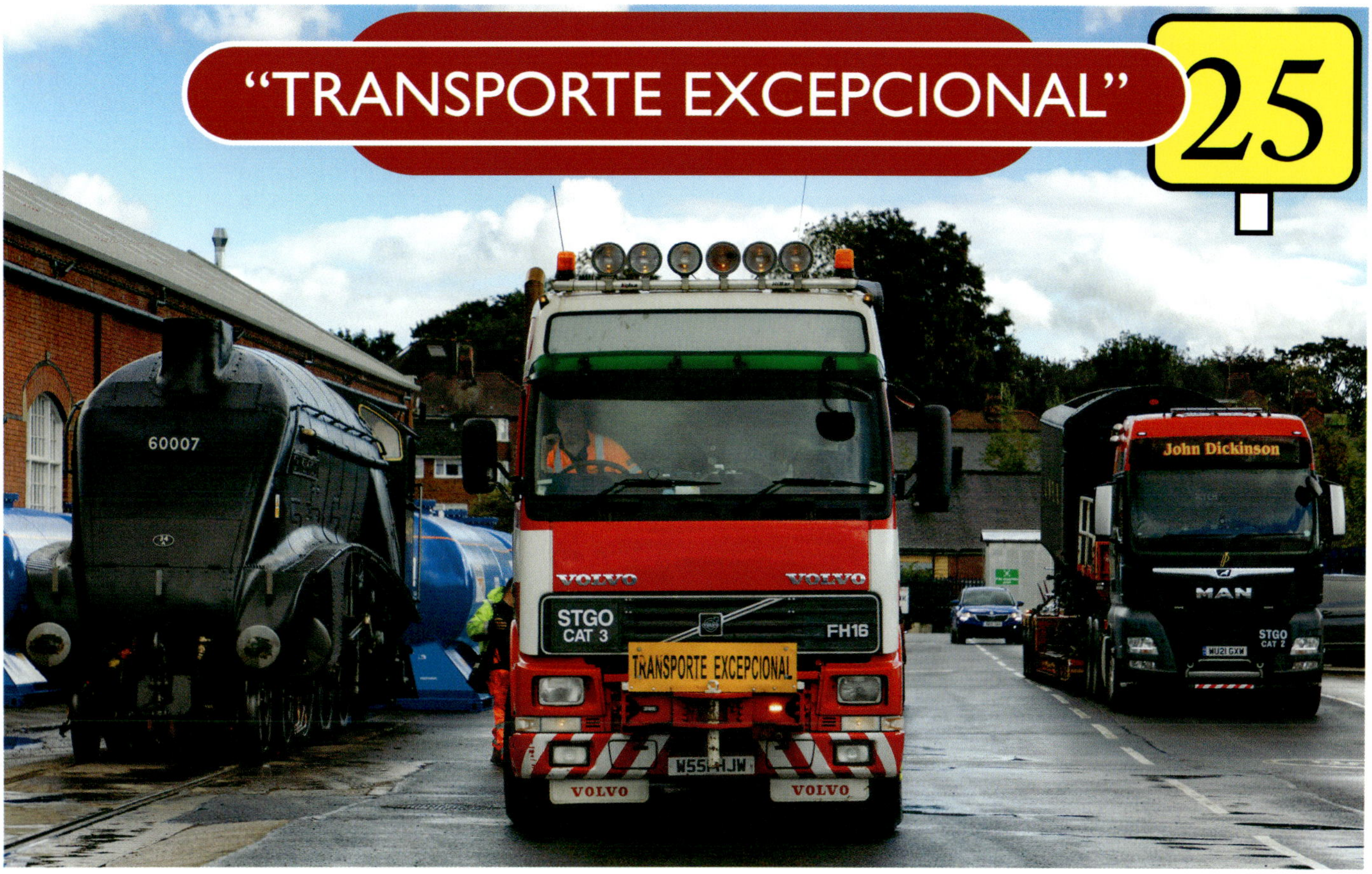

September 2021

Having completed the short rail move from the NRM to Network Rail Holgate Engineering we expected to continue the journey to LSL in Crewe a couple of days later, but there was a delay as one of the transports was stuck on the wrong side of the North Sea and we were postponed a couple of days more. Another 6am start for me and the team as we had agreed to meet up at 8am. Darrin and Richard Swales were joined by Bob Shearman, Andy Barwick, and Tony Bickerstaffe, who formed Darrin's hand picked crew for the job. Peter Ellis came along to watch the day's events as his good lady had told him he had to because she didn't want him under her feet all day!

The forecast was not good, and it was unfortunately correct, as our arrival at Holgate coincided with a cloud burst. It was the type of rain that you would normally associate with water coming out of a shower head, so definitely a day for waterproof clothing, or maybe not as it turned out. The transports had been delayed so we were able to take cover until they arrived late morning, by which time the weather was clearing and Bob, formerly a weatherman, assured us that the front had now passed through so it would get out sunny with maybe a light shower, which it proved to be as this weatherman actually got it right.

With both transports now on site the tender would be

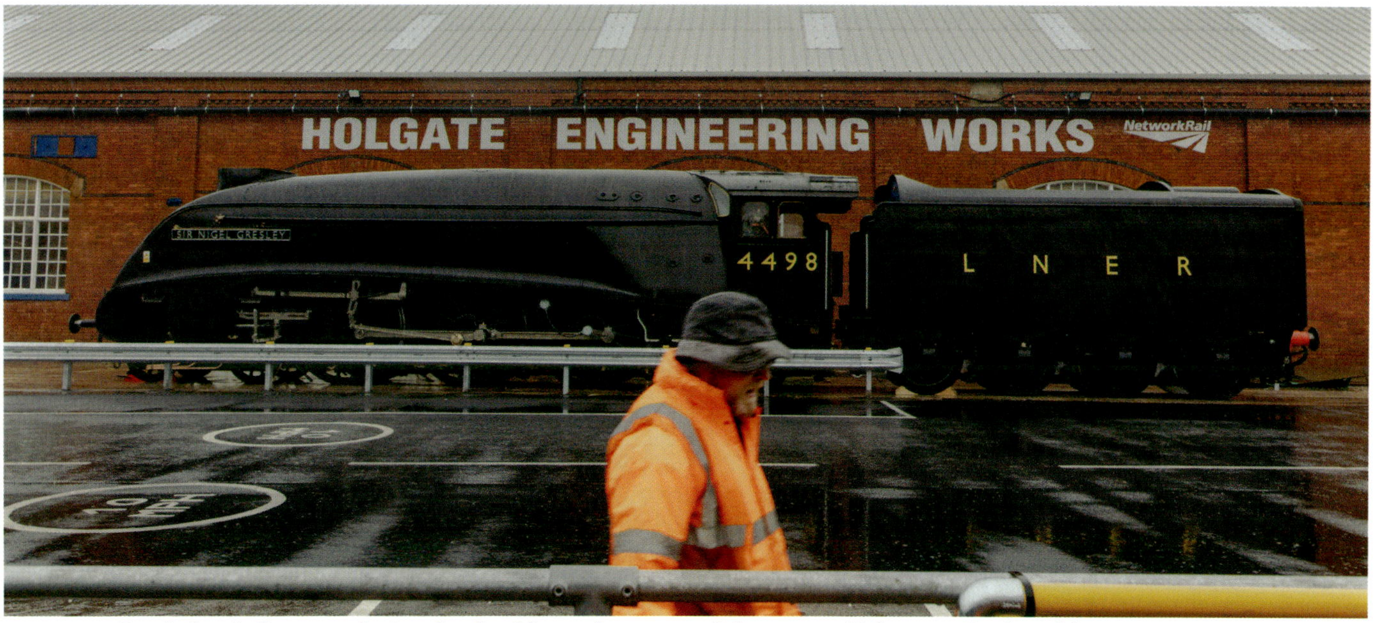

Darrin heads for cover during the cloud burst that greeeted the team at Holgate on the day of the move to Crewe

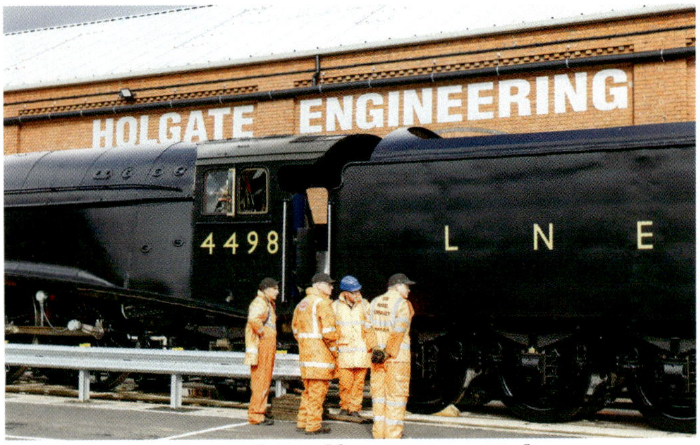

The tender and loco are separated

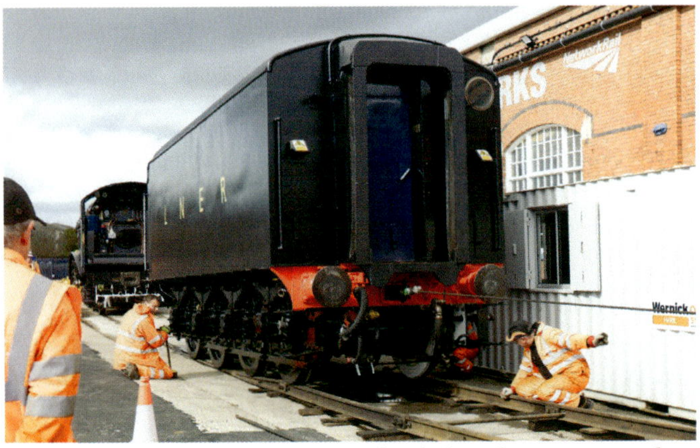

Slowly and carefully the tender is hauled up the ramp onto the trailer

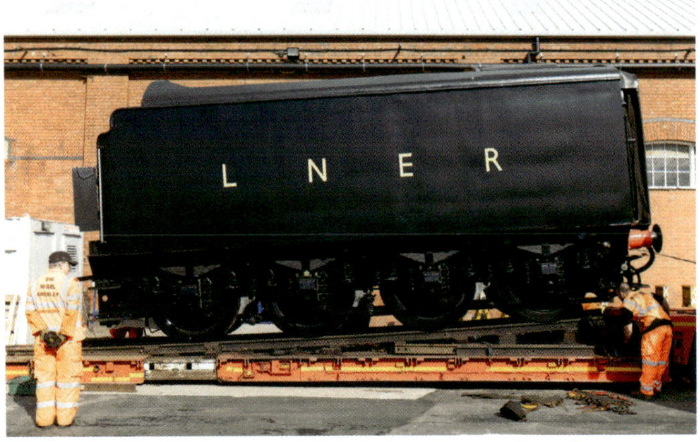

The tender in position on the transport's trailer

loaded onto the smaller one first. I say smaller in relative terms as these transports were huge, the trailer for the loco having 13 axles with 104 wheels and the total length of the tractor and trailer together a colossal 45 metres, that's just a couple of strides short of half the length of an athletics track straight. Standing with Andy staring at this enormous unit he jokingly commented "That would be a top of the range Volvo then." These units needed a lot of space, space for the tractor and trailer and also space to manoeuvre them into careful alignment with the rail tracks. This was completed with excellent precision by the driver and his team using radios and also aided by the fact that the trailer could be steered. The shorter loading ramp for the tender was then assembled and checked for gauge. We certainly did not want a derailment here, and with the tractor's powerful diesels and winch the tender was carefully hauled on to the trailer. The tender actually travels on this trailer at the angle at which it is loaded, so after the RSS team confirmed the tender's brakes were firmly on, it was secured in position and the tractor and trailer moved down the yard. The much larger tractor and trailer for the loco was then brought into position and the larger 22 metre loading ramp was assembled. Everything had gone well so far so now for the loco. The procedure was repeated and with the sun now out and glorious blue sky above, spirits were high. There had to be a problem of course, but we had problem solving Darrin in our team. During the haul onto the ramp No.7, like a child not wanting to go into a car seat, decided she would apply her brakes and stubbornly came to a halt on the ramp. Bob explained what had happened. When the first drive wheels started to ascend the incline of the ramp, which was steeper than you would ever get on a normal railway, the second wheelset dropped in the axle horns to a point that meant the brakes would come on. The solution was to free the brake blocks, which meant that Darrin had to climb underneath and haul himself forward into a position to do so using a 4" spanner, except the required spanner was in a toolbox which had just disappeared out of the yard in the departing tender. To turn the adjustment nuts free it was time for a big hammer to replace the missing spanner and working in the tightest of spaces wielding a very large one, Darrin came to the rescue and the brakes were soon released. With Darrin

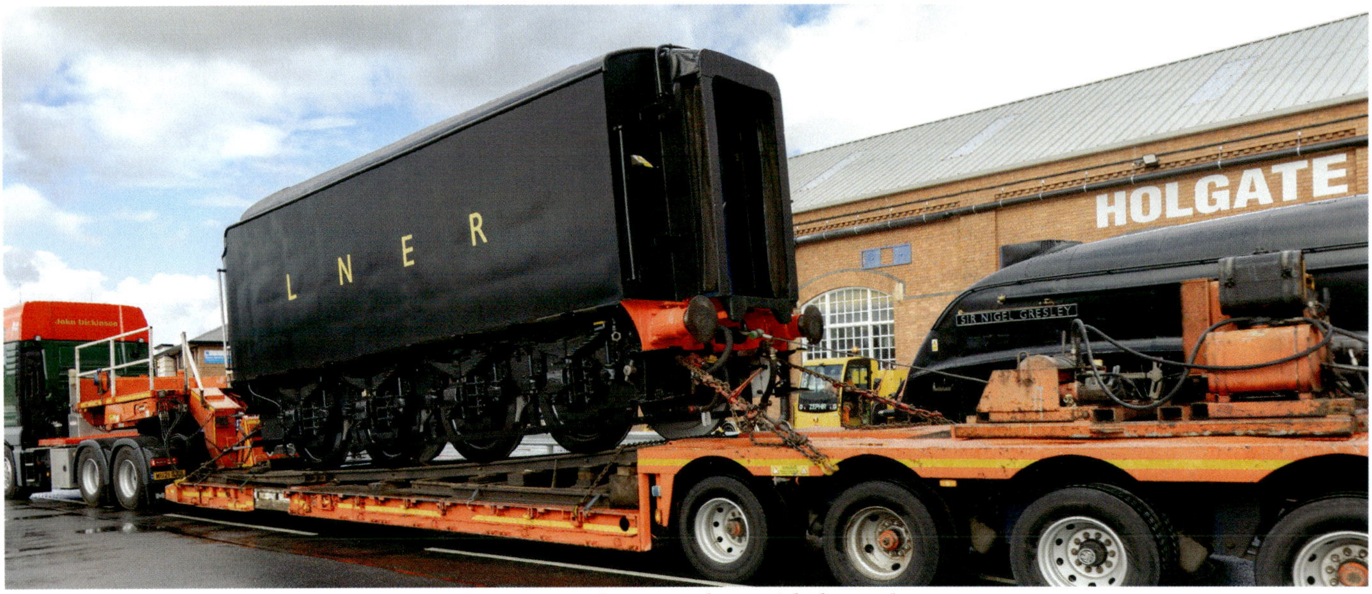

The transport departs Holgate with the tender

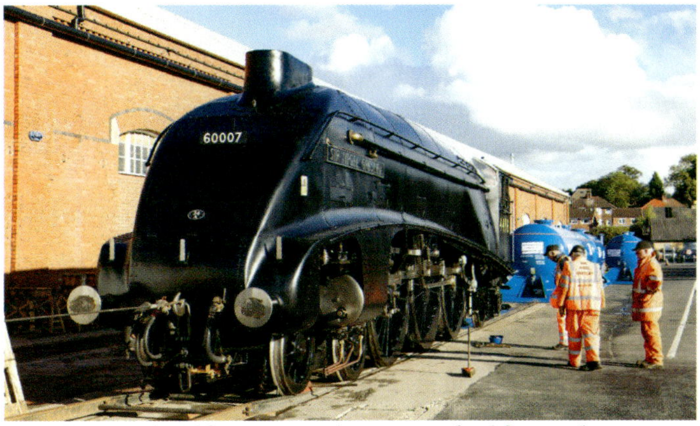

With cable connected No.7 is inched forward towards the transport

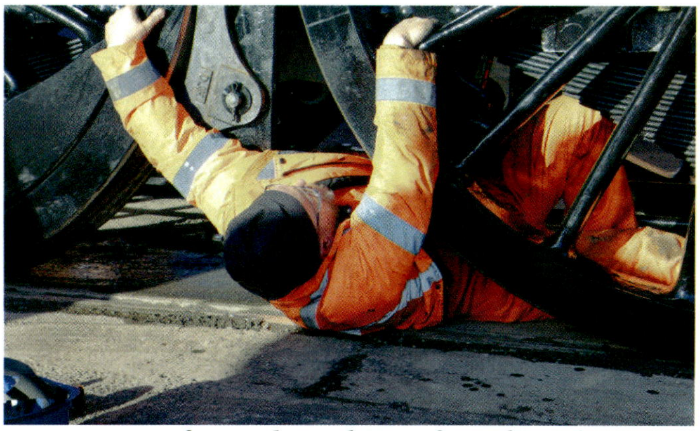

Darrin emerges from underneath No.7 after making adjustments to the brakes

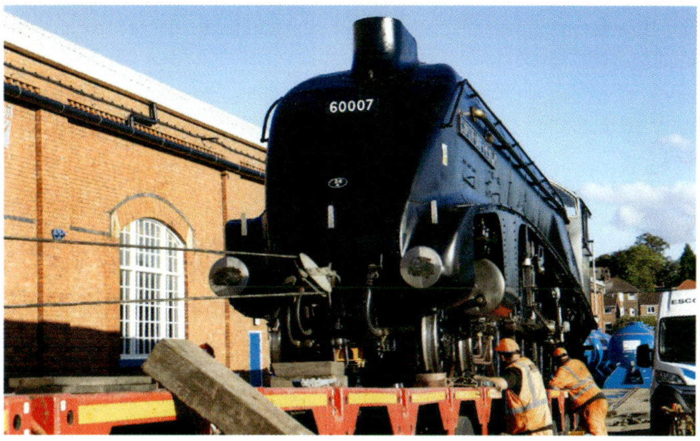

No.7 rises up the ramp onto the huge trailer

"That would be a top of the range Volvo then"

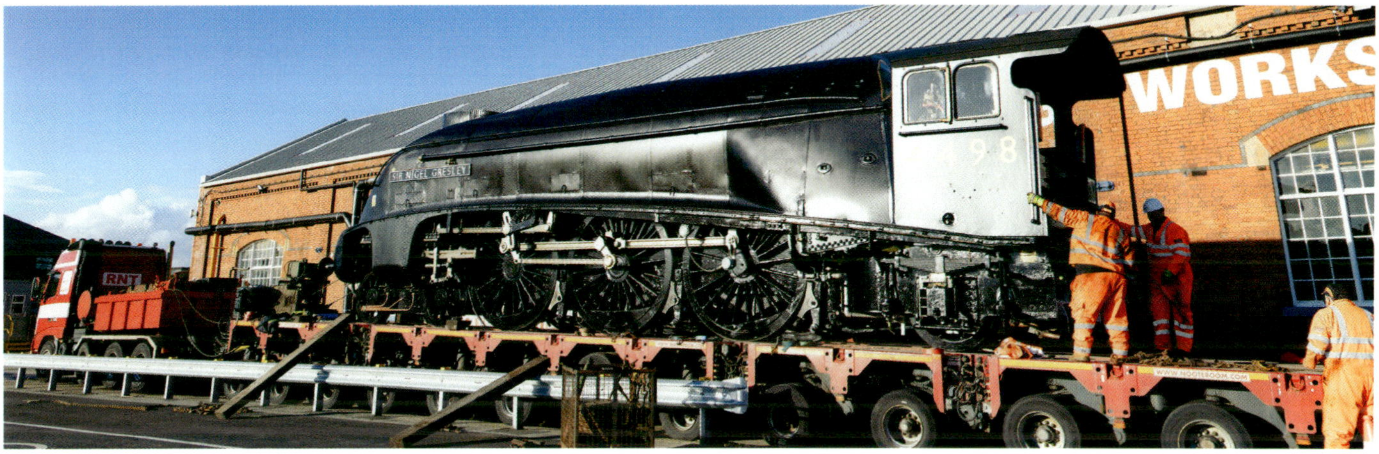

Safely in position on the trailer

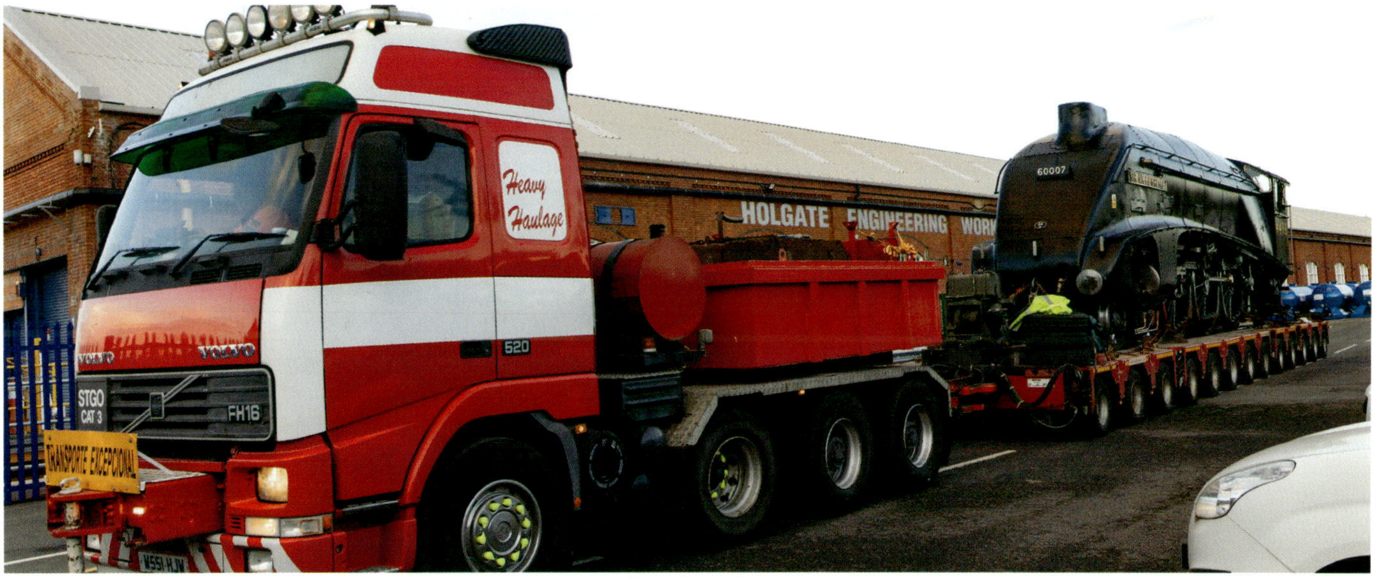

No.7 departs Network Rail's Holgate Engineering yard

extricating himself from his claustrophobic location the haul continued with no other problems. Unlike the tender trailer this trailer had a hydraulic track bed, and the loco could be lowered back to a level elevation before being secured in position.

Day 1 of the road move was finished. The lorries would now park up and continue their journey to Crewe tomorrow where Darrin, Richard, Tony, and myself would be waiting on their arrival, another long day for all of us was in store. Now it was off home for a few hours sleep.

The following day the loco and tender left York around 4am and they were expected to arrive in Crewe around 1pm, so Darrin and Richard arrived in one car around 10:30am and Tony and I in another arrived an hour or so later, which gave us time to go through the LSL induction training for this vast and hugely impressive locomotive depot. The weather had once again deteriorated into very heavy and squally rain but having been allocated an office for the day we were warm and dry as we awaited the lorries' arrival. They had been delayed as the weather was hampering their progress but yet again like yesterday at Holgate Engineering, that delay did us a favour as when they did arrive late afternoon the rain had stopped and besides we were being amused by Tony's wonderful story telling.

Now, you will recall much earlier in my story how impressed I was when the driver who took the boiler to Llangollen reversed off a bridge around a 90 degree T junction and a few hundred yards to the depot at Llangollen, well our drivers today definitely top trumped that manoeuvre. The entrance to LSL comes off the road bridge over the West Coast Mainline next to the railway station in Crewe and again it is at 90 degrees to the road. This was some manoeuvre, particularly as after the corner there was around 300 yards of reversal, the start of which is downhill and although, relatively, a straight road it has a few kinks and is not particularly wide in places. I take my hard hat off

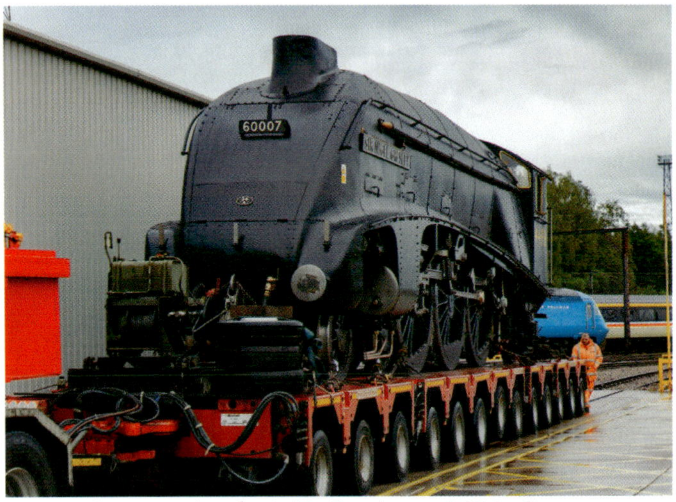

In torrential rain No.7 arrives at LSL in Crewe

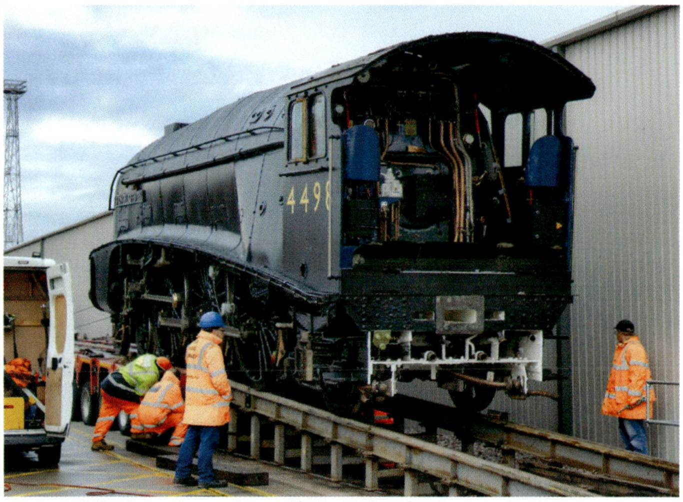
With the rain easing off No.7 slowly descends the ramp

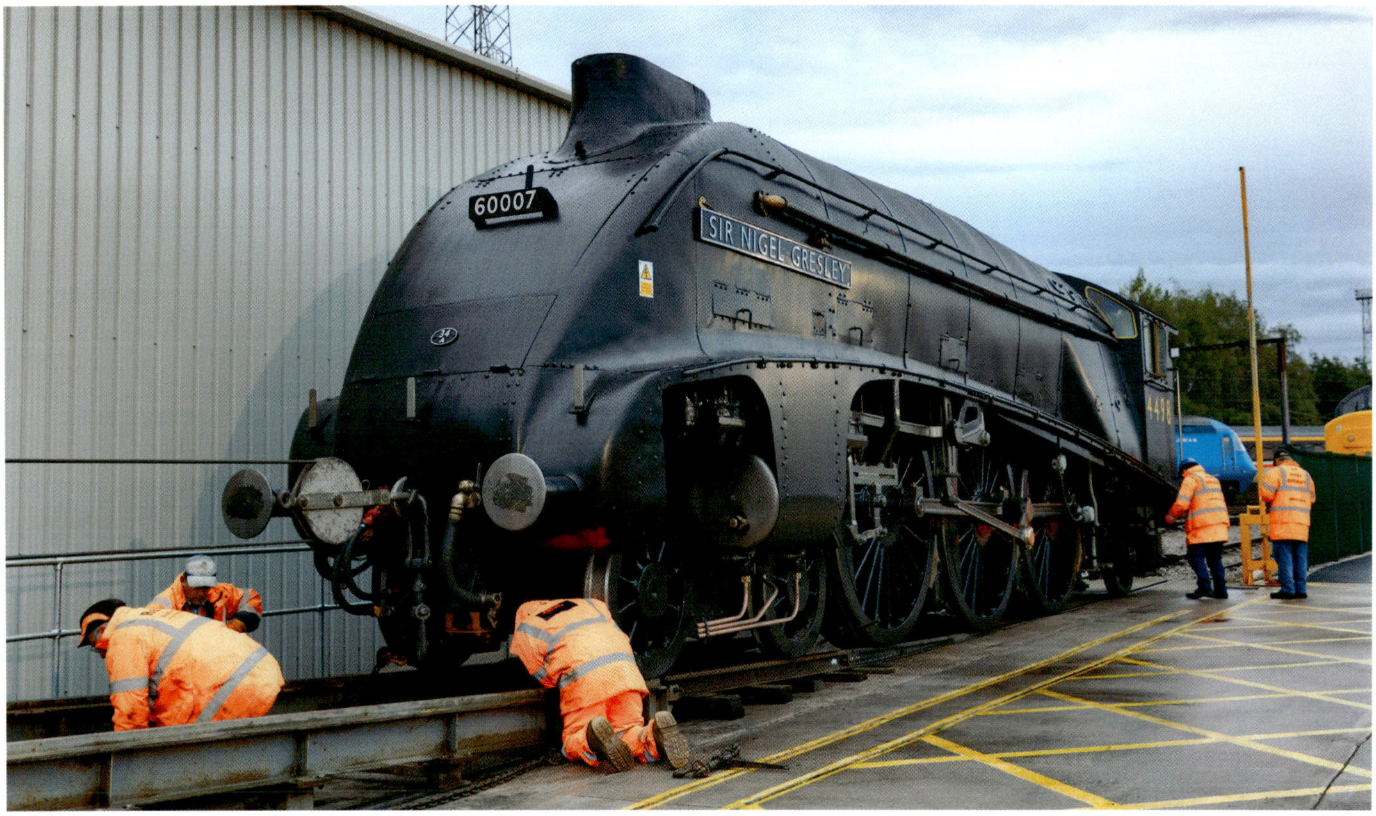

Just a few feet to go and No.7 will be safely on the rail tracks at LSL in Crewe

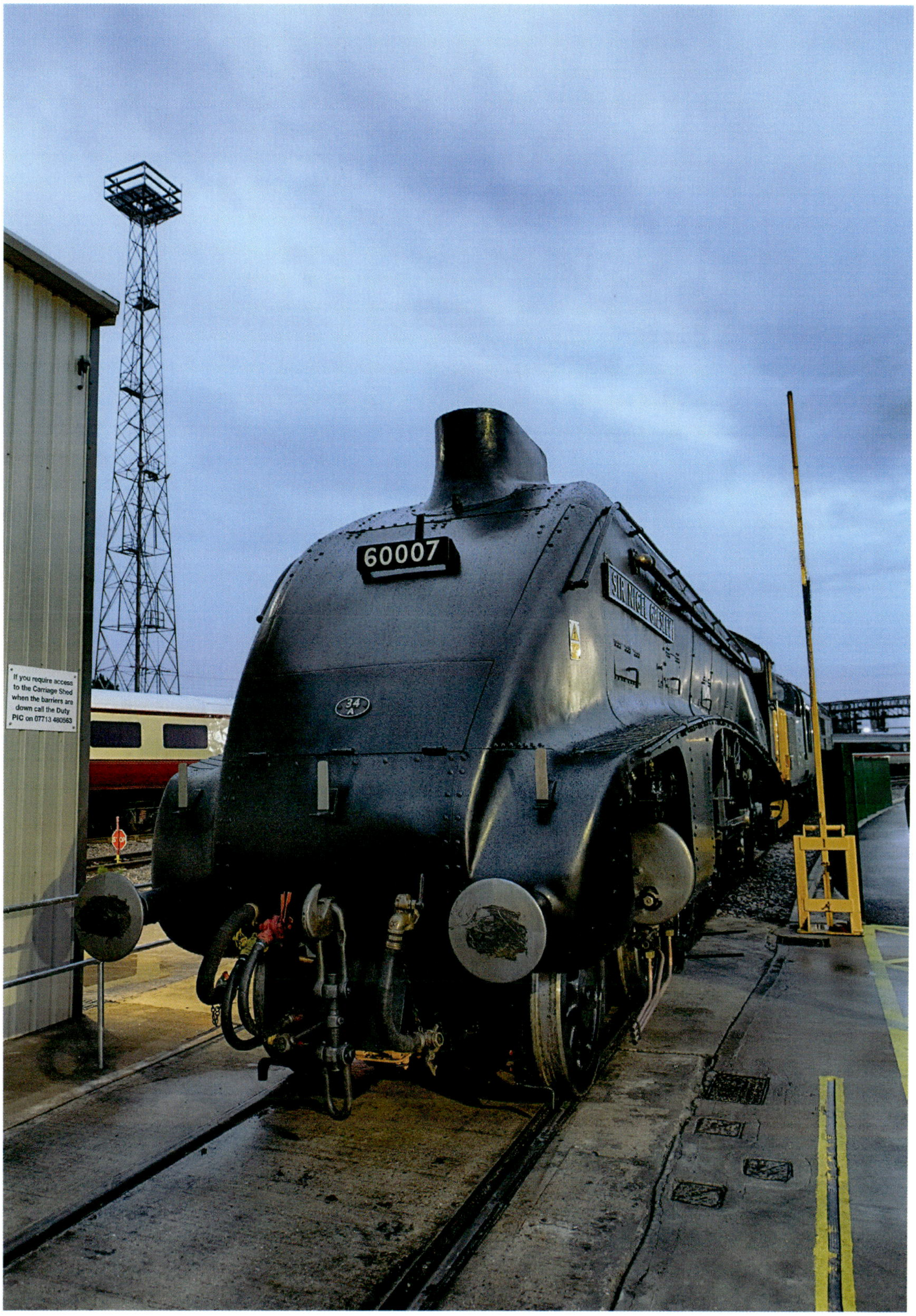

In the gathering gloom an LSL Class 37 hauls No.7 clear of the road transporter

to the RSS drivers and their support team on the radios. Unfortunately, I don't have a picture to show you these manoeuvres as I was waiting at the arrival point.

The loading procedure in reverse now took place, with the loco being unloaded first, and this all went off without any problems. In the early evening autumnal gloom, it was starting to get very dark for photography even though the depot's floodlights were slowly switching themselves on to help. So after consulting with Darrin, it was agreed Tony and I would depart for home and Darrin and Richard would wait for the tender to be unloaded before they also left. Ben Roberts, who was the LSL man in charge of the unloading, had said that they would leave the loco and tender at the unloading location and shunt them into the workshop the following day.

Another long day was coming to an end and as Tony drove us back across the Pennines on the M62 my thoughts drifted to how well things had turned out and what brilliant relationships we had developed with everyone we had met from Network Rail, RSS and LSL.

TIME WARP

October 2021

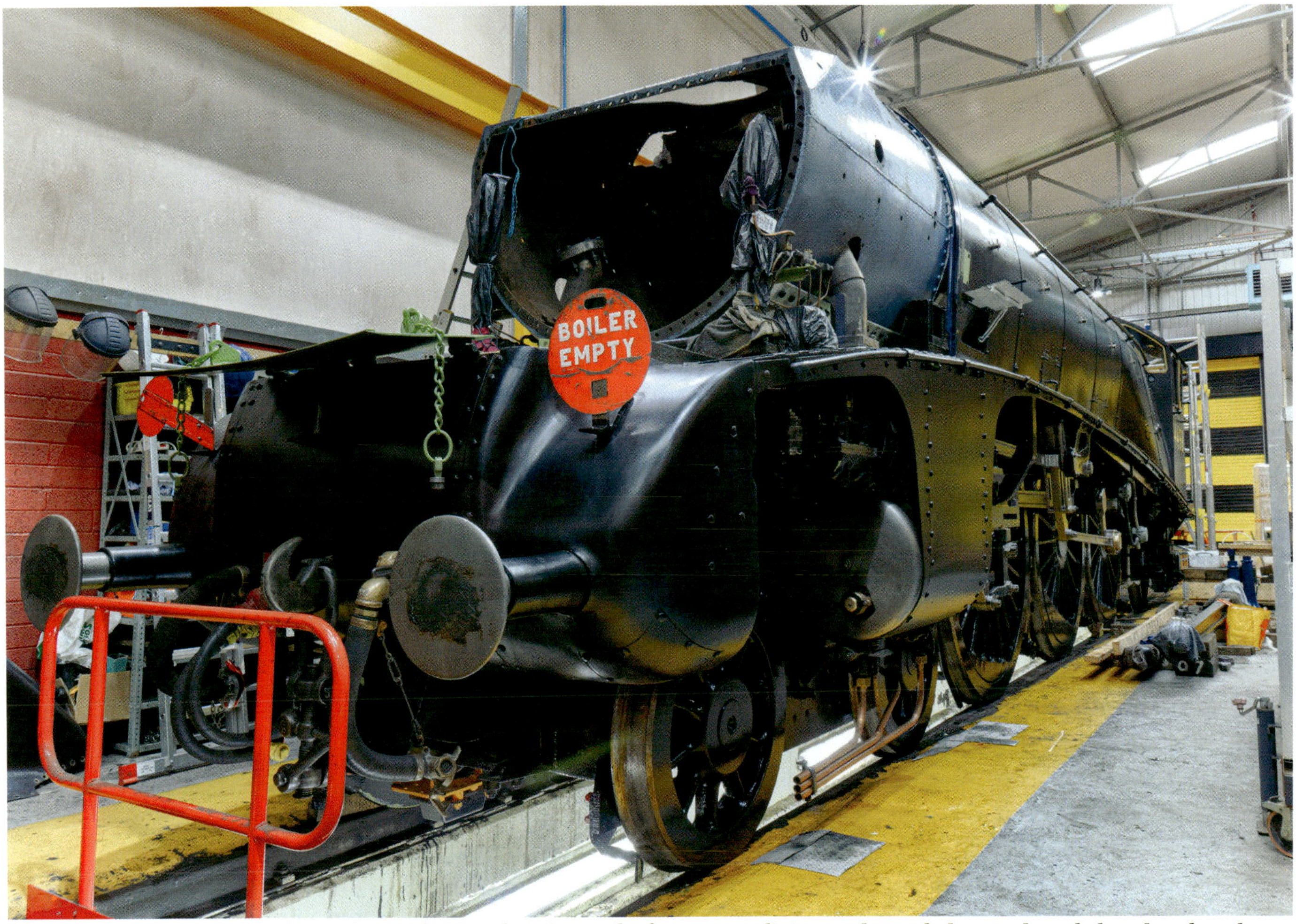

Having been assembled for the road move from York to Crewe the front streamlining, cod mouth door and smokebox door have been removed in preparation for fitting the steam pipes in the smokebox

It was a month since we had relocated to Crewe and the team were starting to settle in. LSL had provided us with excellent facilities including a lock up container which, with multiple trips between York and Crewe, we were starting to slowly fill with our equipment and spares. Darrin was starting to get a better idea of engineer availability; the change of location meant that some would no longer be able to volunteer due to the core of the team living in the north east of England or in and around York. This was a big negative for so many but there were positives, one of which was LSL making three twin rooms available for three nights a week, which allowed some who couldn't do the lengthy return trip in a day the opportunity to stay overnight. Another positive was the chance for new volunteers who did live within reasonble travelling distance of Crewe to now join the team. Having moved from north east England to West Yorkshire last year I was now in that reasonable travelling distance group myself so I wouldn't need the overnight accommodation.

On my first visit back to Crewe since the day we moved loco and tender from York I caught up with one of the new volunteers, Nigel Hoskison, who was a lifelong fan of No.7, having been initially attracted by sharing the same name. He was delighted to be able to volunteer now, as travelling from his West Midlands home to York was previously too far, the same problem others now had but in reverse. Today Darrin had set Nigel a task of removing a plate at the front inbetween the frames which was a little easier as the front streamlining and cod mouth doors had been removed in preparation for the fitting of the main steam pipes. However, as Nigel found out after climbing up on top of the frames, what should be the correct sized socket isn't always, so a climb down to change it and a climb back up again. Whilst Nigel had climbed down to fetch the correct socket I took the opportunity to get some images showing the missing steam pipes sections in the smokebox, steadying myself against one of the buffers which I had forgotten had a fair amount of grease applied to them. Hmm, I'll be popular when I present this shirt for washing I thought!

Also at Crewe today was pipe fitter extraordinaire, Peter Ellis, and his pipe fitter's mate, Barry (Baz) Horton. Travelling from the north east Peter had taken up the opportunity of an overnight stay and was sharing a twin

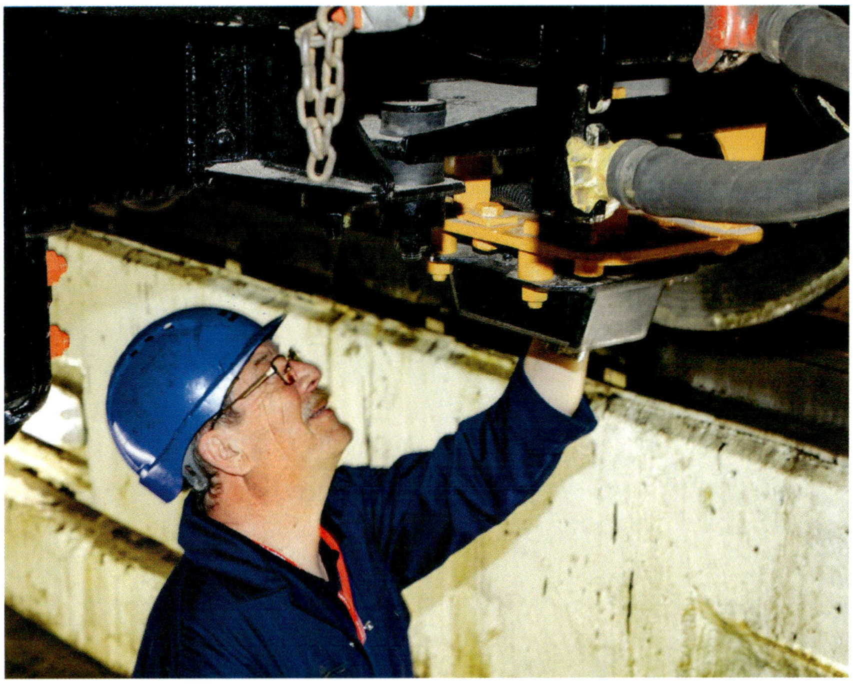

In the pit below No.7 Nigel Hoskison checks the position of bolts between the frames

Nigel Hoskison

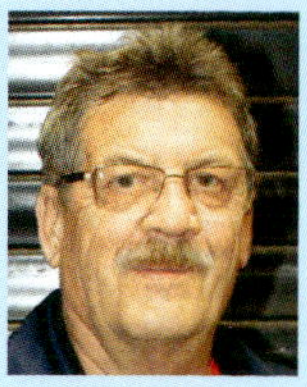

Living in Yorkshire as a child, lifetime Ferroequinologist Nigel stumbled across the rare sight of an A4 in his local railway sheds when it hauled a football special from London for an FA Cup semi-final at Sheffield's Hillsborough stadium in March 1962 between Tottenham and Manchester United. Spurred on by the first glimpse of the loco he got closer and found it carried the same christian name as himself, and so a lifelong devotion to Sir Nigel Gresley ensued.

Now retired Nigel has an extensive collection of model railway locomotives including 24 versions of Sir Nigel Gresley, so you can safely say No.7 is definitely his favourite locomotive.

Previously volunteering for the Trust on the sales stand at the Great Central Railway and the Warley model railway exhibition at the NEC, Nigel has volunteered on the gallery displays held at the NRM. Now living closer to Crewe than York has enabled Nigel to volunteer on a weekly basis as a member of the engineering team which has been a great help to Darrin and the team after the move to LSL at Crewe.

room with Darrin. No doubt the craic that is usual for this team will result in plenty of bedtime stories being told. Barry was, like me, travelling from over the Pennines in Yorkshire, so he was a there and back in a day reasonable travelling distance team member. CME Richard Swales had asked them to relocate a gauge which required a T piece cutting into a section of existing pipe. They were sorting out the required fittings when I caught up with them and Peter showed me the schematic diagram of what was needed which I have to admit was easier to understand than trying to follow it on the loco's maze of pipework. This and the job Nigel was undertaking were just a couple of the thousands of small but crucial tasks the team had to undertake in making this overhaul the most thorough ever undertaken on No.7.

"Big End?" I asked Richard, and "Yes, centre cylinder," was his reply. Not the toughest component to identify but I was nonetheless proud of my correct identification when I interrupted Richard's work on the said component to which he was applying fine adjustment on the working surfaces.

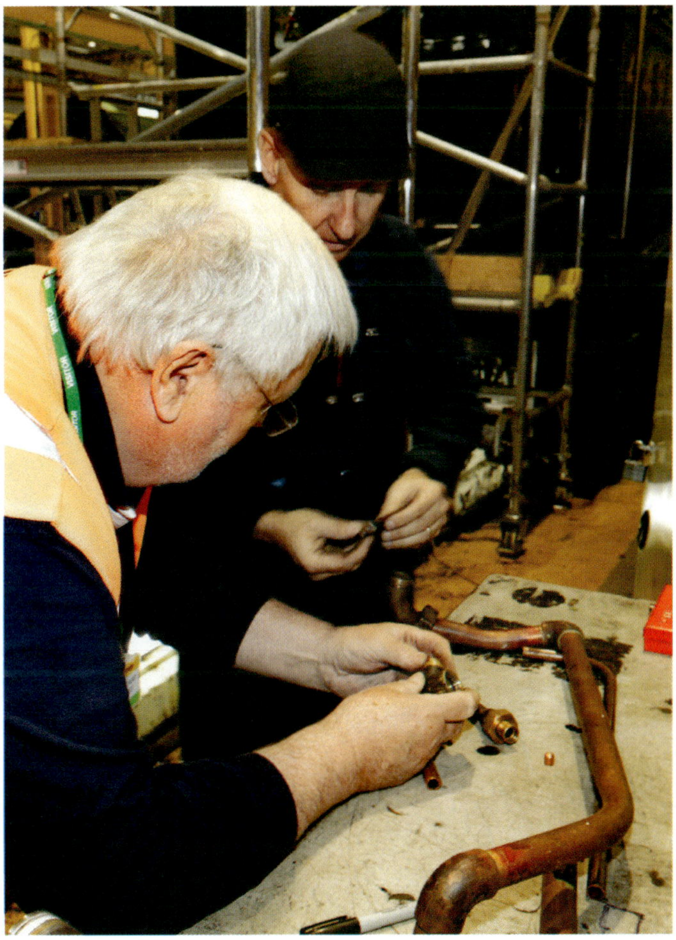

Peter Ellis and Barry Horton sort out pipe fittings

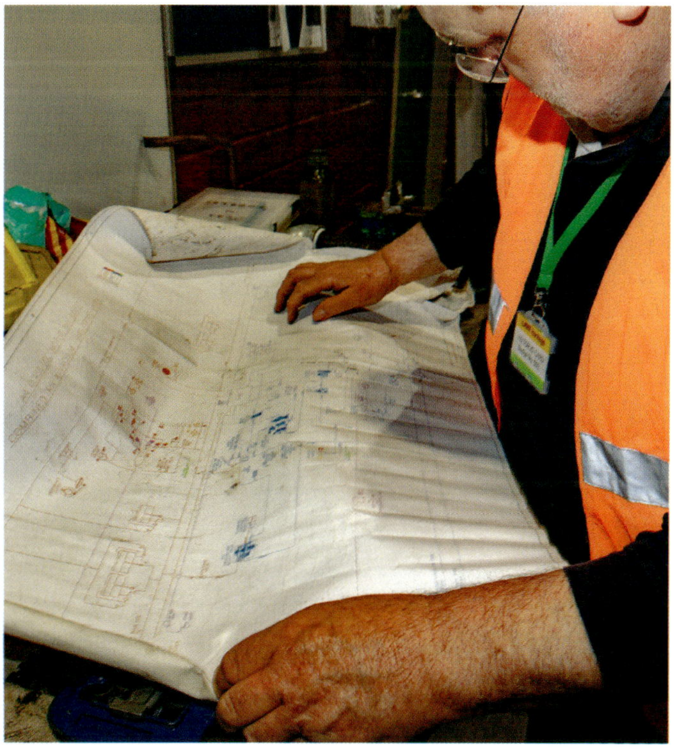

Peter Ellis refers to original LNER pipework plans

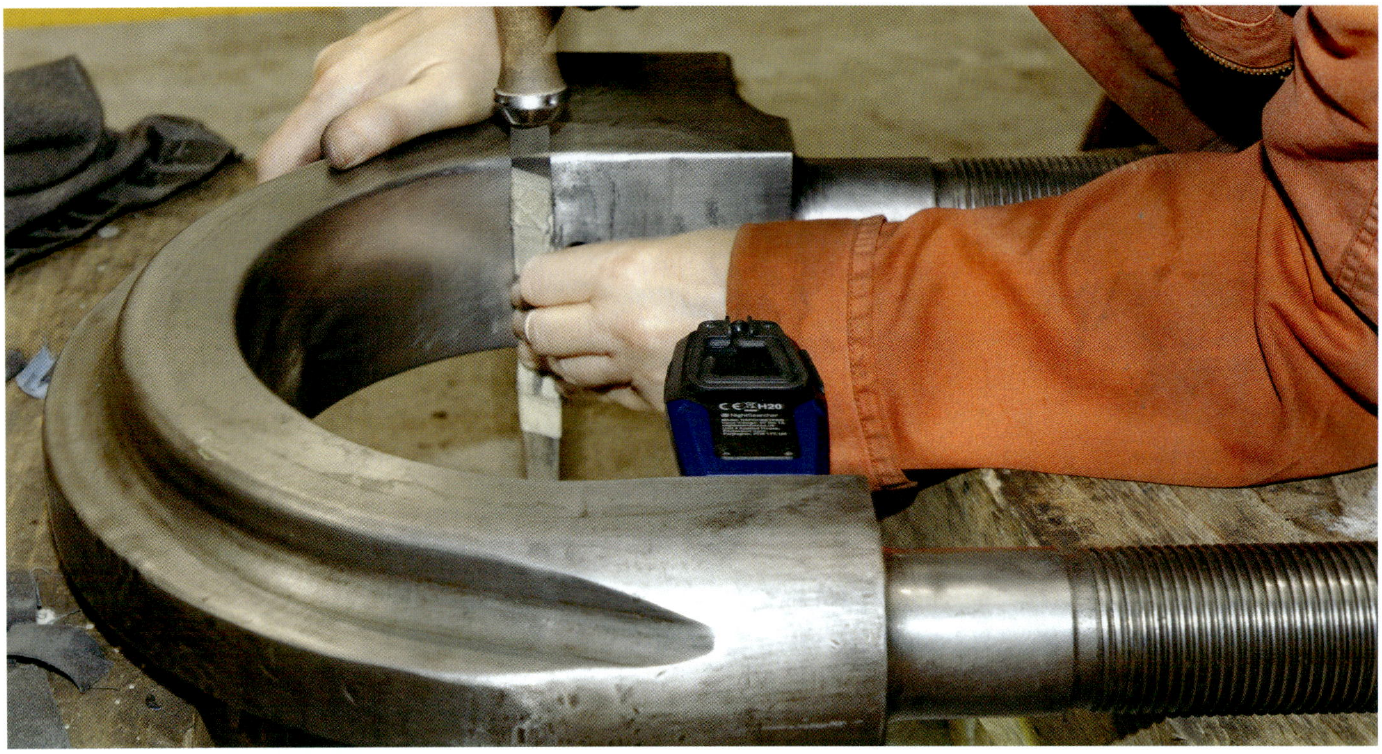

The middle cylinder connecting rod big end is carefully dressed by Richard Swales

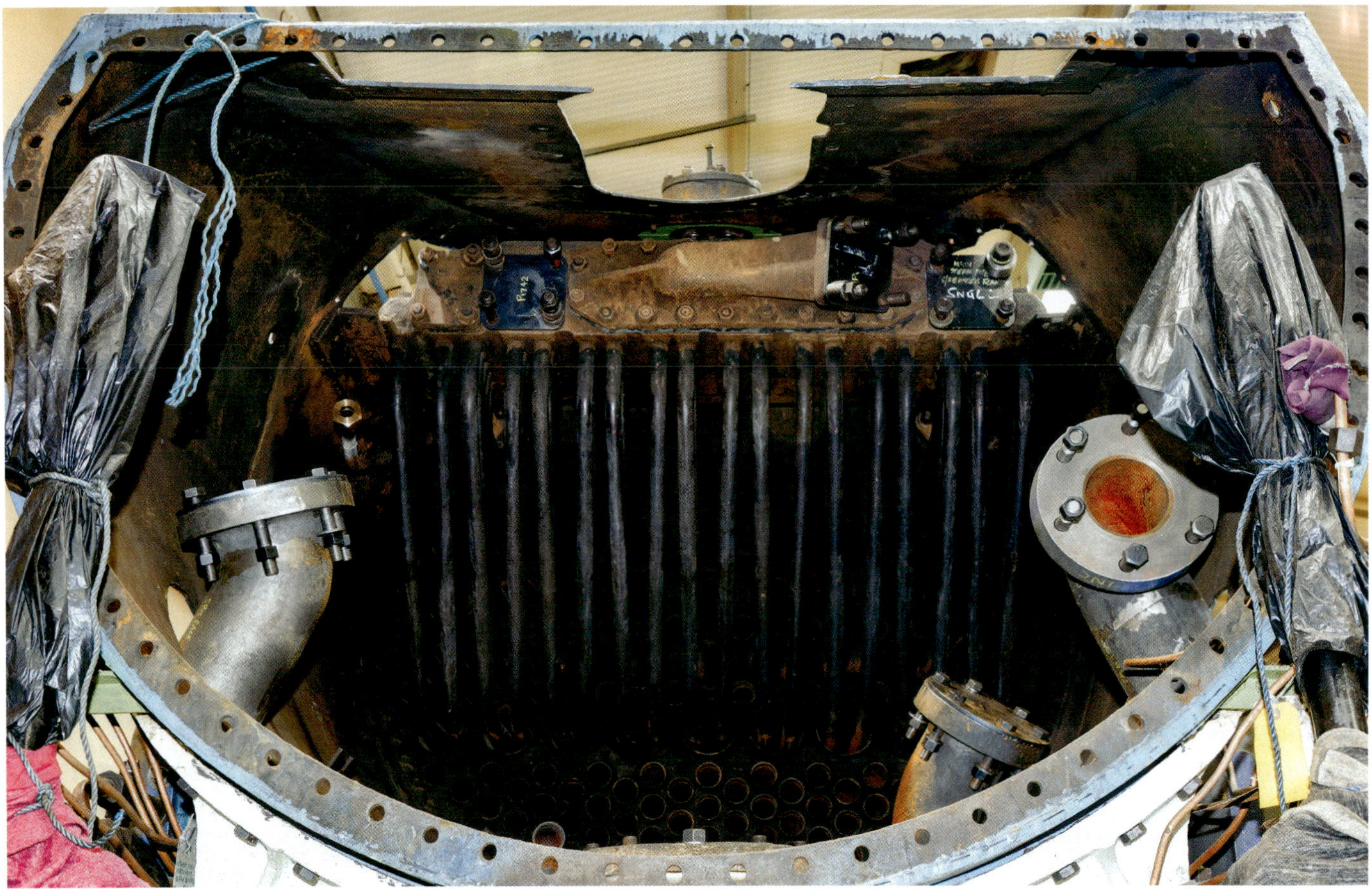

The inside of the smokebox showing the gaps in the steam pipes between the lower sections and the superheater above

Yet another example of the attention to detail and the thoroughness of the overhaul.

I had captured everything I needed for today and more importantly familiarised myself with the procedures on the site, which were very different from the NRM in York as it is a large working motive power depot with steam and diesels, so I packed up my gear, said my goodbyes and left the workshop. Walking outside I heard the unmistakeable sound of a 'whistler' or what I remember from the 1970s as an EE Type 4 later reclassed a Class 40, which along with two tone green Brush Type 4s also outside, took me back many years. It was early afternoon so after lunching in the car I drove away from this time warp site and set off back over the Pennines to try and avoid the worst of the M62 traffic.

November 2021

Now well settled into our new home in Crewe, steady progress was being made by the team as the light at the end of the tunnel was finally coming into view. The next major milepost event was going to be the installation of the main steam pipes between the superheater and the cylinders followed by hydraulic testing. So, I arranged with Darrin that my next visits would be for those events, which was just as well I had been laid low with a particularly nasty winter bug which lasted most of the month of November. I find November a depressing month, with short dull days and long dark nights The only solace, being a photographer, was the autumnal colours of the fauna and, if you were lucky, catching some nice 'golden hour' sunlight to capture the variety of colours that presented. However, I was still busy working from home since the media was now paying more attention as the return to traffic crept ever nearer and my pictures were in demand. Our return to the pages of the steam railway media unfortunately coincided with the premature early retirement of sister locomotive No.9 (Union of South Africa) which had failed whilst working on the East Lancs Railway. Having only a few months left on the boiler ticket and no plans to renew it, the decision was taken to bring forward No.9's retirement date. This meant that with no A4 Pacific currently in traffic our return was all the more eagerly awaited and, with talks underway between the various parties to ensure this would happen as soon as possible, dates for running in and operational runs on the Severn Valley Railway were being pencilled in along with mainline trips hauling Saphos Trains in their 2022 calendar.

Back in Crewe progress was picking up and as time went by more and more of the team were available to Darrin once they had passed the LSL induction training, leading to a comprehensive list of tasks to be undertaken. One notable task had been completed by the piping team when the final piece of piping in the cab was fitted. Team leader, Peter Ellis, and the team are to be rightly congratulated on their wonderful work. There are so many tasks being undertaken by the team, so if you would like more detail, you will find Darrin's reports on the Trust's website very comprehensive.

Since the start of the Covid pandemic, work which could be carried out by volunteer engineers at their homes had helped to keep things moving whilst the NRM workshop was closed by the Covid regulations, and it had worked well. So offsite work continued even after the reopening of the NRM workshop and was still going ahead now. Phil Gillespie was refurbishing the chimney cowl at home, inbetween his other passion of rock climbing and his regular engineering job. I needed to catch up with Phil as, being a relative newcomer to the team I had yet to meet him on any of my visits to York or Crewe, so I planned my next visit with that in mind. Fortunately, this coincided with the final pressure test on a cylinder main steam pipe, the other two cylinders having already been successfully completed.

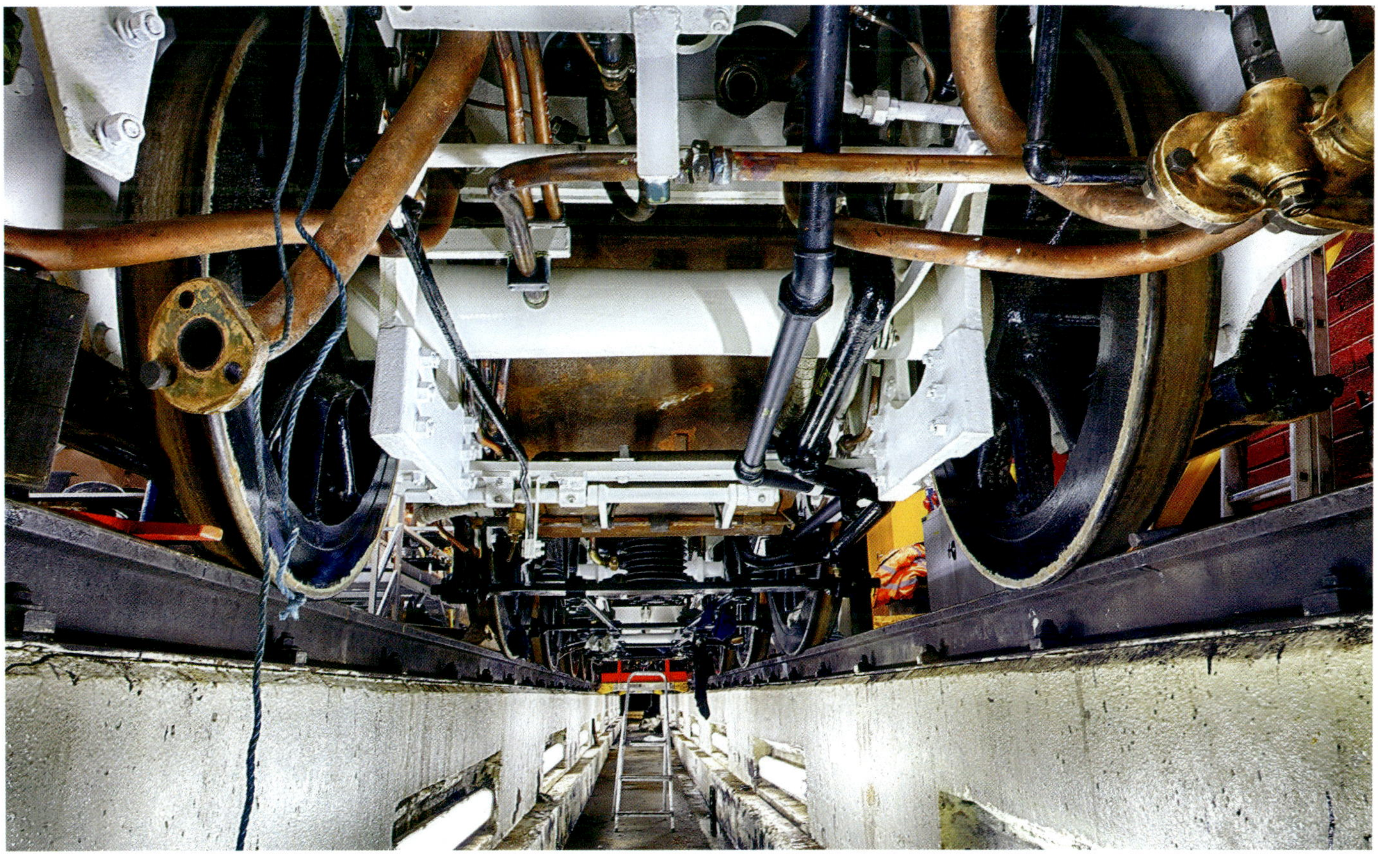

View from the pit showing the clean refurbished components and fresh new paint work

26 PRESSURE ON

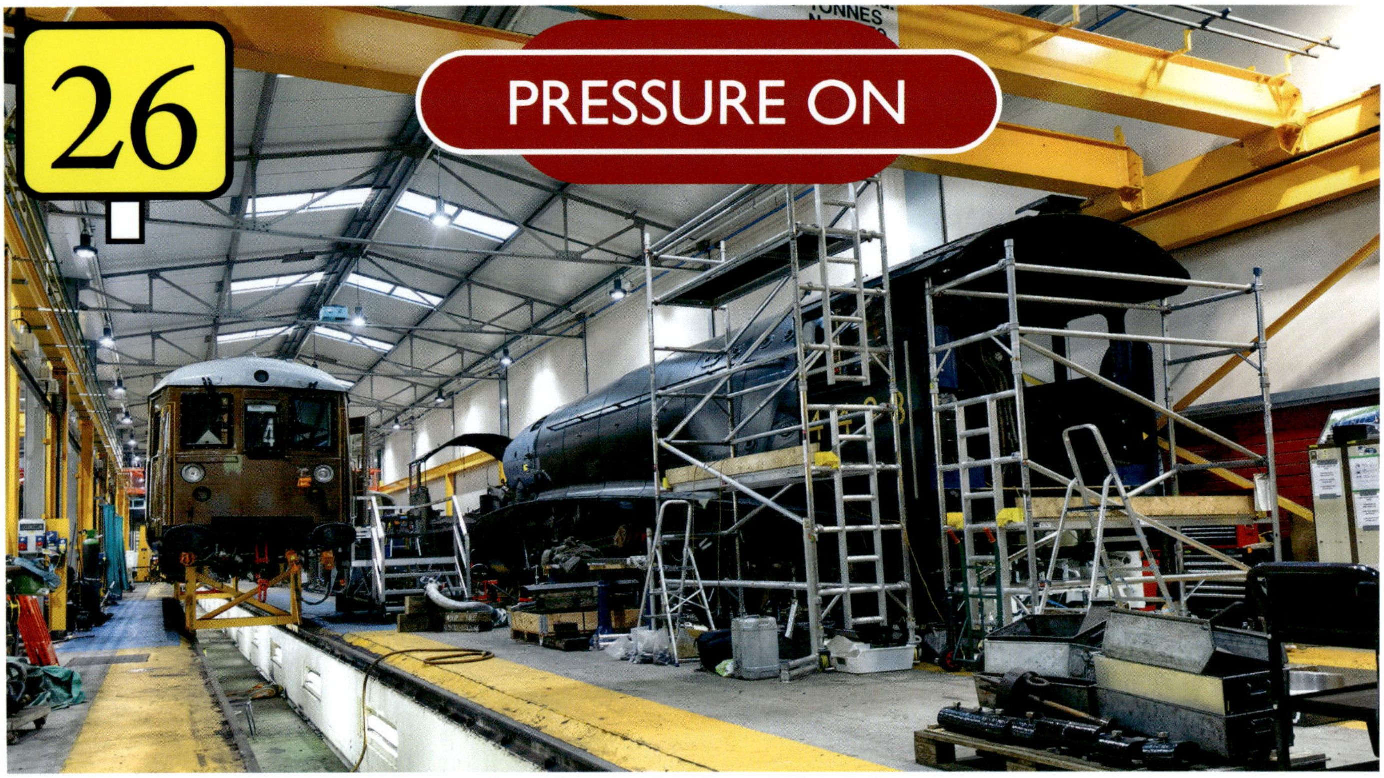

No.7's new home in the LSL workshop at Crewe.

"Other photographers are available!"

December 2021

The response by the volunteer engineering team to travelling and working at Crewe was nothing short of phenomenal, considering that, for the majority, the NRM at York was comparatively local. However, on arriving today Darrin was shorthanded as two stalwarts had rung in sick, so after having plenty of hands on deck recently there were only four working today these being Darrin and Richard Swales, Paul Aston and Phil Gillespie. Still, it worked out well for me as I wanted to catch up with Phil and Paul and also to witness the final steam pressure test on the main steam pipes which was to take place today. After a torturous drive across the Pennines on the M62 and around Manchester's largest car park, the M60, which took two and half hours to cover the 80 miles, I arrived just before lunch.

We made a lasting impression this lunchtime which probably wasn't appreciated by everyone. The pathways in and out of the mess room had been freshly painted and barriered off to stop unwelcome footprints whilst drying. Oh no, I hear you say. Oh yes! We did manage to find a door that had no barrier in front or warning sign and gave access straight onto the wet paint. I'll say no more other than the anti-bacterial wet wipes on the lunch tables did an excellent job at removing the footprints we had left on the dry mess room floor, but the still wet floor beyond the door was another matter!

Returning to the workshop after the embarrassing faux pas, I found Richard and Paul installing steam injector pipes. After talking to Darrin this was less straight forward than it seemed, as Richard had discovered that the flange holes did not line up correctly, but such is the excellent attention to detail at this overhaul, a proper engineering fix was devised and applied by Richard.

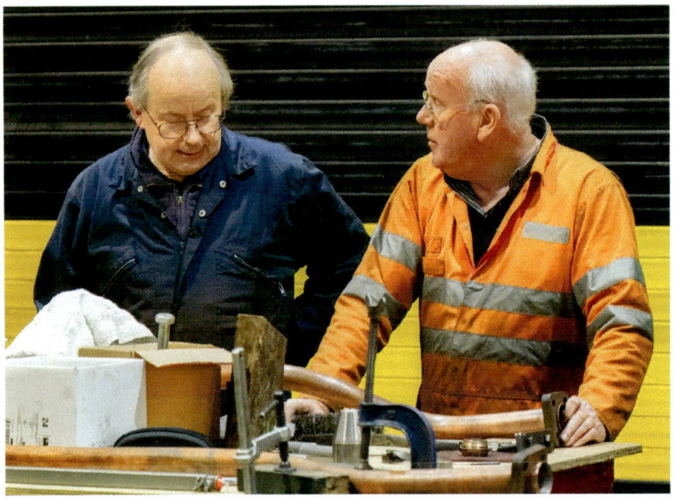

Richard Swales (left) and Paul Aston discuss the planned work

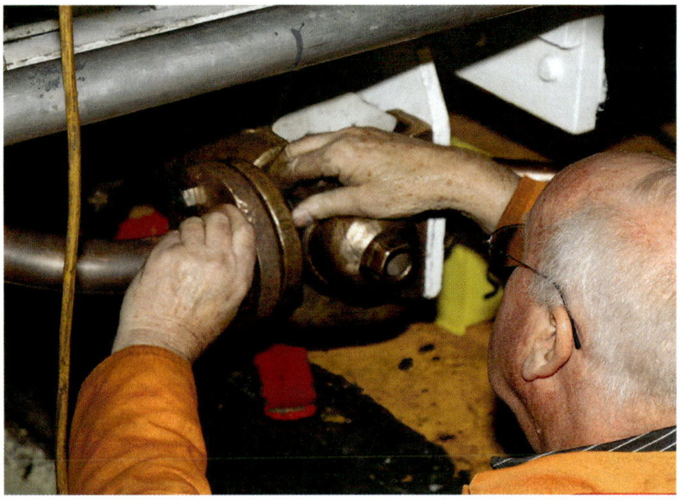

Paul Aston connects up a steam injector pipe

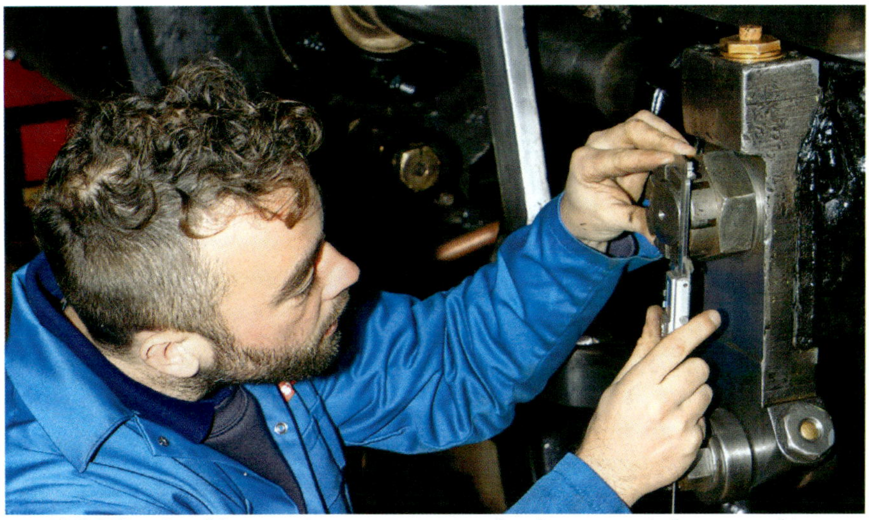

Phil Gillespie taking precise measurements for the cotter pin he was making

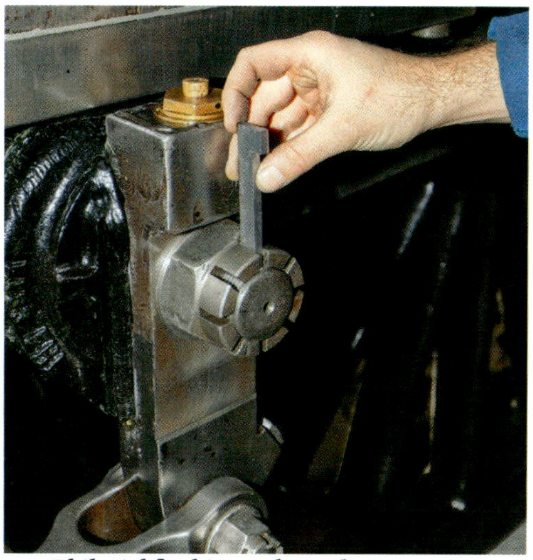

Phil trial fits his newly made cotter pin

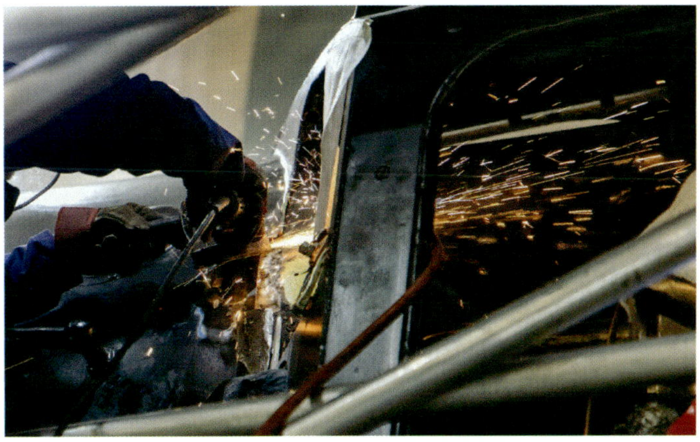

Darrin Crone gets to grips with a particularly awkward section of cladding around the left hand side window

PHIL GILLESPIE

It was a love of railways that led to a career in engineering. I first saw number 7 at Carnforth Steam Town when I was only 2 years old with my parents. Over the years a number of heritage railway visits inspired me to eventually get 'hands on' and I became a volunteer with the Churnet Valley Railway at 15. Spending my weekends wielding spanners it felt only right to apply for an engineering apprenticeship and I have worked as an aircraft fitter with BAE Systems in Lancashire ever since.

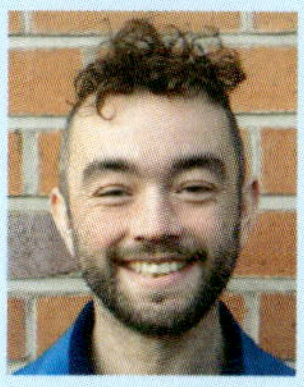

With number 7 in York for overhaul and a change in shift pattern at work I had the opportunity to volunteer on a loco that I had admired for over 30 years. It's been a real privilege to be part of its story and a joy to work with such a great team.

Phil was making a cotter pin to fit in the left hand cylinder's gudgeon pin which links that cylinder to the connection rod. With his own diagram to work to this was precision work requiring a lot of careful filing to finish with a snug fit in the gudgeon pin, as being heavy handed here would result in the pin being too slack and mean him having to start all over again. As with many of the jobs carried out during the overhaul, some parts are no longer available and have to be made, whereas back in British Railway days parts would have been off the shelf or pinched from another loco which was further down the queue awaiting its turn for overhaul.

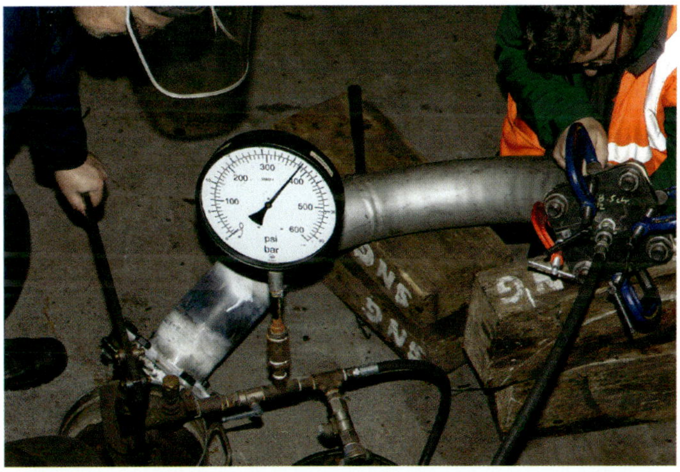

Pressure reaches 375psi during the final test on the steam pipes

A less precise finish than Phil's cotter pin was required on the cladding and boiler band up around the forward facing, left hand side window which Darrin was working on with an angle grinder and mig welder. Whilst Phil was working in thousands of an inch Darrin was working in fractions and I dared to comment on the contrast in the tolerances which received the humorous rebuke from Darrin of "Other photographers are available!"

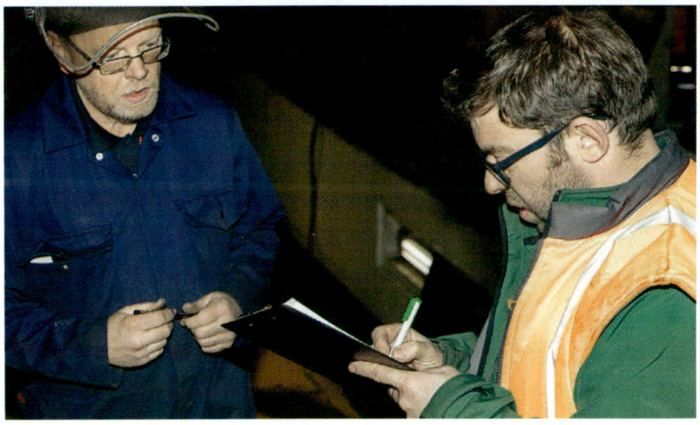

Boilersmith Mark O'Brien signs off the all important steam pipe pressure test paperwork

Only one pic left to capture, which was my main reason for today's visit, and that was the pressure test on a main steam pipe which connects the superheater with the cylinders. This was the last of the three pipes to be done and once signed off by the boilersmith, Mark O'Brien, the pipes could be installed back in the smokebox and the front of No.7 could be reassembled for the last time. The test went well, reaching the required pressure of 375psi, which is 50% higher than the boiler working maximum of 250psi, and the paperwork was duly signed off. The pipes could now be installed next week when I would return to capture the moment.

December 2021

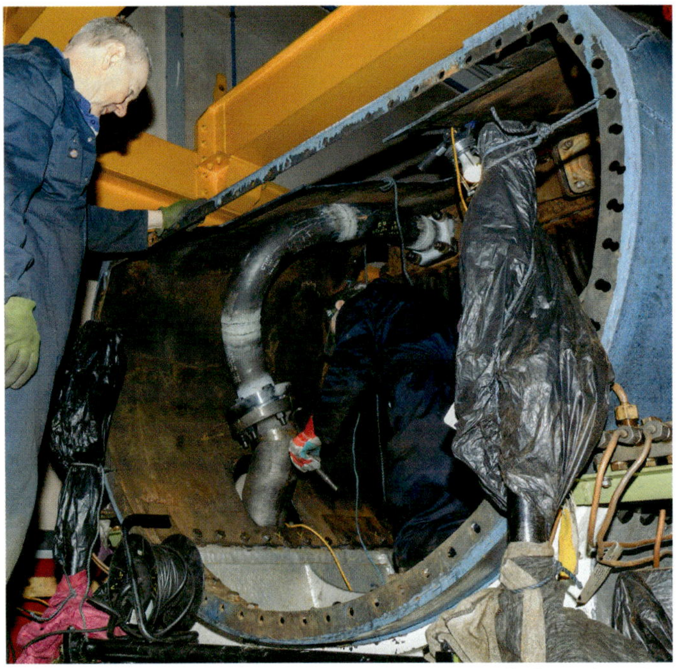

Phil Slater looks on as Dave Lee gets to grips with the right side cylinder steam pipe in the smokebox

I wanted to see the steam pipes being put back in the smokebox so I needed to make sure I was there early on the planned day for the job. Rising from my very comfortable king size mattress at 4:30 and being on the road by 5:30 I reckoned would get me to Crewe by 8:00. Being December it was very dark, with added heavy, sleety snow as I set off on another torturous drive over the Pennines. Early as it was I wasn't early enough as Darrin greeted me when I entered the workshop just after 8:00 with the news that the boys had done way better then he expected on the day before and the pipes had already been trial fitted. However, the right hand cylinder pipe was still in situ so I managed to capture some images of Phil Slater and Dave Lee working on this one. I was a bit deflated having battled my way through the weather and traffic, but the simple fact is I couldn't expect the guys to wait for me to arrive before they pushed on with the work as the clock was well and truly ticking with arrangements and dates being discussed for the return to traffic. There was a silver lining, for me at least, as getting my pictures early morning meant I could return home in the middle of the day when the traffic was lighter. So leaving before lunch time I managed to lop an hour off my return trip, getting 2 hours 35 minutes down to 1 hour 35 minutes.

The relocation to Crewe meant many of the team had more difficult and costly trips, and talking with Phil and Barry Horton at tea break I realised they had even longer journeys but they were sticking with the trains; Phil travelling from Scunthorpe and Barry from Doncaster. Part of Barry's journey was along the picturesque Hope Valley through the Peak District between Sheffield and Stockport, made all the more appealing for real ale fan Barry as Stockport station boasted a real ale pub.

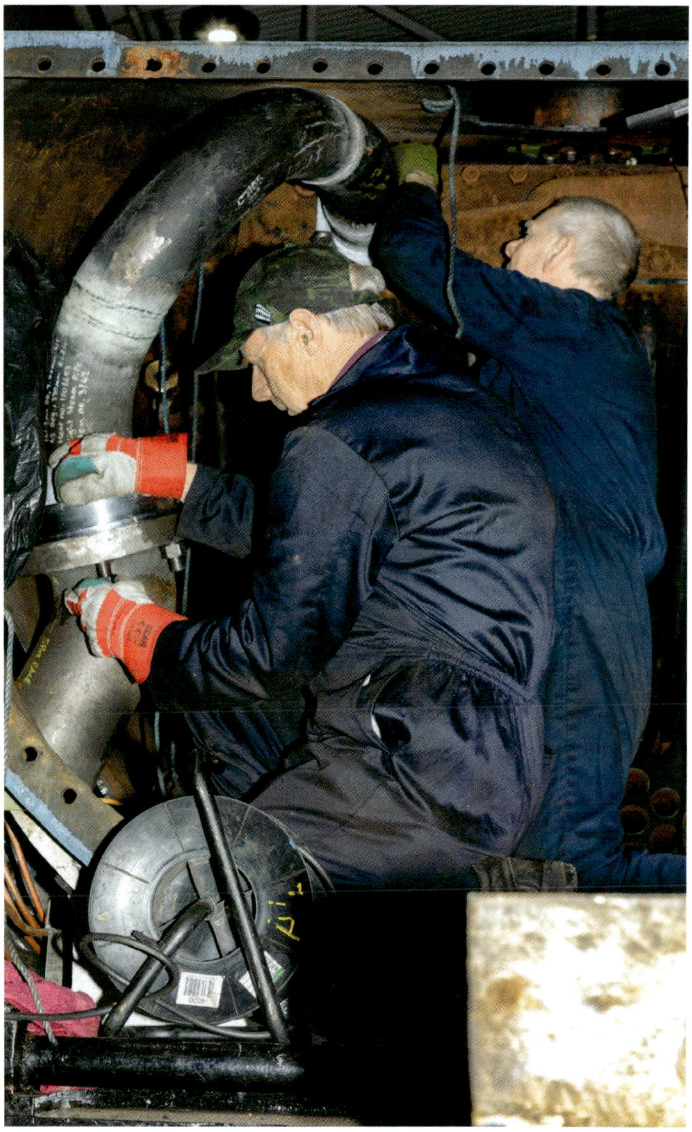

Dave Lee (left) and Phil Slater trial fitting the right side cylinder steam pipe in the tight constraints of the smokebox

Back to the job in hand. Darrin said he hoped to have the pipes back in the smokebox and the front reassembled this side of Christmas so I still had chance to get the images I wanted. However Covid intervened again. The pandemic's Omicron variant was now sweeping through the UK and stricter control measures were being introduced, not directly affecting the overhaul as yet, but my daughter, who lives in Spain, was travelling home for Christmas and would have to quarantine on arrival. Although we didn't have to quarantine with her, I felt the sensible thing to do would be to avoid unecessary exposure, so I reluctantly informed Darrin that I would not get back to Crewe until the New Year.

SLIP UP

January 2020

"Don't forget the copper slip"

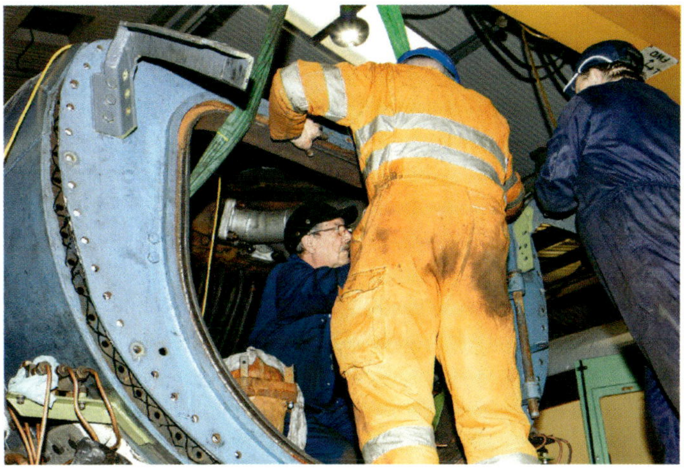

The smokebox front plate is lifted back into position

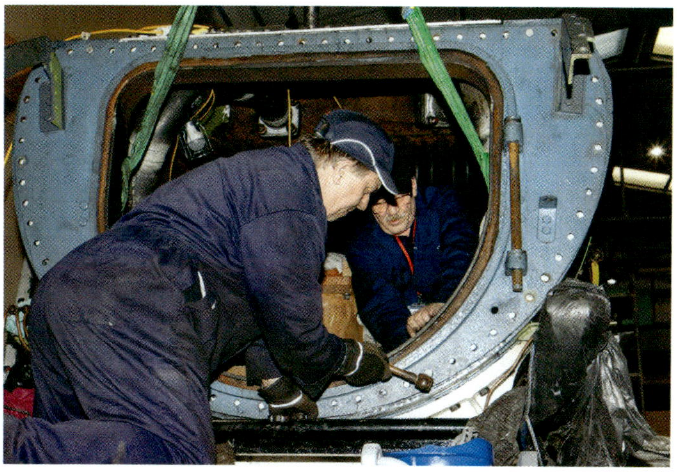

Richard Hawkhead tightens one of the eighty securing bolts on the smokebox frontplate

Christmas had been and gone and the new year had arrived. On my last visit Darrin was hoping to have the steam pipes back in the smokebox along with the blast pipe casting so the rebuild of the front of the loco could begin. However as with all the best laid plans, it didn't quite turn out that way and work was now a little behind. I was hoping to capture a shot of the rebuilt front but after talking with Darrin and realising that would not be the case I nearly postponed my planned trip, but, I was missing the craic and there was always a picture and a story to be had, so I decided to go anyway.

Today, the team were rebuilding the front now the steam pipes had been fitted and sealed in the smokebox and the blast pipe had also been refitted. The smokebox front plate was readied, and with the help of LSL staff operating the overhead crane, it was lifted into position and bolted back onto the smokebox. Richard Hawkhead, Nigel Hoskison and Paul Aston had been delegated to do this job and all went well, with some 80 bolts to be fitted and tightened. The team were delighted when Darrin reminded them not to forget to apply 'copper slip' reasonably early into the fittimg of the eighty, as they actually did forget! Copper slip is anti seizing grease for use in high temperature areas, so I guess they would have been devasted if Darrin had left it until the mid seventies before checking on them. It did however bring a wry smile to the boss's face.

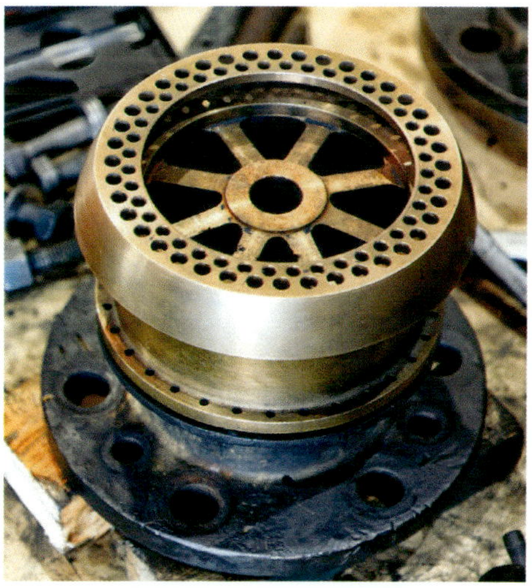

One of the two steam safety valves

Earlier Paul had been working with Richard Swales on the steam safety valves, something he had done at the previous overhaul after Richard had written a procedure which followed the methods of the then Chief Mechanical Engineer, John Graham. I thought this was an excellent example of skills being passed down the generations and I wondered if this was how John had learned the procedure at the previous overhaul. The valves were now ready to be refitted to No.7, but not today. My image shows one of them sitting ready on the workshop table.

Having had my fix of the craic as well as getting my pics for the day I slinked off a little early to miss the worst of the traffic. Some of today's team had shorter trips home than my 180 mile round trip, one with a nonetheless significant 100 mile round trip, and some staying several nights in order to work multiple days or just overnight to carry on the following day, after which they would themselves finish early for their trips home, the longest of these being a round trip of over 400 miles. Now people travelling large distances and staying overnight is not an unusual occurrence but you have to remember that the team are volunteers and were doing this in their own time and, more significantly, at their own expense. This is a testimony to the wonderful team that had evolved over the years and had stayed to complete the job even after the relocation from York to Crewe had significantly increased the travel and expense for the vast majority.

My next visit would see a couple of hundred or so Trust members on site, as in order to enable Trust members to see the black, wartime approximate livery, a Members' Day had been arranged for February 12th which was to be kindly hosted by LSL in the North Yard of the depot. This was logistically challenging with LSL being a working motive power depot, so the Trust were very appreciative of our LSL colleagues for accommodating this event.

27 MEMBERS DAY

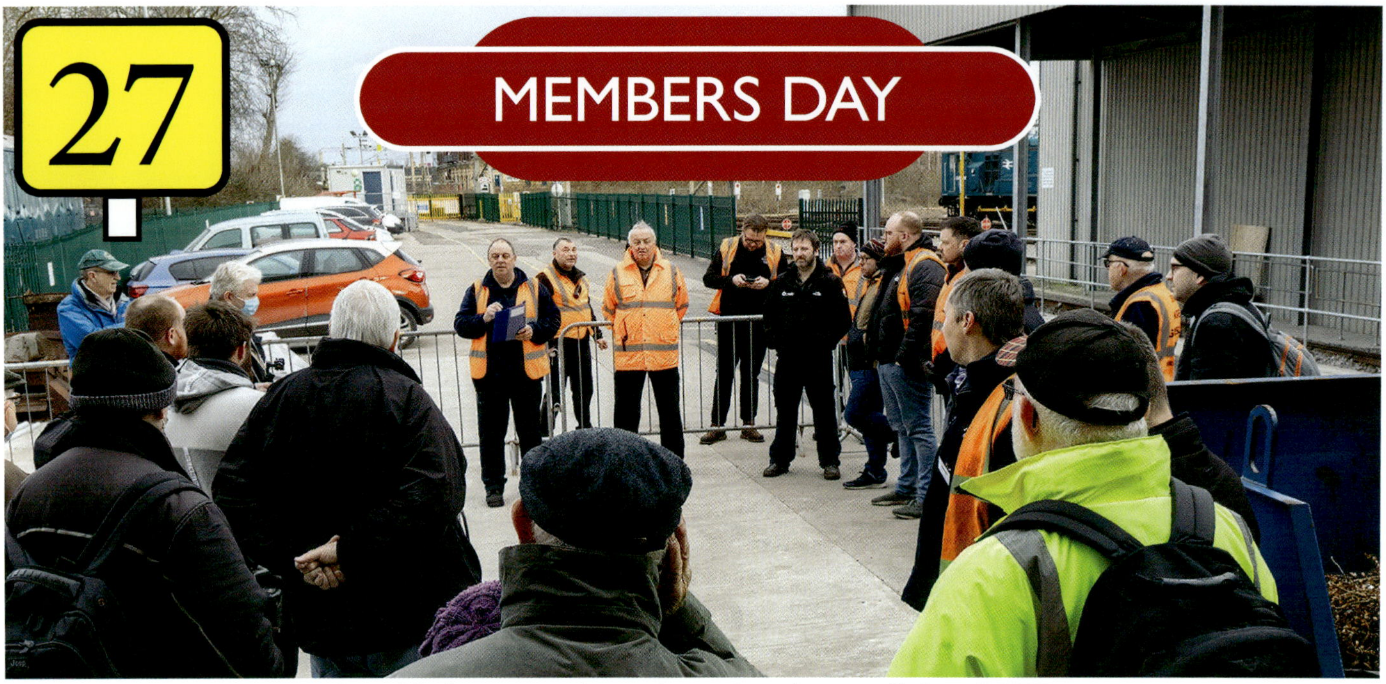

One of the four tour parties receives an introductory talk from SNGLT Operations Manager Richard Hawkhead

February 2022

It was Members' Day and it was going to be a long one for all involved, so the Friday night before resulted in the local hotels being heavily booked by the Trust engineering team and attending trustees - myself and my good lady, Janice, included. What a difference it was getting up and having a relaxed breakfast and still being at LSL by 08:30am in good time for the first batch of members who were due at 10:00am.

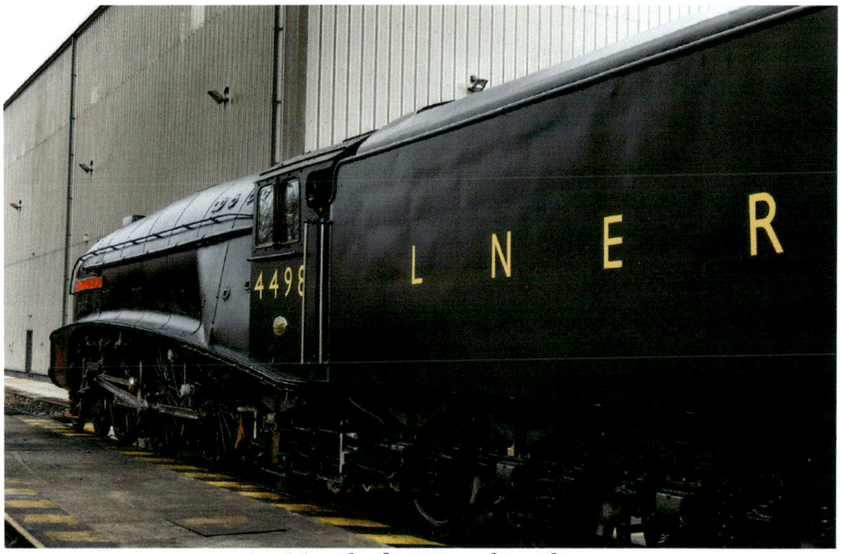

Awaiting the first tour of members

Sporting her temporary wartime black livery number 4498 was being shunted out of the maintenance building into the viewing position when we arrived onsite. The place was a hive of activity already. LSL staff were managing the loco movements and were to take charge of the groups of members for guided tours of the depot. SNGLT engineering team were onsite to answer any questions, trust officials were onsite to support sales of merchandise and I had a 'Meet the Author' stand promoting this very book that you are now reading. So, as I also had to capture the day in pictures, and needed to be in more than one place, Janice was to help by manning the stand whilst I flitted about with my camera. And I can't forget Darrin's wife, Julie, and Richard Swales' wife, Helen, for their sterling work keeping the team supplied with refreshments.

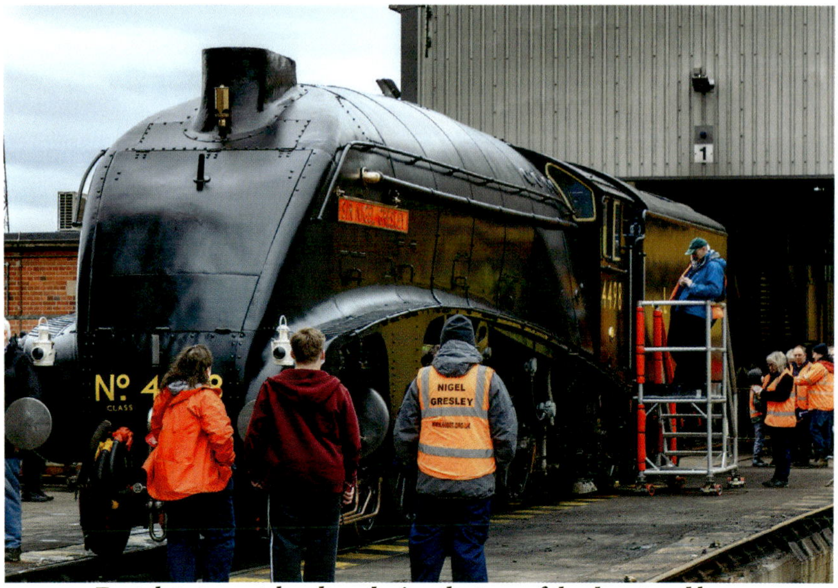

Footplate tours take place during the tour of the depot and loco

It was very cold and, as the day wore on, it became very wet as well, but this, it seems, is no deterrent to our hardy members as the numbers attending were pretty much everyone who had booked a place. Several members introduced themselves to me to say they had pre-ordered copies of this book and I was really pleased to meet them as they had raised

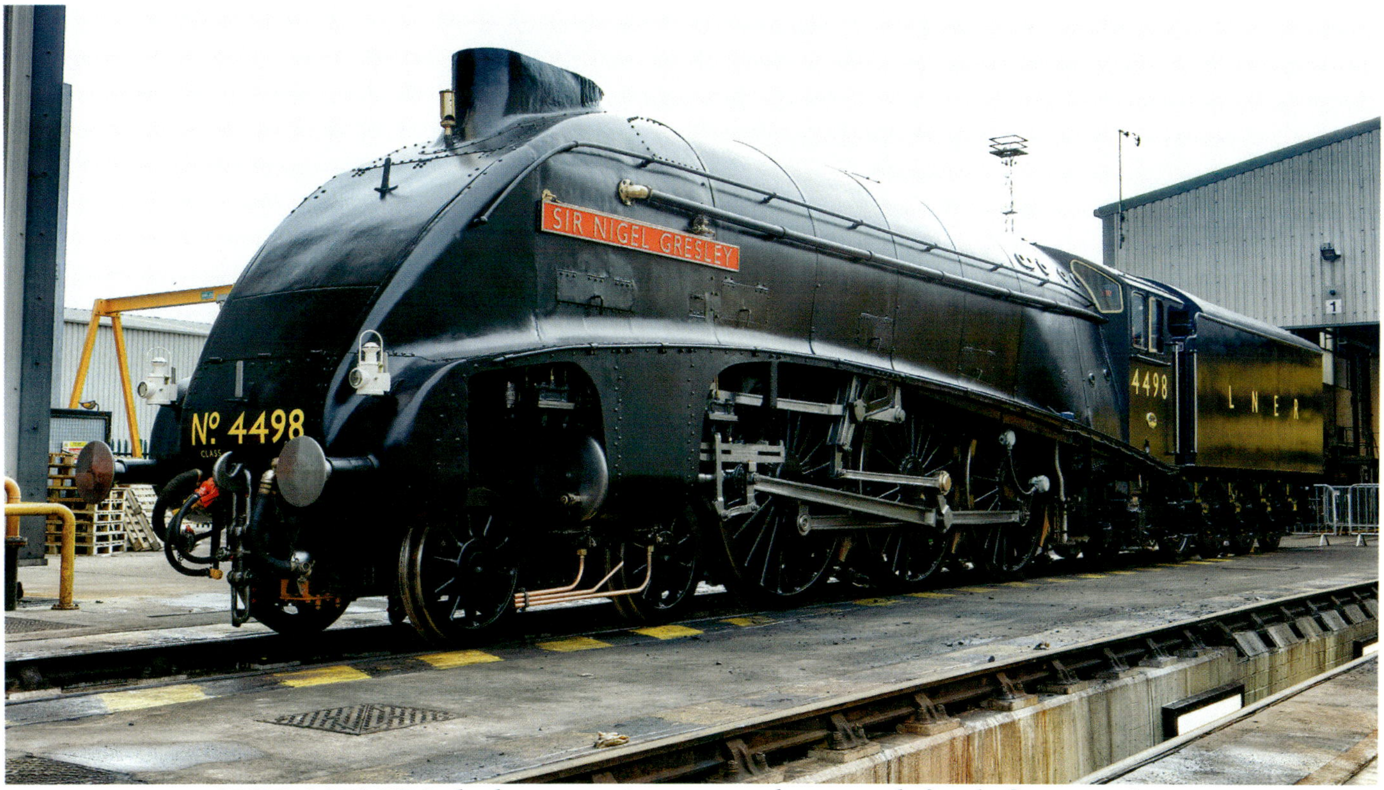

ABOVE & BELOW I take the opportunity to capture the moment before the first tour arrives

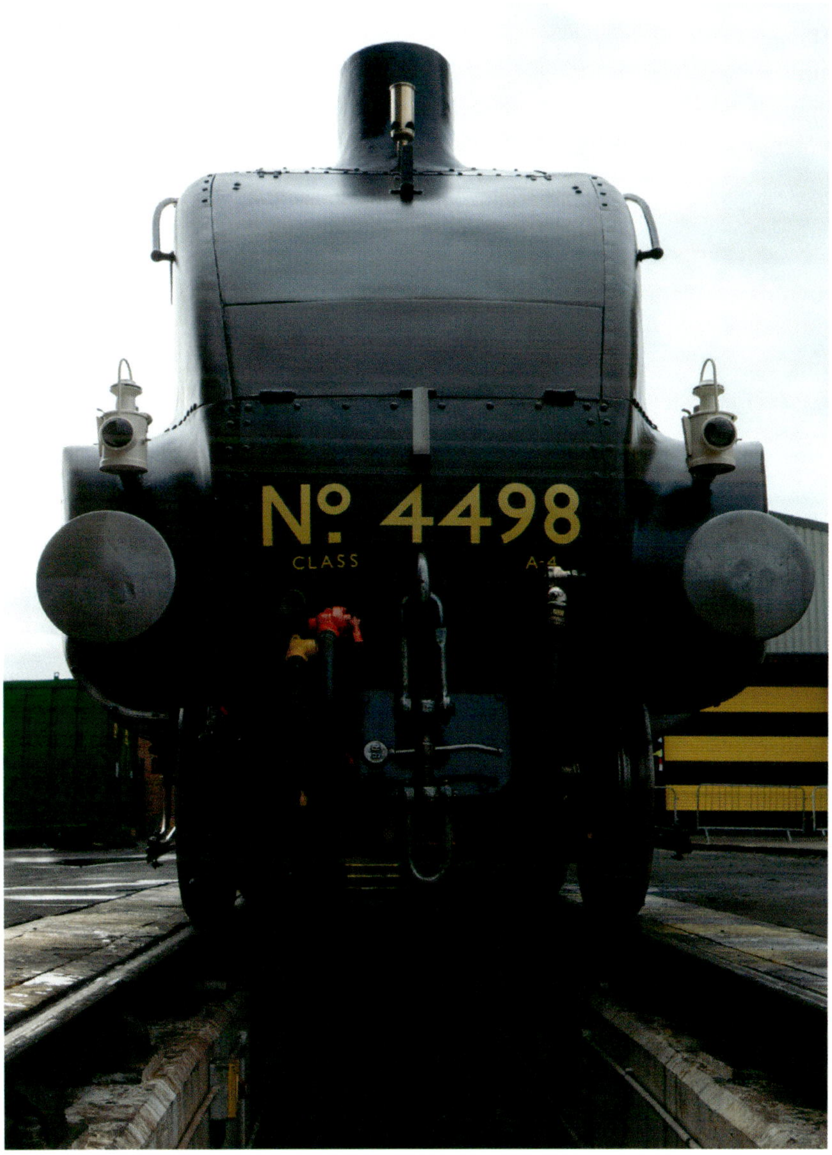

the vital funds to finance the book's first print run without which there was a degree of risk as, at the onset, I had no idea of how it would develop. The day went off brilliantly and the response from all parties and on social media groups was outstanding, prompting Trust chairman, Nigel Wilson, to deliver a huge thank you to LSL for not only allowing the event to take place but providing manpower and assistance over the day.

As mentioned earlier (p159) the temporary black livery was effectively a protective undercoat applied for protection during the move from York to Crewe, but the opportunity had been taken to simulate the wartime appearance by applying temporary decals for the number and lettering. To keep the authenticity correct for today's photo opportunity, the front 60007 number plate had been removed and replaced by gold 4498 numbers and lettering which was applied between the buffers. The wartime black livery was creating quite a debate, particularly on social media and the photos taken will no doubt continue to fuel these discussions with the progress of time.

The success of the day made us all think that if this is the response with the loco not yet in steam what is it going to be like when we actually get a fire lit and she starts running under her own steam; we would soon find out as the final arrangements were now being made for the running in trials on the Severn Valley Railway.

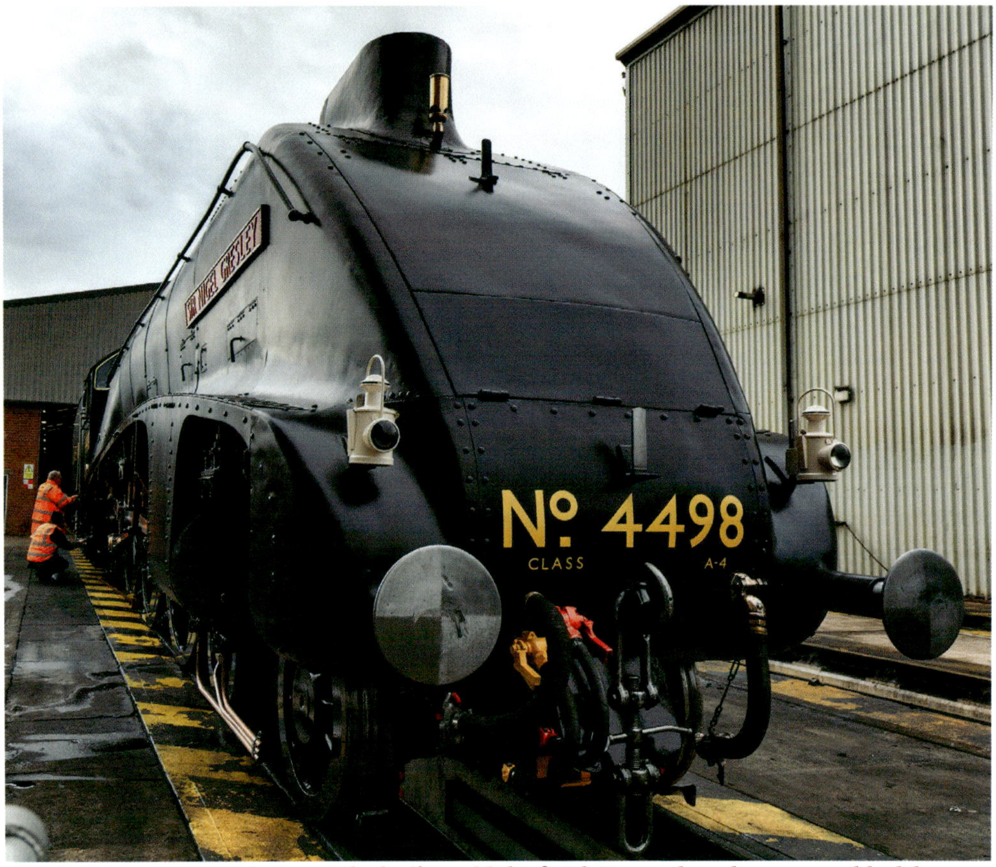

The temporary decals between the buffers add the finishing touch to the wartime black livery

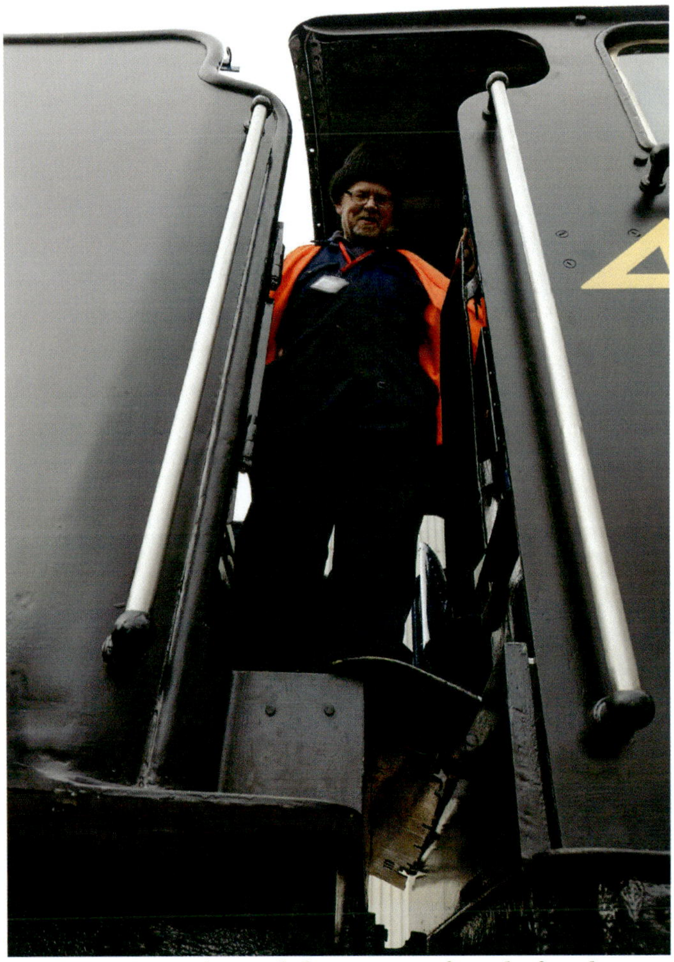
Darrin Crone shares a lighter moment from the footplate

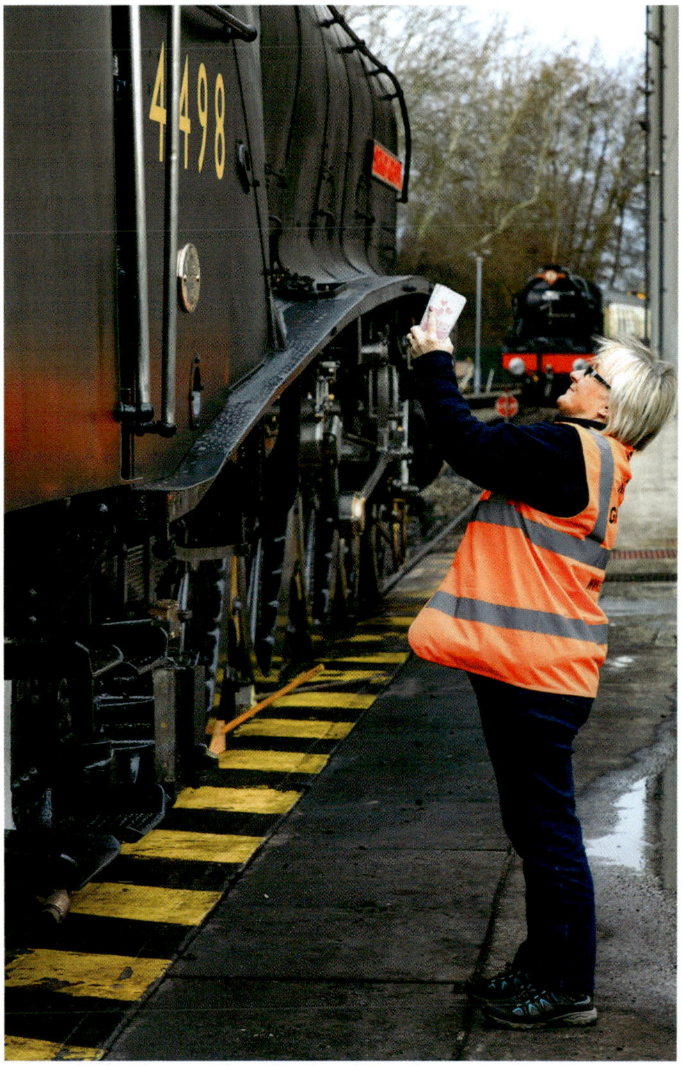

Julie Crone videos husband Darrin on the footplate

BIG END

February 2022

We were getting ready for a big end in more ways than one as I was to find out today when I was introduced to the Tos Varnsdorf horizontal boring machine. The centre con rod's big end bearing needed machining to a fine tolerance before the con rod could be fitted to the loco between the middle cylinder and the main drive wheelset's axle and this machine was going to do the job. With LSL's Tony in charge, our own Tony (Bickerstaffe) was on hand to help but he was happy to admit that this machine was a little daunting, and he was quite content just watching in awe I have to add. I must admit I was quite transfixed watching the machine's tool delicately cut its way through the white metal of the big end's journals. With the job completed successfully the con rod could now be fitted, but not today. It would need to be soon as Darrin had weighing equipment arriving in a few days to weigh the locomotive, so it would have to be in place before that could happen.

Another major task today was putting the firebox brick arch former in place, which Richard Swales had wriggled into the firebox to do. The brick arch is no longer made from bricks but concrete which would be mixed onsite and applied to the wooden former - as with the con rod, not today but soon. I was beginning to wish I had arranged an overnight stay as I really could have done with being there every day at the moment. Getting a picture of the wooden former had me on my hands and knees at the firebox door with the widest lens I had with me today, but I was also on my hands and knees as Richard Hawkhead was showing off his joinery skills creating a new wooden floor for the cab which meant I was crawling across the new floor to avoid putting dirty footprints on his handiwork.

The big end, in the other sense of the word, was racing towards us and it became even more real when Peter Ellis and Barry Horton, who, on fitting a pipe, Barry announced that it was a "last pipe to be fitted" ceremony. The pipe in question runs along the top of the left hand nameplate and carries steam to the atomisers in the smokebox. Now, hopefully, most ceremonies go off without a hitch, but in this case maybe that was just a pipe dream for our intrepid pipe team (sorry I couldn't resist that). The collar that clamps around the pipe where it passes through the steamlining is held in place by small brass nuts and bolts, and the nuts were proving difficult to hold in place and kept being dropped from their lofty location only to fall, not on the smokebox

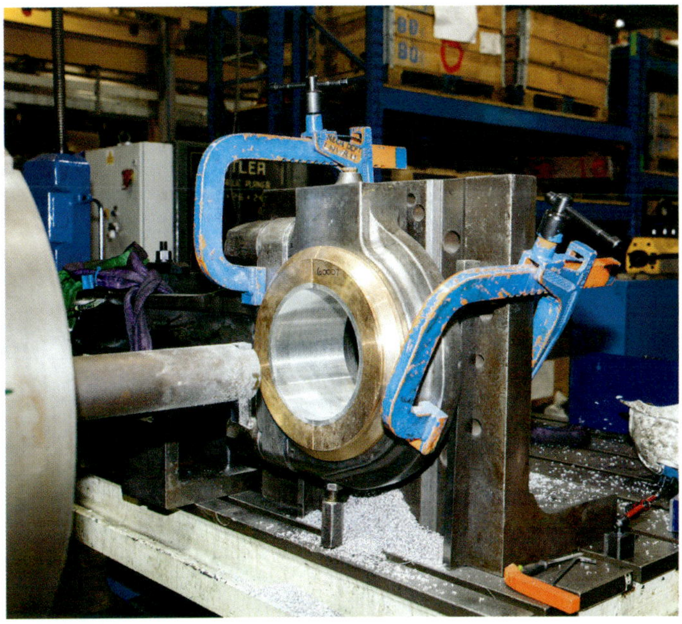

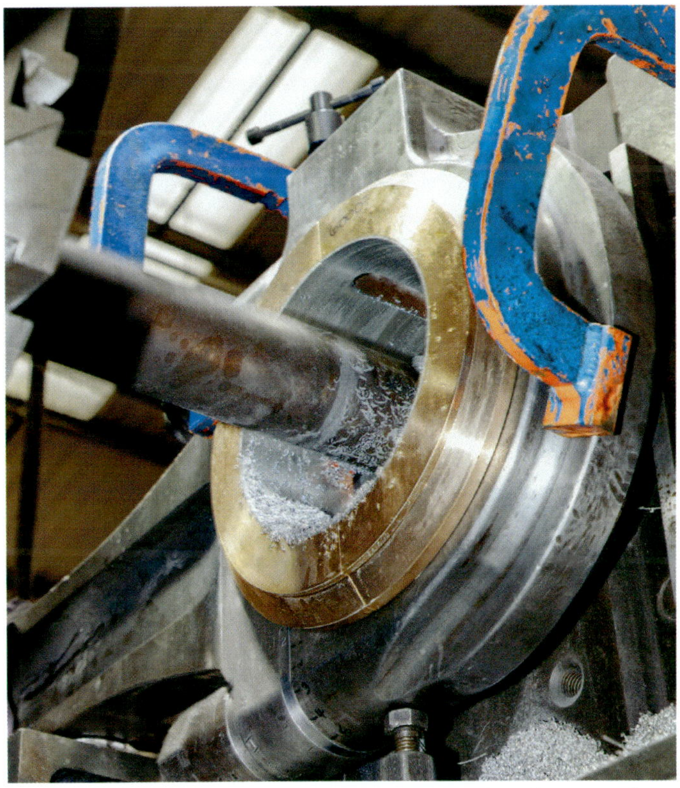

The horizontal borer gets to work machining the white metal on the big end journals of the middle cylinder

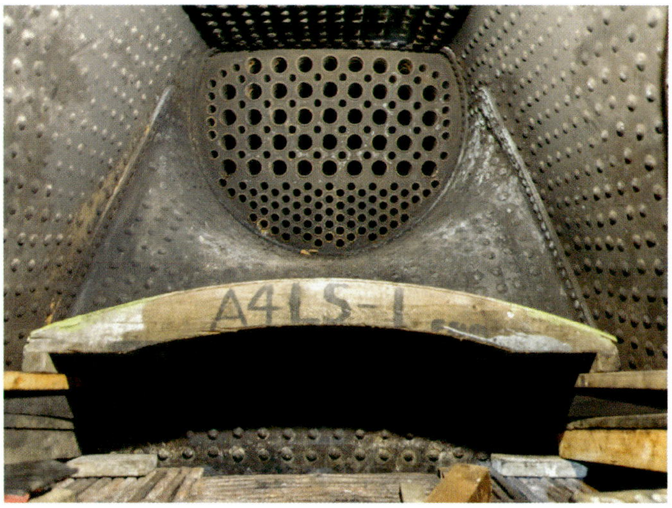

The firebox wooden brick arch former is in place ready to receive the concrete mix which was now replacing the original fire bricks

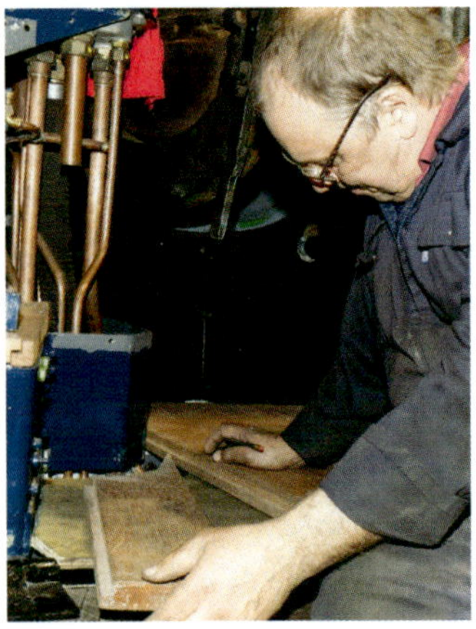

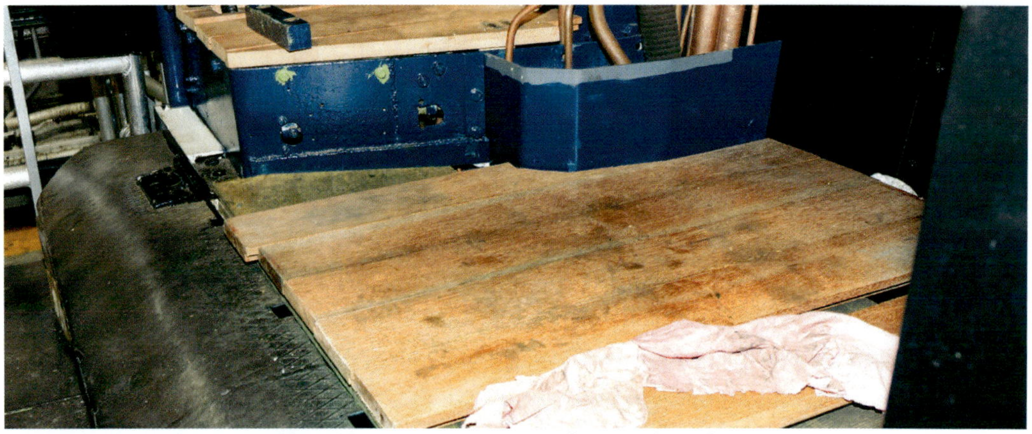

LEFT & ABOVE - Richard Hawkhead fitting a new wooden floor in the cab

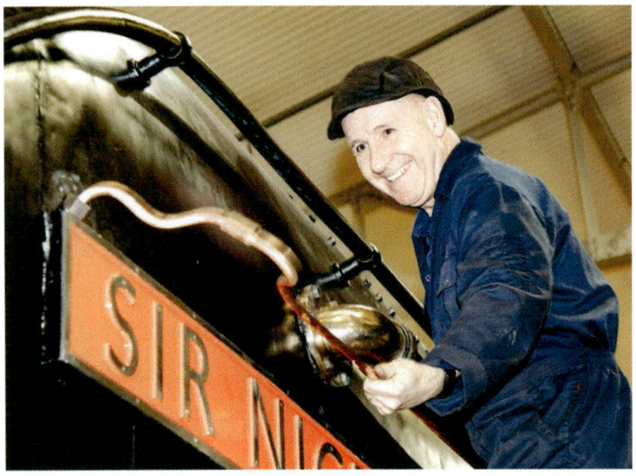

Barry Horton & Peter Ellis (out of picture in smokebox) fit the last pipe of this overhaul

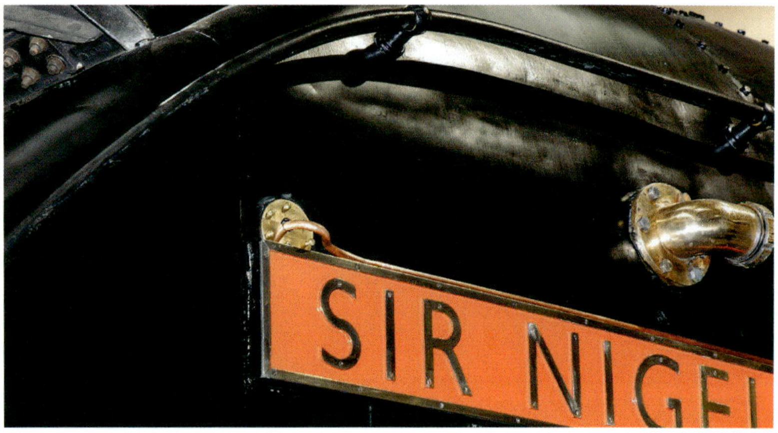

The final pipe to be fitted at the overhaul takes steam to the atomisers in the smokebox and sits along the top of the left hand nameplate

floor, but all the way down into the pit below. Eventually Peter and Barry, with a little help from Darrin, completed the task triumphantly and there were smiles all round.

Darrin and I took a little time reflecting back today and having worked on this locomotive for over six years you do tend to forget that it is world famous, but the reality comes back when visitors, as we had today with a group of teenage prospective engineers, are awed by getting so close to No.7. It is a privilege working on this project and with Darrin announcing an end of overhaul celebratory meal date in mid March and the fact that in 2-3 weeks we would be lighting a fire, it was fast coming to its big end.

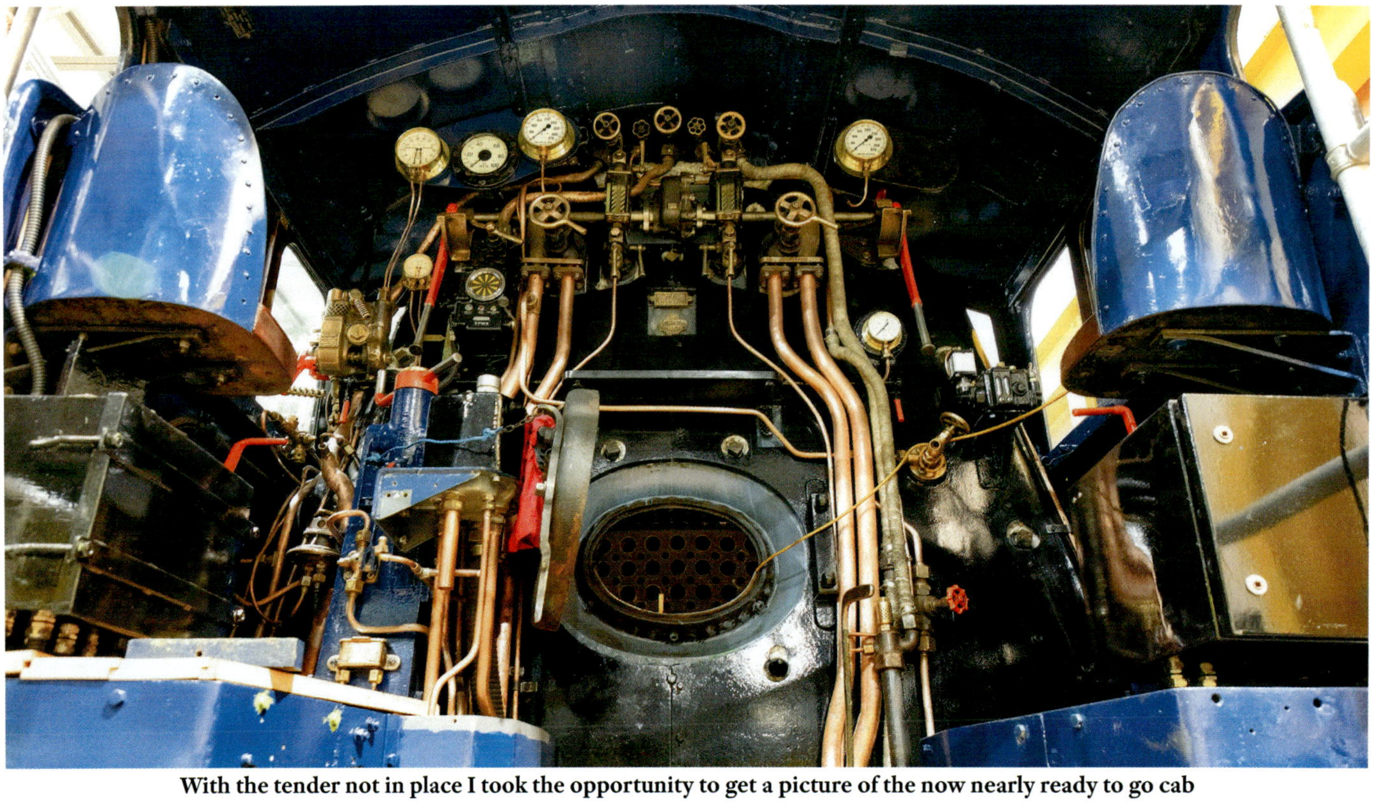

With the tender not in place I took the opportunity to get a picture of the now nearly ready to go cab

WEIGH IN

March 2022

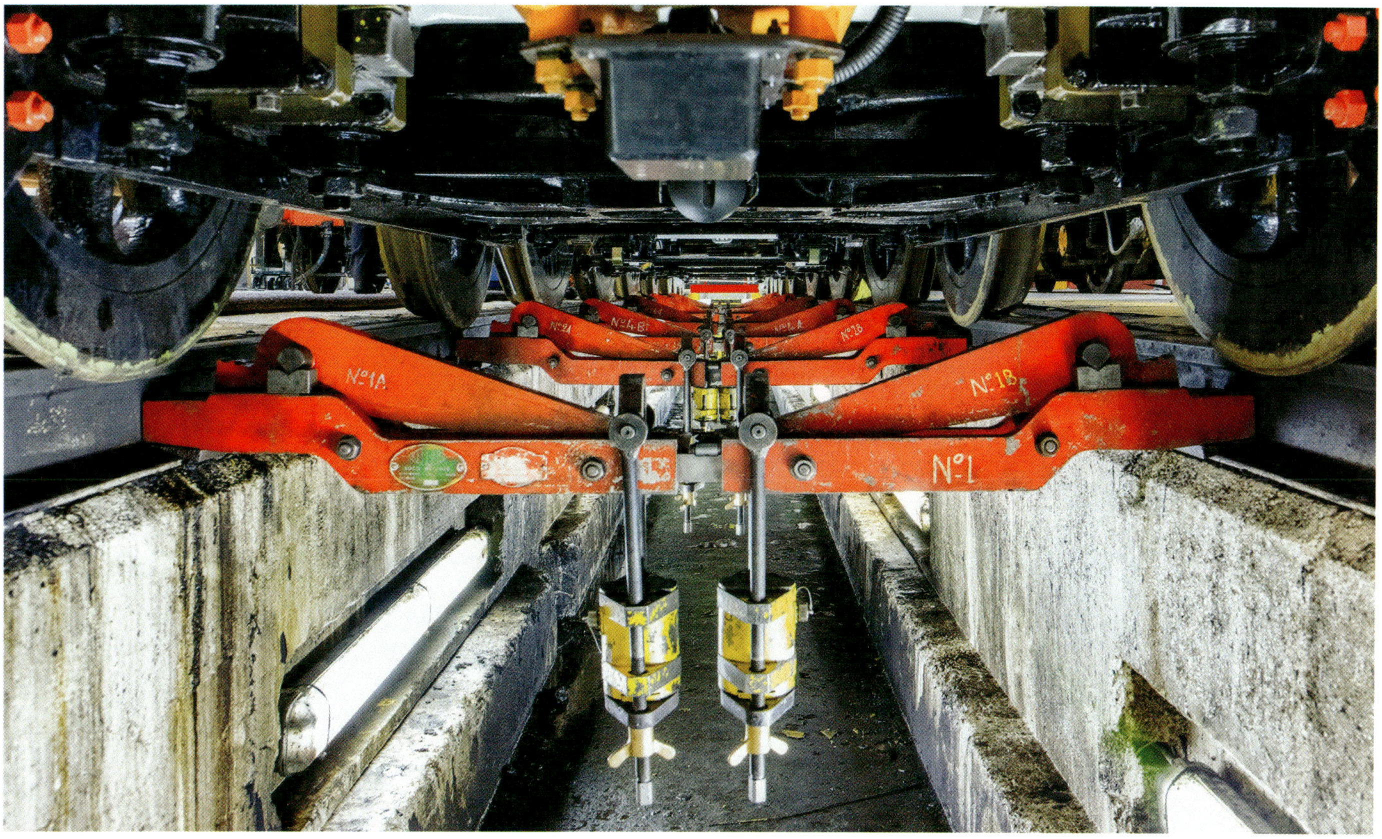

The six hydraulic jacks in situ under the wheel flanges and ready for the lift

It sounded a bit like a prequel to a boxing match when Darrin first said, "We will be doing the weigh in next week," except my interpretation of weigh in was actually weighing. I am blaming his Teesside accent, but nonetheless I still had the wrong idea of what it actually meant until I turned up on the day, saw the kit, and spoke with Richard Swales and Andy Barwick, who would be performing the task. Mark O'Brien from LSL was kindly helping as he had a wealth of experience with this kit which had been kindly supplied by the NYMR. I'm not sure what I expected as I knew the loco needed to be clear of the rails but being suspended at the end of some giant butcher's shop scales was something I was struggling to visualise. Of course, it was nothing like that, but a more sophisticated system as in simple terms a hydraulic jack is placed between the rails under each axle wheelset, six in our case, and jacked up under the flange of each wheelset until each wheel is clear of the rail. The clearance

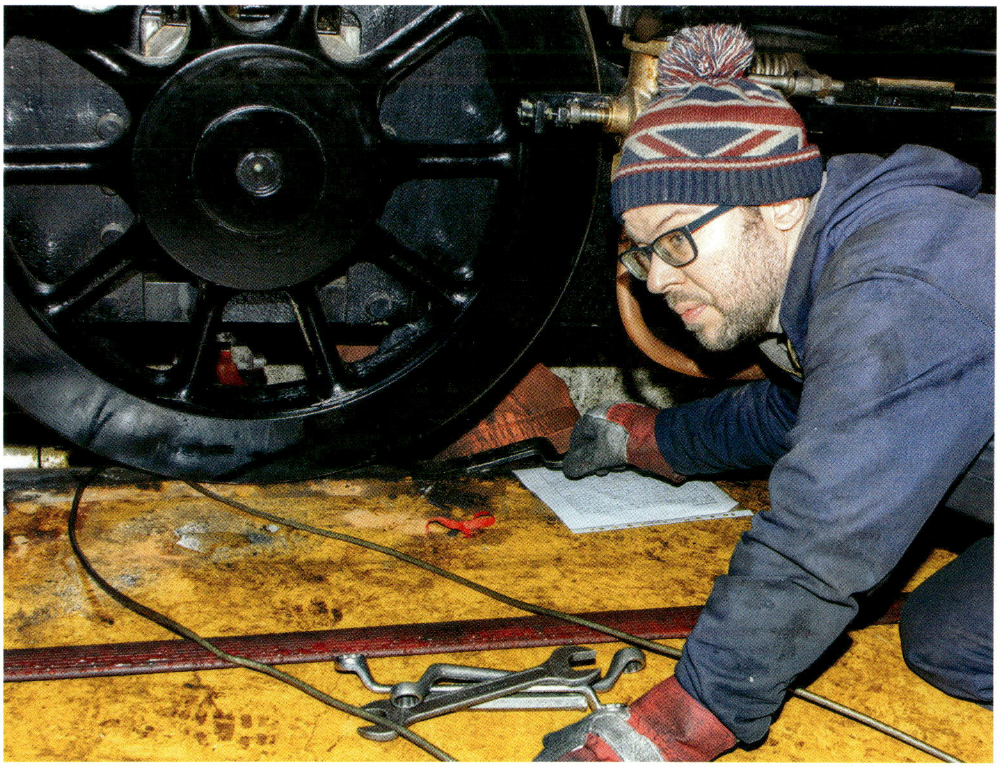

Mark O'Brien checks the first wheel is clear of the rail

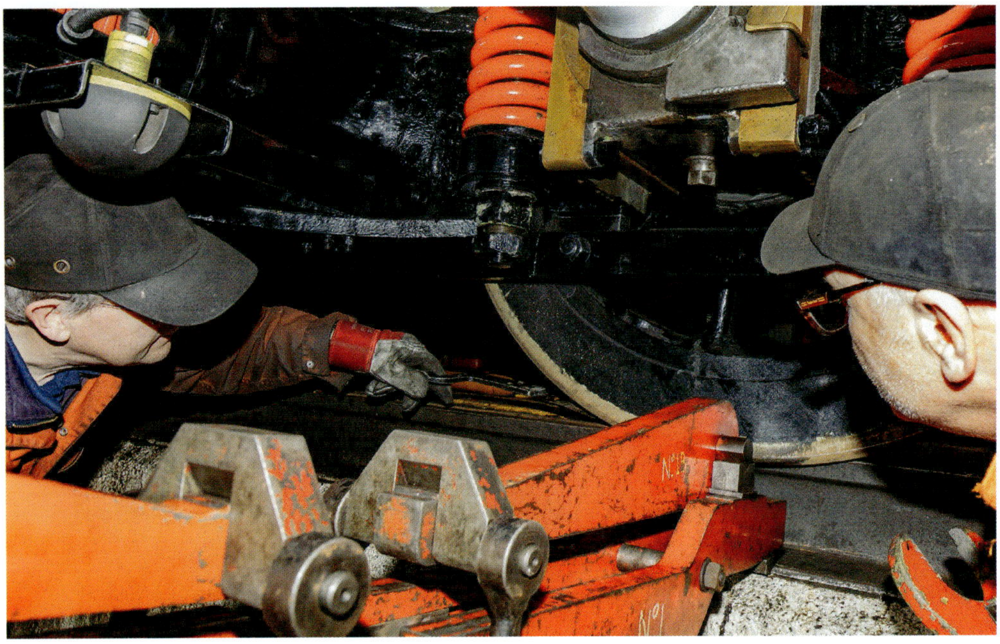

Richard Swales (left) uses a feeler gauge to check the wheel is clear of the track whilst Andy Barwick looks on

only needs to be a few thousands of an inch (sorry, still using old money) to record the weight which the kit electronically transfers to display panels.

That is the simple bit. The point of doing this exercise is not so much to record the locomotive's weight as to ensure each wheel is in contact with the track and none of them are 'floating', i.e., not applying the same amount of weight to the track as all of the other wheels. Getting that done was going to require a lot of hard work from Richard and Andy. With twelve wheels we had twelve readings which needed to be, within a tolerance, the same, and the way to do this was to adjust the axle springs which required the use of a very large 2" socket and a lot of physical effort on the twelve wheels springs. At the end of the day the weighing was not quite completed but Richard said he was very pleased with the results so far.

Due to the high volume of locomotives returning to traffic for the upcoming season of operations, it was a very busy time of the year for the weighing kit, which meant there was only a small window of availability for its use. The loco had to be completed before it could be weighed so the middle cylinder con rod, which was having its big end machined on my last visit, needed to be fitted in place and the concreting of the arch in the firebox had to be carried out, making it quite a race against time for the team. The boiler also needed to be full of water, so this was supervised by boilersmith, Mark O'Brien. Next week the pressure would still be on as once the tender is full of coal and water it will also be weighed before the equipment is returned to the NYMR.

We essentially now had a complete locomotive but there were still lots of jobs being finished by the team and I was not to be excused from assisting them, as putting my camera down for a moment resulted in some rags and white spirit being thrust into my hands. Darrin had promoted me to a cleaner - maybe in a few years I may make it to fireman - a job that would soon be needed as the day for lighting that first fire was almost upon us.

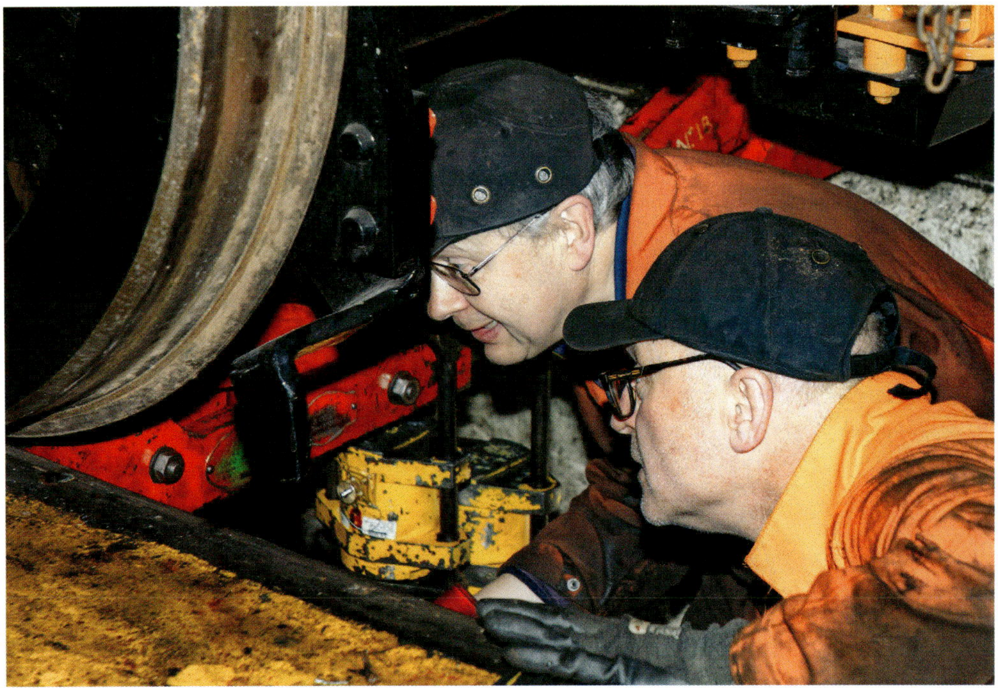

Richard Swales (left) and Andy Barwick check the position of a hydraulic jack

FIRE SMOKE & STEAM

29

March 2022

I have recorded the major events of the last six years by appending a milepost marker, some events have been more important than others but today was to positively trump everything that had gone beforehand as I was eagerly awaiting news from Darrin that the first fire had been lit and that the pressure in the boiler was starting to rise for the first time in over six years. It was planned for Sunday, but the call finally came advising Sunday had now become Tuesday and as I was planning to go the day after the lighting of the fire when the boiler pressure would have risen nicely for the spectacular steam testing to take place, my Monday drive to Crewe had now turned into a Wednesday drive to Crewe. Of course I would have liked to have captured the lighting of the fire, but as flexible as I was, I simply couldn't make that ceremony. I was nonetheless delighted to find out that Peter 'Bracko' Brackstone had the honour of doing it as it was very fitting that lighting the fire to heat the water in the boiler should be done by our top boiling the kettle tea maker Bracko. On a more serious note Bracko was by far the highest attending volunteer over the last six and a bit years and fully deserved the honour.

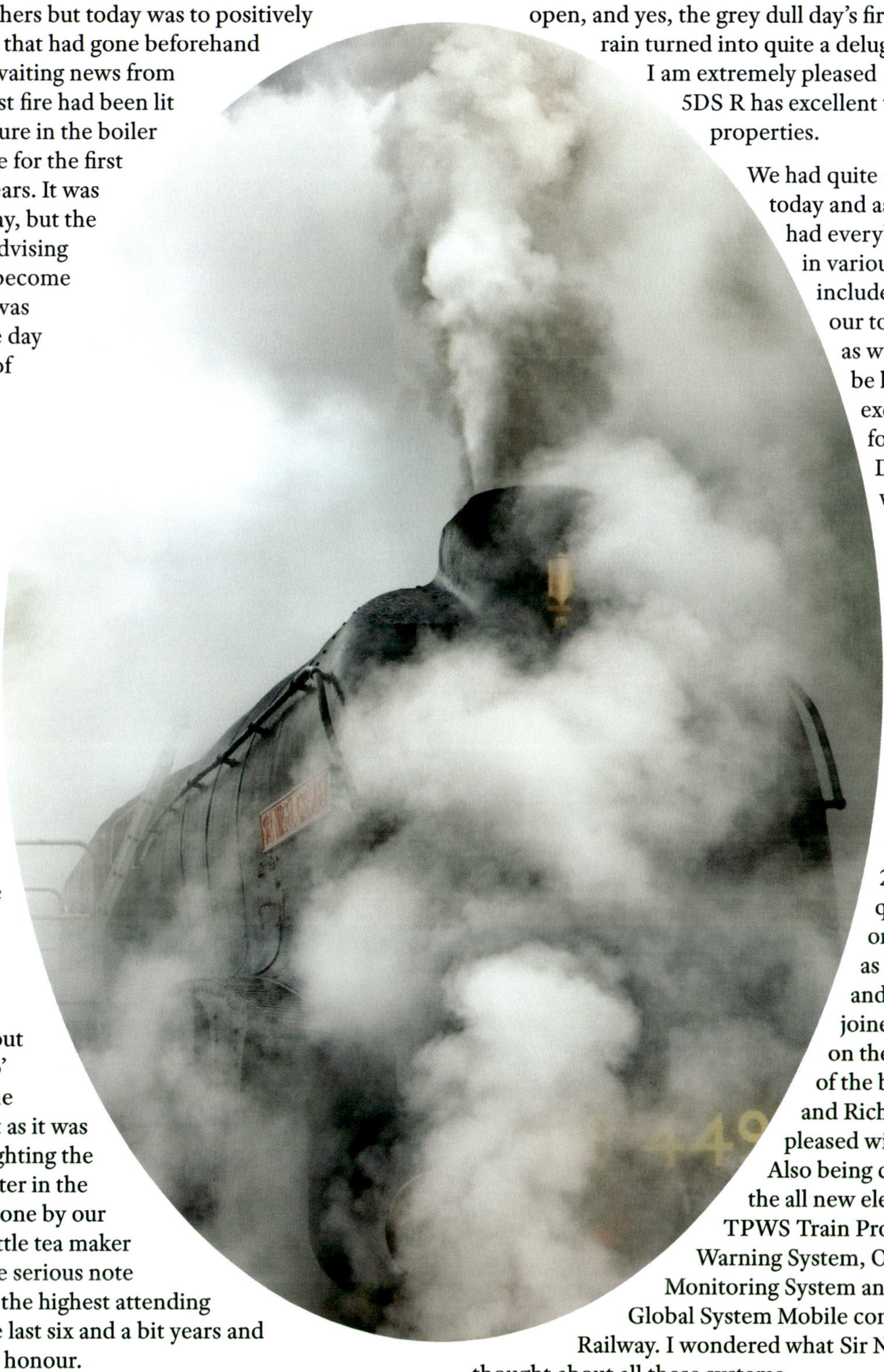

No.7/4498 was simmering away nicely in the prep bays on the north side of the running shed when I arrived on what was a grey overcast day, not ideal for taking pictures of white steam against a slightly less white, grey sky and the images would require some careful editing. Still, I am getting used to that as we have more often than not had dull and wet weather conditions when we have been out in the open, and yes, the grey dull day's first few spots of rain turned into quite a deluge yet again, so I am extremely pleased that my Canon 5DS R has excellent waterproofing properties.

We had quite a team on site today and as always Darrin had everybody organised in various tasks which included collecting all our tools together as we would soon be leaving LSL's excellent facilities for new pastures. Darrin himself was initially busy with the boiler inspector who was checking everything was in order with our documentation before joining Richard Swales on the footplate as the pressure was rising to 250psi. It was quite crowded on the footplate as Andy Barwick and Nigel Hoskison joined us for checks on the water tightness of the backplate. Darrin and Richard were well pleased with the results. Also being checked was the all new electric systems, TPWS Train Protection Warning System, OTMS On Train Monitoring System and the GSMR Global System Mobile communication Railway. I wondered what Sir Nigel would have thought about all these systems.

The steam tests were indeed spectacular and provided me with some great pictures, the first of many more to come as we now move on to the Severn Valley Railway running in trials.

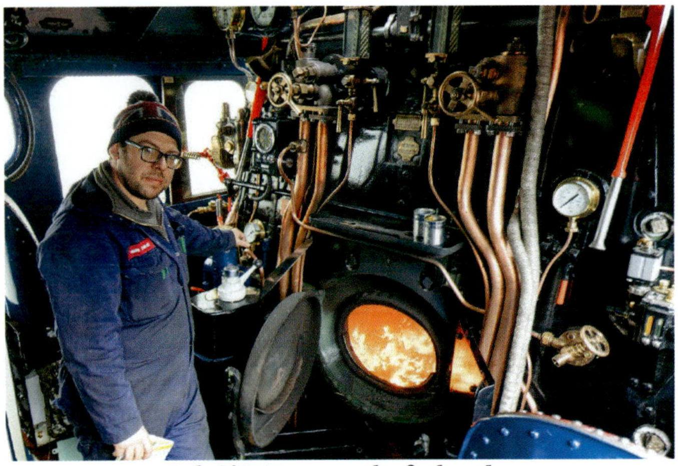

Mark O'Brien opens the firebox door

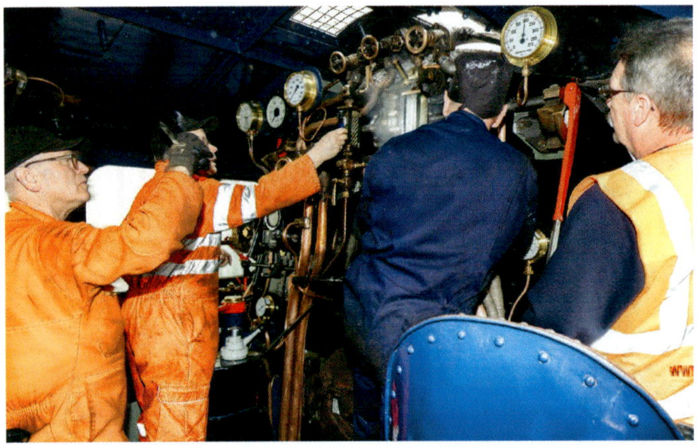

Andy Barwick and Richard Swales hold lights while Darrin Crone checks for leaks as the pressure rises during the steam test, Nigel Hoskison watches on

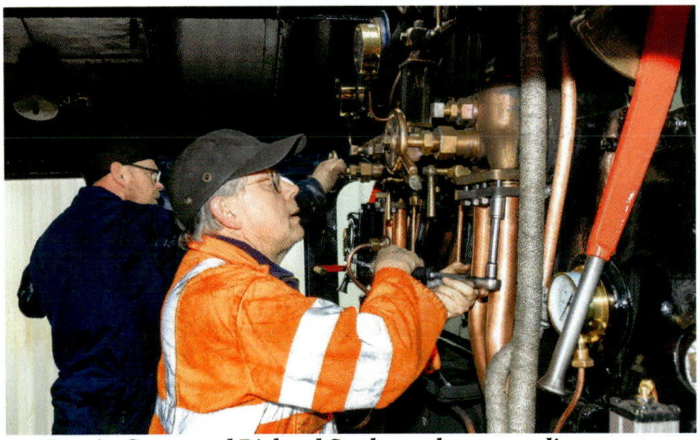

Darrin Crone and Richard Swales make some adjustments

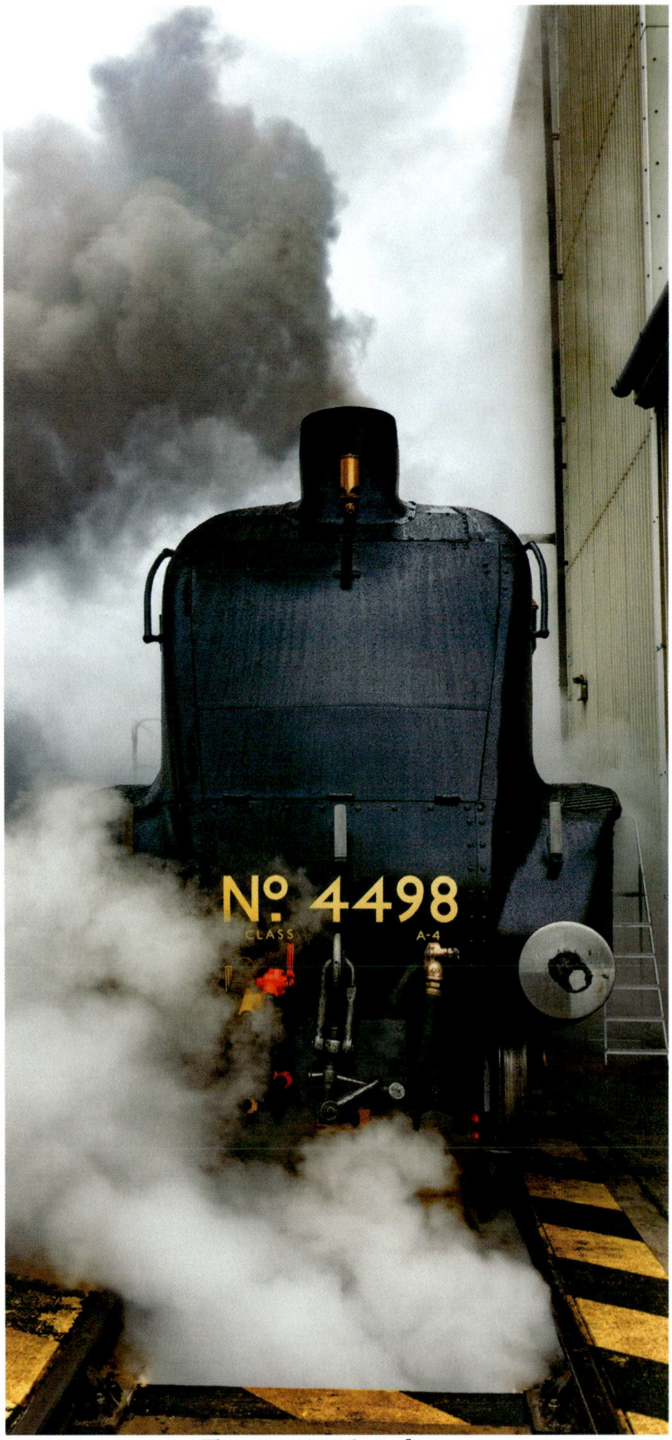

The steam test is underway

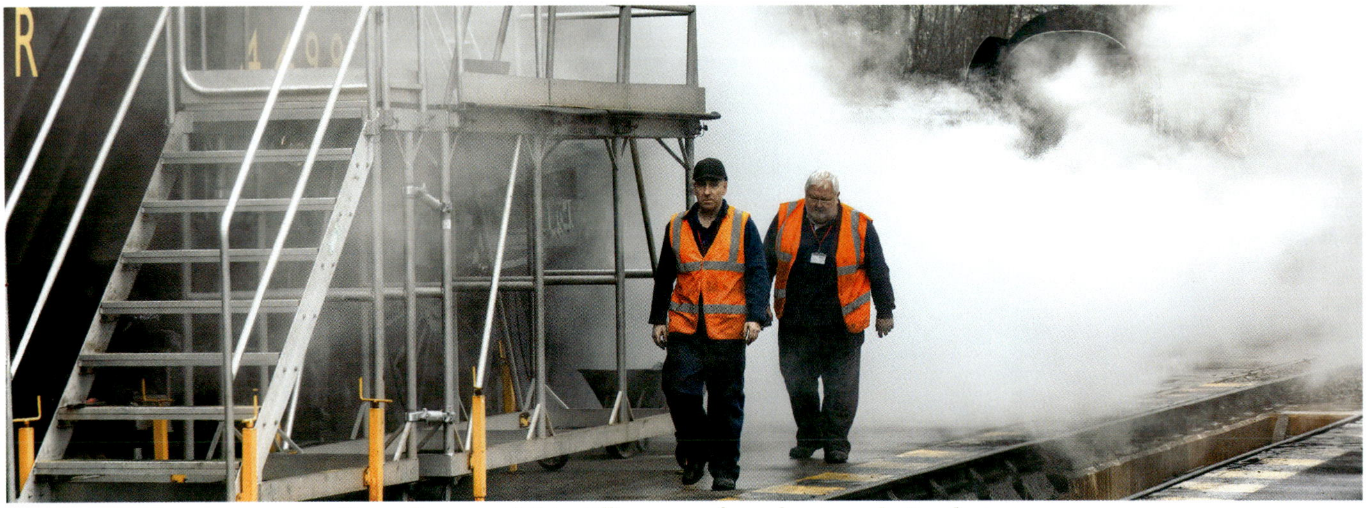

Barry Horton and Peter Ellis emerge from the steam during the tests

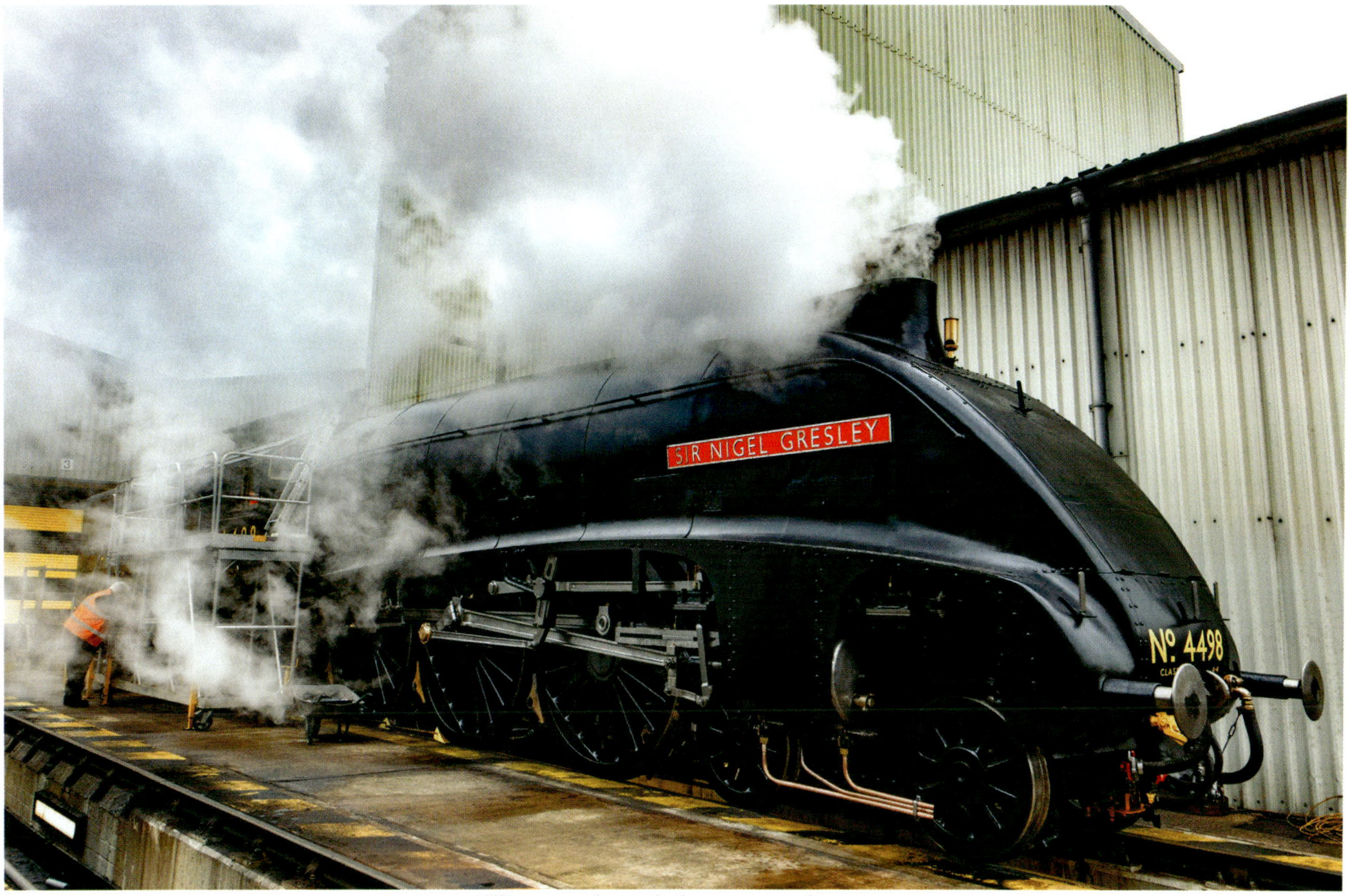

Peter Ellis peers through the steam as the test starts

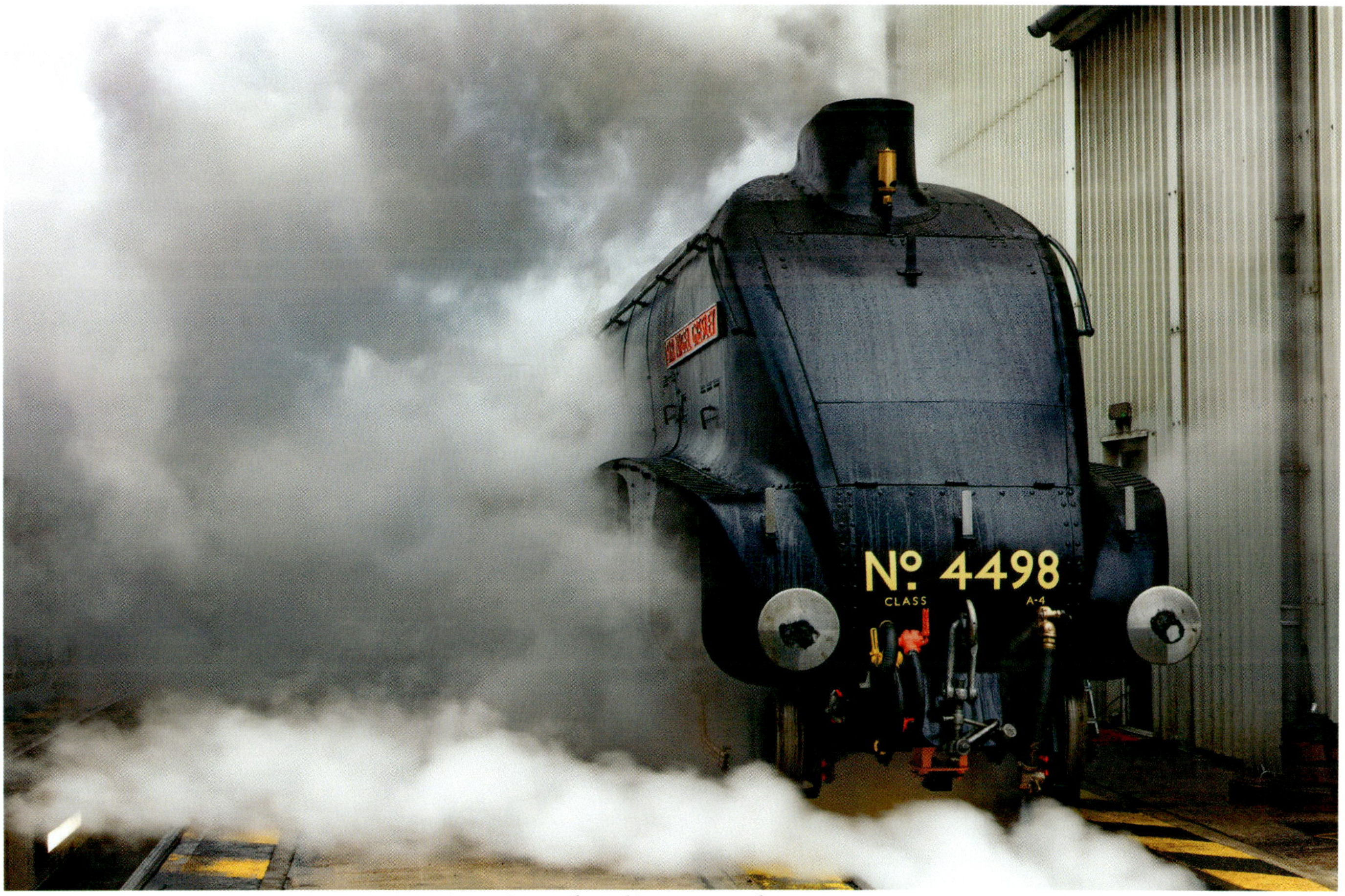

No.7/4498 is enveloped in steam as the test progresses

30

RUNNING IN

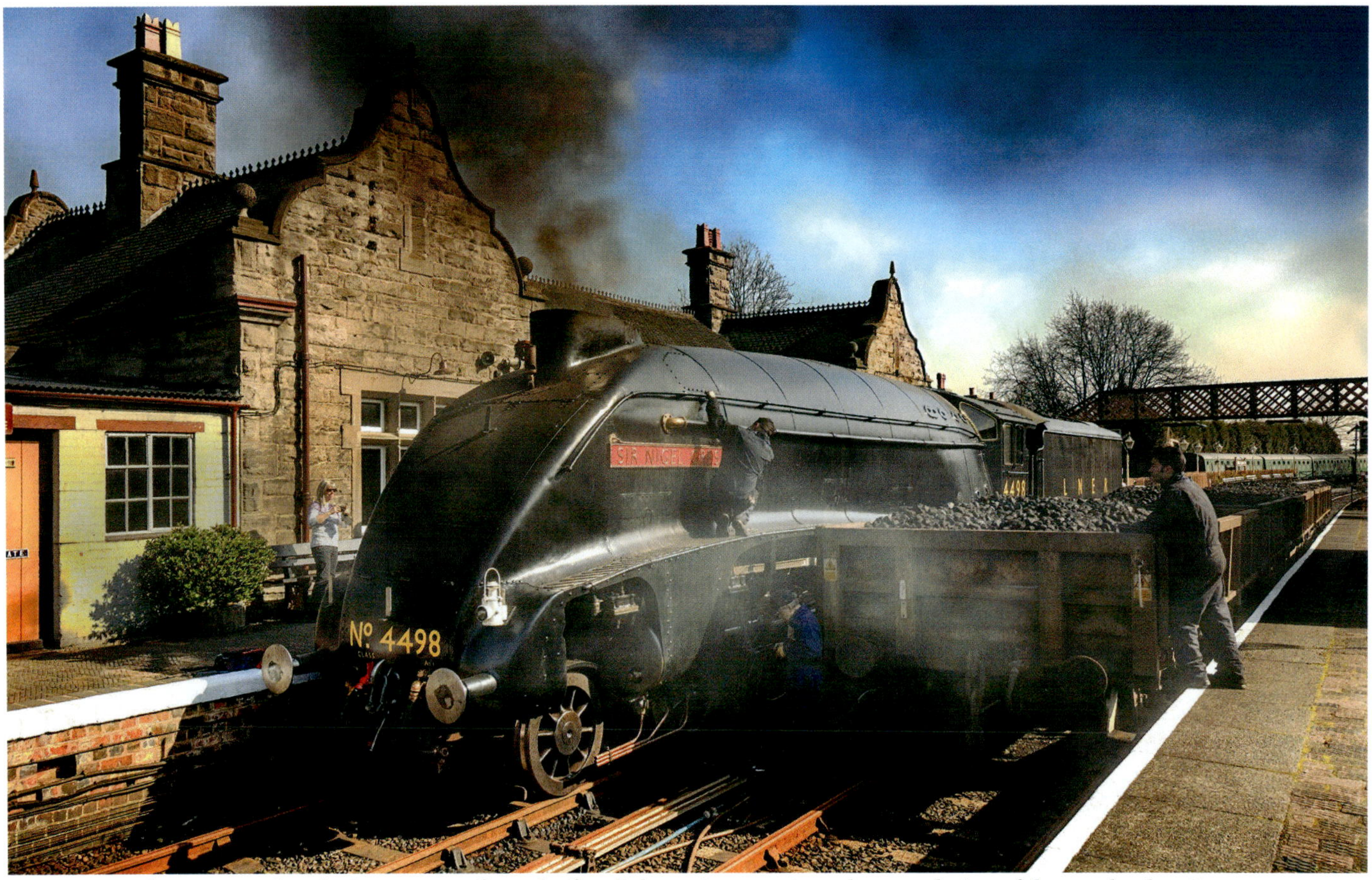

A busy scene after the loco returns from the first light engine run since the completion of the overhaul at Crewe

March 2022

After a road move from Crewe to Bridgnorth and six and a half years after coming out of traffic, No.7 would be moving under her own steam again today on the picturesque Severn Valley Railway. It was a very warm day with no clouds, just wall to wall blue skies and not even the hint of a breeze; the kind of summer's day when all you want to do is laze back and listen to bird song interrupted by the odd voices from distant conversations. Except it wasn't summer. We had only just crept over the line from winter into spring, so this unseasonably warm weather was surely a good omen for our running in trials which was how it turned out to be with the first three days of trials all going very well with just a few snags to resolve.

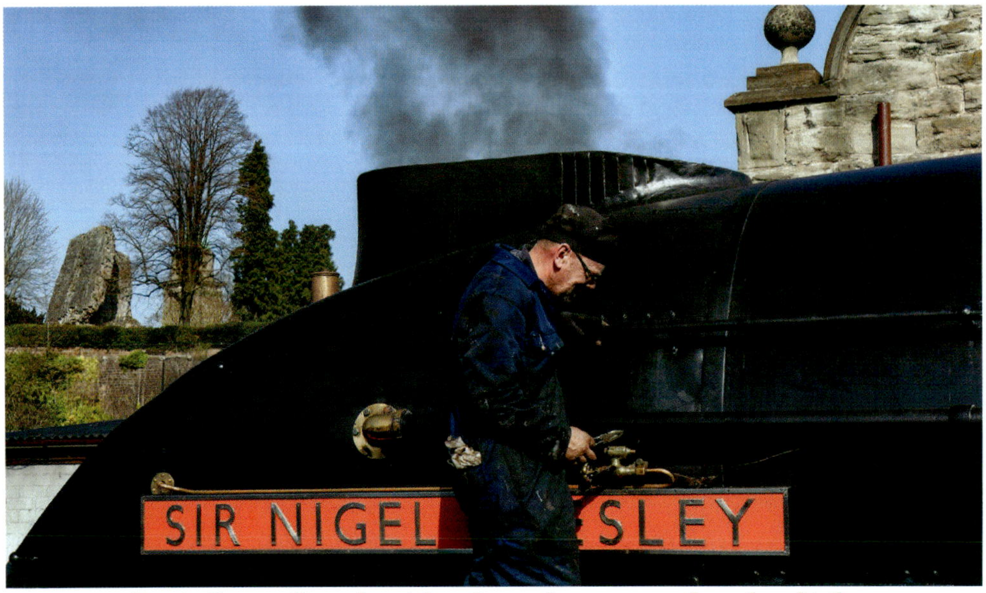

Darrin Crone adjusts the sticky valve on the steam supply to the whistle

I arrived at Bridgnorth station at around midday, parked the car and walked up the bank to the very quiet platform wondering where everyone was. I'd missed the departing loco by a few minutes, but the planned run was only a few miles down the line, so I quickly surveyed the station for vantage points, of which there were several. After catching up with Darrin, whom I found in the support coach, I moved into

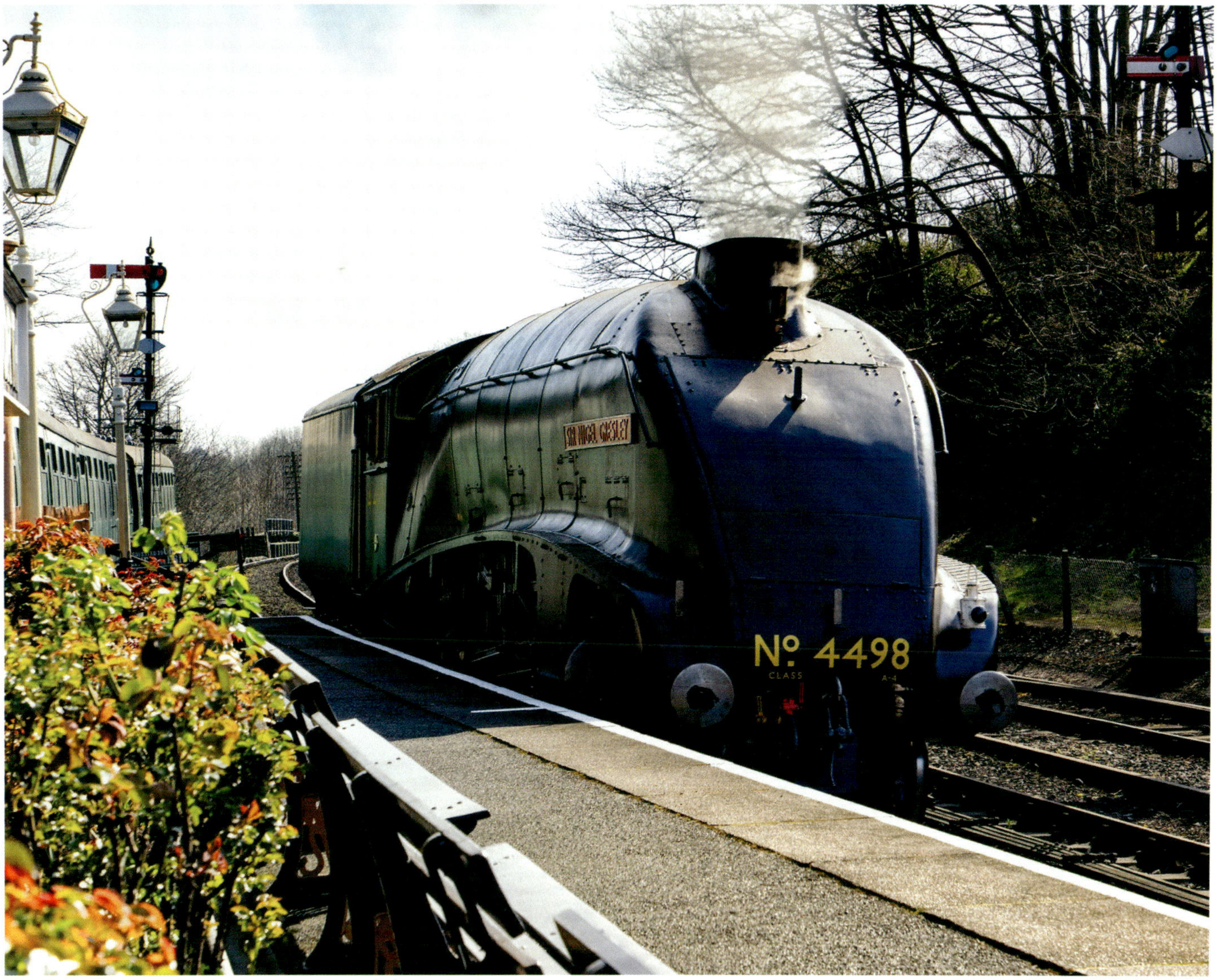

Returning into Bridgnorth from the first light engine run down to Highley on the Severn Valley Railway

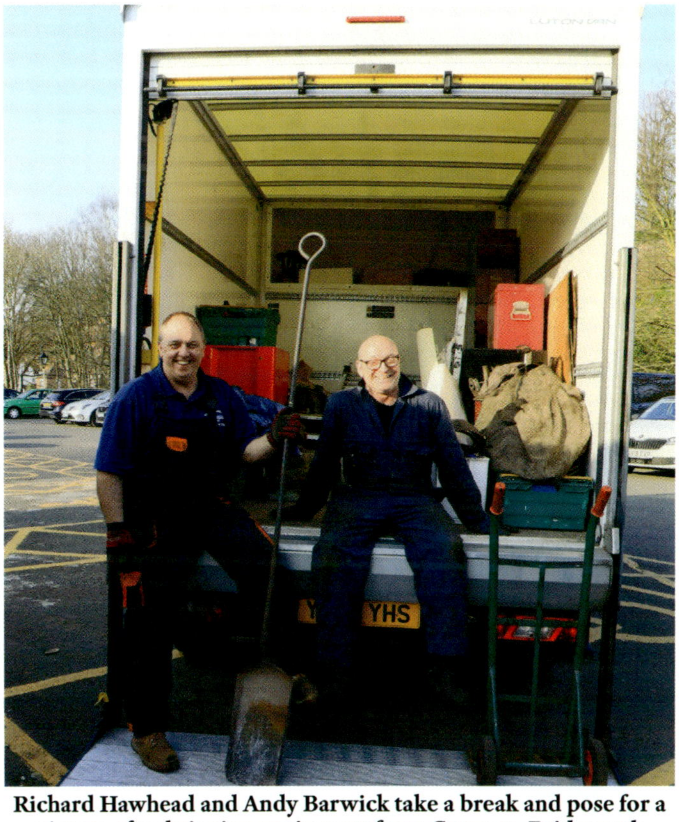

Richard Hawhead and Andy Barwick take a break and pose for a picture after bringing equipment from Crewe to Bridgnorth

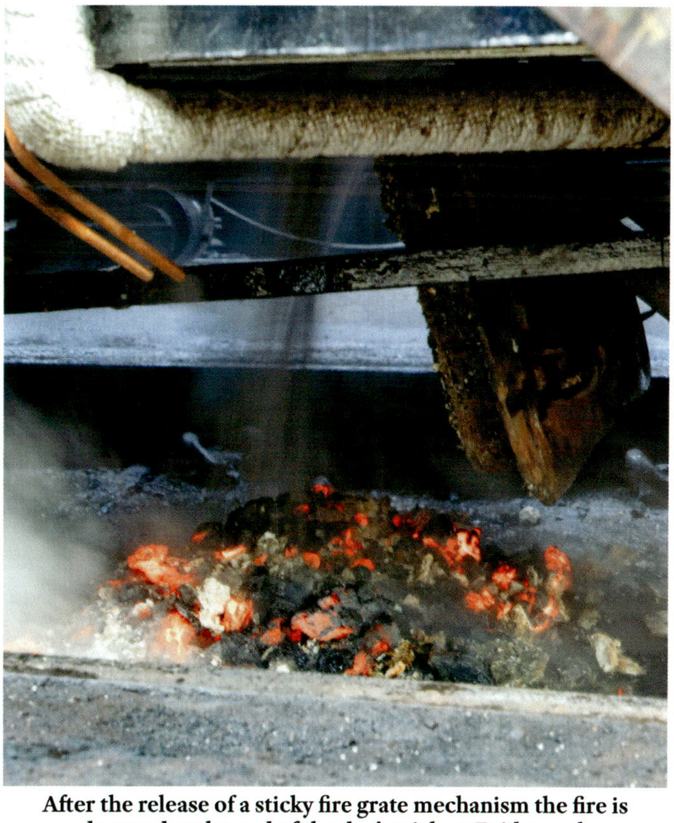

After the release of a sticky fire grate mechanism the fire is dropped at the end of the day's trials at Bridgnorth

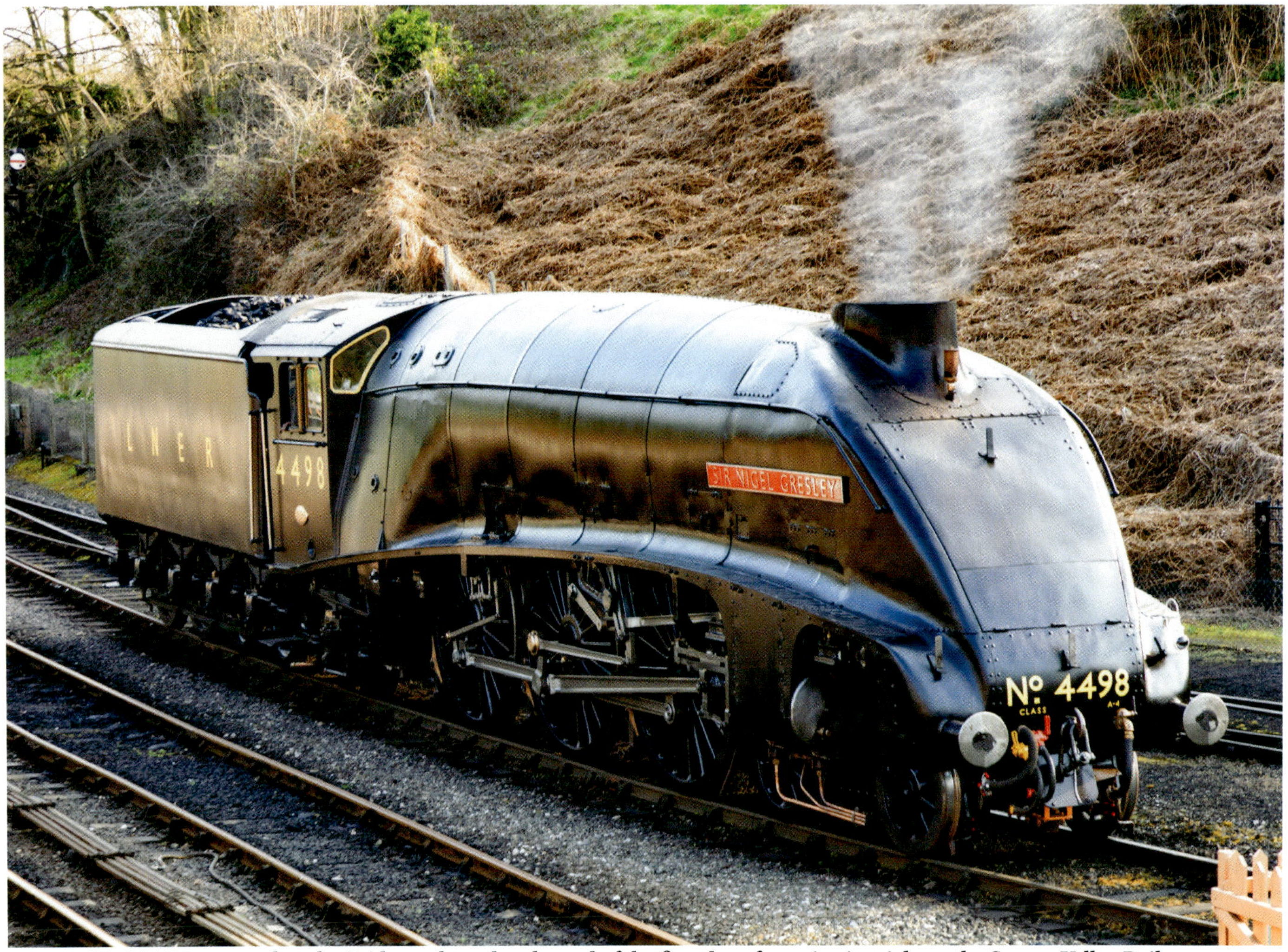

No.7/4498 approaches the Bridgnorth yard at the end of the first day of running in trials on the Severn Valley Railway

position at the end of the platform. After capturing images of the return, I walked down to the alive again loco to mix with the team who had suddenly all appeared.

On the next run I did try to follow by car but not being familiar with the line, all I achieved was a reconnaissance trip. I would have more success the following day when I discovered Arley station and the Victoria Bridge. On this return the loco was routed to a stabling point outside of the sheds as the first day of trials was coming to an end. The fire was dropped after the release of a stubborn fire grate mechanism, and a sticky valve on the whistle that was causing it to stay on was adjusted. However my first day ended with me being hijacked whilst walking past the support coach, as H&B removals, in the form of Richard Hawkhead and Andy Barwick in a hired van, had arrived with a cargo of spares, tools, tea and biscuits which they had brought from Crewe, and all this needed to be transferred to the support coach. The tea and biscuits and even the washing up bowl and cutlery I managed quite easily. However sets of leaf springs for railway vehicles and four feet high metal tool boxes are just a bit heavier and required a trolley jack and a lot of muscle power to move and raise from platform height to coach door height. I really did enjoy my soak in the very comfortable bath at the George hotel later.

Trust chairman, Nigel Wilson, was also staying in the George so we decided to have our evening meal there. Barry Horton and Peter Ellis, along with Richard H and Andy joined us, and the food was excellent as was the ale. Refreshed after a good night's rest and a full English breakfast we all went our separate ways the following morning. Richard H and Andy returned to Crewe to collect more equipment and spares, Barry and Peter E joined up with Darrin, Richard Swales, Peter Brackstone and Phil Gillespie back up at the station, while Nigel and myself headed off down the track to different locations to capture images.

I was heading for the Victoria Bridge between Bewdley and Arley, but first I wanted to catch the loco at Arley station, as she returned north from her first run. I had timed this just right as I only had a few minutes to wait, and during this time I introduced myself to the army of station volunteers who were getting this beautiful little station ready for the start of the season. They gave me a warm welcome and I thank them very much for their hospitality. After the northbound run the local volunteers estimated I had about an hour to get to the Victoria Bridge before the next southbound run, so after chatting a little longer I set off on my mile or so riverside walk to the bridge which was not accessible by car.

Mine and Nigel's paths crossed again as I found him waiting to film on my way to my chosen location but I had to press on to be in position in time so it was a brief "I'll catch up with you after the southbound," and onwards under the bridge to be in a position with the sun as much as possible behind me. Ideally a boat in the middle of the river would have been perfect but I didn't have a boat, so I was as close to the river as I dare go. Southbound shots taken, I rejoined

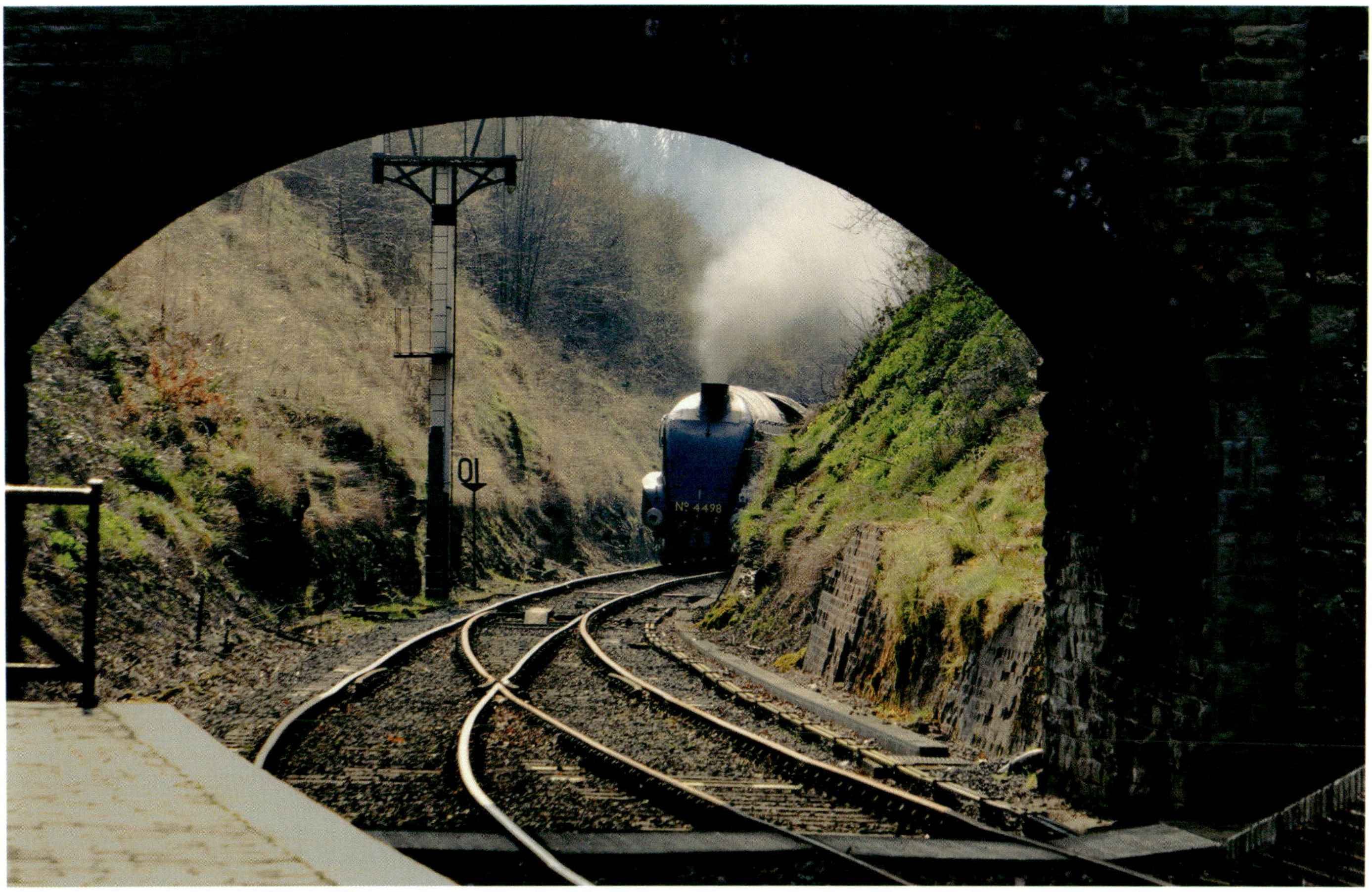

Approaching Arley station on Day 2 of the light engine running in trials at the Severn Valley Railway

Nigel in time to capture the northbound return before we returned to the cars and I called into Arley station to bid farewell to my new friends.

I had all the pics I wanted so I was now heading the 150 miles home to Yorkshire, and as I drove home there was a tinge of sadness in my thoughts that the overhaul, which was now coming to an end with these trial runs, had been a major part of my life for over six years and that was about to change with the upcoming return to traffic. What was I going to do? Only time will tell but sell some books hopefully!

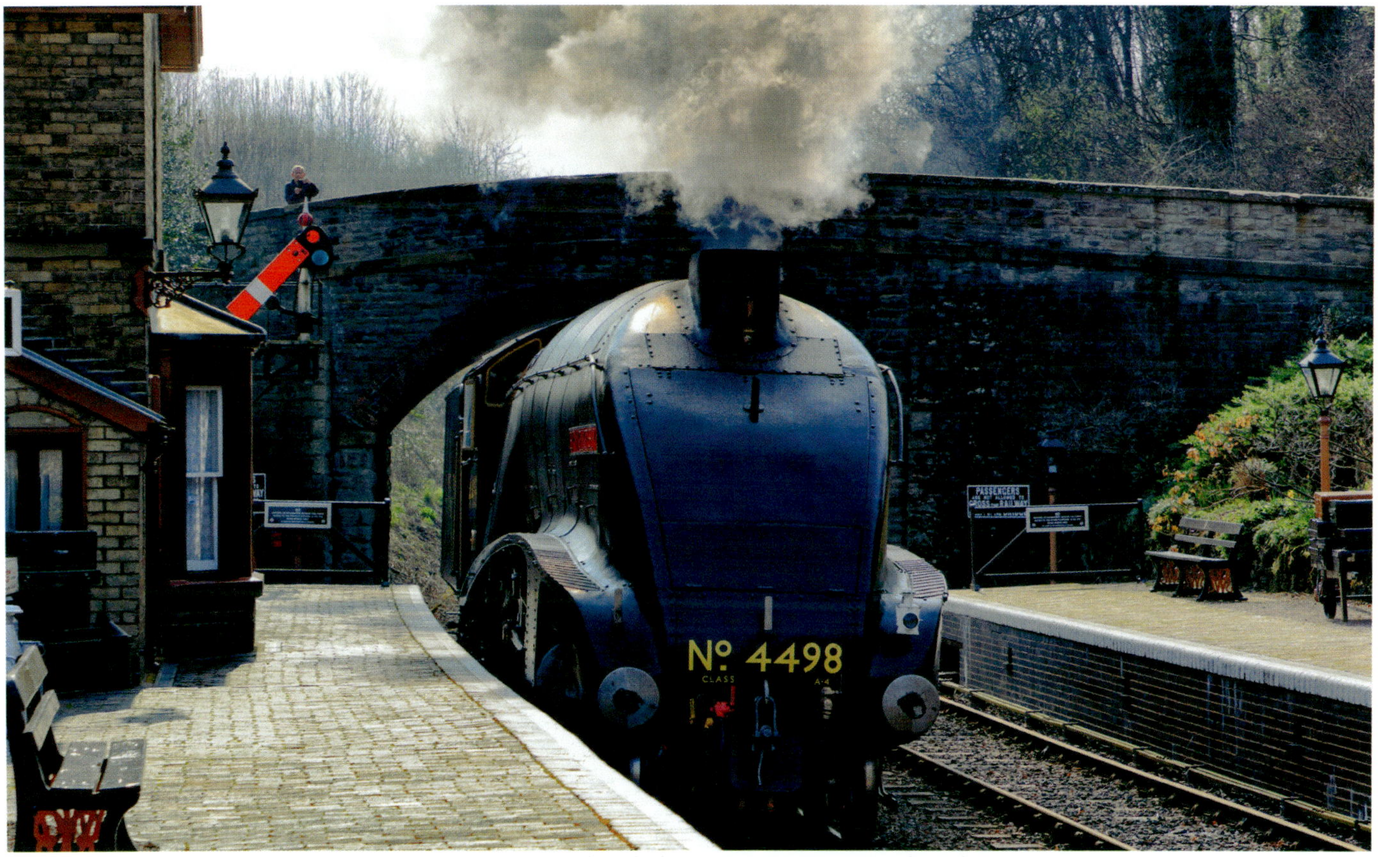

Passing through Arley station on Day 2 of the light engine running in trials at the Severn Valley Railway

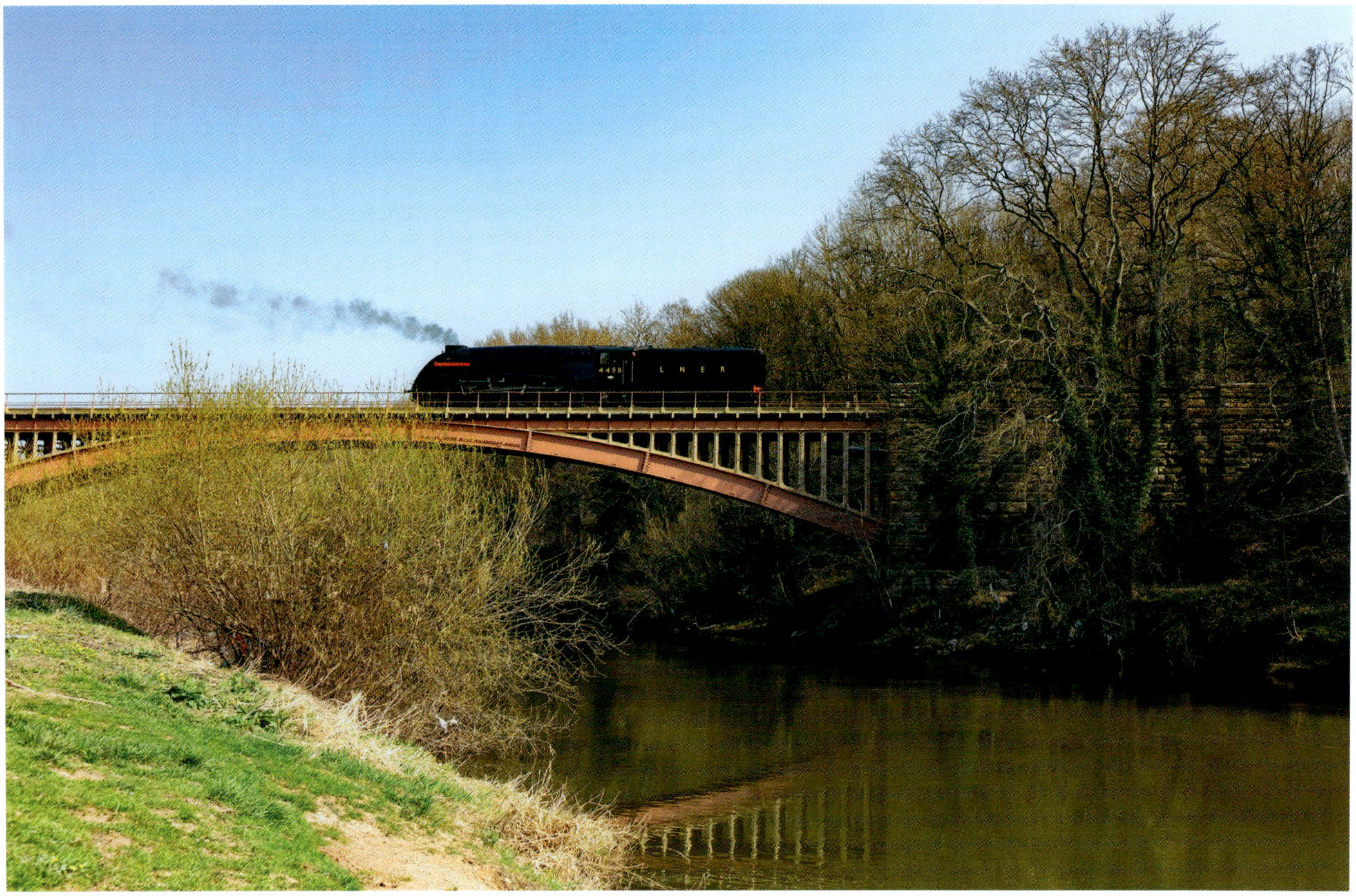

Heading to Bewdley across the River Severn at Victoria Bridge in reverse during Day 2 of the light engine running in at the Severn Valley Railway

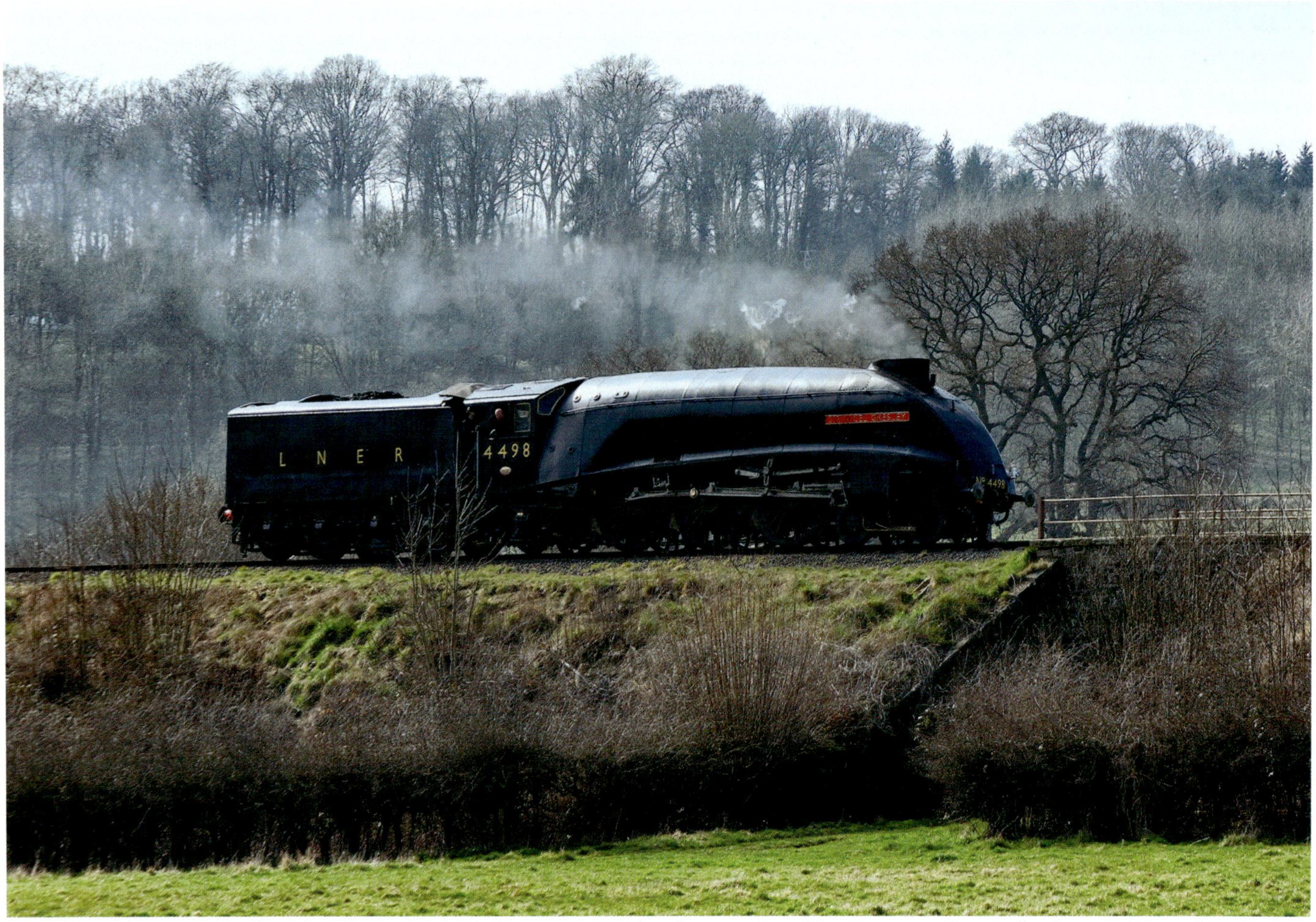

After crossing the Victoria Bridge the light engine trial heads towards Arley station on the run back to Bridgnorth

CHAMPAGNE RETURN

31

April/May 2022

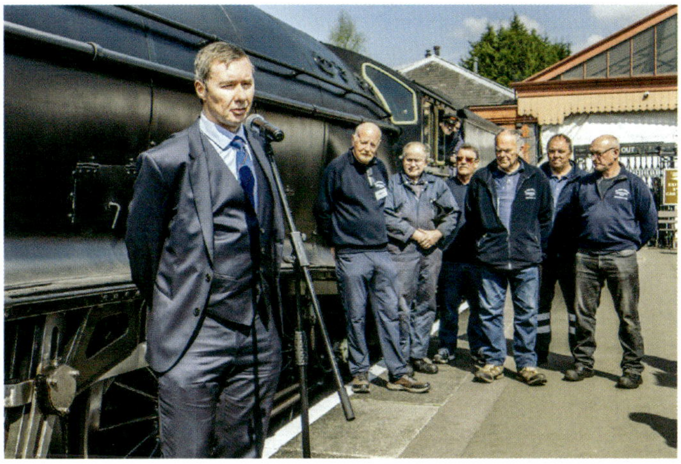

Trust chairman Nigel Wilson announces the return to traffic at Kidderminster on the Severn Valley Railway

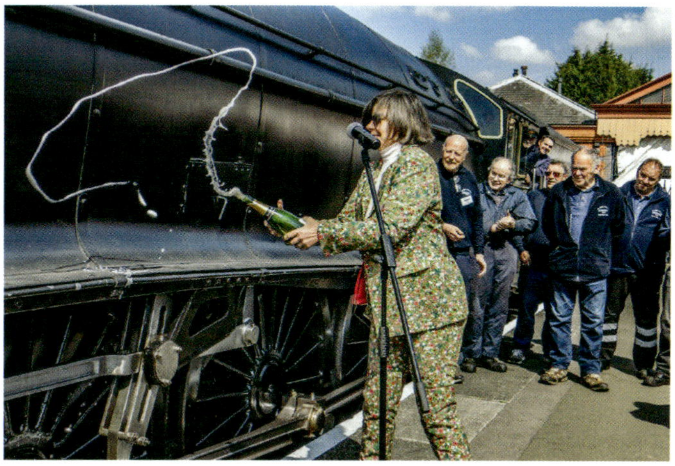

Susan Wilson cracks open the champagne at the return to traffic ceremony at Kidderminster on the Severn Valley Railway

The return to traffic was at the SVR Spring Gala in Kidderminster where the milepost event was marked with a champagne ceremony and speech by Trust chairman, Nigel Wilson. The overhaul was now essentially complete. Darrin still had a list of jobs to be done and we needed a successful loaded test run on the mainline, but we were operational, running up and down the SVR just about every day. My story of the overhaul was coming to an end and would finally conclude with the Trust's members' inaugural run on the mainline which was planned for 21st May 2022. I took a little time off awaiting the conclusion of the trials and during the SVR Gala and ceremony and I thank Michael Holden from Rail Advent for supplying pictures of the event in my absence.

This period did remind me, and not for the first time, of how privileged I had been to have my exclusive access. Following the running in/SVR Gala on social media there were now hundreds, if not thousands, of people taking photographs and videos when previously there was just me. As the lone photographer on Arley station, in a field along the line and on the banks of the river Severn by the Victoria Bridge, I found the inspiration for the phrase "Seven on the Severn", leading to my poster design shown here.

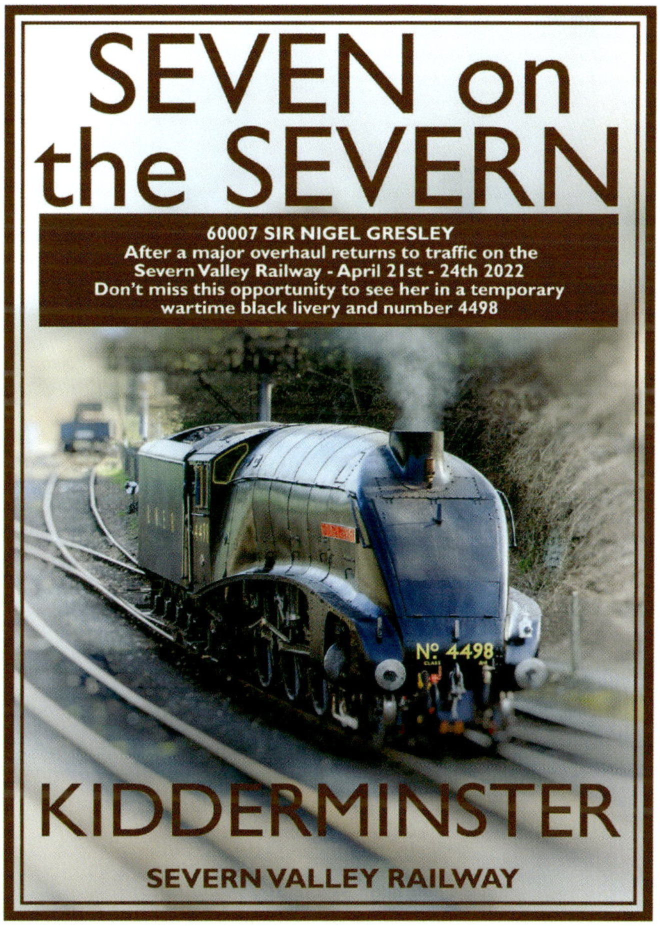

Earlier in the overhaul the Trust members had voted for the loco to return to traffic in her familiar BR Express Blue, but as covered on page 159 and as a result of having to move the loco from York to Crewe, the loco had been painted black and would now be repainted blue at a later date in LSL's paint shop. However due to operational reasons, logistics and paint shop availability, the repaint would not take place before the inaugural Trust members' run, so the return to mainline traffic would now be in the temporary wartime black livery.

There was no break for Darrin, Richard and the engineering team as although the overhaul was effectively complete with the return to operational running on the SVR, there were still quite a few niggles to sort out before the return to mainline duties. I didn't get back into the seat until after the trials, the SVR steam gala and the loaded test runs to approve Network Rail mainline running were complete and that seat was a very comfortable, cushioned one on the return to mainline traffic between Crewe and Carlisle on Saturday 21st May 2022.

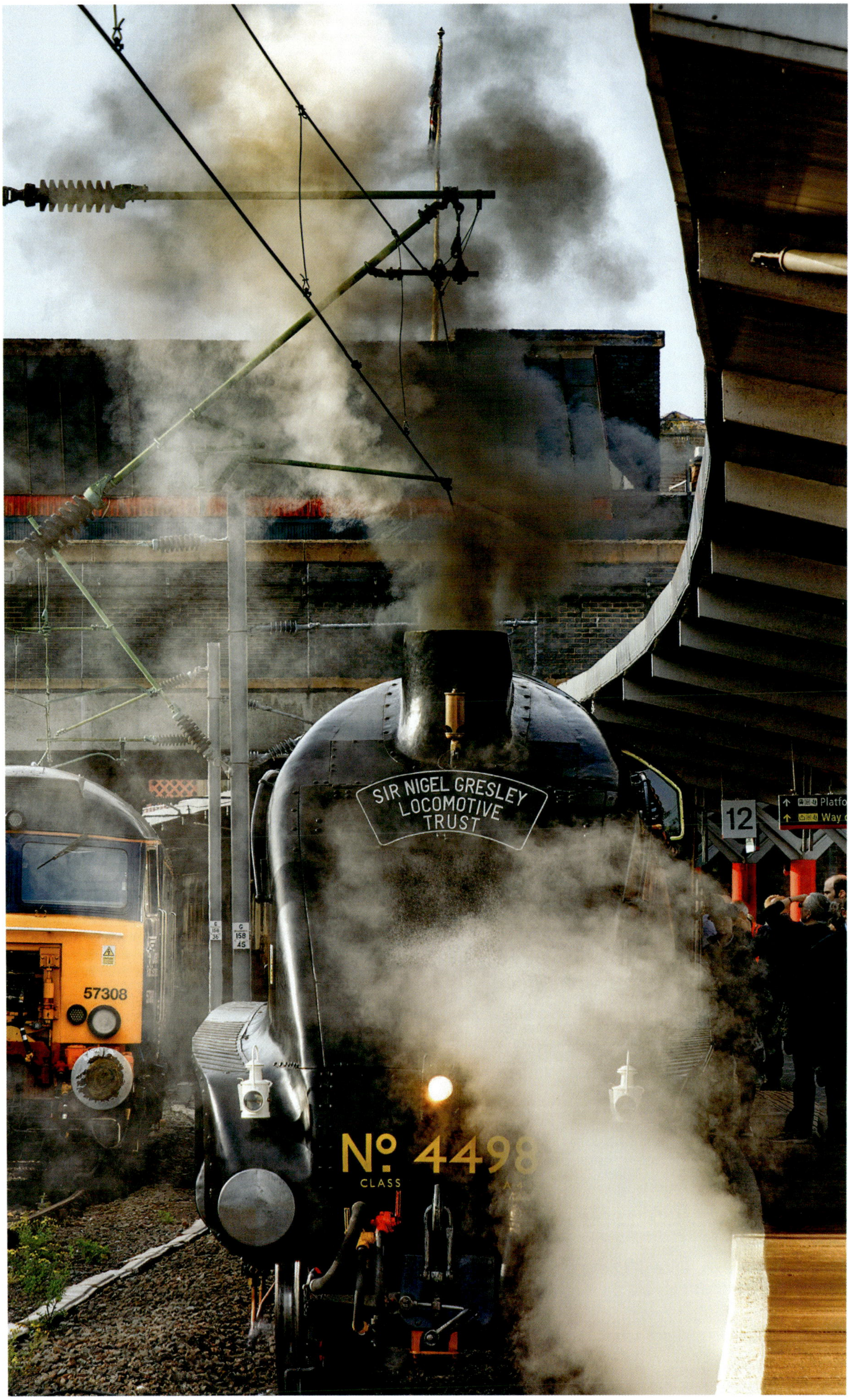

No.7/4498 waits at Crewe's platform 12 while the passengers board the Inagural Run to Carlisle and return

INAUGURAL RUN

32

May 2022

The big day was finally upon us. Saturday 21st May 2022 and the return to mainline traffic with the Trust Members' Inaugural Run from Crewe to Carlisle, a repeat of the very first inaugural run in 1967 after the loco was purchased from British Railways by what is now SNGLT.

Due to the early departure from Crewe many ticket holding members had stayed overnight and were in the various hotels around the area, which resulted in smaller groups getting together the night before. I had a very enjoyable evening meal with trustees and engineering volunteers, in particular two members of 'The Stayers' (Page 170), Malc Bateman and Bob Shearman together with his wife, Jackie, and Trust member and friend, Nick Clennett, resulting in the quality of the desserts being compared with Jackie's excellently baked cakes which were regularly consumed in the blue cabin back in York, provoking yet another fond memory of the last few years.

When we left the hotel by taxi for the station the following morning it was a little overcast, cool and still but not raining, which was a bonus as a wet day had been forecast. We thought we would be early, but we were not the first by some margin, encountering so many familiar faces with the trip being a members' special. The train had also beaten us, so after locating our coach and seats, I was off up the platform, saying, "Hello, Hello, Hello," all the way to the front where I found No.7/4498 gently panting away waiting for the departure. I also found a very proud and smiling Richard Swales on the footplate and the members of our newly formed Operations team by the loco. We now had two teams of volunteers, Engineering and Operations, with an overlap between the two. This was the first outing for the newly formed Operations team, and I managed to group them all together alongside the loco for a team photo before we departed. Wearing their smart new uniforms, they complemented the equally well turned out Saphos Trains team who were our hosts for the day.

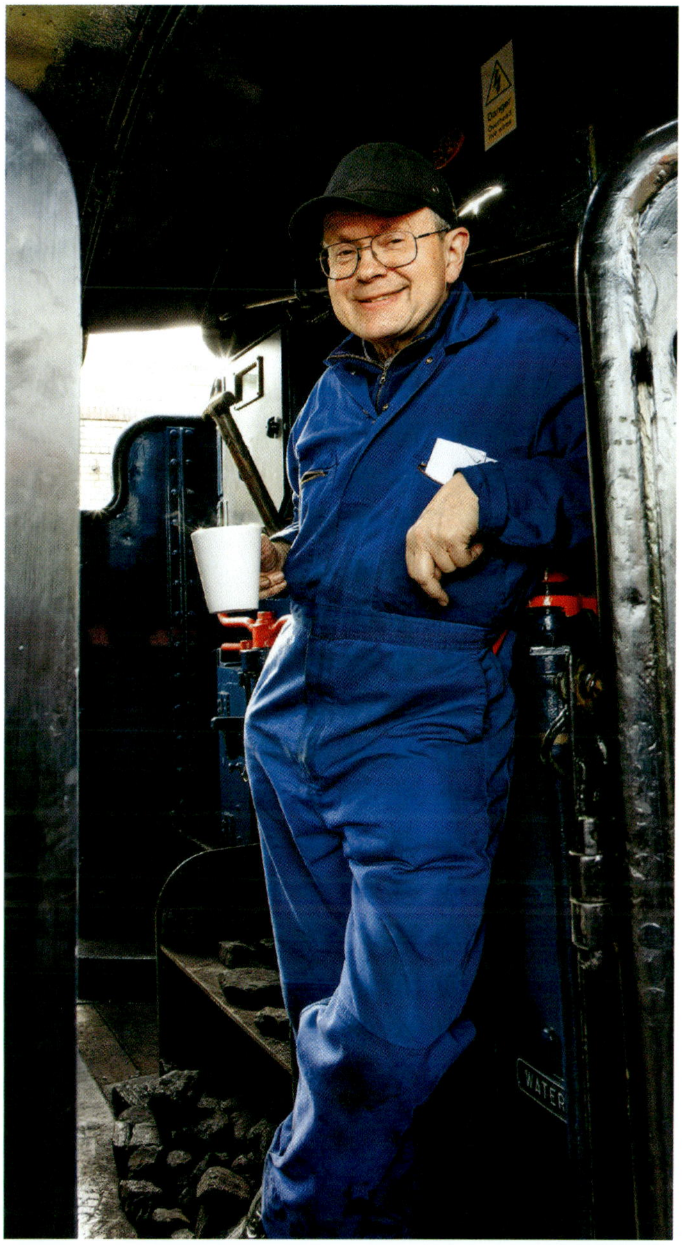

CME Richard Swales savours the proud moment on the footplate

It was a great day out with the loco performing exceptionally well - northbound over the Settle and Carlisle line with crowds all the way particularly at Ribblehead viaduct, and then southbound over the WCML, tackling the climb to Shap summit with ease before racing down the Lune gorge with a water stop in Preston before the sprint back to Crewe.

I hadn't planned to take lots of images today, I wanted to sit back and enjoy the trip, but I had brought my camera as there were a few shots I needed, especially of the notable personalities who were on-board the trip, including Ben Godfrey, grandson of Sir Nigel Gresley himself, and Les Jackson and Neil Cadman, who were members of the train crew on the 1967 runs after the loco was purchased from British Railways in 1966, not to mention Richard Mellish who was also on the 1967 inaugural run. At the Appleby station water stop I did the dash from our coach, second from the back down the platform, over the foot bridge and up to the front of the train to capture some shots. Before I could catch my breath, the footplate crew told me they had been told by the signalman that the stop would be shorter than the planned twenty minutes. So, after getting the shots, it was a rush back and over the footbridge and a jump back on the train at the first available door. Now, 40 to 50 years ago I would have done all this without breaking sweat and would have particularly enjoyed the side-stepping rugby moves needed to weave through the throng of people on the platforms, but today it was a little exhausting and my appearance back at the table was greeted with a few smiles from Nick, engineering team member Alan Pitt and his wife Barbara. Alan and Barbara provided a further smile on the trip back when they were looking for

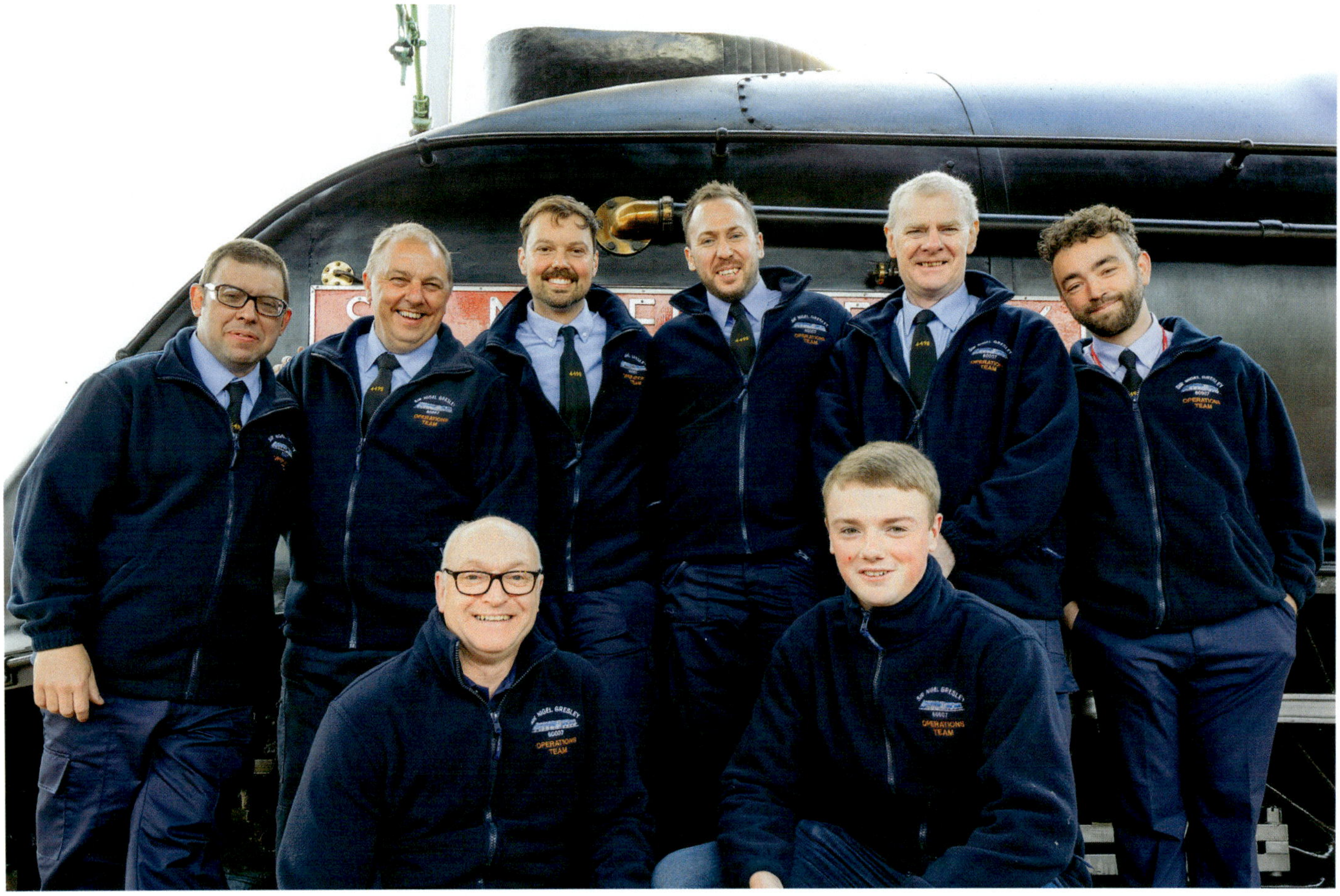

SNGLT Operations Team members BACK left to right Mark O'Brien, Richard Hawkhead, Andi Heald, Chris Whiteley, Phil Slater, Phil Gillespie, FRONT Andy Barwick, Scott Middlemiss

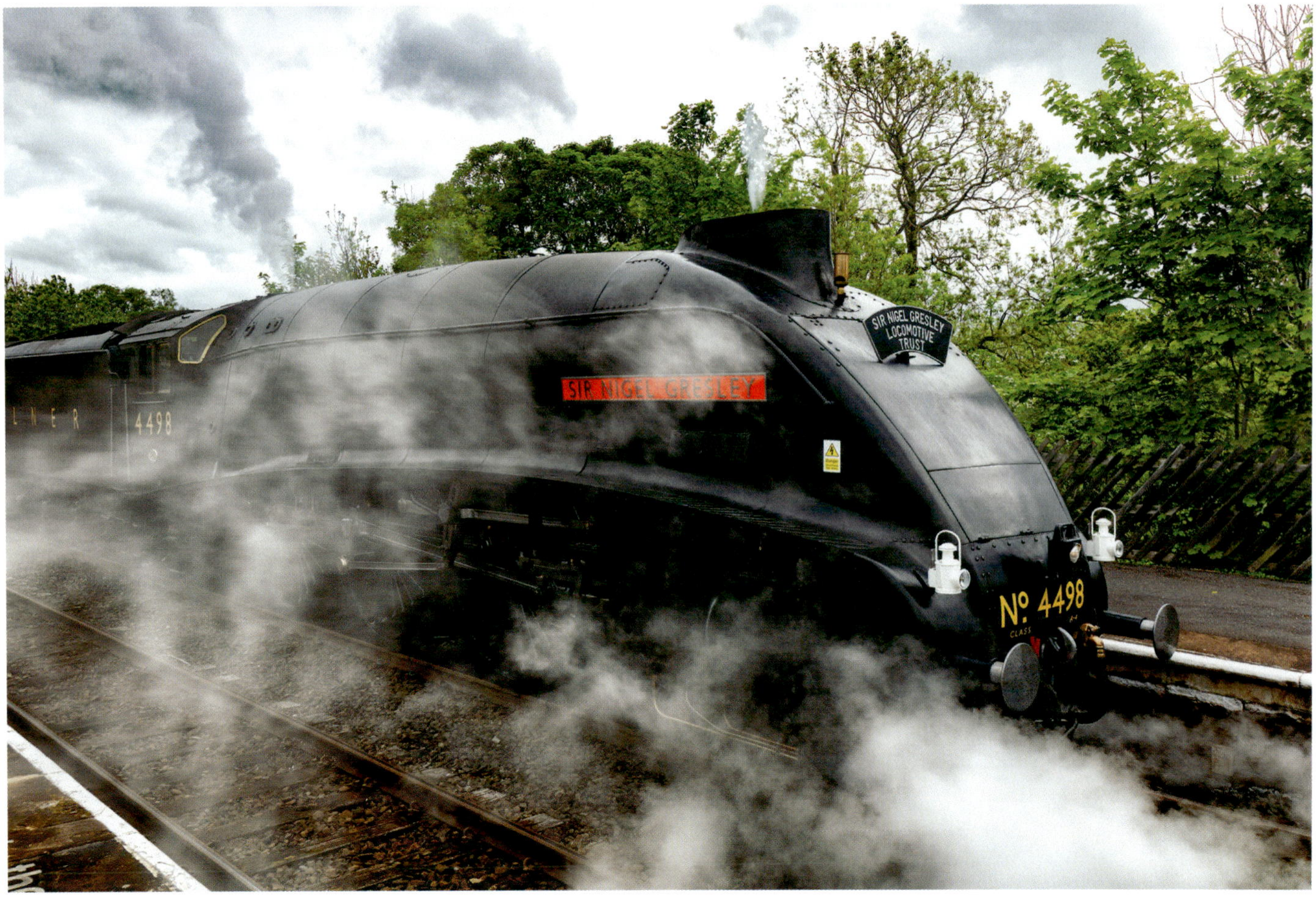

A brief stop for water at Appleby en route to Carlise on the inaugural run

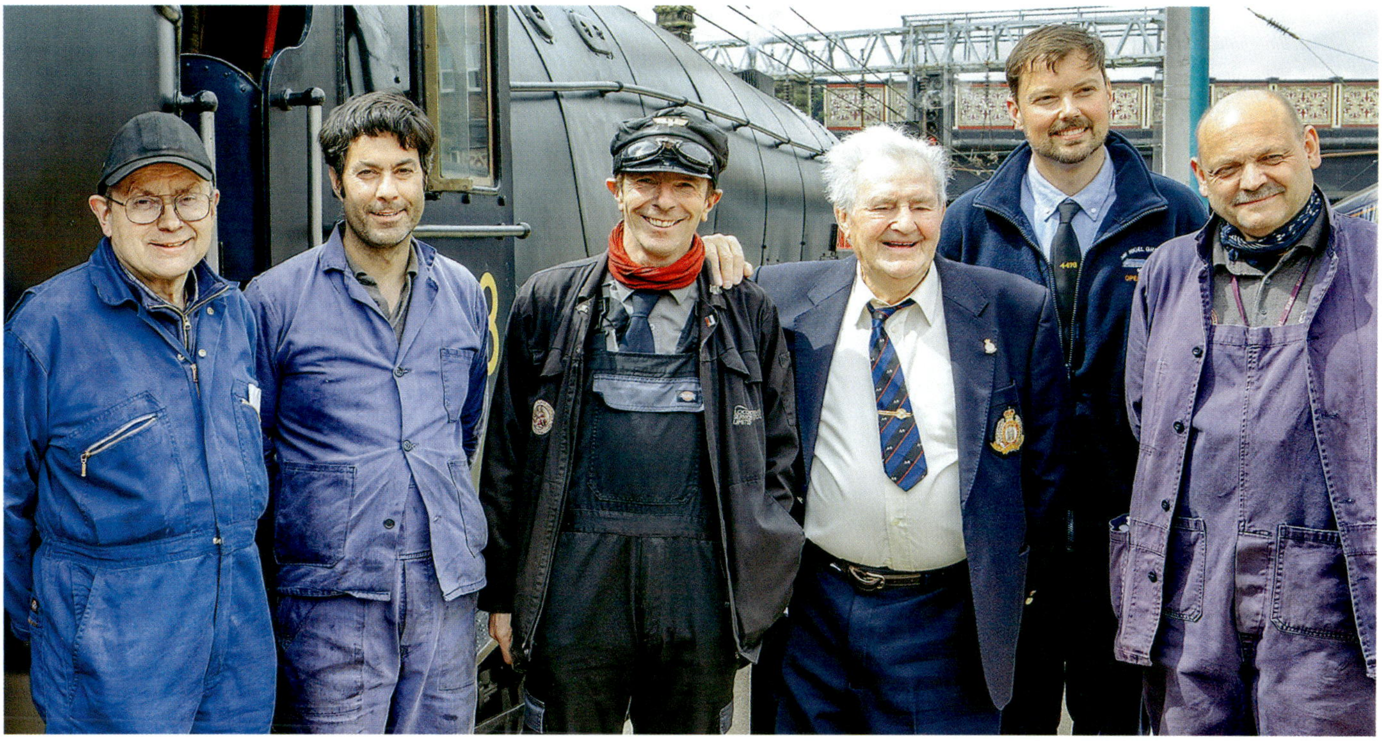

Footplate crew left to right Richard Swales (Chief Mechanical Engineer), Ian Walker (Fireman), Mike Wylie (Driver), Les Jackson (Oct 1967 Fireman) Andi Heald (SNGLT Support Crewe), Sean Levell (Footplate Inspector)

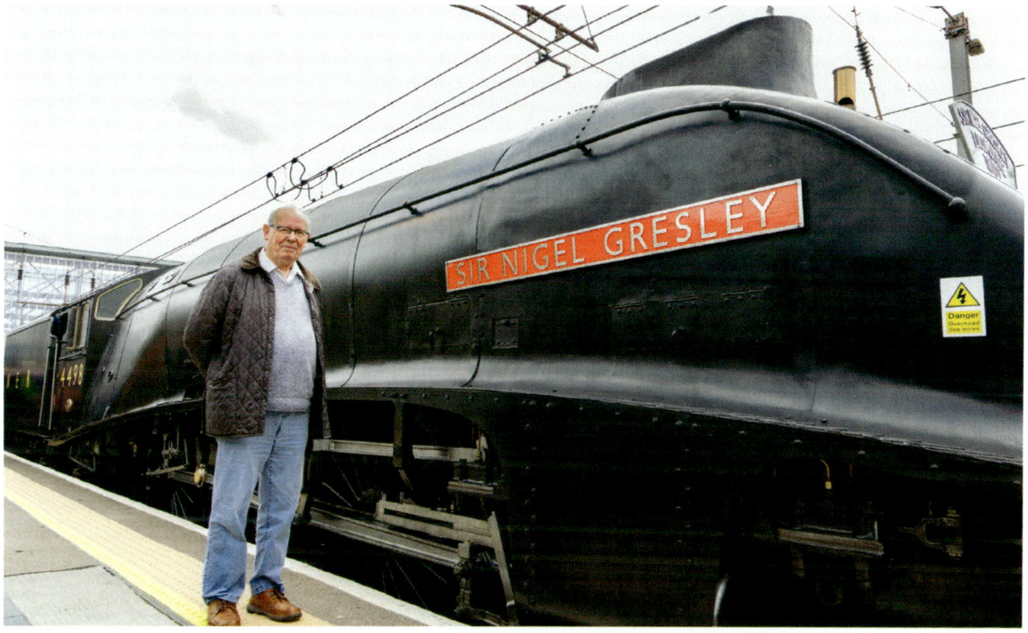

Ben Godfrey, Sir Nigel Gresley's grandson alongside the loco on the platform at Carlisle

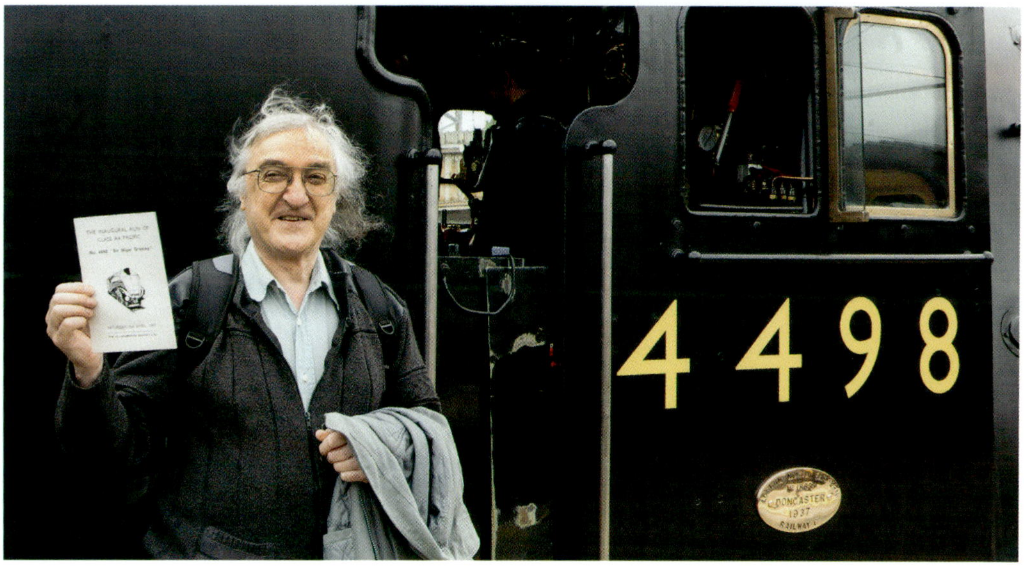

Richard Mellish on the platform at Carlisle with his programme from the first inaugural run after preservation in 1967

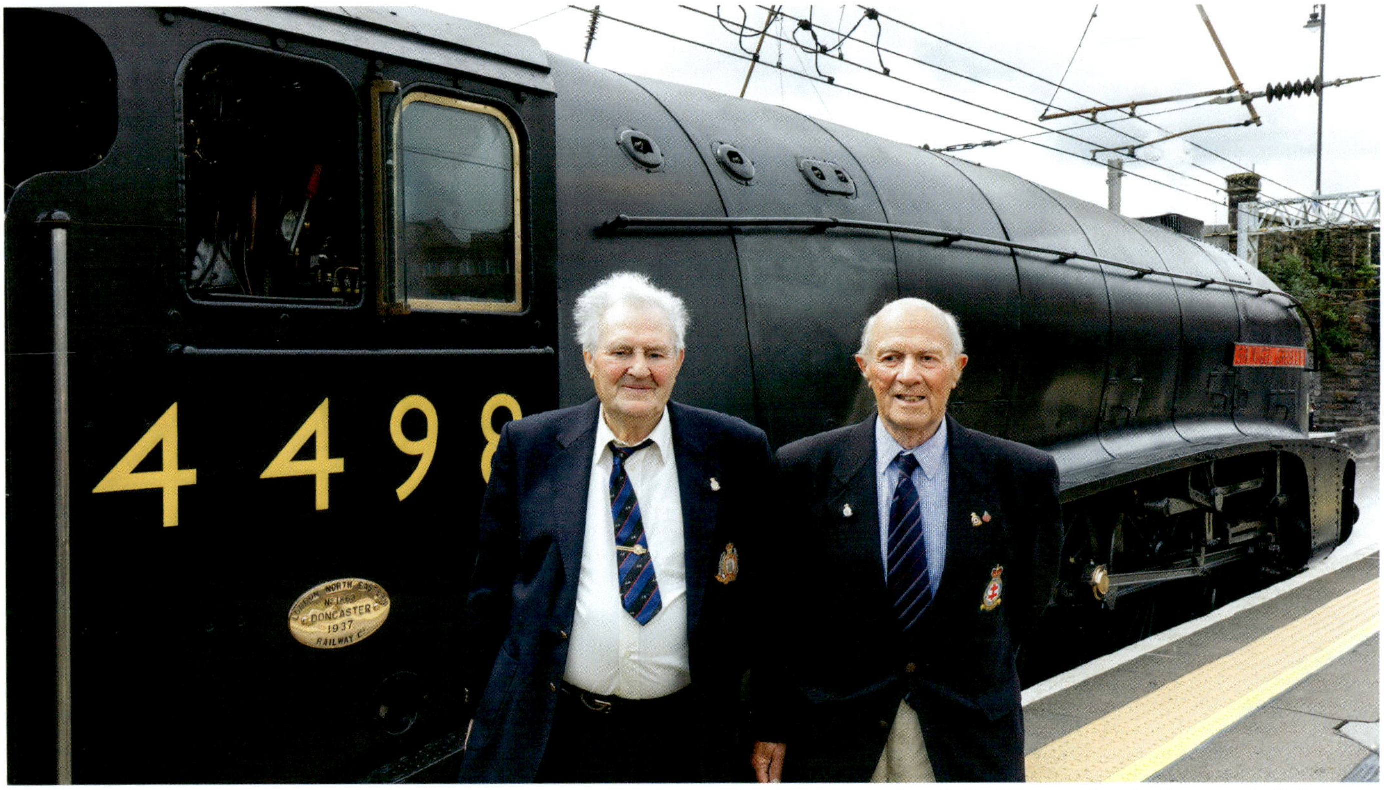

Les Jackson, Oct 1967 fireman and Neil Cadman, Apr & Oct 1967 fireman reunited with the loco on the platform at Carlisle

their daughter, Emma, who was waiting to shoot a video of our passing train at Acton Bridge, only for an Avanti West Coast express to pass on the northbound track at exactly the wrong moment!

We arrived back at Crewe on time, some eleven hours after departing and I was pretty tired after the day of opulence. The Saphos team looking after us, the SNGLT Operations team and of course the train crews, must have been even more exhausted, and they should all be very proud of their achievements over the day. Of course, their day wasn't yet finished as the loco and train had to be put to bed, whilst I, on the other hand, headed back to the hotel for a shower and a couple of very relaxing, winding down drinks in the bar.

All right for some I hear you say.

So, six and half years after 'It started with a ping', here we are. It has been a great experience, and I have made lots of new friends, learned a lot, become an author and also been given a new focus (sorry!) for my photography.

There are far too many memories to recall all of them here, but the outstanding ones for me are seeing the boiler leave through the city walls at York (Page 35), and later watching the lorry reverse round the T junction on the bridge at Llangollen and down to the engineering workshops; the lift off the lorry the following morning in icy sleet and rain, with the best bit left till last - nearly getting stranded in snow

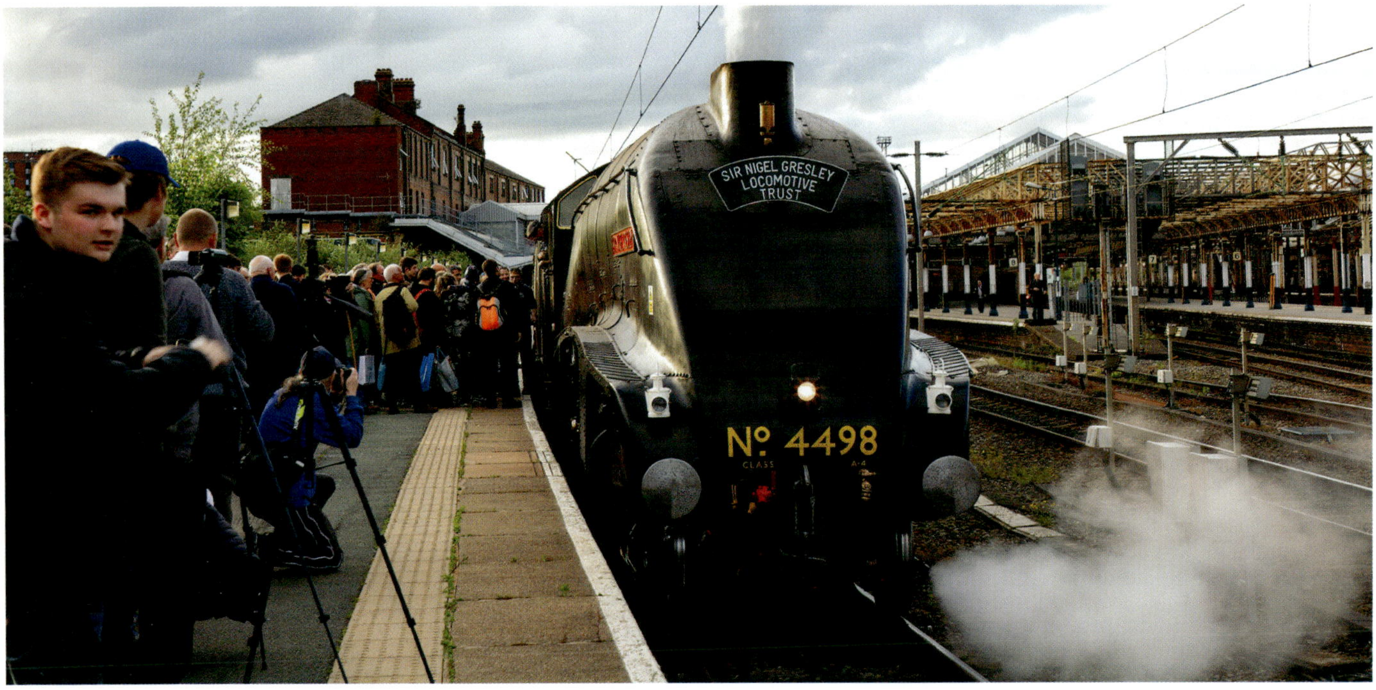

Back at Crewe after completing the inaugural run provided one last opportunity for photos and videos

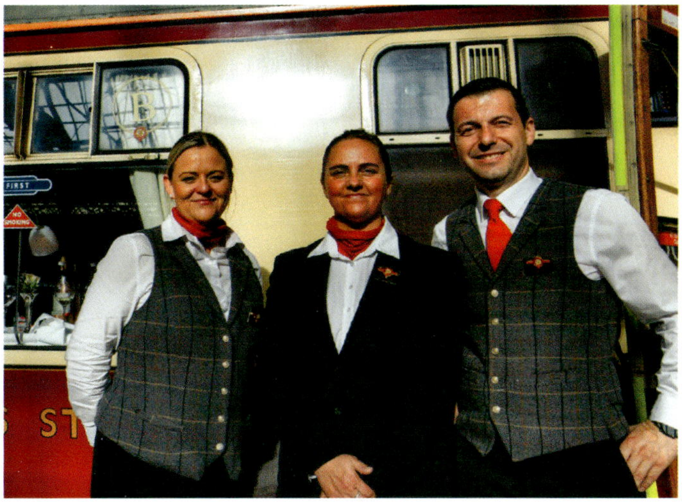

Saphos Trains crew left to right Lisa Barclay, Sandra Jones, Cristian Caraman

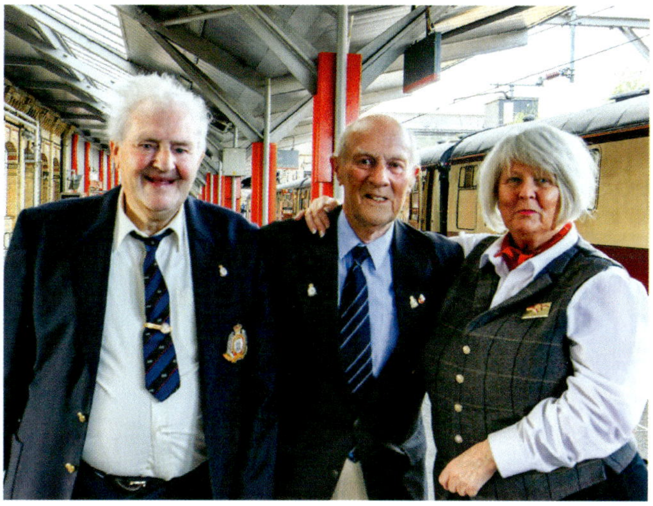

Firemen Les Jackson and Neil Cadman from the 1967 runs with Jan Humpheys from Saphos Trains

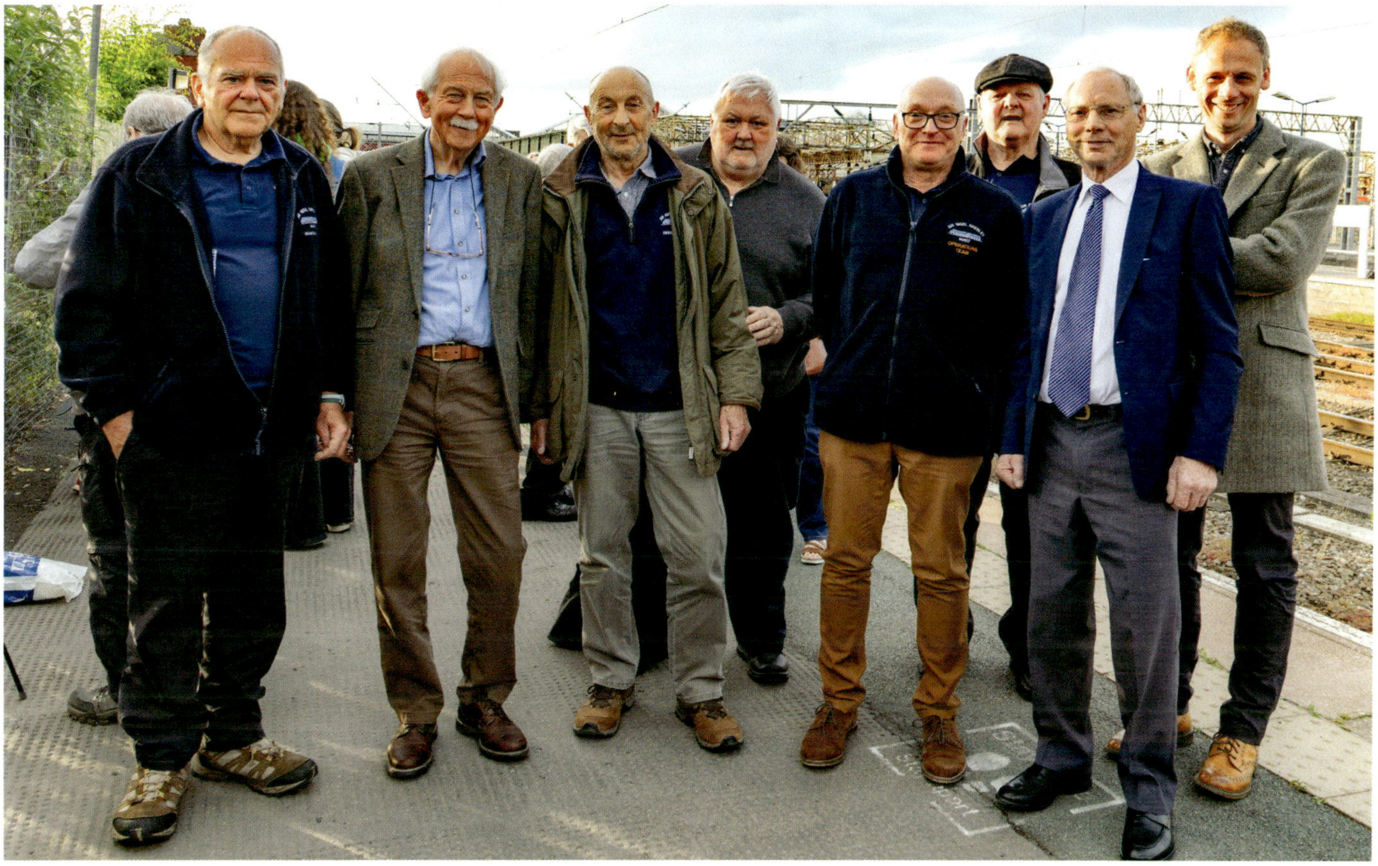

SNGLT Engineering team members left to right Peter Brackstone, Bob Shearman, Mike Page, Peter Ellis, Andy Barwick, Trevor Wilford, Dave Lee, Andy Evans

blocked roads trying to cross the Pennines travelling home with Malc (Page 36). Other early notable events are 'Smoke on the Water' (Page 53) and 'Disco Lights' (Page 66). Later it was the 'Into the Light' leaving the NRM workshop (Page 162) and 'The Convoy' (Page 171) when, without great fanfare, we became the last locomotive outshopped from what was York MPD 50A. The last few indelible memories are feeling proud on seeing the loco outside at LSL on Members' Day and having the lineside all to myself in a field by the river at Victoria Bridge on the Severn Valley Railway during the trial runs before joining up with chairman, Nigel Wilson, who was videoing events from the next field (Page 196). It all seemed a bit surreal being on our own at an historic occasion. Finally, of course, it must be today's Inaugural Run Members' Special.

I'm all out of words and pictures now. I hope you have enjoyed my six and a half year journey and if you have read all of it you have managed around 40,000 words and viewed (I lost count) approximately 650 pictures. Thank you and three cheers for the wonderful team of volunteer engineers led magnificently by Darrin Crone.